Materials Horizons: From Nature to Nanomaterials

Series Editor

Vijay Kumar Thakur, School of Aerospace, Transport and Manufacturing, Cranfield University, Cranfield, UK

Materials are an indispensable part of human civilization since the inception of life on earth. With the passage of time, innumerable new materials have been explored as well as developed and the search for new innovative materials continues briskly. Keeping in mind the immense perspectives of various classes of materials, this series aims at providing a comprehensive collection of works across the breadth of materials research at cutting-edge interface of materials science with physics, chemistry, biology and engineering.

This series covers a galaxy of materials ranging from natural materials to nanomaterials. Some of the topics include but not limited to: biological materials, biomimetic materials, ceramics, composites, coatings, functional materials, glasses, inorganic materials, inorganic-organic hybrids, metals, membranes, magnetic materials, manufacturing of materials, nanomaterials, organic materials and pigments to name a few. The series provides most timely and comprehensive information on advanced synthesis, processing, characterization, manufacturing and applications in a broad range of interdisciplinary fields in science, engineering and technology.

This series accepts both authored and edited works, including textbooks, monographs, reference works, and professional books. The books in this series will provide a deep insight into the state-of-art of Materials Horizons and serve students, academic, government and industrial scientists involved in all aspects of materials research.

Review Process

The proposal for each volume is reviewed by the following:

1. Responsible (in-house) editor
2. One external subject expert
3. One of the editorial board members.

The chapters in each volume are individually reviewed single blind by expert reviewers and the volume editor.

Rekha R. Warrier · Geeta Joshi · A. N. Arunkumar
Editors

Handbook of Asian Rosewoods

Springer

Editors
Rekha R. Warrier
Chemistry and Bioprospecting Division
ICFRE-Institute of Forest Genetics
and Tree Breeding
Coimbatore, Tamil Nadu, India

Geeta Joshi
Indian Council of Forestry Research
and Education
Dehradun, Uttarakhand, India

A. N. Arunkumar
Silviculture and Forest Management
Division
ICFRE-Institute of Wood Science
and Technology
Bengaluru, Karnataka, India

ISSN 2524-5384 ISSN 2524-5392 (electronic)
Materials Horizons: From Nature to Nanomaterials
ISBN 978-981-96-9458-7 ISBN 978-981-96-9459-4 (eBook)
https://doi.org/10.1007/978-981-96-9459-4

This Springer imprint is published by the registered company Springer Nature Singapore Pte Ltd.
The registered company address is: 152 Beach Road, #21-01/04 Gateway East, Singapore 189721, Singapore

Foreword

The genus *Dalbergia* holds immense ecological, economic, and scientific significance, particularly in Asia, where it is well represented across diverse landscapes. These leguminous trees and shrubs, renowned for their high-quality timber, medicinal properties, and nitrogen-fixing abilities, contribute significantly to biodiversity conservation and sustainable forestry. However, the increasing pressure from habitat loss, illegal logging, and climate change poses a serious threat to many species within this genus, some of which are now critically endangered.

This book, *Handbook of Asian Rosewoods*, provides a comprehensive exploration of the taxonomy, ecology, distribution, and conservation of *Dalbergia* species across the continent. Drawing upon the expertise of leading researchers, the book presents the latest scientific insights, from molecular phylogenetics and propagation techniques to sustainable management practices and policy frameworks. Such an interdisciplinary approach is crucial to addressing the conservation challenges faced by *Dalbergia* species and ensuring their continued ecological and economic contributions.

As we move toward a future that demands both conservation and sustainable resource utilization, this book serves as a valuable reference for botanists, foresters, conservationists, policymakers, and researchers working at the intersection of biodiversity and sustainable development. By fostering a deeper understanding of the genus, it is my hope that this work will contribute to the long-term preservation of *Dalbergia* species and the ecosystems they support.

I commend the editors, authors, and contributors for their dedicated efforts in bringing together this wealth of knowledge and trust that this book will inspire further research and conservation action for these remarkable trees.

Kanchan Devi, IFS
Director General
Indian Council of Forestry Research
and Education
Dehradun, Uttarakhand, India

Preface

Rosewoods, belonging to the genus *Dalbergia*, are among the most valuable timber species in the world due to their high-quality wood, aesthetic appeal, and commercial importance. These species are widely distributed across Asia and play a significant ecological role in forest ecosystems. The management of the forest genetic resources of rosewoods, particularly species in the *Dalbergia* genus, is crucial for biodiversity conservation, sustainable utilization, and economic benefits. However, the increasing demand for rosewood has led to overexploitation, habitat loss, and illegal logging, threatening their survival. Sustainable management practices and conservation strategies are essential to preserving these valuable species for future generations.

The species under discussion include *Dalbergia cochinchinensis*, *Dalbergia latifolia*, *Dalbergia oliveri*, *Dalbergia sissoo* and *Dalbergia tonkinensis*. Efforts to cultivate these species in China have been explored, with research focusing on improving growth conditions, adaptability, and propagation techniques.

The wood structure, physico-anatomical properties, and phytochemistry of *Dalbergia latifolia* have been presented. Propagation of rosewood species, especially *Dalbergia latifolia* and *Dalbergia sissoo*, presents significant challenges due to difficulties in adventitious rooting. Hence micropropagation techniques have been extensively reviewed, for *Dalbergia latifolia* and *Dalbergia sissoo*. Similar studies on other species remain sparse, offering a significant scope for future research.

Various diseases and insect pests pose a threat to *Dalbergia* species, impacting their growth and timber quality. Integrated pest management strategies are essential for disease control and the long-term health of these trees. Research on genetic variability and tree improvement programs has been conducted for *Dalbergia sissoo* to enhance stability and adaptability, ensuring that these species can withstand environmental stressors.

Understanding the reproductive biology and pollination ecology of rosewoods is crucial for breeding and conservation. *Dalbergia sissoo* has been extensively studied in this regard, with insights into its breeding strategies, pollination mechanisms, seed maturity indices, and storage conditions. These studies contribute to improved genetic conservation efforts and facilitate the development of more effective breeding programs. By enhancing our understanding of these factors, researchers can develop strategies to improve seed viability and promote successful natural regeneration in forest ecosystems.

Rosewood species play a vital role in agroforestry due to their high economic value and ecological benefits. Their inclusion in agroforestry systems can contribute to sustainable land use while providing economic returns for farmers. By integrating these species into farming systems, communities can benefit from diversified income sources while also contributing to forest conservation. The promotion of sustainable agroforestry practices can reduce pressure on natural forests and support the livelihoods of rural communities.

Researchers have assessed the current and future habitat suitability and vulnerability of Asian rosewoods to multiple threats to guide conservation and restoration efforts. The findings indicate that climate change will significantly impact their distribution, necessitating proactive conservation strategies to ensure their long-term survival. Additionally, restoration ecology and afforestation strategies using *Dalbergia* species remain underexplored, highlighting the need for further research in this area. Prioritization of conservation areas, considering ecoregional diversity, is essential for the effective management of forest genetic resources.

Collaboration among researchers, policymakers, and local communities is crucial to safeguarding the genetic resources of rosewood species for future generations. Additionally, increased investment in research and development can lead to innovative solutions for addressing the challenges faced by these species.

This handbook spans multiple aspects of *Dalbergia* species, including distribution, cultivation, genetic resources, habitat modeling, wood properties, phytochemistry, propagation, pests, diseases, and conservation. The inclusion of regional studies from India, China, Myanmar, Sri Lanka, Indonesia, and Vietnam ensures a broad geographical perspective, making this a comprehensive reference for researchers and practitioners alike.

Perspectives on varied topics have made this book a state-of-the-art reference material. Given the economic importance of rosewood species, the knowledge compiled in this handbook is conducive to exploring and protecting the genus. Readers may wish to focus on specific chapters rather than reading the book in its entirety. Bearing this in mind, we have retained some overlaps between chapters rather than relying solely on cross-referencing between sections.

This book has been conceived for a varied readership, including foresters from different countries, researchers within and beyond forestry, academics from various disciplines, and plantation growers. By fostering information exchange and collaboration, this book aims to inspire scientists to conduct further research on this iconic genus across the globe. Through collective efforts, we can ensure the sustainable conservation and utilization of rosewood species, preserving their ecological and economic value for future generations.

Coimbatore, India — Rekha R. Warrier
Dehradun, India — Geeta Joshi
Bengaluru, India — A. N. Arunkumar
March 2025

Contents

Editors and Contributors

About the Editors

Dr. Rekha R. Warrier is Scientist-G, heading the Chemistry and Bioprospecting Division and co-ordinating the Environment Information, Awareness, Capacity Building and Livelihood Programme (EIACP) at the ICFRE-Institute of Forest Genetics and Tree Breeding, Coimbatore, India. She has put in more than 25 years in the field of Forest Biotechnology and is currently serving as Head of the group in Plant Tissue Culture. She has been working on different aspects of medicinal plants and their productivity, specifically, in the field of metabolite production. Some of her major activities include tissue culture facility for commercial production of plants and in vitro production of secondary metabolites through transformation, propagation and cultivation of commercially important tree species, genetic improvement of medicinal trees and participation in extension and training programs for foresters, research staff, students, and farmers. She has to her credit several research papers and books/chapters. She is the Chair of the Asia Pacific Forest Genetic Resources (APFORGEN), the Assistant National Country Co-ordinator of the APFORGEN and Liaison to the Central Secretariat of International Society of Tropical Foresters. She is a member of the IUCN SSC Western Ghats Plant Specialist Group and Seed Conservation Specialist Group.

Dr. Geeta Joshi is Scientist-G and Assistant Director General (Media and Extension) at the Indian Council of Forestry Research and Education, Dehradun, India. She has 25 years of research experience related to seed collection, handling, germination, storage, standardization of protocol for mass production of quality planting material in tree species. Dr. Joshi has also worked on seed-related aspects of threatened and endangered tree species in the Western Ghats, the propagation and cultivation of bamboo, and the morphological and genetic diversity of Teak and *Melia dubia*. Dr. Joshi has published several research papers, books, and chapters. She was involved in the production of quality planting material for end users, and was associated with conducting trainings for farmers, State Forest Departments, plantation companies, and NGOs on seed handling, nursery, and plantation technology. Dr. Joshi is currently coordinating the extension of ICFRE research findings to various stakeholders.

Dr. A. N. Arunkumar is Scientist-G in the Silviculture and Forest Management Division at the Institute of Wood Science and Technology, Bengaluru, India, and has more than 25 years of experience in forestry research. His research interest is in documenting growth and variability in different timber species such as Indian Sandalwood, Red Sanders, Teak, *Hardwickia binata, Chloroxylon swietenia, Dalbergia latifolia, Melia dubia*, etc. He has been involved in establishing clonal, seedling seed orchards, and progeny trials of sandalwood, teak, and Melia dubia. As a Project Investigator, Dr. Arun has carried out various research projects and published several research papers, along with multiple books and book chapters. He has been providing scientific input on promoting tree cultivation and is invited to deliver lectures in various forums. He was also an organizing member in conducting national and international conferences. Presently, he is a member of the Sandalwood Development Committee, constituted by the National Institution for Transforming India (NITI Aayog), a premier policy think tank of the Government of India.

Contributors

A. N. Arunkumar ICFRE-Institute of Wood Science and Technology, Malleswaram, Bengaluru, India

R. Athulya Forest Protection Division, ICFRE-IWST, Bengaluru, India

Tika D. Atikah Research Centre for Ecology and Ethnobiology, National Research and Innovation Agency, Bogor, West Java, Indonesia

Mu Mu Aung Forest Research Institute of Forest Department, Ministry of Natural Resources and Environmental Conservation, Nay Pyi Taw, Myanmar

Tin Hnaung Aye Forest Department, Ministry of Natural Resources and Environmental Conservation, Nay Pyi Taw, Myanmar

Sabyasachi Banerjee Forest Pathology Discipline, Forest Protection Division, ICFRE-Forest Research Institute, Dehradun, India

Kulvir Singh Bangarwa Department of Forestry, CCS Haryana Agricultural University, Hisar, India

Nitika Bansal Forest Pathology Discipline, Forest Protection Division, ICFRE-Forest Research Institute, Dehradun, India

Maneesh S. Bhandari Division of Genetics and Tree Improvement, ICFRE-Forest Research Institute, Dehradun, Uttarakhand, India

Chaloun Bounithiphonh Forest Research Center, National Agriculture and Forestry Research Institute, Vientiane, Laos

K. Charishma Forest Protection Division, ICFRE-FRI, Dehradun, India

Nguyen Minh Chi Forest Protection Research Centre, Vietnamese Academy of Forest Sciences, Hanoi, Vietnam

Bernard Dell Agriculture and Forest Sciences, Murdoch University, Murdoch, Australia

P. Devamanikandan Forest Genetic Resource Management Division, ICFRE-Institute of Forest Genetics and Tree Breeding, Forest Campus, Coimbatore, India

Payal Dhyani Division of Genetics and Tree Improvement, ICFRE-Forest Research Institute, Dehradun, Uttarakhand, India

Fifi G. Dwiyanti Department of Silviculture, Faculty of Forestry and Environment, IPB University, Bogor, West Java, Indonesia

Tobias Fremout Bioversity International, Lima, Peru

Hannes Gaisberger Bioversity International, Rome, Italy

Phi Hong Hai Vietnamese Academy of Forest Sciences, Hanoi, Vietnam

Ida Hartvig Department of Geosciences and Natural Resource Management, University of Copenhagen, Copenhagen, Denmark;
Smithsonian Environmental Research Center, Smithsonian Institute, Washington, USA;
Center for Evolutionary Hologenomics, Globe Institute, University of Copenhagen, København K, Denmark

Tin Hang Hung Department of Biology, University of Oxford, Oxford, UK

Riina Jalonen Bioversity International, Penang, Malaysia

Geeta Joshi Indian Council for Forestry Research and Education (ICFRE), Dehradun, Uttarakhand, India

R. Kalaiselvi ICFRE-Institute of Forest Genetics and Tree Breeding, Coimbatore, India

Chris Kettle Bioversity International, Rome, Italy

Shweta B. Kukreti College of Forestry, Veer Chandra Singh Garhwali Uttarakhand University of Horticulture and Forestry, Ranichauri, Uttarakhand, India

Rakesh Kumar Wood Properties and Processing Division, ICFRE-Institute of Wood Science and Technology, Bengaluru, Karnataka, India

Shailendra Kumar Forest Products Division, Forest Research Institute Dehradun, Dehradun, Uttarakand, India

S. Lalitha Forest Genetic Resource Management Division, ICFRE-Institute of Forest Genetics and Tree Breeding, Forest Campus, Coimbatore, India

Soulivanh Lanorsavanh Department of Biology, Faculty of Natural Sciences, National University of Laos, Dong Dok Campus, Saythany District, Vientiane Capital, Lao PDR

Yongxiang Li Institute of Forestry, Chinese Academy of Forestry, Haidian District, Beijing, China

John MacKay Department of Biology, University of Oxford, Oxford, UK

Deepak Kumar Mahanta Forest Protection Division, ICFRE-FRI, Dehradun, India

Shristi Manikpuri Forest Pathology Discipline, Forest Protection Division, ICFRE-Forest Research Institute, Dehradun, India

Rajendra Kumar Meena Division of Genetics and Tree Improvement, ICFRE-Forest Research Institute, Dehradun, Uttarakhand, India

Leonard J. Mills Coorg Rosewood Timbers, Kattugollahalli, Bengaluru, Karnataka, India

Kamal Mishra ICFRE-Institute of Wood Science and Technology, Bengaluru, India

A. Muthu Kumar Forest Protection Division, ICFRE-IWST, Bengaluru, India

Ismita Nautiyal Research and Planning Division, Indian Council of Forest Research & Education, Dehradun, Uttarakhand, India

J. S. Nayal Uttarakhand State Seed and Organic Production Certification Agency, Dehradun, Uttarakhand, India

Nitika Negi Forest Pathology Discipline, Forest Protection Division, ICFRE-Forest Research Institute, Dehradun, India

Nong Phuong Nhung Vietnamese Academy of Forest Sciences, Hanoi, Vietnam

Amit Pandey ICFRE-Institute of Forest Productivity, Ranchi, Jharkhand, India

Shailesh Pandey Forest Pathology Discipline, Forest Protection Division, ICFRE-Forest Research Institute, Dehradun, India

S. S. Phartyal School of Earth Science and NRM, Mizoram University, Aizawal, Mizoram, India

Huang Ping Research Institute of Forestry, Chinese Academy of Forestry, Beijing, China

Bayu A. Pratama Research Centre for Ecology and Ethnobiology, National Research and Innovation Agency, Bogor, West Java, Indonesia

Henti H. Rachmat Research Centre for Ecology and Ethnobiology, National Research and Innovation Agency, Bogor, West Java, Indonesia

R. Rajathewa Bandaranayakapura, Indigolla, Gampaha, Sri Lanka

Ramkrishna Forest Pathology Discipline, Forest Protection Division, ICFRE-Forest Research Institute, Dehradun, India

K. R. Sasidharan Forest Ecology and Climate Change Division (Retired), ICFRE-Institute of Forest Genetics and Tree Breeding, Forest Campus, Coimbatore, India

Anil Kumar Sethy ICFRE-Institute of Wood Science and Technology, Bengaluru, Karnataka, India

Titiek Setyawati Research Centre for Ecology and Ethnobiology, National Research and Innovation Agency, Bogor, West Java, Indonesia

Rohit Sharma Wood Properties and Processing Division, ICFRE-Institute of Wood Science and Technology, Bengaluru, Karnataka, India

Siya Sharma Forest Pathology Discipline, Forest Protection Division, ICFRE-Forest Research Institute, Dehradun, India

Sushant Sharma Genetics and Tree Improvement Division, ICFRE-Forest Research Institute, Dehradun, India

Vishwajeet Sharma Indian Council of Forestry Research and Education (ICFRE), Dehradun, India

Prashant Shaw Department of Botany, University of Delhi, Delhi, India

Satish Kumar Sinha College of Forestry, Navsari Agricultural University, Navsari, Gujarat, India

Thea So Institute of Forest and Wildlife Research and Development, Phnom Penh, Cambodia

Nileshkumar J. Sohagiya College of Forestry, Navsari Agricultural University, Navsari, Gujarat, India

Sineath Sreng Forest Development and Botanical Research Centre, IRD, Phnom Penh, Cambodia

Syneath Sreng Institute of Forest and Wildlife Research and Development, Phnom Penh, Cambodia

S. M. C. U. P. Subasinghe Department of Forestry and Environmental Science, University of Sri Jayewardenepura, Nugegoda, Sri Lanka

Rajesh Tandon Department of Botany, University of Delhi, Delhi, India

Ajay Thakur Genetics and Tree Improvement Division, ICFRE-Forest Research Institute, Dehradun, India

D. Thangamani Chemistry and Bioprospecting Division, ICFRE-Institute of Forest Genetics and Tree Breeding, Forest Campus, Coimbatore, India

R. C. Thapliyal ICFRE-Forest Research Institute, Dehradun, Uttarakhand, India

Ida Theilade Department of Geosciences and Natural Resource Management, University of Copenhagen, Copenhagen, Denmark

Pham Quang Thu Forest Protection Research Centre, Vietnamese Academy of Forest Sciences, Hanoi, Vietnam

Pramod Kumar Tiwari Genetics and Tree Improvement Division, ICFRE-Tropical Forest Research Institute, Jabalpur, M.P., India

G. S. Uma Forest Protection Division, ICFRE-IWST, Bengaluru, India

T. Vamadevan EIACP Programme Centre Resource Partner on Forest Genetic Resources and Tree Improvement, ICFRE-Institute of Forest Genetics and Tree Breeding, Forest Campus, Coimbatore, India

R. Vasudevan Chemistry and Bioprospecting Division, ICFRE-Institute of Forest Genetics and Tree Breeding, Forest Campus, Coimbatore, India

Preeti Vats College of Forestry, Navsari Agricultural University, Navsari, Gujarat, India

Praveen Verma ICFRE-Forest Research Institute, Dehradun, India

Wita Wardani Research Centre for Biosystematics and Evolution, National Research and Innovation Agency, Bogor, West Java, Indonesia

Rekha R. Warrier Chemistry and Bioprospecting Division, ICFRE-Institute of Forest Genetics and Tree Breeding, Forest Campus, Coimbatore, Tamil Nadu, India

Shambhavi Yadav Genetics and Tree Improvement Division, ICFRE-Forest Research Institute, Dehradun, India

Zheng Yongqi Research Institute of Forestry, Chinese Academy of Forestry, Beijing, China

Li Yongxiang Institute of Forestry, Chinese Academy of Forestry, Beijing, China

Kusumadewi S. Yulita Research Centre for Ecology and Ethnobiology, National Research and Innovation Agency, Bogor, West Java, Indonesia

Vivi Yuskianti Research Center for Biota Systems, National Research and Innovation Agency, Bogor, West Java, Indonesia

Part I
Management of Forest Genetic Resources (FGRs) of Rosewoods

Chapter 1
Rosewoods of Asia: A Precious Timber Resource

Rekha R. Warrier, Geeta Joshi, and A. N. Arunkumar

Asian rosewoods, or *Dalbergias*, are a genus of flowering plants in the legume family Fabaceae, renowned for their richly hued, durable hardwood. The wood is highly prized for crafting musical instruments, fine furniture, and decorative pieces due to its striking reddish-brown hue and unique grain patterns. Native to tropical and subtropical forests regions across Asia, including India, Southeast Asia, and China, *Dalbergia* species are valued for their strength and resistance to decay. Notable species include *Dalbergia latifolia* (Indian rosewood) and *Dalbergia sissoo* (shisham), which are widely used in high-end woodworking. *Dalbergia cochinchinensis* (Vietnamese rosewood) is another prized species known for its durability and rich colour. Under the label "*Hongmu*," meaning "Red wood" in Mandarin, the common term for Rosewood in China and other Asian markets (Table 1), around 30 species from the genera *Dalbergia*, *Pterocarpus*, *Diospyros*, *Millettia*, and *Cassia* are sought after by importers for luxury furniture and flooring [1].

Rosewood holds a special place in Asian culture due to its durability and exceptional quality. Moreover, furniture crafted from rosewood is highly prized not only for its functional benefits but also for the cultural symbolism and aesthetic appeal it embodies. This deep-rooted cultural affinity has fuelled an insatiable demand for rosewood, leading to flourishing luxury furniture markets in China and Vietnam. However, this demand has also triggered an illegal rosewood trade, as traffickers seek to capitalize on the high value of the wood. Rosewood has become one of the most trafficked natural products globally, with the majority flowing into China's

R. R. Warrier (✉)
ICFRE-Institute of Forest Genetics and Tree Breeding, Coimbatore, India
e-mail: rekha@icfre.org

G. Joshi
Indian Council of Forestry Research and Education (ICFRE), Dehradun, India

A. N. Arunkumar
ICFRE-Institute of Wood Science and Technology, Bengaluru, India

R. R. Warrier et al. (eds.), *Handbook of Asian Rosewoods*, Materials Horizons: From Nature to Nanomaterials, https://doi.org/10.1007/978-981-96-9459-4_1

Table 1 *Dalbergia* species categorized as "Hongmu" *Adapted from [1, 2]

S. No.	Species	Common names	Geographic range
1.	*Dalbergia bariensis*	Burmese Rosewood, Tamalan	Vietnam, Thailand, Cambodia, Myanmar, and Laos
2.	*Dalbergia cochinchinensis*	Siamese Rosewood	Vietnam, Laos, Cambodia, and Thailand
3.	*Dalbergia cultrata*	Burmese Blackwood	Myanmar, India, China (Yunnan), Vietnam
4.	*Dalbergia latifolia*	Indian Rosewood	India, Indonesia
5.	*Dalbergia odorifera*	Fragrant Rosewood	China
6.	*Dalbergia oliveri*	Burmese Rosewood, Tamalan	Thailand, Myanmar, Laos
7.	*Dalbergia tonkinensis*	Yellow Rosewood	Vietnam

booming luxury furniture market. The allure of rosewood furniture, which has a long-standing historical and cultural tradition, makes it challenging to curb consumption in a short period [3].

1 Characteristics of Dalbergias

Morphology: All rosewoods are medium to large deciduous trees, with *Dalbergia cultrata* being a medium-sized evergreen tree. The leaves of the genus are pinnate with leaflet numbers varying from 3 to 9 and shapes ranging from ovate to elliptic and lanceolate depending on the species. The bark shows light variations ranging from dark grey and black to reddish-brown and light brown; hence distinguishing based on the bark is difficult. The flowers are small, fragrant, and typically arranged in clusters or panicles, exhibiting the papilionaceous (butterfly-like) structure, characteristic of the Fabaceae family. The flower colour ranges from pale yellow (*D. cochinchinensis*), white (*D. latifolia*), pinkish-white (*D. oliveri*), to pale purple (*D. cultrata*) which helps distinguish the species. The flowers are bisexual and capable of self-pollination, although they also attract pollinators like bees. The flowering periods vary by species and region, often occurring in spring or early summer. The pods are mostly flat and oblong, with some being narrow and thin or having curved edges. The wood colour, a major distinguishing feature, varies from dark brown with purple streaks, dark reddish-brown, and golden brown to dark brown (Table 2). *Dalbergia tonkinensis* seeds are yellowish-brown, thin, and oval-shaped.

Wood characteristics: Table 3 compares five *Dalbergia* species based on their density, scent, grain characteristics, rot resistance, workability, and uses. *Dalbergia latifolia* (density: 830 kg/m^3) has a distinct rose-like scent, interlocked grain, and medium texture. It is very durable but challenging to work with and is used in luxury

Table 2 Morphological features of common rosewood (*Dalbergia*) species

Feature	*Dalbergia latifolia*	*Dalbergia cochinchinensis*	*Dalbergia cultrata*	*Dalbergia oliveri*	*Dalbergia sissoo*	*Dalbergia tonkinensis*
Growth habit	Medium to large deciduous tree	Medium to large deciduous tree	Medium-sized evergreen tree	Medium to large deciduous tree	Medium-sized deciduous tree	Medium-sized deciduous tree
Leaf type	Pinnate, 5–7 leaflets	Pinnate, 5–7 leaflets	Pinnate, 5–9 leaflets	Pinnate, 5–7 leaflets	Pinnate, 3–5 leaflets	Imparipinnate, 11–17 leaflets
Leaf shape	Ovate to elliptic	Oblong to elliptic	Oblong to ovate	Elliptic to obovate	Lanceolate to ovate	Elliptic
Bark colour	Dark grey to black	Reddish-brown	Dark brown to black	Greyish-brown	Light brown to grey	Yellow-brown or pale-grey
Flower colour	White to pale pink	Pale yellow	Pale purple to white	Pinkish-white	Yellowish-white	White
Pod shape	Flat, oblong with thin margins	Flat, narrow, and elongated	Thin, linear and flat	Flat, oblong with curved edges	Flat, narrow, and thin	Flat, narrow, and thin
Wood colour	Dark brown with purple streaks	Dark reddish-brown	Dark brown to black	Reddish-brown to purple	Golden brown to dark brown	Dark reddish-brown with black streaks

furniture and musical instruments. *Dalbergia cochinchinensis* (density: 1,035 kg/m^3) has a mild scent, straight to interlocked grain, and fine texture. It is durable and easier to work with, making it suitable for furniture and small specialty items. *Dalbergia cultrata* (density: 1,060 kg/m^3) has an irregular grain, high density, and excellent durability but is difficult to work with. It is used in carvings and percussion instruments. *Dalbergia oliveri* (density: 940 kg/m^3) has a rose-like scent, straight to interlocked grain, and good workability. It is used in flooring and furniture. *Dalbergia sissoo* (density: 770 kg/m^3) has a mild scent, straight grain, and good workability, making it ideal for veneer, flooring, and boatbuilding. *Dalbergia tonkinensis*, known for its fine texture, workability, and durability, is used in premium furniture and musical instruments.

2 Global Trade and Market Trends

Dalbergia timber is primarily exported from India, Myanmar, Thailand, and Laos, with China being one of the largest importers. The European Union and the United States are significant markets for luxury wood products made from *Dalbergia* species. Due to limited supply and increasing restrictions, the price of *Dalbergia* timber has surged in recent decades. Despite regulatory measures, the rising demand for luxury furniture and musical instruments continues to drive the market.

Historically, China has been a major rosewood importer, primarily for its traditional furniture industry. The rosewood resources have long been exhausted. China relied on rosewood imports to meet domestic demand. Prior to 2020, China was a major importer of rosewood species, primarily sourced from Southeast Asian countries such as Myanmar, Laos, and Cambodia. The rosewood imports into China were mainly logs and sawn timber [4]. China emerged as the primary destination for most international rosewood trade, establishing itself as the world's largest importer of rosewood [5, 6]. Between 2014 and 2019, China imported 5.50 million m^3 of rosewood logs and 0.83 million m^3 of rosewood-sawn timber [7].

However, overexploitation led to the depletion of native rosewood populations in these regions. This unsustainable harvesting prompted both national and international regulatory measures to control the trade of endangered rosewood species. In 2014, rosewood imports peaked at over $2 billion. Subsequent enforcement actions and regulatory measures led to a significant decline, with imports dropping to one-fifth of the 2014 levels by 2020.

It was demonstrated that the import of rosewood to China was notably impacted by export restrictions implemented by source countries and international conventions, such as the Convention on International Trade in Endangered Species of Wild Fauna and Flora (CITES) [4]. It was observed that CITES was ineffective in curbing the illegal trade of rosewood from Thailand, specifically *D. cochinchinensis* and *D. bariensis* [8]. After genera *Dalbergia* was listed in CITES Appendix II in 2016, local hoarding of genera *Dalbergia* expanded in Madagascar and an average annual increase in Malagasy rosewood logging of 127.6% [9, 10]. Similarly, the trade of *D.*

Table 3 Wood characteristics of common rosewood (*Dalbergia*) species

Species	Average density (kg/m^3)	Scent	Grain/ texture	Rot resistance	Workability	Uses
Dalbergia latifolia	830	Distinct rose-like scent when worked	Grain narrowly interlocked, Medium texture with good natural lustre	Very durable Good termite resistance	Challenging to work with due to its interlocked grain and high density. Wood may contain chalk-like deposits that quickly dull cutting edges.	Luxury furniture, musical instruments, veneer, turned objects, and other speciality wood objects.
Dalbergia cochinchinensis	1,035	Mild scent	Grain straight to slightly interlocked, Medium to fine uniform texture with good natural lustre.	Very durable	Fairly good	Luxury furniture, turned objects, musical instruments, and small speciality wood items.
Dalbergia cultrata	1,060	Mild scent but not resembling rosewoods	Grain is irregular, Uniform medium texture and very good natural lustre.	Resistant to termites	Challenging to work with due to high density. Turns and finishes nicely, and takes a high polish.	Turned objects, carvings, furniture, inlay, percussion instruments, and small speciality wood items.
Dalbergia oliveri	940	Distinct, rose-like scent when worked	Grain is usually straight or slightly interlocked, Medium-fine texture and a good natural lustre	Very durable	Easy to work with both hand and machine tools. Wood may contain chalk-like deposits that quickly dull cutting edges. Natural oils in the wood can disrupt the drying process. Turns and polishes well.	Furniture, flooring, percussion instruments, turned objects, and other small specialty wood items.

(continued)

Table 3 (continued)

Species	Average density (kg/m^3)	Scent	Grain/ texture	Rot resistance	Workability	Uses
Dalbergia sissoo	770	Distinct yet mild rosewood scent	Straight grained. Texture is medium to coarse with a good natural lustre	Very durable	Good working characteristics, and responds well to nearly all machining. Wood may contain chalk-like deposits that quickly dull cutting edges. Turns, glues, and finishes well.	Veneer, plywood, percussion instruments, furniture, flooring, boatbuilding, carving, and turned objects.

Source https://www.wood-database.com; [3]

cochinchinensis has been limited under the CITES restrictions since 2013. *Dalbergia oliveri* belongs to the same genus as *D. cochinchinensis*. The woody texture of both species *D. oliveri* and *D. cochinchinensis* being similar, while the price of *D. oliveri* is much lower than that of *D. cochinchinensis*, the countries started to stock *D. oliveri* as a substitute for *D. cochinchinensis* [11] leading to a price jump for *D. oliveri* also [12]. In contrast, India, as the leading producer of *D. latifolia*, severely restricted its exports as early as 1980 [13]. Including *D. latifolia* to CITES Appendix II in 2016 had little effect on its price following the designation [12]. Thailand has banned rosewood trading, but smuggling persists through the Cambodian and Lao borders. Traders circumvent the ban by smuggling Thai rosewood into Laos and falsely labelling it as Lao rosewood. According to EIA [2], around 60% of rosewood traded in Laos originates from Thailand. The country is the hub of high-risk tropical sawn timber from neighbouring countries such as Laos, Malaysia, Myanmar, and Cambodia. However, the case of the import is presently controlled largely through regulations.

3 Conservation Challenges

The overexploitation and illegal logging of Asian rosewoods have emerged as critical environmental and socio-economic challenges in recent years [3]. Renowned for their striking reddish hue and durability, rosewoods are in high demand for applications such as furniture manufacturing, musical instruments, and traditional medicines. However, this surge in demand has led to the depletion of rosewood populations, particularly in Laos, Thailand, Cambodia, and Vietnam [14].

One of the primary consequences of rosewood overexploitation is the significant loss of habitats. The population of most *Dalbergias* is on the decline across its natural range due to overexploitation for timber and illegal harvest and trade, as well as the conversion of lowland habitats for agriculture and settlement space [11]. Further, pressure on the harvest of wild populations of one species increases due to rules and regulations imposed on another resulting in the numbers in the wild becoming increasingly rare due to exploitation [14]. All rosewoods continue to be exploited across their natural ranges despite logging bans and depleted populations and are subject to illegal logging and harvesting [1].

One of the primary drivers of habitat destruction and deforestation of Asian rosewoods is the illegal logging industry. Three valuable Asian rosewood species—Siamese rosewood (*Dalbergia cochinchinensis*), Burma blackwood (*Dalbergia cultrata*), and Burmese rosewood (*Dalbergia oliveri*)—are at significant risk across 75% of their native ranges due to critical threats from overexploitation, habitat conversion, and climate change [11]. Efforts to combat illegal logging have yielded mixed results. Most countries have introduced stringent regulations and monitoring mechanisms to curb illegal activities, while others have established protected areas and conservation programmes to safeguard remaining rosewood populations. However, enforcement remains a significant challenge due to the high profitability of the illegal timber trade and the limited resources available for monitoring and law enforcement.

Efforts to combat habitat destruction and deforestation of Asian rosewoods must involve a multi-faceted approach. This includes strengthening laws and regulations related to logging and timber trade, enhancing monitoring and enforcement mechanisms, and promoting sustainable forest management practices. Additionally, raising awareness about the importance of preserving Asian rosewood forests and the impact of their destruction is critical in garnering public support for conservation efforts.

International cooperation is also essential in addressing the issue of habitat destruction and deforestation of Asian rosewoods. Given that demand for these trees often comes from global markets, collaboration among countries to regulate and monitor the trade in rosewood timber is crucial. The enforcement of international treaties and agreements, such as the CITES, can play a significant role in curbing illegal logging and ensuring the sustainable management of Asian rosewood forests.

4 CITES Regulations and Efforts to Protect Rosewoods

CITES is an international treaty designed to protect endangered species from the threats posed by international trade. Established after a 1963 IUCN resolution, the agreement was finalized on 3 March 1973, by representatives from 80 countries. CITES officially came into force on 1 July 1975, aiming to prevent trade from endangering species' survival.

CITES oversees international trade in selected species through a licensing system that manages their import, export, re-export, and introduction. The inclusion of various rosewood species in the CITES Appendices has evolved over time in response to growing threats from overexploitation, driven largely by their high value in the timber trade. A chronology of key events related to the listing of rosewood species under CITES is detailed below.

1992	Brazilian Rosewood (*Dalbergia nigra*): One of the earliest rosewood species to be listed. *Dalbergia nigra* was listed in CITES Appendix I due to significant declines in its population from logging and habitat loss, which severely restricted its international trade
2013	Madagascar Rosewood (*Dalbergia* spp.): Several species of rosewood from Madagascar, including *Dalbergia baronii* and *Dalbergia louvelii*, were listed in CITES Appendix II. This listing aimed to regulate the trade of these species to ensure that their exploitation did not threaten their survival
2016	All *Dalbergia* spp. (Global Listing): At the 17th meeting of the Conference of the Parties (CoP17) in Johannesburg, South Africa, the genus *Dalbergia* was listed in CITES Appendix II. This broad listing was implemented to address the issue of species misidentification and to provide comprehensive protection against unsustainable trade, particularly for species targeted for high-end furniture, musical instruments, and other luxury goods
2019	Further Listing Clarifications and Adjustments: During the 18th meeting of the Conference of the Parties (CoP18) held in Geneva, Switzerland, exemptions and annotations were made to the listing of *Dalbergia* species, to facilitate trade in finished products like musical instruments while maintaining protection for the species

Including all *Dalbergia* species in CITES Appendix II mandated that international trade in these species requires permits, ensuring legality and sustainability. This regulation significantly impacted China's rosewood imports, stabilizing the market between 2018 and 2019. To promote the sustainable harvest and trade of rosewood tree species while safeguarding their survival in the wild, CITES released a report at the 27th meeting of the CITES Plants Committee. This report serves as a guide for CITES Parties involved or intending to participate in the sustainable trade of rosewood specimens.

CITES plays a vital role in ensuring the sustainable trade and conservation of Asian rosewoods by developing management plans, harvesting quotas, and traceability systems. By collaborating with range countries, industry stakeholders, and conservation organizations, CITES promotes legal and sustainable trade practices while protecting these valuable species. Its regulations provide a legal framework for international trade, fostering cooperation, scientific research, and habitat conservation. Through compliance mechanisms and collective efforts, the global community can safeguard Asian rosewoods for future generations. The principles of CITES are essential in securing a sustainable future for these species and other endangered wildlife.

5 Improving Timber Identification Methods

Species-level separation within *Dalbergia* has been a long-standing goal, but with a few exceptions, it remains unachievable by wood anatomy alone [1]. Seized logs often lack distinctive morphological and descriptive features, making it nearly impossible to determine the species and source of origin [15]. This challenge in identifying specimens at checkpoints facilitates the illegal timber trade, contributes to deforestation, and devastates local communities [1].

Timber identification and tracking have undergone significant advancements through the application of scientific methods to enhance the verification of timber legality. The primary techniques employed for wood identification encompass visual identification, chemical methods, and DNA approaches [16]. Visual identification relies on wood anatomy and dendrochronology, which help determine the age and characteristics of timber. Anatomical differences between species or groups of species within the genus provide prospects for wood identification at the species level. Variations of axial parenchyma patterns and vessel size potentially can be used for species discrimination. Chemical methods include mass spectrometry, near-infrared spectroscopy, stable isotopes, and radiocarbon dating, all of which contribute to precise timber authentication [13]. In the context of rosewood identification, visual methods such as wood anatomy analysis play a crucial role in distinguishing species based on structural characteristics [17]. Chemical techniques, such as near-infrared spectroscopy, have proven effective in differentiating rosewood species due to their unique chemical signatures [18]. However, the chemical stability of the compounds varies, being influenced by factors such as tree age or habitat [15, 16].

DNA barcoding has also emerged as a powerful tool for accurately identifying rosewood species and tracing their geographic origin [19]. DNA barcoding and other molecular identification techniques can be valuable tools for identifying and tracking rosewoods and other high-value hardwoods. In recent years, multiple studies have showcased the effectiveness of DNA barcoding in species identification and forensic analysis of protected plant species [20].

Chloroplast (cp) DNA markers, such as matK, rbcL, and trnL (UAA), have been proposed as standard DNA barcoding markers [21] and have been effectively used in plant studies for species identification, phylogenetic analysis, and determining sample origin. Research on Asian *Dalbergia* species has demonstrated the potential of these markers for species identification [14, 22, 23]. Additionally, phylogenetic studies have highlighted the utility of standard barcodes like nrITS, matK, rbcL, and trnH-psbA [14, 19, 22, 24–27].

The recent application of DNA barcoding using plastomes has emerged as a powerful method for plant identification and distinguishing closely related taxa [28, 29]. *Dalbergia* plastomes exhibit a diverse array of repeat sequences in terms of number, type, and distribution [30]. These distinctive types, locations, numbers, and abundances of SSRs offer the potential to develop markers for species identification. The integration of scientific methodologies has greatly improved the traceability and accountability of timber harvesting [31].

The LogPro system in Suriname showcases the power of digital technology in timber tracking. By requiring detailed timber data and using independent processors for verification, it reduces manipulation and ensures data integrity. This enhances tracking accuracy, supports efficient forest management, and improves fee collection. Introduced in 2000, LogPro served for two decades before being replaced by SFISS, the Sustainable Forest Management Information System Suriname, which continues advancing forest monitoring and management. Indonesia's approach of combining forest certification with a national legality verification system has set a benchmark for other countries. The involvement of independent auditors ensures transparency and compliance with legal standards. Similarly, Thailand's extensive checkpoint system and mandatory travel permits for timber transportation have proven effective in deterring illegal logging.

6 Conclusion

Asian rosewoods represent a rich natural heritage, combining strength, beauty, and versatility. However, their future depends on responsible trade and conservation efforts. Protecting these species is essential to maintaining ecological balance and sustain traditional craftsmanship in the region.

The high demand for these woods in the global market has created a thriving industry that supports local economies and provides livelihoods for many communities in Asia. The sale of Asian rosewoods generates significant revenue for countries where these trees are grown and harvested, providing a valuable source of income for individuals and businesses involved in the woodworking industry. The sustainable management and responsible harvesting of Asian rosewoods can help ensure the long-term viability of this industry and its economic benefits for local communities.

The importance of Asian rosewoods lies in their exceptional beauty, durability, and versatility. These woods are prized for their aesthetic appeal, acoustic properties, and environmental benefits, making them a valuable resource for artisans, craftsmen, and musicians around the world. The preservation and sustainable management of Asian rosewoods are essential in ensuring the continued availability of these precious trees for future generations. By recognizing and appreciating the unique qualities of Asian rosewoods, we can help promote the conservation of these valuable species and ensure their ongoing contribution to the woodworking industry and the natural environment.

References

1. United Nation Office on Drug and Crime (UNODC) (2020) World wildlife crime report 2020: trafficking in protected species. United Nations, New York, p 132. https://www.unodc.org/documents/dataandanalysis/wildlife/2020/World_Wildlife_Report_2020_9July.pdf
2. Environmental Investigation Agency (EIA) (2014) Routes of extinction: the corruption and violence destroying Siamese rosewood in the Mekong. London
3. CITES Secretariat (2024) Report on the conservation and trade of CITES-listed rosewood tree species [Leguminosae (Fabaceae)]. https://cites.org/eng/topics/flora/cites-and-forests
4. Bian Q, Zhou Y, Zhang H (2020) Research on the status quo and countermeasures of mahogany furniture industry. China For Econ 5:67–69
5. John W (2019) Rosewood trade: China's economic exploitation of Nigeria a case study of Taraba state. IOSR J Humanit Soc Sci 24:28–40
6. Treanor NB (2015) China's Hongmu consumption boom: analysis of the Chinese rosewood trade and links to illegal activity in tropical forested countries. Forest Trends, Washington, DC, USA, p 35
7. Yin JP, Yin YF, Tang ZQ, Jiang XM, Bin LYU (2021) Dynamics of international governance of Hongmu species and analysis of Hongmu imports in China in recent years. Chin J Wood Sci Technol 35(6):77–82
8. Siriwat P, Nijman V (2018) Online media seizure reports: a tool to monitor CITES implementation in regulating the international rosewood trade. For Policy Econ 97:67–72. https://doi.org/10.1016/j.forpol.2018.09.004
9. Cargill N (2022) Assessing the efficacy of CITES restrictions on Malagasy rosewood (Honors theses). University of Richmond, Virginia, USA
10. Waeber PO, Schuurman D, Ramamonjisoa B, Langrand M, Barber CV, Innes JL, Lowry PP, Wilmé L (2019) Uplisting of Malagasy precious woods critical for their survival. Biol Conserv 235:89–92. https://doi.org/10.1016/j.biocon.2019.04.007
11. Gaisberger H, Fremout T, So T, Thammavong B, Bounithiphonh C, Hoa TT, Yongqi Z, Kanchanarak T, Changtragoon S, Sreng S, Ping H et al (2022) Range-wide priority setting for the conservation and restoration of Asian rosewood species accounting for multiple threats and ecogeographic diversity. Biol Conserv 270:109560. https://www.sciencedirect.com/science/article/pii/S0006320722001136?via%3Dihub6
12. Ding Y, Yin Z (2024) The study on the effect of restrictive rosewood trade policies on China's rosewood import prices. For Econ Rev 6(1):58–73
13. Winfield K, Scott M, Grayson C (2016) Global status of Dalbergia and Pterocarpus rosewood producing species in trade. In: Convention on international trade in endangered species 17th conference of parties. Johannesburg, South Africa, pp 1–246
14. Hartvig I, Czako M, Kjær ED, Nielsen LR, Theilade I (2015) The use of DNA barcoding in identification and conservation of rosewood (*Dalbergia* spp.). PLoS One 10:e0138231. https://doi.org/10.1371/journal.pone.0138231
15. Espinoza EO, Wiemann MC, Barajas-Morales J, Chavarria GD, McClure PJ (2015) Forensic analysis of CITES-protected Dalbergia timber from the Americas. IAWA J 36(3):311–325. https://doi.org/10.1163/22941932-20150102
16. Shang D, Brunswick P, Yan J, Bruno J, Duchesne I, Isabel N, VanAggelen G, Kim M, Evans PD (2020) Chemotyping and identification of protected *Dalbergia* timber using gas chromatography quadrupole time of flight mass spectrometry. J Chromatogr A 1615:460775. https://doi.org/10.1016/j.chroma.2019.460775
17. Gasson P, Baas P, Wheeler E (2011) Wood anatomy of CITES-listed tree species. IAWA J 32(2):155–197. https://doi.org/10.1163/22941932-90000050
18. McClure PJ, Chavarria GD, Espinoza E (2015) Metabolic chemotypes of CITES protected *Dalbergia* timbers from Africa, Madagascar, and Asia. Rapid Commun Mass Spectrom 29(9):783–788. https://doi.org/10.1002/rcm.7163

19. Hassold S, Lowry PP, Bauert MR, Razafintsalama A, Ramamonjisoa L, Widmer A (2016) DNA barcoding of Malagasy rosewoods: towards a molecular identification of CITES-listed *Dalbergia* species. PLoS One 11(6):e0157881. https://doi.org/10.1371/journal.pone.0157881
20. COBOL Plant Working Group (2009) A DNA barcode for land plants. Proc Natl Acad Sci USA 106:12794–12797. https://doi.org/10.1073/pnas.0905845106
21. Hollingsworth P, Graham S, Little D (2011) Choosing and using a plant DNA barcode. PLoS One 6:e19254. https://doi.org/10.1371/journal.pone.019254
22. Bhagwat RM, Dholakia BB, Kadoo NY, Balasundaran M, Gupta VS (2015) Two new potential barcodes to discriminate *Dalbergia* species. PLoS One 10:e0142965. https://doi.org/10.1371/journal.pone.0142965
23. Phong DT, Van Tang D, Hien VTT, Ton ND, Van Hai N (2014) Nucleotide diversity of a nuclear and four chloroplast dna regions in rare tropical wood species of *Dalbergia* in Vietnam: a DNA barcode identifying utility. Asian J Appl Sci 2:116–125
24. Li Q, Wu J, Wang Y, Lian X, Wu F, Zhou L, Huang Z, Zhu S (2017) The phylogenetic analysis of *Dalbergia* (Fabaceae: Papilionaceae) based on different DNA barcodes. Holzforschung 71(12):939–949. https://doi.org/10.1515/hf-2016-0236
25. Rakotonirina TJ, Viljoen E, Rakotonirina AH, Leong Pock Tsy JM, Radanielina T (2023) A DNA barcode reference library for CITES listed Malagasy *Dalbergia* species. Ecol Evol 13(3):e9887. https://doi.org/10.1002/ece3.9887
26. Rather SA, Wang T, Liu H, Schneider H (2023) Characterization of the complete chloroplast genome of *Dalbergia congesta* (Fabaceae), an endangered legume endemic to the Nilgiri Hills of Tamil Nadu, India. Funct Integr Genom 23(2):126. https://doi.org/10.1007/s10142-023-01047-7
27. Sotuyo S, Pedraza-Ortega E, Martínez-Salas E, Linares J, Cabrera L (2022) Insights into phylogenetic divergence of *Dalbergia* (Leguminosae: Dalbergiae) from Mexico and Central America. Front Ecol Evol 10:910250. https://doi.org/10.3389/fevo.2022.910250
28. But GWC, Wu HY, Siu TY, Chan KT, Wong KH, Lau DTW, Shaw PC (2023) Comparison of DNA extraction methods on CITES-listed timber species and application in species authentication of commercial products using DNA barcoding. Sci Rep 13(1):151. https://doi.org/10.1038/s41598-022-27195-7
29. Qin M, Zhu CJ, Yang JB, Vatanparast M, Schley R, Lai Q, Zhang DY, Tu TY, Klitgård BB, Li SJ, Zhang DX (2022) Comparative analysis of complete plastid genome reveals powerful barcode regions for identifying wood of *Dalbergia odorifera* and *D. tonkinensis* (Leguminosae). J Syst Evol 60(1):73–84. https://doi.org/10.1111/jse.12702
30. Song Y, Zhang Y, Xu J, Li W, Li M (2019) Characterization of the complete chloroplast genome sequence of *Dalbergia* species and its phylogenetic implications. Sci Rep 9(1):20401. https://doi.org/10.1038/s41598-019-56917-8
31. Stäuble T, Nogueron R, Knorr-Evans M, Schneider T, Foddy L (2023) Timber traceability: a diagnostic tool for practitioners and policymakers: a report. World Resources Institute, Washington, DC. https://doi.org/10.46830/wrirpt.21.00067

Chapter 2
Dalbergia cochinchinensis Pierre

Chaloun Bounithiphonh and Soulivanh Lanorsavanh

1 Introduction

Indochina is home to 28 indigenous species of *Dalbergia* genus. *Dalbergia cochinchinensis* is a species in the family Fabaceae. It is commonly referred to as Siamese or Thailand Rose Wood, and the vernacular name is *Kha Gnoung* in Lao, *Payung* in Thai, and *Trac* in Vietnamese [1]. This high-value species is in demand in domestic and international markets. It is native to Cambodia, Laos PDR, Thailand, and Vietnam [1, 2], However, despite its broad geographic distribution, its area of occupancy is limited due to fragmented populations. This fragmentation is primarily caused by unsustainable harvesting practices throughout its range.

It is found in lower Mekong countries, namely Thailand, Laos, Vietnam, Cambodia, and Myanmar. The species can grow in any type of soil at an elevation of 1,200 m, growing up quickly at a young age and slowing down over the years [3, 4]. *D. cochinchinensis* can be found mainly in mixed deciduous forests, evergreen forests, and dry dipterocarp forests. Reports state that in Thailand, the number of *D. cochinchinensis* trees dropped from 300,000 in 2005 to 80,000 to 2011 [5]. In Vietnam, even within the protected areas, *D. cochinchinensis* density was as low as 1–10 trees per hectare [5], similar to 1.37 per hectare in Cambodia [6]. In Laos, field surveys in 2012 confirmed no mature trees could be found in the natural populations of *D. cochinchinensis* [7].

C. Bounithiphonh (✉)
Nongviengkham Village, Saythany District, Vientiane Capital 7170, Lao PDR
e-mail: chalounb@yahoo.com; k.jular28@gmail.com

National Agriculture and Forestry Research Institute, Vientiane, Laos

S. Lanorsavanh
Department of Biology, Faculty of Natural Sciences, National University of Laos, Dong Dok Campus, Saythany District, Vientiane Capital, Lao PDR

R. R. Warrier et al. (eds.), *Handbook of Asian Rosewoods*, Materials Horizons: From Nature to Nanomaterials, https://doi.org/10.1007/978-981-96-9459-4_2

According to the Environmental Investigation Agency (EIA), between 2000 and 2014, more than 3.5 million cubic meters of rosewood were illegally logged in Southeast Asia [8]. Thailand alone reported over 100,000 cases of rosewood smuggling between 2009 and 2019, resulting in the confiscation of nearly 50,000 cubic meters of illegal timber. In Cambodia, deforestation rates linked to rosewood have soared, with forests declining by 20% between 2001 and 2020 [9, 10]. The high demand in China for luxury "Hongmu" furniture has driven market prices of rosewood to over $50,000 per cubic meter, making it more valuable than gold. Imports took place despite the fact that harvesting in some cases had been illegal or that exports were in contravention of national laws [8]. The rapid depletion of Siamese rosewood reflects broader challenges in combating illegal logging and preserving biodiversity in the Mekong region.

2 Habitat and Ecology

This evergreen tree species grows up to 15–35 m tall, with trunk diameters ranging from 60 to 120 cm at breast height [11]. It is commonly found in lowland mixed deciduous or dry evergreen forests [12] and sparsely distributed in semi-deciduous forests at 400–500 m above sea level. At higher elevations, its presence diminishes, but it can occasionally be found in riparian or evergreen forests [5]. Known as an intermediate pioneer species [13], its seedlings are light-demanding yet require forest cover for sustained growth. The species grows slowly, with a few trees reaching maturity. It reproduces through coppicing and root suckering [4, 13, 14]. Flowering typically occurs from March to June, with fruiting completed by December in most regions [15]. However, mature trees capable of flowering and fruiting are rare [12]. The species is likely insect-pollinated, while its flattened seedpods suggest wind dispersal [3, 12].

3 Morphology and Taxonomy

Consider as trees with average height 10 to 30 meters tall, bark greyish to brownish yellow, young shoots puberulous; *stipules* small, caducous; *petioles* 2.3–5 cm long; *leaves* imparipinnate, spiral; rachis 6.3–15.2 cm long, glabrous; *leaflets* 7–9, alternate; lamina ovate to narrowly ovate or ovate-lanceolate, 3–8.5 × 2–5 cm, base obtuse or rounded, margin entire, apex obtuse or acute, upper surface dark green, lower surface greyish, both of surfaces glabrous; secondary veins 7–9 pairs; petiolules 2–5 mm long (Fig. 1). Inflorescence's panicles, terminal and in leaf axils, 10–20 cm long; Bract and bracteoles early caducous. *Pedicels* ca. 1 mm long. *Flowers* white or whitish, 5–6 mm long. *Calyx* 5-teethed, ca. 5 mm long, yellowish, glabrous or sparely hairy. *Corolla* 5 petals, white; limb of standard petal orbicular, 3.5–4.5 × 2.5–3.7 mm, claw ca. 1.5 mm long; wing petals oblong, ca. 3.5 × 1 mm, claw

Fig. 1 Profuse fruiting (left) and imparipinnate leaves (right) of *Dalbergia cochinchinensis*

ca. 1.5 mm long; keel petals ca. 3.5 × 2 mm, claw ca. 1.5 mm long. *Stamens* 10, monadelphous (5 + 5); 5–6 mm long. *Ovary* glabrous, 2–4-ovuled; stipe hairy, ca. 3 mm long; style ca. 1.5 mm long; stigma small. *Pod* linear-oblong, glabrous, 4–8 × 0.8–1.2 cm, pericarp very thin, light brown when dry, indehiscent, 1–4-seeded. *Seeds* reniform, ca. 4 × 6 mm, brown or reddish [14, 15].

4 Distribution

Lao PDR: In Lao PDR, *Dalbergia cochinchinensis* thrives in diverse forest types, including mixed deciduous, evergreen, and Dipterocarp forests. It grows at elevations ranging from 400 to 600 m and is distributed across several provinces: Xayabouly, Luang Prabang, Vientiane, Vientiane Capital, Bolikhamxay, Khammouane, Savannakhet, Saravan, Attapeu, and Champassak [3, 4].

Cambodia: *Dalbergia cochinchinensis* is widely distributed in provinces such as Kampong Thom, Kampung Speu, Preah Vihear, Ratanakiri, Pursat, Siem Reap, Kratie, Koh Kong, Stung Treng, and Modulkiri [16]. However, its population size remains uncertain and is believed to have drastically declined. Mature individuals are now considered extremely rare outside strictly protected areas [17].

Vietnam: *Dalbergia cochinchinensis* is sparsely distributed across central and southern provinces, including Da Nang city and Quang Nam (Hien and Phuoc Son districts), Kon Tum (Dak Ha, Sa Thay, Ngoc Hoi, and Kon Plong districts), Gia Lai (Krong Pa, Chu Pah, Ia Pa, Ayun Pa, Mang Yang, Dak Doa, Ia Grai, K'Bang, and Duc Co districts), Dak Lak (Buon Don, Ea Sup, Ea Kar, Krong Nang, and Krong Pong districts), Lam Dong (Cat Tien, Bao Lam, and Da Teh districts), Binh Thuan (Ham Thuan Bac district), Dong Nai (Dinh Quan, Trang Bom, Vinh Cuu, Tan Phu, and Thong Nhat districts), Tay Ninh (Tan Bien district), and Kien Giang (Phu Quoc district) [18, 19]. Within Vietnam's protected areas, *Dalbergia cochinchinensis* is found in Dak Uy Special Use Forests, Chu Mom Ray National Park, Kon Plong watershed protection forests, Yok Don National Park, Ea So Nature Reserve, Krong

Nang, Ia Rsai, Nam Song Ba, Tan Phu, Ham Thuan—Da Mi, and Duc Co watershed protection forests. It also exists in Kon Ka Kinh National Park, Kon Chu Rang Nature Reserve, Dong Nai Nature Reserve, Cat Tien National Park, and Phu Quoc National Park [18].

Thailand: The species has been recorded from 14 provinces, scattered in the northeast region at an altitude below 800 m, the southeast region, Pisanuloke, Petchaboon, Lopburi, Nakorn Nayok, and Saraburi.

5 Present Status

Laos PDR: Historical surveys, such as those by Vidal (1956–1960), identified this species in semi-evergreen forests on *terre rouge* soils (red soils) derived from basaltic parent material. Sampling locations included Phou Ba Chieng, east of Pakse; Dong Houa Sao, south of Pakse; and Phang Ham in Salavane Province. Vidal's study highlighted that *D. cochinchinensis* is notably dominant in various forest types, such as evergreen forests, mixed deciduous forests, semi-evergreen forests, degraded montane forests, and deciduous Dipterocarp forests and woodlands. However, over the last five decades, significant forest destruction, particularly of evergreen forests, has led to a decline in suitable habitats. The National Forest Inventory survey by the Department of Forestry, in 2019 [20], documented the presence of *D. cochinchinensis* and related species. In Nakai-Nam Theun National Park, Khammouane Province, *D. cochinchinensis* was recorded in a single sample plot. In contrast, *D. oliveri* was identified in 27 records across 15 sampling plots distributed in Houaphanh, Xiengkhouang, Khammouane, Savannakhet, Sekong, and Champassak provinces. The survey spanned 511 plots, covering a total area of 100 ha, with comprehensive tree species documentation. Further field surveys on the occurrence, distribution, and population status of *D. cochinchinensis* were conducted in 2022, by the National Agriculture and Forestry Research Institute (NAFRI) and the National University of Laos (NUOL). This effort focused on six provinces—Bolikhamxai, Khammouane, Savannakhet, Salavan, Attapeu, and Champassak—and involved 24 districts and 52 villages. The survey identified *D. cochinchinensis* in 183 plots across 23 districts and 51 villages, with most occurrences concentrated in Bolikhamxay, Khammouane, Savannakhet, and Salavan provinces. A total of 6,697 individual *D. cochinchinensis* trees were inventoried during the survey, emphasizing the importance of these forests for conserving this valuable species [23, 24].

Cambodia: The largest known population of *D. cochinchinensis* is a seed source in Siem Reap Province. The second largest population, found in the Leap Kuy Community Forest in Kampong Speu Province, consists of approximately 200 trees within a 107-ha natural forest. Additional populations include the Damrey Chak Thlork Community Forest (15,000 ha) in Kampong Speu, O Soam Community Forest in Kampong Thom Province, which hosts 50–100 trees with dbh of 10–15 cm, and Tbeng Lech Community Forest in Siem Reap Province, where only about 10 trees remain after the largest tree was illegally cut in 2017 [17]. The Cambodia Tree Seed Project (2003) reported an average of just 1.34 *D. cochinchinensis* trees per hectare in the natural forests of Sre Nauy commune, Siem Reap Province. A 2007 survey in Stung Treng Province's lowland forests revealed the species had been extirpated locally due to illegal logging. Surveys in Samkos within the Central and Eastern Cardamom Mountains in 2012 reported only one individual *D. cochinchinensis*—a root sucker surviving after the mother tree was felled. Rangers in the Southern Cardamom Mountains similarly noted that all trees in the area had been felled for the rosewood trade. Field guides reported in 2015 that many populations documented between 2010 and 2012 no longer existed due to deforestation and logging [17, 20].

Vietnam: Field surveys estimate approximately 8,000 timber trees (DBH1.3 > 6 cm) in Dak Uy Special Use Forest (SUF) and 8,000–10,000 in Yok Don National Park [21]. A similar population exists in Kon Ka Kinh National Park, although it contains fewer large trees (Hoang Thanh Son, pers. comm.). Additionally, about 1,000 seed-bearing trees were reported in the Tan Phu watershed protection forest (Le Viet Dung, pers. com.). Smaller populations of *Dalbergia cochinchinensis* are sparsely distributed in areas like Chu Mom Ray National Park, Nam Song Ba, and Ia Rsai watershed protection forests. They are also found in plantations, forest gardens, production forests, and local villagers' crop fields in central provinces. However, comprehensive surveys to quantify the number of timber trees in these locations are lacking [20, 22].

Despite government efforts in all the countries to conserve the species through laws and regulations, illegal logging and habitat destruction persist. Key threats include land conversion for agriculture and infrastructure projects. Surveys reveal heavy harvesting, with 619,641 m^3 of timber lost, fragmenting habitats and hindering natural regeneration [23, 24].

Propagation

Its propagation can be achieved through various methods, including seeds, vegetative techniques, and natural regeneration means such as coppicing and root suckers. The seeds require pretreatment for enhancing germination. Pre-soak the seeds in water for 24 h before sowing in sand, covering them lightly with soil. Water regularly, ensuring the sand remains moist but not overly saturated. Seeds typically germinate within 7 days. Transfer seedlings to poly bags 10–14 days after germination. When transplanting, use a 4 × 6-inch grid arrangement. The growing medium should consist of alluvial soil, sand, rice husk ash, and compost in a 5:2:2:1 ratio. After 3–5 months, the seedlings should reach a height of approximately 30 cm and be ready for out-planting.

Pests and Diseases

Fungi isolation usually happens with the leaves of the seedlings as well as the big trees, where the isolated fungal species are *Alternaria alternata*, *Colletotrichum siamense*, *Curvularia aeria*, and *Maravalia peterocarpi* [25]. The heartwood is hard and resistant to insect damage (including termites).

Uses

Siamese rosewood, prized for its durable red heartwood, is a sought-after timber for luxury furniture, carvings, and instruments. The timber is also used in construction activities [6].

Genetic Resources and Conservation

Cambodia: Cambodia has implemented significant measures to protect Siamese rosewood (*Dalbergia cochinchinensis*), recognizing it as a priority species for conservation. The species is protected under the Cambodian Forestry Law of 2002, with a ban on its processing, use, and export since 2013. Networks of conservation stands have been established and protected by Ministerial Declaration to preserve genetic diversity. These stands serve as potential seed sources for future replanting and restoration programmes. A notable restoration effort has been taken up at Khun Ream commune, Siem Reap, where a natural stand of mature trees is strictly preserved. Trial plantations were initiated in 2004 in the Kbal Chay watershed area, Sihanouk province, to study and expand the species' growth. A grafted seed orchard covering 0.45 ha was established in Pursat province with support from the UK Darwin Initiative, ensuring sustainable seed production for propagation.

Laos PDR: The Government of Laos has enacted laws and regulations to guide sustainable forest management and conservation. In 2000, research identified *Dalbergia cochinchinensis* for inclusion in conservation efforts, focusing on seed source establishment, improvement, and procurement in Lao PDR [26]. In 2007, the Ministry of Agriculture and Forestry issued Decision No. 0116/MAF, dated 7/5/2007, which stipulates three tree species lists with the following three categories (i.e. MAF Appendix I, II, and III).

MAF-Appendix I: refers to rare, medicinal, endangered, endemic, slow-growing, enduring species, some of which are listed in the International Convention on Trade in Endangered Species of Wild Fauna and Flora (CITES).

MAF-Appendix II: refers to the species of trees that grow and reproduce in some areas and grow slowly in nature, with moderately durable woods.

MAF-Appendix III: refers to species that grow, propagate, and grow well in natural, low-tolerance woods.

Efforts to conserve *Dalbergia cochinchinensis* have been guided by government policies, conservation projects, and research initiatives. A national order in 2008 banned harvesting of *D. cochinchinensis*, with Decision No. 0448/MAF (2021) listing it under Appendix I, prohibiting wild harvest without government approval. Logging in natural forests was suspended in 2013, limiting legal timber harvesting to land conversion projects. Conservation efforts focus on creating protected areas, such as those spanning Bolikhamxay to Attapeu provinces. Under the NAFRI-Darwin Dalbergia project, *in situ* and *ex situ* conservation units were established in Savannakhet province. Seed sources for *D. cochinchinensis* were identified. Currently, a single viable seed source of *D. cochinchinensis* remains at NAFRI's Forest Research

Center, housing 62 breeding trees. Restoration efforts have included natural regeneration and enrichment planting, though documentation on planted areas remains sparse. Experimental plantations have also been established, with notable successes in Vientiane Capital, demonstrating effective conservation and genetic resource preservation [23, 24].

Vietnam: The country has undertaken substantial measures to protect *Dalbergia cochinchinensis* (Siamese rosewood) through legislation, conservation initiatives, and international cooperation. The country has introduced strict regulations to control the use of Siamese rosewood. A national logging ban, first implemented in 1992 and extended in 2006, prohibits the harvesting of logs and sawn wood from natural forests, except for plantations. In 2014, the government further prohibited the commercial harvesting of *D. cochinchinensis*, limiting its use exclusively to scientific activities such as breeding and artificial propagation [27]. Additionally, Vietnam is a signatory to the European Union Forest Law Enforcement, Governance, and Trade Voluntary Partnership Agreement (EU FLEGT VPA). This initiative aims to promote sustainable forest management, combat illegal logging, and ensure the legality of timber exports to the EU. Through this agreement, Vietnam has strengthened efforts to trace the origins of timber, including Siamese rosewood, demonstrating its commitment to sustainable forestry practices.

Efforts for Conservation and Maintenance of Genetic Diversity

There are country-level conservation practices and policies in place for *Dalbergia cochinchinensis* (Siamese rosewood), implemented to varying degrees across its native range.

Cambodia: In Cambodia, the species is designated as a priority for conservation and protected under the Cambodian Forestry Law of 2002 [14]. A ban on the processing, use, and export of the species has been in effect since 2013. To preserve genetic diversity, conservation stands have been protected by Ministerial Declaration, which could serve as future seed sources for replanting and plantation programmes. A restoration programme is underway at Khun Ream commune, Siem Reap, where a natural stand of mature trees is strictly preserved. In 2004, trial plantations were established in the Kbal Chay watershed area, Sihanouk province. Additionally, a grafted seed orchard spanning 0.45 hectares was set up by a farmer in Pursat province with technical and financial support from the UK Darwin Initiative to produce seeds for future propagation efforts.

Lao PDR: Lao PDR banned the logging of *D. cochinchinensis* in 2008 and prohibited its timber export in 2011. The 2007 Lao Forestry Law emphasizes enriching non-timber forest products, including surveying and conserving threatened species [6]. The revised Forestry Law of 2019 has intensified conservation actions, listing *D. cochinchinensis* in Tree List I of prohibited species [28]. Efforts continue to strengthen legal frameworks and implementation to ensure the species' survival [29]. The National Strategic Conservation plan proposed 2 action plans with 6 focused targets to strengthen the conservation efforts for *Dalbergia cochinchinensis*.

Thailand: In Thailand, *D. cochinchinensis* is classified as restricted timber. A logging ban on natural forest specimens has been in place since 1989 [6]. The species remains protected within designated conservation areas, ensuring its preservation in natural habitats [5].

Vietnam: Vietnam has stringent policies governing the species' use, limiting it exclusively to scientific purposes, such as breeding and artificial propagation. Commercial harvesting of *D. cochinchinensis* was banned in 2014 [30, 31]. Logging prohibitions initially established in 1992 were expanded in 2006 to cover logs and sawn wood from natural forests, with exceptions for plantations. Vietnam is a signatory to CITES and has initiated the EU Forest Law Enforcement, Governance, and Trade Voluntary Partnership Agreement (FLEGT VPA) to combat illegal logging [32]. However, challenges remain. Between 2016 and 2017, timber trade quotas were introduced, facilitating cross-border trade with Cambodia, where protected areas were targeted for timber harvesting [31]. Discrepancies in timber import–export records highlight gaps in transparency and the need for improved enforcement to curb illegal logging activities.

Conclusion

The conservation of *Dalbergia cochinchinensis* is a pressing issue that requires immediate attention. Despite the implementation of various conservation measures, these species continue to face significant threats from overexploitation and habitat loss. To ensure the long-term survival of these valuable species, a multi-faceted approach is necessary. This includes strengthening law enforcement, promoting sustainable forest management practices, and investing in scientific research. Community engagement is crucial, as local communities play a vital role in conservation efforts.

By fostering international collaboration and sharing knowledge, we can enhance our understanding of these species and develop effective conservation strategies. Ultimately, the success of these efforts will depend on a concerted effort from government agencies, non-governmental organizations, and local communities.

In conclusion, *D. cochinchinensis* represents a significant component of Indochina's rich biodiversity. This highly valued hardwood species faces immense pressure due to overexploitation and habitat degradation, underscoring the urgent need for conservation efforts. Sustainable forest management practices, coupled with stricter enforcement against illegal logging and habitat preservation, are crucial to ensuring the long-term survival of this valuable resource. The ecological and economic significance of *D. cochinchinensis* necessitates a comprehensive and collaborative approach involving local communities, governments, and international organizations to safeguard this species for future generations.

References

1. Niyomdham C (2018) *Dalbergia* L. f. In: Flora of Thailand 4 (3.1). The Forest Herbarium, Bangkok, pp 221–487
2. POWO (2024) Plants of the world online. Facilitated by the Royal Botanic Gardens, Kew. https://powo.science.kew.org/taxon/urn:lsid:ipni.org:names:490142-1. Accessed 10 Dec 2024
3. Hartvig I, So T, Changtragoon S, Tran HT, Bouamanivong S, Theilade I, Kjær ED, Nielsen LR (2017) Population genetic structure of the endemic rosewoods *Dalbergia cochinchinensis* and *D. oliveri* at a regional scale reflects the Indochinese landscape and life-history traits. Ecol Evol 8(1):530–545
4. Hartvig I, So T, Changtragoon S, Tran HT, Bouamanivong S, Ogden R, Senn H, Vieira FG, Turner F, Talbot R, Theilade I (2020) Conservation genetics of the critically endangered Siamese rosewood (*Dalbergia cochinchinensis*): recommendations for management and sustainable use. Conserv Genet 21(4):677–692. https://doi.org/10.1007/s10592-020-01279-1
5. CITES (2013) Convention on international trade in endangered species of wild fauna and flora. CoP16 Prop 60. https://cites.org/sites/default/files/eng/cop/16/prop/E-CoP16-Prop-60.pdf
6. Winfield K, Scott M, Graysn C (2016) Global status of *Dalbergia* and *Pterocarpus* rosewood producing species in trade. In: Convention on international trade in endangered species 17th conference of parties. Johannesburg, South Africa, pp 1–246
7. Hung TH (2022) Ecological genomics and adaptation of rosewoods *Dalbergia cochinchinensis* and *D. oliveri* for conservation and restoration. Doctoral dissertation, University of Oxford
8. Wiedenhoeft A (2016) Best practice guide for forensic timber identification. united nations office on drugs and crime: international consortium on combating wildlife crime. Vienna, Austria: Laboratory and Scientific Section, Global Programme for Combating Wildlife and Forest Crime, New York, NY, USA, pp 1–226
9. Siriwat P, Nijman V (2018) Using online media-sourced seizure data to assess the illegal wildlife trade in Siamese rosewood. Environ Conserv 45(4):352–360
10. Siriwat P, Vincent N (2023) Quantifying the illegal high-value rosewood trade and criminal trade networks in the Greater Mekong Region. Biol Conserv 277:109826
11. Eiadthong W, Tangmitcharoen S (2015) Ecological and geographical distributions and demographic status of Siamese rosewood (*Dalbergia cochinchinensis* Pierre) in Thailand. In: FORCONS proceedings, 2015, pp 1–3
12. Moritsuka E, Chhang P, Tagane S, Toyama H, Sokh H, Yahara T, Tachida H (2017) Genetic variation and population structure of a threatened timber tree *Dalbergia cochinchinensis* in Cambodia. Tree Genet Genomes 13(6):1–11
13. So T, Theilade I, Dell B (2010) Conservation and utilization of threatened hardwood species through reforestation—an example of *Afzelia xylocarpa* (Kruz.) Craib and *Dalbergia cochinchinensis* Pierre in Cambodia. Pac Conserv Biol 16(2):110–116
14. Cheang D, Burgess S, Sloth A, So T, Sok T (2004) Cambodia tree species monographs. Forestry Administration, Cambodia Tree Seed Project, DANIDA
15. Jøker D (2000) *Dalbergia cochinchinensis* Pierre. Seed leaflet, DANIDA Forest Seed Centre, No. 26, pp 1–2
16. Khorn S (2003) Distribution of selected tree species for gene conservation in Cambodia. In: Forest gene conservation strategy, CTSP/DANIDA, Phnom Penh, Cambodia, pp B2.1–B2.7
17. UNEP-WCMC (2014) Non-CITES timber species from South Asia (Leguminosae) potentially warranting further protection. UNEP-WCMC, Cambridge, UK
18. La QT, Bui TT, Nguyen MH, Van DTK, Nguyen HN (2021) Development of distribution maps for *Dalbergia cochichinensis* and *Dalbergia oliveri* in Vietnam. Center for Nature Conservation and Development, Ha Noi, Vietnam
19. Ha NM, Van DTK, Trung LQ, Hiep NT (2021) A review of the current harvest control and monitoring of *Dalbergia cochinchinensis* and *Dalbergia oliveri* in Vietnam. Center for Nature Conservation and Development, Hanoi, Vietnam, p 18
20. Kameda C, Thammavong V, Phouphasouk S (2019) Forest plants of central Laos: a field guide for forestry. Faculty of Forest Science, National University of Laos, Vientane

21. Dinh TKV, La QT, Ta QT, Hoang TS, Nguyen MH (2021) Report on the systematic field survey of the population distribution, abundance and stocking of *Dalbergia cochinchinensis* and *Dalbergia oliveri* in four key protected areas of Dak Uy, Bu Gia Map, Yok Don and Cat Tien in Vietnam. Center for Nature Conservation and Development, Ha Noi, Vietnam
22. Ha NM, La QT, Dinh TKV, Ban DV, Hiep NT (2021) Identification manual for *Dalbergia cochinchinensis* Pierre and *Dalbergia oliveri* Gamble ex Prain. Center for Nature Conservation and Development, Hanoi, Vietnam p 34
23. NAFRI (2022) Developing CITES non-detriment findings for rosewoods in Lao PDR. Report on the status quo of rosewood species of *D. cochinchinensis* and *D. oliveri* in Lao PDR. NAFRI, Lao PDR
24. DoF—Department of Forestry (2021) The implementation of five year report (2016–2020) and plan of forest sector in Lao PDR (2021–2025)
25. Jitjak W, Chairop W, Sanoamuang N (2021) Molecular identification of fungal species causing brown circular leaf spot disease in seedlings of Siamese Rosewood (*Dalbergia cochinchinensis* Pierre ex Laness). Sci Technol Asia 26(3):156–166
26. Theppavong B, Khamphan K, Vonghachack S (2001) Conservation and management of forest genetic resources in Lao DPR. Asia Pac For Genet Resour 1(1):29–34
27. Environmental Investigation Agency (EIA) (2016) Red alert: how fraudulent Siamese rosewood exports from Laos and Cambodia are undermining CITES protection. EIA, London p 6
28. Government of Lao PDR 2019 (2019) Planting Promotion and Forest Restoration Sectoral, Department of Forestry, Ministry of Agriculture and Forestry. Summary of the 2019 Forest Planting Promotion and Forest Restoration Program and 2020 Plan No. 455/PFRS dated December 23, 2019. Vientiane, Lao PDR
29. Ministry of Agriculture and Forestry (MAF) (2021) The classification of prohibited tree species. Laos PDR
30. Environmental Investigation Agency (EIA) (2012) Rosewood robbery: the case for Thailand to list rosewood on CITES. EIA, London, p 4
31. Environmental Investigation Agency (EIA) (2018) Serial offender. Vietnam's continued imports of illegal Cambodian timber. EIA, London, p 15
32. Environmental Investigation Agency (EIA) (2017) Repeat offender. Vietnam's persistent trade in illegal timber. EIA, London, p 20

Chapter 3
Cultivation Prospects of *Dalbergia cochinchinensis* Pierre in China

Yongqi Zheng and Yongxiang Li

1 Introduction

Dalbergia cochinchinensis Pierre is a large tree species in the Fabaceae family and the *Dalbergia* genus. This species was named by French botanists Pierre and De Lanessan in 1886. *D. cochinchinensis*, commonly known as Siam rosewood, is primarily distributed in Laos, Thailand, Vietnam, and Cambodia. It does not occur naturally in China and is only limited in cultivation (Fig. 1). Current research on *D. cochinchinensis* mainly focuses on species identification, seedling cultivation, wood structure analysis, and germplasm collection and conservation. To ensure the sustainable use of this species, *D. cochinchinensis* has been included in the revised Convention on International Trade in Endangered Species of Wild Fauna and Flora (CITES) list.

2 Botanical Characteristics Description

Dalbergia cochinchinensis grows up to 30 m tall with a 60–120 cm diameter. Covered in fine hairs, its bark is light yellow (Fig. 2a) [1], with pinnate leaves, sparse white flowers (Fig. 2b) [2], in terminal or upper axillary panicles (Fig. 2c) [3], and leguminous, elongated fruits with 1–3 seeds (Fig. 2d) [1]. Flowering occurs in February–June and fruiting during April–August [4].

Y. Zheng (✉) · Y. Li
Institute of Forestry, Chinese Academy of Forestry, Haidian District, Beijing, China
e-mail: zyq8565@126.com

R. R. Warrier et al. (eds.), *Handbook of Asian Rosewoods*, Materials Horizons: From Nature to Nanomaterials, https://doi.org/10.1007/978-981-96-9459-4_3

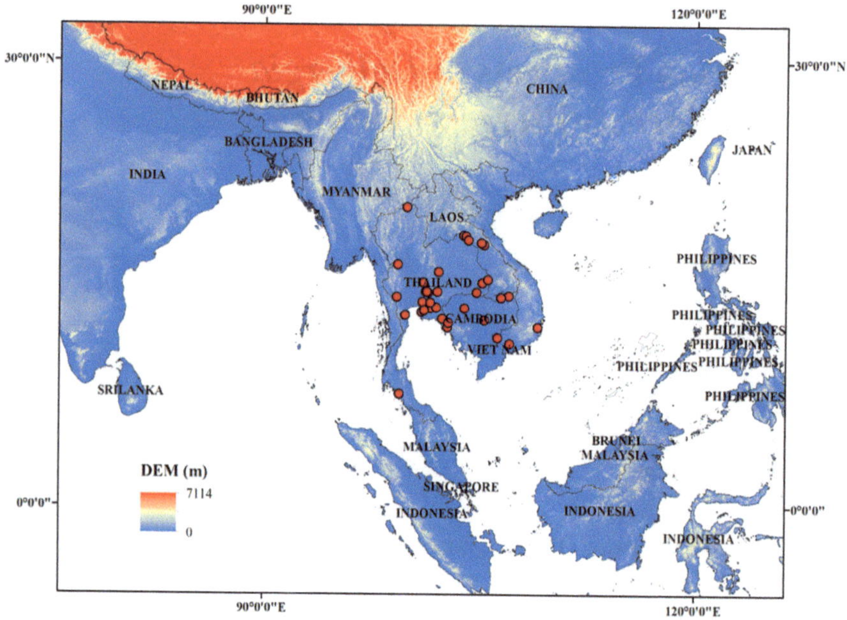

Fig. 1 Geographical distribution of *Dalbergia cochinchinensis*

3 Habitat Description

Dalbergia cochinchinensis thrives in tropical rainforests at 100–500 m elevations, requiring 20–32 °C and 1200–1700 mm annual rainfall. Shade-tolerant in the initial stages, saplings demand more sunlight. It grows slowly but adapts to various soils, including clay, sandy, and calcareous, tolerates drought, and fixes nitrogen in nutrient-depleted soils, improving forest conditions [5].

4 Taxonomy and Identification

Given the numerous species within the *Dalbergia* genus and their similar morphological traits, accurately distinguishing between different rosewood species is essential for identifying legally traded timber and preventing illegal logging and trade. Traditional taxonomic methods primarily rely on external morphological characteristics of the tree, such as leaves, bark, inflorescences, and fruits. Microscopic analysis of wood structures, such as vessels, fibres, and rays, is another effective identification method. Microscopic examination of *D. cochinchinensis* wood slices and near-infrared spectroscopy effectively differentiate its unique wood structure (Fig. 3) [6]. Wood has interlocking grain patterns, with a relatively low amount of brown gum in the vessels.

Fig. 2 The partial morphological photos of *Dalbergia cochinchinensis*

The primary cell type in the vessels is cylindrical, comprising 44.56% of the vessel cells, while the number of balloon-shaped cells is minimal at 3.96% [7].

With the advancement of molecular biology techniques, DNA barcoding and other molecular markers have become widely used in plant taxonomy [8]. These techniques analyse specific DNA sequences to identify species accurately. For *D. cochinchinensis*, commonly used genetic markers include chloroplast genes (e.g., rbcL, matK) and ribosomal genes (e.g., ITS). Additionally, specific chemical compounds in *D. cochinchinensis* wood, such as flavonoids [9], serve as chemical fingerprints for identification. Techniques like gas chromatography-mass spectrometry (GC-MS) [10] and liquid chromatography-mass spectrometry (LC-MS) [11] are frequently used to detect these chemical components.

5 Cultivation Practices

The seeds of *D. cochinchinensis* mature between November and December each year, measuring approximately 6.3 cm long, 4.4 cm wide, and 1.2 cm thick, with a thousand-seed weight of 25.2 g. To protect the seeds from pests and diseases, collection is typically done when the pods turn from green to brownish-yellow, followed

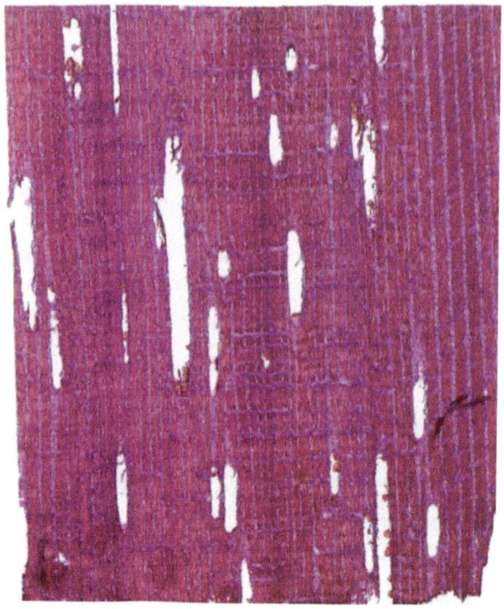

Fig. 3 Radial longitudinal section of *Dalbergia cochinchinensis* wood

by drying and threshing [12]. Seed germination requires specific temperature and humidity conditions, with research indicating that temperatures of 25–30 °C and relatively high humidity promote germination [13]. Additionally, pre-soaking and light scarification of seeds can improve germination rates. The soil should be moist but not waterlogged during sowing to prevent seed rot [14, 15] (Fig. 4).

The highest survival rate for *Dalbergia cochinchinensis* vegetative propagation is achieved using spring cuttings. Optimal results are obtained by maintaining cutting lengths at 12 cm and treating them with growth regulators, mainly NAA. Applying a 300 mg/L multi-growth regulator solution for subsequent management ensures a survival rate of 94.87%. After treatment, seedlings exhibit optimal growth when transplanted into nutrient-enriched bags, supporting their early development and vigour [16].

Fig. 4 Seedling rearing of *Dalbergia cochinchinensis*

Improvement Programmes in China

Growth differences were observed among *D. cochinchinensis* families maintained in a state-owned forest farm in Longhai, Fujian Province. This highlights the need for selecting high-performing families in improvement programmes for *D. cochinchinensis* [17].

To encourage the cultivation of *D. cochinchinensis* in the south subtropical region of China, seeds from 10 provenances and 25 families from Thailand and Cambodia were introduced to Pingxiang, Guangxi, and their germination and growth performance were evaluated. The trial established in 2016 revealed that by the end of three years, growth performance among provenances showed no significant differences, while significant variation was observed among families with Thai provenances and families demonstrating superior adaptability [18].

A provenance cum progeny trial evaluation of 10 provenances from Thailand and Cambodia, across three sites for provenance differences, environmental interactions, geographic variation, and genetic stability, revealed that Southern China, especially tropical zones especially Ledong (Hainan), Guangzhou (Guangdong), and Zhangzhou (Fujian) offer suitable conditions for *D. cochinchinensis* cultivation, with Chachoengsao Province from Thailand and Siem Reap Province from Cambodia ideal for genetic breeding and introduction efforts [15, 18].

In an evaluation of growth differences and genetic stability of 26 *D. cochinchinensis* introduced from Cambodia and Thailand across three sites in Zhangzhou, Fujian Province, it was observed that a mixed planting model of *D. cochinchinensis* with species such as acacias, large-flowered eucalyptus, and giant eucalyptus, followed by management practices such as pruning and thinning, can effectively promote the growth and result in large diameter trees. Thailand provenances performed better in the sites [19].

The heritability of each growth trait in the young forest of *D. cochinchinensis* is generally high and is under firm genetic control. The mixed-forest model supports better than a monoculture block model. The early growth traits vary significantly in different parts of the country with different provenances from Thailand and Cambodia. Hence it is mandatory to promote families with strong adaptability and high productivity for general promotion, while families with high yield and poor stability are to be deployed only in suitable areas showing moderate growth.

Successful *D. cochinchinensis* plantations have been established in China, Thailand, Vietnam, and other regions, showing good growth. To gradually build sustainable stands, these efforts combine scientific research, community involvement, and government support through seed orchard establishment, nursery cultivation, and ecological monitoring (Table 1).

Table 1 Existing provenance or progeny trials of *D. cochinchinensis* across their species range

Country	Location	Number of provenances	Provenances covered	Number of families	Year of establishment
Cambodia	Siem Reap	7	Cambodia	100	2016
Thailand	Khao-Yai Field Station	7	East Thailand	85	1988
Vietnam	Chumomray NP and buffer zone	2	–	100	2003

6 Genetic Diversity

Studying genetic diversity in germplasm resources is a crucial prerequisite for research on tree breeding and selection of superior varieties, playing a decisive role in advancing breeding processes. The natural distribution of *D. cochinchinensis* is limited, primarily concentrated in Southeast Asian countries and regions such as Cambodia, Laos, Thailand, and Vietnam. According to existing records of natural forest distributions, the species is often found in scattered or localised patches and is relatively sparse. There is limited research on the population genetics of *D. cochinchinensis*. To develop effective resource conservation strategies and provide a scientific basis for the exploration and utilisation of superior germplasm resources, researchers have employed molecular markers such as isoenzymes, Random Amplified Polymorphic DNA (RAPD), inter-simple sequence repeat (ISSR), simple sequence repeat (SSR) (Table 2), and (Single Nucleotide Polymorphisms (SNP) to conduct comprehensive studies on the genetic diversity of *D. cochinchinensis* populations in Thailand, Laos, Vietnam, and Cambodia [20, 21]. Studies indicate that it has a moderate level of genetic diversity, high degrees of differentiation, and extremely high levels of outcrossing within populations. The genetic distance between populations is proportional to their geographic distance.

Table 2 Genetic characteristics of *Dalbergia cochinchinensis* populations

Country	R	N	Na	Na (rare)	Ho	He	Fis	T_2	Mm
Laos	0.83	20.57	2.89	2.21	0.41	0.41	0.00	0.46	0.42
Thailand	1.00	28.50	5.00	2.88	0.49	0.53	0.06	−1.40	0.35
Cambodia	0.66	18.23	5.43	3.59	0.67	0.64	−0.01	0.31	0.31
Vietnam	0.77	14.00	3.50	3.01	0.50	0.49	0.01	0.37	0.36

Note Results of genetic diversity study on SSR markers. R = clonal richness estimate, N = number of unique genotypes used for further analyses, Na = mean number of alleles, Na (rare) = mean number of alleles corrected for sample size differences, Ho = observed heterozygosity, He = expected heterozygosity, FIS = Fixation index, T_2 = bottleneck statistic representing deviation between observed and expected heterozygosity based on number of alleles, and Mm = Garza-Williamson index representing ratio between number of alleles and allelic range

7 Threats and Risks

Dalbergia cochinchinensis faces significant threats in its native regions due to over-exploitation, with local governments and environmental agencies implementing resource management measures (Fig. 5) and strictly limiting the export of this species' timber (Table 3). Over the past decade, the market price for timber has surged, particularly for large-sized timber, reflecting a substantial increase in demand (Table 4).

However, the growth of this tree species requires specific environmental conditions, such as high elevation, well-drained soils, and mature forest ecosystems, making its habitat very limited and fragile. Human activities pose severe threats to its survival, especially illegal logging (Fig. 6) which leads to forest degradation and destruction that significantly impedes the species' regeneration capacity and lowers its survival rate. Additionally, drought-induced forest fires present another major threat. Research indicates that approximately 60% of study plots experience fires during the dry season each year, particularly in mixed and dry deciduous forests. Ground-level seedlings and resprouting plants, including *D. cochinchinensis*, suffer severe damage from these annual fires, adversely affecting their survival rates. Fires further disrupt the natural regeneration and replenishment of these trees.

Habitat fragmentation is another factor contributing to the risks faced by this species. Agricultural expansion, road construction, and urbanisation have fragmented their habitats into smaller forest patches. The reproductive cycle and patterns of *D. cochinchinensis* are uncertain, with low seed dispersal efficiency and slow, irregular regeneration due to natural barriers and genetic characteristics. Overall,

Fig. 5 Illegally logged *Dalbergia cochinchinensis* confiscated in Thailand

Table 3 Conservation actions of *Dalbergia cochinchinensis* by the countries in their species range around early 2000s [23]

Thailand	Lao	Cambodia	Vietnam
(1) Very high priority species (2003) (2) Recommended to establish more ex situ conservation areas (3) 5 genetic conservation stands, 56 ha, 38 plus trees (1999) (4) 43 plus trees (2003–2007)	(1) Priority species (1999) (2) Category A (Endangered) (3) 200 ha plantation area in Bolikhamxai Forest (1992) (4) Used mainly for ex situ conservation and demonstration	(1) Threat level 5 (2003) (2) 50 ha seed source (2002) (3) 5 kg seed demand supplied from CTSP and local people (2003) (4) 50 ha in situ conservation area (5) 3 ha production area (6) 2 genetic conservation stands, 69 ha, 147 trees (2006)	(1) Priority species (2003) (2) 1 ex situ conservation site (3) 2 seed sources with 2,600 trees planted

Table 4 Wood volume (m^3) of *D. cochinchinensis* reported as illegally harvested and/or traded in the Choam Ksant District, 2013–2019 [22]

Particulars	2013	2014	2015	2016	2017	2018	2019	Total
Sawn wood	6.19	1.36	0.20	0	0	0	0	7.75
Hardwood (log)	2.49	1.76	6.21	2.80	9.23	0.60	0	23.09
Total	8.68	3.12	6.41	2.80	9.23	0.60	0	30.84

illegal logging, forest fires, habitat fragmentation, and reproductive risks collectively threaten the survival of *D. cochinchinensis*, highlighting the urgent need for enhanced conservation measures to ensure the long-term persistence of this species [22].

8 Future Outlook

Dalbergia cochinchinensis is a rare and highly valuable timber species, with growing demand in China. Research and cultivation efforts face considerable challenges due to the long rotation period of over 50 years. Tree improvement research focuses on germplasm collection, improved clonal seedling technologies, developing superior genotypes, innovative afforestation models, and improving ecological efficiency. Though the production of plant material through cuttings and grafting has been meeting existing demand, it is not sufficient for large-scale afforestation. The development of tissue culture techniques is critical as it can ensure uniformity,

Fig. 6 Protected wild tree of *Dalbergia cochinchinensis*

improve propagation rates, and support automated processes to achieve industrial-scale seedling production, resulting in the species expansion. Breeding efforts aim to assemble diverse germplasm resources and observe the adaptability and growth rates of *D. cochinchinensis* from various provenances. These studies help in estimating the inherent genetic diversity, providing a scientific basis for identifying and selecting superior genotypes having better growth traits for higher yield and stress resistance and ensuring maintenance of a stable broad genetic base. Continuous stress-resistance trials and monitoring are vital for sustainable utilisation and improvement of breeding programmes. Considering its long rotation period, at present it may be less suitable for pure plantations. Mixed planting with species of shorter rotation periods is an ideal approach, promoting diversity, enhancing ecological stability, and reducing risks associated with monocultures, such as pest outbreaks and soil degradation. The research efforts must focus on long-term trials, improving propagation methods, and refining afforestation strategies. The integration of advanced genetic analysis and tissue culture technologies can accelerate progress in breeding and cultivation. Partnerships between research institutions, local governments, and forestry enterprises are essential to scaling up operations and ensuring sustainable harvesting, utilisation, and meeting the country's rosewood demand.

References

1. Zeng S (2024) https://ppbc.iplant.cn/tu/19123912
2. NParks Flora & Fauna Web (2022) https://www.nparks.gov.sg/florafaunaweb/flora/2/8/2835
3. Sudarat Homhual (2012) https://phargarden.com/main.php?action=viewpage&pid=249
4. Jøker D (2000) Seed leaflet—Danida Forest Seed Centre. No.26, 2, Danida Forest Seed Centre, Humlebaek, Denmark
5. Narong N, Sobon K (2014) Review of biological and silvicultural characteristics of timber trees planted in Cambodia, 96p. https://www.irdfa.org/wp-content/uploads/2017/11/Review-of-biological-and-silviculturalcharacteristics-of-timber-trees-planted-in-Cambodia.pdf
6. Plants of the World Online (2024) https://powo.science.kew.org/taxon/urn:lsid:ipni.org:names:490142-1/images
7. Zhang FD, Xu CH, Li MY, Chen XD, Zhou Q, Huang AM (2016) Identification of *Dalbergia cochinchinensis* (CITES Appendix II) from other three *Dalbergia* species using FT-IR and 2D correlation IR spectroscopy. Wood Sci Technol 50:693–704
8. Hartvig I, Czako M, Kjær ED, Nielsen LR, Theilade I (2015) The use of DNA barcoding in identification and conservation of rosewood (*Dalbergia* spp.). PLoS One 10(9):e0138231
9. Liu RH, Wen XC, Shao F, Zhang PZ, Huang HL, Zhang S (2016) Flavonoids from heartwood of *Dalbergia cochinchinensis*. Chin Herb Med 8(1):89–93
10. Jiang T, Li K, Liu H, Yang L (2016) The effects of drying methods on extract of *Dalbergia cochinchinensis* Pierre. Eur J Wood Prod 74(5):663–669. https://doi.org/10.1007/s00107-016-1041-2
11. Xiang Z, Chen X, Zhao Z, Xiao X, Guo P, Song H, Yang X, Huang M (2018) Analysis of volatile components in *Dalbergia cochinchinensis* Pierre by a comprehensive two-dimensional gas chromatography with mass spectrometry method using a solid-state modulator. J Sep Sci 41(23):4315–4322. https://doi.org/10.1002/jssc.201800636
12. Seng M, Cheong EJ (2020) Comparative study of various pretreatment on seed germination of *Dalbergia cochinchinensis*. For Sci Technol 16(2):68–74
13. Sun Z, Pang W, Wu C, He G, Lu S, Deng J, Huang W, Liu C, Xie H, Xu Y, Wei S, Zhu Y, Zhang T (2024) Research progress on cultivation techniques of *Dalbergia cochinchinensis* in China. J Green Sci Technol 26(11):96–101
14. Zeng XZ (2017) Seedling nurture technology of in *Dalbergia cochinchinensis*. J Fujian For Sci Technol Trop Crops 42(2):39–41
15. Mai B, Hong Z, Xu D, Luo M, Zhang N, Huang X (2019) Variation of seed germination and seedling growth among different families of *Dalbergia cochinchinensis*. J Nanjing For Univ 43(02):153–160
16. Peiyan WU (2023) Effect of cutting season and plant growth regulator on cutting effect of *Dalbergia cochinchinensis*. J Fujian For Sci Technol 50(2):53–57. https://doi.org/10.13428/j.cnki.fjlk.2023.02.010
17. Chun ZH, Peiyan WU, Wenqing CH, Junqiang ZH, Zhouyang CH, Yulin WU (2024) Comparison of growth in cutting orchard of the Family of *Dalbergia cochinchinensis*. J Fujian For Sci Technol 51(1):116–119. https://doi.org/10.13428/j.cnki.fjlk.2024.01.022
18. Mei LF, Ling WJ, Jiang PS, Yan JH, Zhou H, Hui CH (2019) Seed vigour of *Dalbergia cochinchinensis* Pierre ex Laness from different provenances and families and its introduction performance in Pingxiang, Guangxi. J Southern Agric 50(1):110–117
19. Hong Z, Wu P, Zhang J, Wang W, Xu L, Shen W, Xu D (2020) Early growth performances and adaptability of *Dalbergia cochinchinensis* in Zhangzhou, Fujian Province. J Nanj For Univ 44(6):118–124
20. Hien VTT, Phong DT (2012) Genetic diversity among endangered rare *Dalbergia cochinchinensis* (Fabaceae) genotypes in Vietnam revealed by random amplified polymorphic DNA (RAPD) and inter simple sequence repeats (ISSR) markers. Afr J Biotech 11(35):8632–8644
21. Moritsuka E, Chhang P, Tagane S, Toyama H, Sokh H, Yahara T, Tachida H (2017) Genetic variation and population structure of a threatened timber tree *Dalbergia cochinchinensis* in Cambodia. Tree Genet Genom 13:1–11

22. Sinly S, Dany C, Sothea H, Phanit P, Sopheap L, Sobon K (2021) CITES non-detriment findings report on *Dalbergia cochinchinensis* and *Dalbergia oliveri* in the Choam Ksant District, Preah Vihear Province. Forestry Administration, Phnom Penh, Cambodia
23. APFORGEN (2018) Conserving rosewood genetic resources for resilient livelihoods in the Mekong—Project Inception Workshop Report

Chapter 4
Dalbergia latifolia: A Legacy of Strength and Sustainability

A. N. Arunkumar, Rekha R. Warrier, and Geeta Joshi

1 Introduction

In 1791, younger Linnaeus established the genus *Dalbergia* as a tribute to Nicholas Dalberg, a Swedish Botanist. In 1798, William Roxburgh (1751–1815) a Scottish botanist popularly known as founding father of Indian Botany, added three new species to the genus *Dalbergia*, and one of them was *Dalbergia latifolia* which has stood the test of subsequent scrutiny and examination both for its generic position and specific rank [47]. *Dalbergia* Linn. f. has a distribution worldwide with diverse habitats in the tropics having 250 species falling under the category of trees, shrubs and climbers/lianas [35]. Being an iconic genus, the following three species are prominently known in India: *D. sissoo* (broadly distributed in Northern parts of India), *D. latifolia*, (broadly distributed in southern and Central parts of India) and its ally *D. sissoides* (restricted to Western Ghats from the hills of Karnataka and southwards to Pulney and Kerala) [62]. *Dalbergia* is entirely a diploid genus with $2n = 20$ chromosomes, and these three species are considered to be closely knitted forming one genetic assemblage [19]. Considering the ecological, economical and heritage value, *D. latifolia* is considered to be one of the prized trees of India's forests.

A. N. Arunkumar (✉)
ICFRE-Institute of Wood Science and Technology, Malleswaram, Bengaluru, India
e-mail: arun@icfre.org

R. R. Warrier
ICFRE-Institute of Forest Genetics and Tree Breeding, Coimbatore, India

G. Joshi
Indian Council of Forestry Research and Education (ICFRE), Dehradun, India

R. R. Warrier et al. (eds.), *Handbook of Asian Rosewoods*, Materials Horizons: From Nature to Nanomaterials, https://doi.org/10.1007/978-981-96-9459-4_4

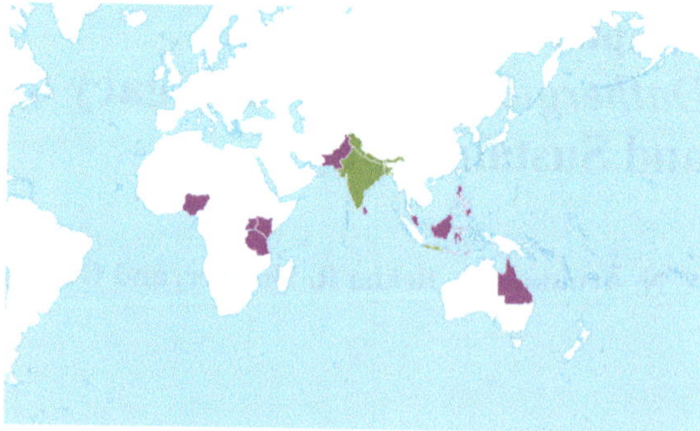

Fig. 1 Worldwide distribution map of *Dalbergia latifolia* (green colour indicates natural population while the purple colour indicates introduced species) [46]

2 General Distribution

Dalbergia latifolia is naturally distributed in India and naturalised in Indonesia (Fig. 1). It has also been introduced to other countries such as Africa and Asia like Kenya, Malaysia, Mauritius, Myanmar, Nepal, Nigeria, Pakistan, the Philippines, Réunion (France), Singapore, Sri Lanka, Sulawesi, Uganda, Tanzania and Vietnam [12, 67]. It is mentioned to be indigenous to Java in Indonesia; however, there are reports indicating that it has been introduced to Indonesia [40] by the colonial government around 1870, and first planted in the Keling area [69]. It was introduced to Ceylon (present Sri Lanka) in 1923–1924 and planted in Anuradhapura [23]. The conflicts related to the global distribution also find mention [12]. The vernacular and common names are collated in Table 1.

3 General Features

Dalbergia latifolia Roxb. popularly known as Indian Rosewood, a large, nearly evergreen or deciduous tree, mostly scattered in the natural habitat, has a full round crown and a straight and somewhat cylindrical bole depending on the local conditions in which it grows (Fig. 2). The smooth grey bark is relatively thin, having irregular short cracks, exfoliating and fibrous. Depending on the tree's age, longitudinal flakes are observed. The root system is a developed tap root that runs deep along with long lateral roots. The roots that are prevalent near the surface or subsurface of the soil produce root suckers when injured. These root suckers can gradually develop into trees if undisturbed [6]. The specific epithet *latifolia* refers to the wide leaflets which is one of the characteristics of *D. latifolia*. The compound leaves are broadly ovate to oblong and leathery, and the flowers are white and the pods are flat and 1–3 seeded [23, 34, 66] (Fig. 3). The tree can reach a total height and clear bole height of about 45 m and 21 m, respectively, with a girth of 6 m and having a large buttress when grown in ideal conditions of Western Ghats in the Southern part of India [6], specially in Coorg, Coimbatore, Malabar and Travancore [44]. It has been reported that trees

Table 1 Vernacular and common names of *Dalbergia latifolia*

Language	Vernacular name
Bengali	*Satsor, Sitsal, Swetasal*
Gujarati	*Kalaruk, Shisham*
Hindi	*Bhotheula, Beete, Bide, Kala shisham, Pai, Sitsal*
Kannada	Agaru *Beete, Bete, Beete, Belbiti, Bide, Ibadi, Ibri, Kareevyaadhi, Karevyadi, Karimbeetti, Kariya ibbadi, Shushmaru, Thodagatti*
Malayalam	*Cholavitti, Itti/Eetti, Kariveeti, Ulmalavitti, Veeti*
Marathi	*Bhotheula, Kalarukh, Sisham, Sissu, Sisva*
Oriya	Sitsal, Sissua
Sanskrit	*Shishapa, Simsapa*
Tamil	*Eruvadi, Iridi, Irupottu, Itti/Eetti, Karundovira, Nukku, Palkonda, Tawadi, Thodagathi, Totakatti*
Telugu	*Chava, Chittegi, Iridi, Iruduchava, Jitegi, Viroogooduchawa, Virugudu, Zitregi*
Urdu	*Shisham, Sissua*
French	*Palisander de l'Inde*
German	*Indischer Rosenholzbaum*
Indonesia	*Sonokeling, Sonobrits*
Java	*Sonokeling, Sonobrits, Sonosungu*
Nepali	*Satisal*
Vietnamese	*Trắc*
Common names	Blackwood Tree, Black Rosewood, Blackwood, Bombay Blackwood, East Indian Rosewood, Indian Palisandre, Java Palisandre, Sonokeling, Roseta Rosewood

of 20 ft girth, 70 ft clear bole and 90 ft height have been found in the semi-moist deciduous forests of South Coimbatore [9]. The best growth of the trees attaining a girth of 10 ft was found in the forests of Kanara and Belgaum (earlier Bombay Presidency) in Karnataka. The logs extracted from them ranged from 15 to 25 ft and girth of 6–8 ft [44]. However, the tree size can be lower while growing in Northern part of India.

4 Distribution in India

In India, it is distributed from the foot hills of Himalaya to tip of Southern India. The states in which rosewood is found are Andhra Pradesh, Assam, Bihar, Chhattisgarh, Gujarat, Jharkhand, Karnataka, Kerala, Madhya Pradesh, Maharashtra, Odisha, Rajasthan, Tamil Nadu, Telangana, Uttarakhand, Uttar Pradesh and West Bengal. Though it was mentioned earlier that Andaman and Sikkim had rosewood, the report on non-detrimental findings by Botanical Survey of India in 2018 indicated that it

Fig. 2 A giant *Dalbergia latifolia* tree at Aanamalai Tiger Reserve, Tamil Nadu, India

Fig. 3 Leaves, flowers and pods of *Dalbergia latifolia*

is not present now, and has also been introduced in Jammu Kashmir, Punjab and Sikkim [10, 61].

Some of the important habitats of rosewood have been mentioned here considering that several states were not yet formed during that period. Generally, it is considered to be a typical species found in different types of deciduous forests and rarely found in evergreen forests [9]. In Uttar Pradesh, it is found in a scattered manner at an altitude of 1100 to 1500 ft in dry deciduous forests of Bundelkhand, rarely along Gonda and Bahraich districts mostly in moist deciduous forests. In the Rajpur forests of North Lalitpur and Manikpur, it is found sparsely. In Orissa, it is in Angul division, Sambalpur, where it is associated with *Shorea robusta* (Sal). In earlier Bihar (which is bifurcated now to Jharkhand), it was commonly found on cooler aspects of Chota Nagpur from 800 to 2500 ft. In Madhya Pradesh, it attains larger proportion in the Chanda Forest (presently in Dindori district). In Maharashtra, it was found in moist deciduous forests at various elevations up to 3000 ft, commonly found in dry savannah in Satpura range of Khandesh, north and south Thana, Surat Forest (presently in Gujarat) and good trees in Nagargali (in Belgaum, Karnataka), and Kanara and Dharawad divisions (presently in Karnataka). In Tamil Nadu (earlier Madras), the best trees were found in north western and western slope of Nilgiris, Anaimalai hills of south Coimbatore, with large dimensions in north and south Malabar (presently in Kerala state), south Kurnool hills and Guntur division (presently in Andhra Pradesh) and Kollegal divisions (presently in Karnataka). The mixed deciduous forests of Mysore plateau, also known as south Karnataka plateau, is bounded on the west and south by Western Ghats (presently in Karnataka, comprising some of the districts such as Hassan, Kodagu, Tumkur and Bengaluru) and moist deciduous forests on the western boundary of Coorg (presently Kodagu) varying from 2600 to 4000 ft above mean sea level. In Kerala, it is found in Travancore areas [23]. Though it is found in different elevations from mean sea level to as high as 4500 ft, the ideal elevation is ~1000 ft in which the best size trees are found [11]. However, it is rarely found in evergreen forests [23]. A detailed documentation of the distribution in several states of India has been provided in the Non Detriment Findings report published by Botanical Survey of India [10].

It has been found that *Dalbergia sissoides* is another closely related species that has been found in the southern part of India, which grows along with *D. latifolia*. This species is also sometimes traded as Indian rosewood. Both the species have been clearly distinguished based on morphological traits such as leaf, inflorescence, bark and microscopic structures such as epidermal cells and stomata [6, 16]. Some of the prominent associate species that grow along with *D. latifolia* across its distribution range are *Careya arborea, Garuga pinnata, Phyllanthus emblica, Shorea robusta, Terminalia tomentosa, Anogeissus latifolia, Acacia catechu, Ougenia dalbergioides, Zizyphus jujube, Hardwickia binata, Boswellia serrata, Terminalia paniculata, Lagerstroemia lanceolata, Xylia xylocarpa, Pterocarpus marsupium, Haldina cordifolia, Tectona grandis* and some bamboo species [23, 52, 66].

5 Geology, Soil and Rainfall

Rosewood is not an exacting species when it comes to geology and soil. It grows on diversified geological formations such as gneiss, trap, laterite, boulder deposits and alluvial formations [23, 66]. However, it has found to grow better in deep well-drained soil adjoining the streams. It grows better in loamy and clayey soils having more lime content, but grows fairly well in black cotton soils; however, drainage is important. It also survives in poor, dry and stony soils, with severely stunted growth. Some of the best trees found in the south Kanara area of Karnataka are in gneissic soil; however, it does not grow well in water-logged soils [23, 61, 66].

Considering the natural habitats in which the tree grows, the minimum and maximum temperature range is 0–15.6 °C and 35–47.8 °C, respectively. The annual rainfall which occurs predominantly in monsoon (May–June–September) ranges from 760 to 5000 mm with 40 to 100% relative humidity [6, 23, 34] with 6 months of dry season.

Being a deciduous and nearly evergreen tree, in moist areas, the leaf fall is not significant and the tree is found with leaves. In drier areas, shedding of leaves occur during February–March, and the new flush of leaves are observed during April–May. However, there can be a bit of early or late leaf shedding depending on the location it grows. For instance, in south Kanara, leaf shedding occurs in early March and by the end of April new leaves are formed. The leaves are imparipinnately compound (10–25 cm long) and are orbicular to broadly ovate to oblong, leathery, dense and darker, and the glabrous rachis length varies from 12.8 to 18.6 cm [23, 34, 63, 66]. The leaf length ranges from 5.24 to 9.61 cm and leaf breadth varies from 4.95 to 8.86 cm [63].

6 Pollination Ecology and Breeding

Not many studies have been conducted on pollination ecology and breeding in *D. latifolia*. It has papilionaceous flowers which is adapted for highly efficient and specific pollination mechanisms with pollinating agents and pollen vectors. Various events that occur during pollination including the different timings have been described [53] and are as follows: anthesis from 5.00 to 8.00 h; anther dehiscence from 7.00 to 9.00 h and nectar secretion from 9.00 to 12.00 h. It was found that the sugar concentration was 6.49% in the nectar. The entomophilic type of pollination happens by the presence of bees and butterflies along with a few birds that visit to feed the nectar in the flowers. Indian honeybee (*Apis cerana indica*) was documented to be the major pollinator as it moves from flowers across the trees to collect nectar and pollen. The common butterflies that are active during pollination are *Neptis hylas* L. (Common sailer) *Euploea klugii* Moore & Horsfield (Common crow), *Hypolimnas misippus* L. (Danaid eggfly), *Prosotas nora* Felder (Common lineblue) and *Rapala manea* Hewitson (Slate Flash). The crimson-backed Sunbird (*Nectarinia minima* Sykes) is a common visitor for nectar feeding. It has been reported by Damaiyani and Prabowo (2019) that in Indonesia also some of the bee species that help in pollination are *Apis cerana* (Asiatic honey bee) and carpenter bees like *Xylocopa confusa, X. fenestrata, X. iridipennis,* and a wasp *Polistes metricus.* Of these visitors, *Apis cerana* and *Xylocopa confusa* were the most common frequent visitors.

Breeding system studies carried out on *D. latifolia* reveal that 5.24% of flowers end up setting fruits under open pollination, and it was two percent under self-pollination (autogamy). When the pollen grains of one flower are transferred to the stigma of the other flower, within the same plant (geitonogamy), fruit setting was not observed. Similar observation was documented in the case of apomixis. However, when the pollen grains of one flower was transferred to the stigma of the other flower from a different plant (xenogamy), fruit setting was observed in 4% of flowers [53].

7 Flowering and Fruiting

Flowering season also varies across its natural habitat and locality in which it grows. It is found to occur during February–March in Travancore (Kerala), February–April in south Kanara (Karnataka) and April along with new leaves in Maharashtra [23, 66]. In Java (Indonesia), flowering occurs from October to November, a month early in West Nusa Tenggara. In some areas in Java such as Rembang, Cepu and Bojonegoro, flowering is observed as early as May [69]. The inflorescence is a corymbose panicles with pale yellowish to white flowers that are scented, and appear clustered (varying from 23 to 74) at the end of a stem resulting in round or oval shape. The inflorescence length varied from 5.00 to 21.6 cm [63]. The bisexual flowers are 6–8 mm long, with five petals which are veined. The bell-shaped calyx is five mm long and five toothed; the white corolla is double the length of calyx and has nine stamens. The filaments in the stamens are united (monadelphous), and the sheath covering them is 5–7 mm long. The flowering intensity varies between trees and can vary from dense to little flowering. Flowers open from 10.30 to 14.00 h and the peak flowering is between 11.00 and 13.00 h [63]. The superior ovary has three to five ovules that contain minute stigma. The flat indehiscent pods are oblong to oblong-lanceolate with 4–9 cm length and 1.5–2.6 cm width with a narrowed base into a stalk. The glabrous, greenish pod turns brown and dries at maturity. The pod generally ripens from December to April and can vary depending on the locality. The pods usually fall once the monsoon sets. The pods commonly have one to three seeds, sometimes four and rarely five. The average percentage number of seeds in a pod varies as depicted in Fig. 4. The average seed weight of 7.36 ± 0.16 g was minimum in four seeded pods, while the single seed pod had the maximum seed weight of 9.37 ± 0.15 g. A study found that these pods remain on the tree until the monsoon's onset to facilitate good germination. The shining seeds are slightly hard, smooth, compressed and brown to black. It is oval to slightly kidney-shaped, 6–7 mm in length and 4–4.5 mm wide [23, 66]. The seeds are not produced regularly and abundantly; alternate years are considered good seed years. Germination, seedling growth and survival do not have any impact on the

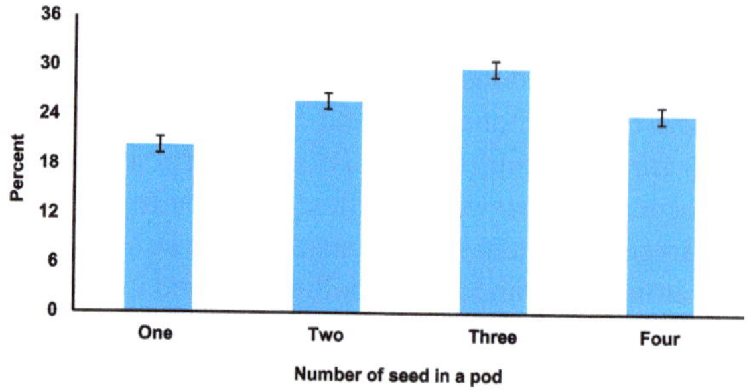

Fig. 4 Variation in average percentage for number of seeds in a pod in *Dalbergia latifolia* (Adopted with modifications [63]). (Error bar indicates standard deviation value)

Table 2 Leaf fall, leaf flush, flowering and fruit maturation in *Dalbergia latifolia*

Sl. No.	Phenological event	Months
1	Leaf fall	February–March
2	Leaf flush	April–May
3	Flowering	January–April
4	Fruiting	November–April

tree size, age and seed-bearing conditions [23]. A general reference for leaf fall, leaf flush, flowering and fruit maturation summarised [53, 56, 66] is provided in Table 2.

8 Regeneration

Natural regeneration: Regeneration is an essential aspect of the survival of the species in its natural habitat. Several factors, either biotic, abiotic or both, impact the reproduction and establishment of tree species in their natural settings [28]. Under natural conditions, the seed fall coincides with the monsoon; germination occurs soon after the rain sets in. In those locations where the pods fall and are protected by the soil or by the moderate growth of grass, seeds tend to germinate easily as the conditions are favourable. Fairly moist soil not compacted with moderate shade results in better regeneration success through seedling emergence [66]. The wind easily carries away the seeds and facilitates seed dispersion to a certain distance. If the conditions are favourable in the natural habitat, regeneration is abundant [45]. Fire, drought, grazing by wild animals, weed infestation and insect damage can severely impact regeneration [18]. Another form of natural regeneration commonly found in the habitat has been through root suckers. An injury or wound to the root bark or lateral roots can stimulate the formation of root suckers [61]. These suckers thrive particularly in areas with an open canopy, where ample light reaches the ground, promoting shoot development. Additionally, fire lines, which cause minor root damage, create favourable conditions for root sucker growth [34]. Similarly, root damage along the edges of forest roads and boundary lines has also been observed to encourage the development of root suckers.

An interesting observation regarding the development of root suckers in rosewood trees was recorded by Kadambi [23]. At a distance of 15 feet from a rosewood tree, a circular elephant trench measuring six feet in width and four feet in depth was excavated. Within this trench, 26 root suckers, approximately 10 years old, were discovered. Of these, 20 had developed from severed, disconnected roots at the outer edge of the trench, while the remaining six emerged from the inner periphery or the trench bottom [23]. To promote enhanced root sucker formation, Coode [13] suggested thoroughly pickaxing the ground within a radius of 10–20 feet from the base of a standing rosewood tree. This should be done to a width and depth of one foot, with trenches extending 8–16 feet in various locations within the perimeter.

Dalbergia latifolia is known for its strong coppicing ability. Coppice shoots regenerating from felled trees are an effective means of natural propagation. Seasonal variations influence the development of these shoots. In north Chanda, Kadambi (1954) reported that coppice shoots were highest—around 100—between April and July, peaking before the monsoon. However, this number declined to 80 in August and dropped to 25 by September, coinciding with the peak monsoon season.

Artificial regeneration: It can be artificially propagated by direct sowing, transplanting the seedlings, stump planting, and macro and micro propagation techniques. Details related to macro and micro propagation have been discussed in other chapters of this book. For direct sowing and raising of seedlings, seed collection, handling and storage play an important role.

9 Seed Collection and Storage

Rosewood produces seeds every alternate year, with pod collection timing varying based on local conditions during the fruiting period. In the wild, seed collection is challenging and labour-intensive due to the lack of synchronised pod maturity and seed predation by monkeys. However, in most regions, the ideal time for pod collection is between February and March. The tree begins fruiting at around 6–10 years of age. Wind disperses the seeds, which can be extracted from the pods by rubbing [22].

For collection, branches bearing dark brown pods are lopped, and ripe pods are gathered. Unripe pods should be avoided [23, 34, 61, 66]. The harvested pods are sun-dried for 2 days before breaking them to extract seeds. Pods may also be broken into small segments containing one or two seeds. The extracted seeds are then stored in gunny bags, sealed tins or earthen pots for future use. While seeds remain viable for up to a year, those sown within 6 months of storage show better germination. Prolonged storage beyond a year significantly reduces viability. Notably, seed viability and germination rates are not significantly affected by the age of the parent tree [23], indicating that seeds can be collected from trees of varying maturity. A kilogram of rosewood seeds contains approximately 18,000–35,000 seeds.

Rosewood seeds are non-dormant and typically do not require pretreatment. However, soaking them in water for 12–24 h can enhance germination. Studies indicate that broken pod segments containing one or two seeds yield better germination rates [48]. A study in Dehradun reported that soaking seeds in cold water for 24 h resulted in a 61% germination rate. Conversely, a study in Tamil Nadu reported germination rates of 5% for seeds treated with boiling water, 29% with hot water, 26% with cold water and 31% for untreated seeds [23]. Overall, germination rates can range from 45 to 80%, producing approximately 10,000–14,000 saplings per kilogram of seeds [22, 33, 34].

Broken pod segments containing a single seed can be directly sown, as they support better germination than clean seeds [20]. This may be due to reduced seed predation by ants and other soil-borne pathogens. The natural habitat of *Dalbergia latifolia*, predominantly dry deciduous forests, is frequently affected by smoke and fire, which influence its growth and regeneration. Laboratory experiments show that exposing *D. latifolia* seeds to smoke and fire resulted in 74 and 125% germination increases, respectively, compared to untreated seeds. This suggests that such natural disturbances may play a crucial role in enhancing seed germination in the field [57].

10 Raising of Seedlings

Freshly collected segmented pods containing one or two seeds are soaked overnight before sowing. They are then dibbled at 5 cm intervals into a porous sandy loam medium in a raised seedbed in a shaded area. Depending on local conditions, the seedbed is watered once daily. During epigeal germination, the radicle emerges from one end of the seed and curves downward into the soil, while the hypocotyl pushes the cotyledons above the soil surface. Germination begins within seven days and is completed within 21 days, with rates varying from 45 to 80%, depending on the seed source.

Once seedlings reach the two- to three-leaf stage, they must be pricked and transplanted into polybags. This process must be done carefully, as seedlings develop a strong taproot, which may be damaged during transplantation. To avoid risk, direct sowing in polybags is preferred. Initially, seedling growth is slow, as root development is prioritised over shoot growth. However, once the root system is established, shoot growth accelerates. By 12–14 months, seedlings can reach a height of 80–100 cm and are ready for field planting [34, 48].

Seedling growth is negatively impacted by deficiencies in trace elements such as copper (Cu), zinc (Zn), manganese (Mn), molybdenum (Mo) and boron (B). Studies have documented the specific deficiency symptoms in seedlings [24] and the associated physiological changes [1]. However, seedlings show a positive response when supplemented with these trace elements [25], as illustrated in Fig. 5.

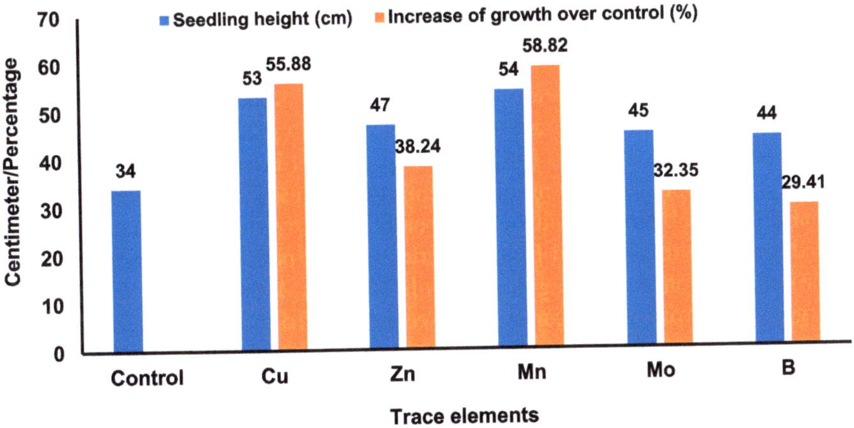

Fig. 5 Impact of application of trace elements on seedling height and percentage increase in growth over control (Adopted with modifications [25])

11 Stump Planting

For stump planting, seedlings that are grown in the nursery bed for 5–7 months are used. Stumps are prepared depending on the age of the seedlings. If the seedlings are smaller, then the stems are cut down to two inches and root to 6 inches. In the case of two-year-old seedlings, the pruning of stems can be about 12 inches and the root to about 2 inches [66]. Stumps are placed in polythene bags of 15 × 22 cm during October–November. These stumps are ready for transplanting in the coming monsoon. Stumps with a collar diameter of 12–15 cm showed better result. It is generally believed that seedlings from stumps have vigorous growth compared to normal bareroot seedlings.

The taproot, which grows longer, has a few side roots. The nodules are formed in the seedlings which facilitate nitrogen fixing. Five-month-old seedlings raised in a polythene bag (13 × 20 cm) had an average of 22 cm root length and 10 root nodules [48]. Considering the root growth, it is always suggested that seedlings be grown in raised beds. Another method of raising seedlings is to collect the root suckers developed preferably from superior trees. The lateral roots of the mother tree are damaged, resulting in the development of root suckers. These suckers are collected along with five-centimeter root bit and planted in polythene bags. These suckers grow well, attaining a height of 60–70 cm by the next planting season and when planted, grow better than bareroot seedlings. Interestingly, in West Java, the people of Cikongga considered rosewood as "*the plant which drags the soil*". This connotation is because, in the area dominated by rosewood trees, there will be profuse root suckers that emerge, resulting in high density of saplings. Depending on the number of the trees in the plantation, it is estimated that the planting density can vary from 1,000 to 200,000 saplings, indicating the innate naturally propagating ability [69].

12 Plantation Establishment and Silviculture

Rosewood thrives in deep, moist soils and benefits from shade during its early growth stages. It is shade-tolerant in the initial years but gradually requires overhead light, typically after 10 years [48]. Seedlings planted in areas with water stagnation are susceptible to damping-off, leading to suppressed growth, making proper drainage essential. Under unfavourable conditions, growth remains stunted. Bare-rooted seedlings transplanted from nurseries often experience drying from the tip to the collar region, weakening their growth. However, if they survive the first year, their growth improves significantly in the second year. To enhance plantation success, planting stumps prepared from 1.5- to 2-year-old seedlings is recommended [34, 48]. During the first year, seedlings are highly susceptible to drought and require adequate soil moisture. However, as they mature, they become drought-resistant and can withstand dry seasons lasting over 6 months. Side shade is beneficial during early growth stages, and while rosewood tolerates overhead shade initially, it requires direct overhead light as it matures, making it a moderate light demander. Without sufficient diffused light, trees tend to grow crooked and branchy, resulting in suboptimal development. Trees have shown better growth in areas where thin, diffused light filters through canopy openings.

Rosewood benefits from soil aeration, and studies indicate that trees achieve greater girth and height when the soil is properly ploughed. The species develops a deep taproot, with some surface-feeding main roots that, when injured, can produce root suckers. It can survive light frost but is highly susceptible to intense frost. Additionally, it has low fire resistance, and crown fires can be fatal, making fire protection a crucial management practice. For successful plantation establishment, proper soil preparation, weeding and climber removal are essential. Young seedlings are highly vulnerable to browsing and require protection from wildlife, including deer, bison, elephants and wild boars, as well as livestock such as cattle and goats [3, 23, 34, 48]. Generally, rosewood is not cultivated as plantations unlike teak in India. However, in Indonesia, it has been extensively cultivated as plantations and four types of planting systems have been identified—monoculture (>80% of the total trees comprises rosewood), mixed plantation (when grown along with teak and mahogany), agroforestry (grown along with the annual crops, on the bunds, and as boundary planting) and grown in savannahs [69].

13 Management of Rosewood

Generally, rosewood occurs as a scattered population in its natural habitat. Depending upon the other highly valued associate such as teak, the management operations are carried out. In case of high forests, the silviculture system used is selection system with a fixed girth size varying from 1.8 to 2.1 m depending on locality or as improvement felling. Coppice system with or without standards is followed if the trees are of coppiced origin. Though clear felling is also followed, it is not preferred in the present situation [61].

14 Growth Rate

Under natural conditions, rosewood exhibits a slow growth rate, which typically remains consistent over time. However, growth rates can vary significantly, with faster growth observed in well-managed plantations, such as those established in Indonesia. In its natural habitat, growth is influenced by site conditions, leading to considerable variation. Reports indicate that the tree reached a girth of 175 cm over a span of 150 years in high forests of north Kanara (Ankola high forest and Kali Nadi slopes) in present Karnataka (Fig. 6) [42]. In a revised working plan for the Mount Stuart Forest, South Coimbatore division, growth statistics such as tree height (m), diameter (cm) and volume (cubic ft) of *D. latifolia* over a period of 120 years reported that the trees attain a girth of 240 cm by the end of 120 years (Fig. 7) [32]. The author suggests that "*D. latifolia appears to grow considerably faster than teak in middle age and slower when it is either very young or very old*".

In another study [55], following were the mean annual girth increment in different aged plantations in Punachi working circle, Coimbatore: 10 years (1.0 cm), 11 years (1.6 cm), 12 years (1.3 cm), 16 years (0.9 cm) and 18 years (1.1 cm), respectively. Considering the growth, it can be safely assumed that the mean annual girth increment in and around Coimbatore Division was ~1.18 cm. As *D. latifolia* is a good coppicer, a study carried out in Bihar, India, suggests that average height (m) and girth (cm) at the end of 14 years was 8.08 m and 26.16 m, respectively, Fig. 8 [66].

It has been reported that annual increment of *D. latifolia* was 0.89 cm and to reach 58 cm diameter, it takes around 80, 110 and 160 years in different natural habitats such as Mysore, Kurnool and north Kanara, respectively [4] and 80 years to reach 1.05 m girth in Dangs forest in Gujarat [3]. However, in forest types such as southern tropical semi-evergreen and moist deciduous forests of north Kanara, the trees tend to grow slow and to attain 60 cm DBH, the tree takes 238 years (Rai, 1978). In an

Fig. 6 Growth rate in girth for *Dalbergia latifolia* in the high forests of North Kanara (Adopted with modifications [42])

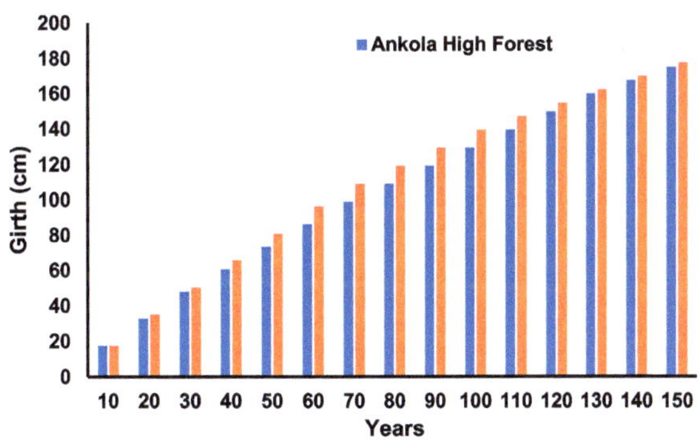

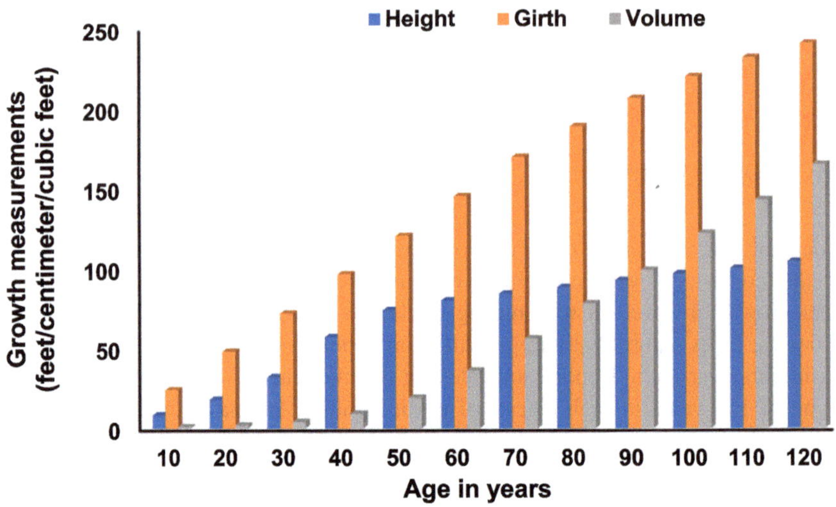

Fig. 7 Growth statistics of a rosewood tree grown in Mount Stuart forest of South Coimbatore (Adopted with modifications [32])

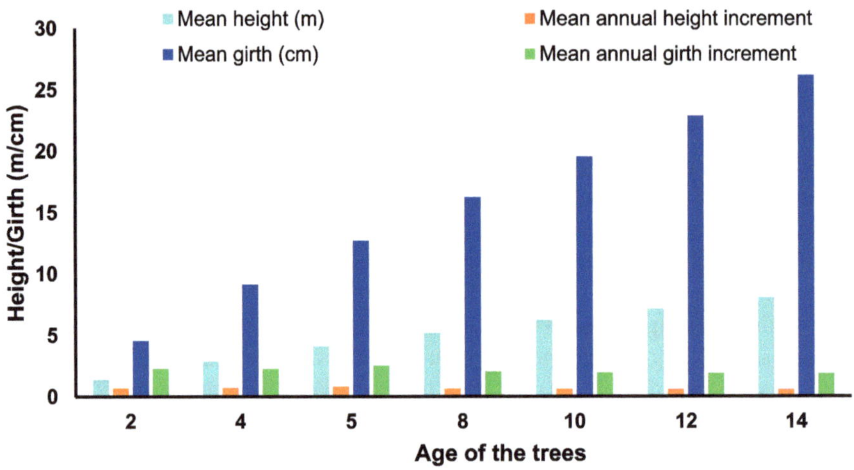

Fig. 8 Growth rate in different aged coppiced plantation of *Dalbergia latifolia* in Kolhan, Bihar (Adopted with modifications [66])

8 year old sandalwood plantation, rosewood (as a host for sandal) had a mean tree height and GBH of 8.72 m and 29.18 cm, respectively indicating an annual average growth of 1.08 m height and 3.64 cm girth, respectively [5].

15 Heartwood and Sapwood

The sapwood in older trees is typically narrow, pale yellowish-white, and often exhibits a pinkish tinge, making it easily distinguishable from the heartwood. It is perishable, less durable and highly vulnerable to powder post beetles. In contrast, the heartwood varies in colour from light brown, dark brown or purplish-brown to nearly golden brown, with hues of light rose purple and darker streaks, sometimes deepening to dark purple with intricate grain patterns. Over time, it darkens further, enhancing its aesthetic appeal and giving the wood a mottled appearance, making it one of the most globally recognised decorative timbers. The distinctive red tint in the heartwood is attributed to red crystalline quinone, latinone and naturally occurring dalbergiones [14]. The wood has a medium to coarse grain texture, is hard and heavy (specific gravity ~0.75 to 0.88) and features shallow to straight interlocked grains that contribute to its ability to take a fine polish. Freshly cut wood emits a faint but distinct aroma. Structurally, it is approximately 34% stronger transversely and more than three times harder than teak in cross-section [7, 23, 41, 43, 61, 66]. The highly durable wood is categorised as class I based on the graveyard test. The time duration of durability ranged from 276 to 288 months [2, 31, 59]. The heartwood is considered to be highly resistant to wood decay fungi as it contains three important secondary metabolites such as latifolin, 4-methoxy dalbergione and dalbergiphenol [51, 54]. Freshly cut wood loses its colour. The wood considered to be moderately refractory needs to be seasoned either through air or kiln seasoning. The wood is susceptible for centre heart or hollowness in the centre of wood due to the presence of white calcareous deposits which is rich in $CaCO_3$. Therefore, the pith needs to be boxed out as it hampers the sawing process resulting in blunting the cutting edges of the saw [61, 68]. Considering the superior wood quality, timber strength was reported in comparison to the strength of teak providing a score of 100 to teak for all the wood properties (Table 3).

As a structural timber, *D. latifolia* is considered to be "very strong and handsome" and "extremely handsome furniture wood and probably the fines furniture wood India possesses" [64]. It is evident from the table 3 that some of the wood traits of D. latifolia are better than teak. "Wood traits such as shock resisting ability and shrinkage, are 30% better than teak. It is 25% higher for weight and shear strength and 20% higher for hardness than teak (Table 3)".

Table 3 Timber strength of *D. latifolia* depicted in percentage compared to strength of teak [64]

Weight relative to teak (unseasoned)	Strength as a beam	Stiffness as a beam	Suitability as a post or strut	Shock resisting ability	Shear	Hardness	Shrinkage
125	85	85	75	130	125	120	130

16 Uses

Considered to be one of the high quality wood in the Indian forests, the wood has diversified uses and has been known for its popularity as cabinet wood and valued for furniture. It is believed that the true blackwood furniture used by the Chinese was initially of *D. latifolia*; as the price increased, other inferior woods were used as substitutes. To appreciate the durability of the wood, British rulers in 1850s had used rosewood for the gun carriage wheels that was utilised in their campaign against Afghan and Egyptians. When employed as posts in houses and temples, it can endure for over a century, and as furniture, it can last even longer [44]. Its extensive use in various categories is provided in Table 4.

17 Medicinal Properties of *D. latifolia*

Similar to most of the tropical trees, *D. latifolia* has various medicinal uses. Some of medicinal properties for which it is used are to treat diarrhoea, dyspepsia, obesity, and skin and worm infections. The pounded mixture of the tree bark along with the bark of *D. paniculata* when applied externally reduces body pain. In Ayurveda, the root, bark and leaves have been considered as a substitute for *D. sissoo.* It is used for treating polyuria, chronic ulcer, leprosy, leukoderma, urinary bladder disorders, burning sensation, oedema, as a brain tonic and as potential for antidiabetic agents [21, 26, 27, 33, 38, 60].

18 Trade

Trade in Indian rosewood dates back to the time of Portuguese and Dutch invaders. By 1907–1908, government depots across various provinces held substantial volumes of rosewood. The Bombay Presidency, particularly the south circles, recorded 36,131 cubic feet, with logs sourced primarily from the Belgaum, north, south, east and west Kanara forest divisions. High-quality logs were priced between 60 and 80 rupees per ton.

In the Madras Presidency, where rosewood was considered among the most valuable timbers [7], the total outturn was 22,861 cubic feet, with south Malabar contributing the largest share. The estimated cost of delivering one ton of rosewood from Malabar (also known as Malabar Blackwood) to Madras, including royalty fees, was 150 rupees per ton. In the central provinces, the Chanda division alone produced 3,000 cubic feet, while the forests of Coorg were capable of yielding 2,000 cubic feet of large logs and 4,000 cubic feet of smaller ones. Logs exceeding 5 feet in girth were sold for over 1.50 rupees per cubic foot, highlighting their value [65].

Table 4 List of diverse uses of D. *latifolia* wood

Sl. No.	Category	Suitable uses
1	Agricultural implements and machinery	Ploughs and yokes
2	Boat and ship building	Knee timber
3	Building	Doors, window frames and house posts
4	Carts and carriages	Bodies of carts, shafts, axles, yokes
5	Carving and turning	Carving, toy making
6	Wooden combs	Combs
7	Coopers works (Casks and vats)	Tubs, buckets and casks
8	Domestic utensils and other articles	Measures for grain, oil, bowls, jars, snuff boxes, spoons
9	Furniture and joinery	High class furniture and cabinet making
10	Gunstocks	Sporting guns
11	Ordnance work and military stores	Spokes and felloes of wheels of gun-carriages, transport wagons, side bars of saddles, planking for foot boards of ammunition wagons, platform-boards, ammunition boxes, carriages, axle-beds for gun carriages and ends of transport boxes, brake box, foot and platform boards, rammer heads
12	Railway carriages, wagons, general fittings	Washers, pulleys, knaves, handles, shelves, internal carriage panels, decorative mouldings
13	Railway sleepers	Highly valuable wood, but can be used for railway sleepers
14	Tools and tool handles	Handles of axes, chisels, gimlets
15	Musical instrument	Guitar
16	Sports goods	Carrom coins, clubs, wall bars, discus, chess pieces, tennis and badminton rackets
17	Miscellaneous	Backs of brushes, walking sticks, handles of Nepalese kukris, knife handles, well curbs and other parts of wells, drawing equipment, artificial limbs, printing blocks, umbrella handles, walking sticks, dobby lags, pegs, barrels, box back blanks, boards for slay races for cotton and jute looms and hessian looms and shuttles for sacking and hessian looms, bookstands, teapots, wall hangings, jewellery boxes, trays, brush backs, and calico printing boxes Nitrogen fixing tree, potential agroforestry species, as shade tree in coffee plantations, leaves are good fodder, as fuel wood with calorific value ranging from 5049 to 5149 k cal The tree yields a gum, oil is occasionally extracted from the heartwood which is used for skin and foot diseases of cattle. The seed oil is believed to cure rheumatism

Source [9, 15, 17, 29, 39, 49, 50, 61, 66]

Britain became the primary market, especially from 1920 onwards. In 1915, the royalty fees for rosewood trees in the United Province (present-day Uttar Pradesh) were 3 rupees for trees with a girth of 36 inches, 1 rupee for those between 25 and 36 inches, and 8 annas (half a rupee) for trees with a girth of 16 to 24 inches at the butt end. In Bihar and Orissa, royalty rates varied from 3 to 7 annas per cubic foot, depending on tree size and quality.

Local market rates in the Central Provinces and Berar (now Maharashtra, Madhya Pradesh and Chhattisgarh) ranged from 2 to 2 rupees 8 annas per cubic foot in Nagpur Wardha, 1 rupee to 1 rupee 2 annas in Buldana and 2 rupees 8 annas in Nimar. In Bombay Province (encompassing Maharashtra, Gujarat and parts of Karnataka), north Thana priced high-quality 15 cubic foot logs at 35 rupees, while 9 cubic foot logs ranged from 10 to 15 rupees. In south Thana, the rate was 8 to 12 annas per cubic foot. In Surat, the best quality wood cost 100 rupees per ton, while second-grade wood was priced between 40 and 64 rupees per ton.

In north and west Kanara, good-quality logs were sold for 60–80 rupees per ton, whereas rates in south Kanara, Dharwad and Belgaum were 5–6 rupees, 10–16 rupees and 15 rupees per khandy (12.5 cubic feet), respectively. In the Madras Province, prices varied by region: Ganjam and Guntur recorded rates of 12 annas and 1 rupee per cubic foot, respectively; south Coimbatore ranged from 2 rupees 3 annas to 3 rupees 4 annas; and central Coimbatore charged 40 to 50 rupees per kandy (13 cubic feet).

In Krishna, the cost was 1 rupee 8 annas per cubic foot, while in south Kurnool, it ranged from 1 rupee 8 annas to 2 rupees. North Malabar priced rosewood at 1 rupee 2 annas per cubic foot, while in Salem, it was 1 rupee 8 annas. In Tinnevelly (now Tirunelveli), rates ranged from 2 rupees 8 annas to 3 rupees 8 annas per cubic foot. In Coorg, the average cost of one cubic foot of dressed logs was 2 rupees.

The estimated annual rosewood outturn in 1915 (Table 5) reflects the significant production potential of the early twentieth century. South Malabar, a major supply centre, was heavily overworked, leading to a decline in tree numbers in the region. London market prices for rosewood varied between £7 and £15, depending on log size and quality [9].

After World War II, France began importing Indian rosewood, followed by Italy, which became a major consumer after 1950. The export of rosewood saw a significant rise during the 1960s (Table 6). In the 1968–1969 Indo-Italy trade, rosewood accounted for 6.7% of total exports. This growing demand also attracted trade with other European countries, including Germany, Switzerland, Holland and the Scandinavian nations, where rosewood was primarily used in high-end furniture making. Indian rosewood soon gained a reputation as a premium choice for aristocratic furniture, second only to walnut. The peak period of rosewood trade in India is considered to be between 1955 and 1970, with European timber importers actively investing in rosewood stocks as a safe and profitable option due to a sustained bullish market (Sivarajan, 1969). The 1960s, therefore, are often regarded as the "golden period" for Indian rosewood trade. However, over time, demand declined as European preferences shifted towards other rosewood varieties and lighter-coloured woods. As exports to Europe decreased, Japan emerged as a key market. Of the total rosewood

Table 5 Estimated annual outturn of rosewood (cubic ft) in various locations during 1915 in British India [9]

Sl. No.	Location	Estimated annual outturn (cubic ft)
1	North Thana	7000
2	South Thana	8000
3	Surat	6000
4	North Kanara	50,000
5	South Kanara	2500
6	West Kanara	2500
7	Dharwad	1250
8	Belgaum	4000
9	North Malabar	10,000
10	South Coimbatore	15,000
11	Guntur	1700
12	Coorg	16,600

Table 6 Rosewood export from India to Italy in 1960s [8, 58]

Sl. No.	Year	Rupees in lakhs
1	1962–1963	208
2	1963–1964	154
3	1964–1965	64
4	1967–1968	130
5	1968–1969	121
6	1969–1970	113.4

exports from India in 1965–1966 (7,742 m^3) and 1966–1967 (11,582 m^3), Japan alone accounted for 3,000 m^3 and 6,592 m^3, respectively, highlighting the shifting global demand for Indian rosewood [58].

19 Classification of Rosewood Logs for Trade

For facilitating better trade and market value, rosewood is classified similar to teak wood but with certain modifications. In Kerala, rosewood logs have been classified based on three important criteria—length, girth and quality. While the first two are quantitative, the quality is based on visual observations.

Length—Based on the length, the logs are classified into three types—short length logs (between one meter to below 2.5 m length) and lengthy logs (2.5 m length to 7.37 m length) and long length logs (>7.37 m logs). Usually, only short and lengthy logs are in vogue.

Girth—Based on girth, the logs are classified into four classes—Class I (above 185 cm mid girth); Class II (between 125 and 184 cm mid girth); Class III (between 75 and 124 cm mid girth); and Class IV (between 60 and 74 cm mid girth). The Class I logs without any defect are also considered as Export class. If the logs are less than 60 cm mid girth, they are considered as rosewood below specification.

Quality—Based on quality, the logs are classified into three classes—A class (logs straight and sound without any defects); B class (fairly straight and sound logs); C class (defective and crooked logs).

Considering all these three aspects, the logs can be categorised as follows:

Short Log I	—	A, B, C
Short Log II	—	A, B, C
Short Log III	—	A, B, C
Short Log IV	—	A, B, C
Lengthy Log I	—	A, B, C
Lengthy Log II	—	A, B, C
Lengthy Log III	—	A, B, C
Lengthy Log IV	—	A, B, C

20 Rosewood Production

The two states that predominantly contribute to the total supply of rosewood are Karnataka (50%) and Kerala (37%). The other states contributing in smaller quantities are Tamil Nadu (12%), Gujarat and Madhya Pradesh (<1%) [36]. The production of rosewood for the past 14 years from 2009–2010 to 2023–2024 in Karnataka and Kerala is provided in Figs. 9 and 10. In Karnataka, two districts, Kanara and Kodagu, are the top two rosewood producing regions; sometimes rosewood is also obtained from Shimoga district. On an average, 82% of rosewood production is from these two districts. During year 2017–2018 there was an unusual spike in production (17,021 m^3) with Shimoga district alone contributing 78 and 20% from Kodagu and Kanara. It is difficult to attribute any specific reason for this increased production. In Kerala, the average rosewood production was 198 m^3 and the lowest production of 20 m^3 was during the year 2011–2012. Keeping aside that particular year, the average production up to 2016–2017 was 259 m^3 and post 2017 was 180 m^3 indicating a decline in production. The Seigniorage rate of rosewood (in m^3) fixed in Karnataka for the years 2018–2019, 2022–2023 and 2024–2025 was Rs. 88,440, 96,400 and 106,040, respectively. The Seigniorage rate of rosewood (in m^3) fixed in Kerala for the different girth classes for short and long logs is provided in Fig. 11. When compared with lengthy and short logs, the cost of Class I logs in quality A and B of lengthy logs is more than 100% of the cost of Class I logs in quality A and

B short logs indicating the demand for the quality logs is higher. However, it is only 10% increase for lengthy logs over short logs in case of A and B quality for Class IV logs (Fig. 11). It is interesting to note that a 100-year-old rosewood tree in Waynad, Kerala, harvested in 2018, with a girth of 246 cm and length of 310 cm easily fit into the Class I category. The auction was conducted by Kerala Forest Department, and the log was sold for 7.15 lakhs INR [37].

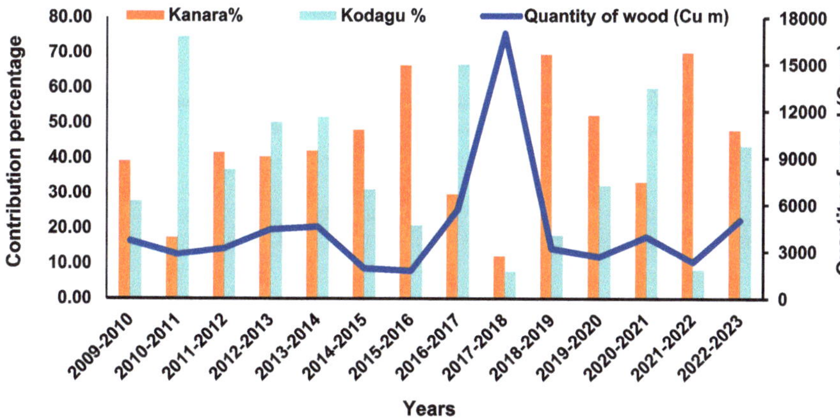

Fig. 9 Quantity of rosewood production in Karnataka (2009–2023) and percentage contribution from Kanara and Kodagu Districts (*Source* Different annual reports of Karnataka Forest Department)

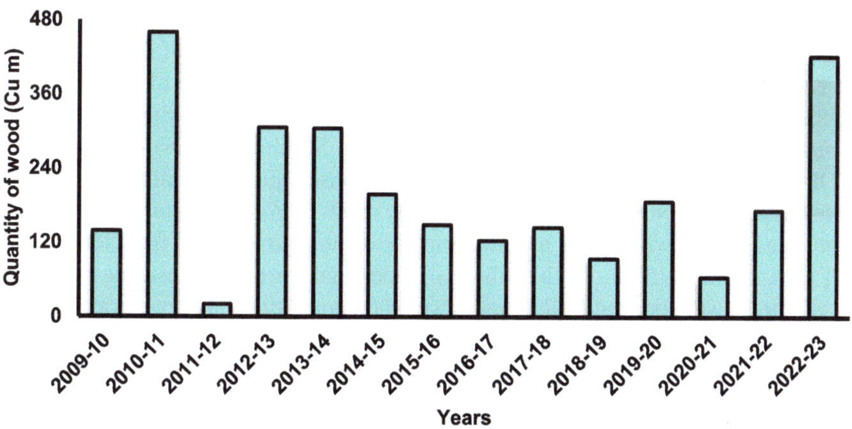

Fig. 10 Quantity of rosewood production in Kerala for the period 2009–2023 (*Source* Different annual reports of Kerala Forest Department)

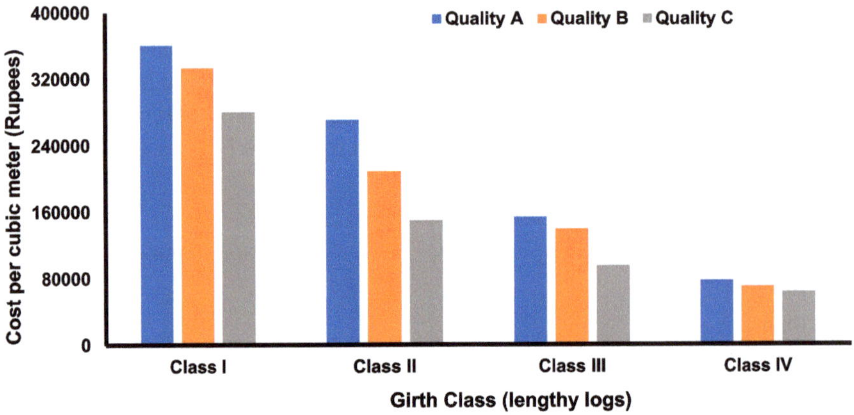

Fig. 11 Seigniorage rate fixed by Kerala Forest Department for different girth class logs under different quality class in *Dalbergia latifolia* (*Source* Kerala Forest Department)

21 Present Status

The huge demand for rosewood has resulted in over-harvesting of mature trees along with the immature trees in the forests of Mysore (present Karnataka) and Kerala since early 1970s. The availability of trees in the natural population has considerably dwindled in both the states. It was suggested that establishing plantations is one of the strong measures to support the demand. If such initiatives not being taken, then the sustained availability of wood will be in peril [58]. Not much improvement was observed over a period considering the constant escalation of rosewood price in Karnataka and Kerala. However, the over-harvesting and trade of rosewood tree species in general and *D. latifolia* in specific resulted in bringing the species under Convention on International Trade of Endangered Species (CITES) Appendix II at COP17 in 2017. The International Union for Conservation of Nature (IUCN) has categorised it as "vulnerable". Current status of *D. latifolia* in terms of population status and the growing stock are not properly documented except for a few studies [6]. It was estimated more than three decades ago, in the homesteads of Kerala 21,921 trees were present [30]. Another study on population and regeneration status in Tamil Nadu and Kerala indicated that less than two trees (girth of more than one metre) were found for every 1.25 ha and on an average only four seedlings (less than 10 cm girth) were present for every 100 m^2 area [53]. A study was carried out during 2011–2017 to assess the population structure and natural regeneration status of *D. latifolia* in Kerala and Tamil Nadu. Out of the 14 forest divisions of Kerala, only in six divisions natural regeneration was good while in the other eight divisions, seedlings were documented and the saplings were not recorded. In Tamil Nadu, the study was carried out in seven divisions and in one division *D. latifolia* was totally absent. Out of the six divisions, in one division the regeneration was good, in three divisions it was fair and in two divisions it was absent. Considering the girth-wise tree population structure, in Kerala, out of 14 divisions, nine divisions had healthy population structure having trees that were distributed from lower to higher girth classes corresponding to their age groups. In the remaining five divisions, as there was presence or absence of trees in specific girth classes/age groups and also the trees were not healthy resulting in the poor population structure. The study inferred that there was poor contribution of seedlings gradually shifting towards sapling stage. In case of tree density, there was a conspicuous absence of small girth trees across various divisions and the prevalence of only higher girth sized trees resulted in poor density as well as indicating the absence of seedling and sapling stage plants gradually becoming pole or young aged trees. The dwindling population in the natural habitat is attributed to forest fires, grazing, habitat loss, presence of exotic species/invasives and climate change [52] apart from the harvest of matured trees for timber [10]. Overall, there are sufficient evidences indicating that population of *D. latifolia* in Kerala and Tamil Nadu is facing critical survival prospects in its natural habitat.

It is also pertinent to realise that there is huge lack of interest in raising plantations of *D. latifolia*. This is because of the inherent slow growth of the species and issues related to obtaining felling and transit permission. As the species is considered as a scheduled/reserved tree or restricted species in most of the states where it is naturally distributed, the complex procedure for harvesting the trees has severely discouraged growing rosewood plantations [10].

With *D. latifolia* coming under Appendix II of CITES, to obtain the actual scenario of rosewood in terms of trade, in 2018, Botanical Survey of India carried out Non Detriment Finding (NDF) study supported by Ministry of Environment, Forest and Climate Change. As per CITES, NDF is defined as follows—"*In accordance with Articles III and IV of the Convention, export permits for specimens of species included in Appendices I and II shall be granted only when a scientific Authority of the State of export has advised that such export will not be detrimental to the survival of the species (a determination known as a 'non-detriment finding' or NDF)*". The NDF report mentions that populations in some parts of western, central and eastern India have dwindled, but is comparatively stable in Karnataka, Kerala and Tamil Nadu. However, in general natural regeneration is a cause of concern in most of the states. Considering the status tree population *per se*, it was opined that as major population is found to be in the protected areas, with the laws enforcing strict prohibition of harvesting in these areas, the natural population as such is safe protected and conserved and it will not hamper the trade if the harvest permission is granted for the trees grown in reserve forest, social forestry plantations and private plantations grown outside protected areas. The number of trees per hectare based on the samples selected and the regeneration rate (number of individuals reproduced or regenerated/number of individuals at the age of reproduction) in percentage based on the samples selected have been provided in Figs. 12 and 13. The report also indicated that considering the indiscriminate management practises as well as identity crisis in demarcating the original *D. latifolia*, assessing the actual quantity of harvest, and the impact on trade and export cannot be determined easily. However, it was concluded that as most of the trees grown and harvested are from private lands and the source

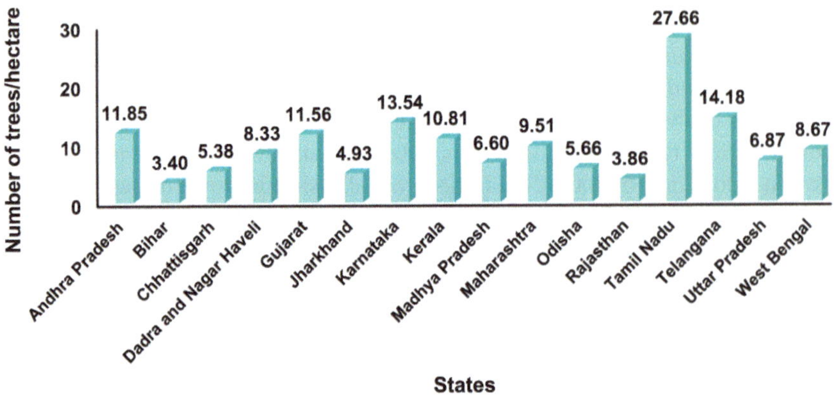

Fig. 12 Number of trees per hectare of *D. latifolia* across different states based on randomly selected sample populations (Adopted with modification [10])

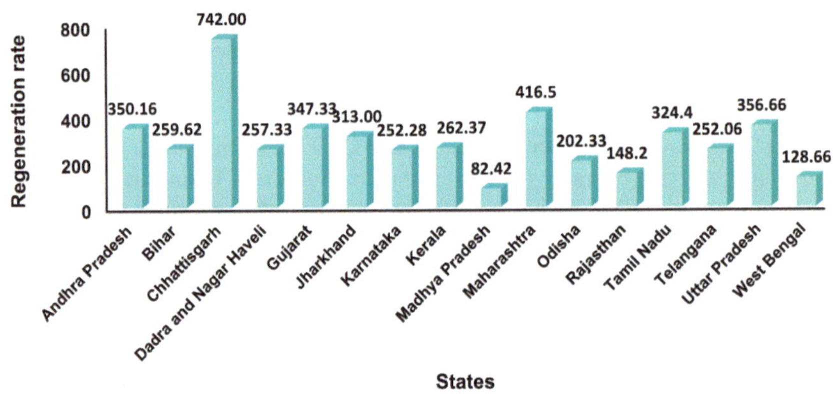

Fig. 13 Regeneration rate of *D. latifolia* across different states based on randomly selected sample populations (Adopted with modification [10])

is properly documented, there will not be any detrimental impact on the survival of wild populations.

22 Need for Conservation and Sustainable Utilisation

Considering the rich value and importance of Indian rosewood as a part of culture and tradition in India, efforts have to be channelised in conserving and its sustainable utilisation. Some of the crucially important broad steps needed to be addressed are as follows:

- Generating base line information about the current population and regeneration status across India.
- Need for identifying and developing suitable practises/techniques for in situ and ex situ conservation.
- Developing simple and rapid micro and macro propagation techniques for facilitating reintroduction and rehabilitation of the species in various areas of its natural habitat.
- Initiating tree improvement programmes that facilitates in identifying superior genotypes and assembling a national as well as regional level germplasm banks for research, utilisation and as repositories for posterity.
- Assessing the inherent genetic variability of the species using advanced molecular tools and identifying region-specific genotypes for better survival and establishment.
- The selected superior genotypes can be used for growing extensive plantations and popularising among tree growers indicating the financial benefits for the growers.
- Developing and demonstrating suitable agroforestry models to popularise the cultivation.
- Framing of suitable legal provisions by the Government that encourages cultivation of rosewood, and facilitates easy harvest, transport and marketing and also strict monitoring of illegal harvest, trade and export.

Nature provides bountiful of benefits to humankind and trees are one of the greatest gifts. *Dalbergia latifolia* popularly known as Indian rosewood has its presence both nationally and internationally. Valuing this non-renewable resource both scientifically and ecologically will help in sustainable utilisation and conservation.

References

1. Angadi VG, Kamala BS, Rai SN, Parthasarathi K (1988) Effect of deficiency of trace elements on leaf area chlorophyll level and photosynthetic efficiency in tree seedlings. My Forest 24:124–128
2. Anonymous (1982) Code of practice for preservation of timber. IS:401, Bureau of Indian Standards, New Delhi, India, p 30
3. Anonymous (1973) Indian timbers rosewood. Forest Research Institute and Colleges, Dehradun, 13p
4. Anonymous (1952) The wealth of India, vol III. Publications and Information Directorate, C.S.I.R., New Delhi, p 236
5. Arunkumar AN, Joshi G, Warrier RR (2022) Growth assessment of young plantation of *Santalum album* with *Dalbergia latifolia* as host. Indian For 148(8):785–790
6. Arunkumar AN, Warrier RR, Kher MM, Teixeira da Silva JA (2022) Indian rosewood (*Dalbergia latifolia* Roxb.): biology, utilisation, and conservation practices. Trees 36(3):883–898
7. Balfour E (1870) The timber trees, timber and fancy wood. Higginbotham Company, Madras, p 370
8. Bansal GL (1970) Prospects of Indo-Italian trade. Foreign Trade Rev 5(3):237–249
9. Benskin E (1915) Note on Blackwood *Dalbergia latifolia* Roxb. Forest Bulletin No.17, Superintendent Government Printing, Calcutta, India, p 12

10. Bhattacharjee A, Krishna G, Kumar A, Sengupta S, Chakraborty S, Dhanavate R, Sarkar S, Sahu RK, Soni AK, Chakraborty O, Mallick B, Prasad K (2018) Report on non-detriment findings (NDFs) of *Dalbergia latifolia* Roxb. in India. Botanical Survey of India, Kolkata, p 70
11. Bourdillon TF (1893) Report on the forests of Travancore. Travancore Government Press, Thiruvananthapuram, Kerala, India
12. CITES Secretariat (2024) Report on the conservation and trade of CITES-listed rosewood tree species [Leguminosae (Fabaceae)]. https://cites.org/eng/topics/flora/cites-and-forests
13. Coode J (1925) Working plan (revised) of the Punachi working circle of South Coimbatore District. Working Plan Office, Oatacamund, Madras
14. Criodain TO, O'Sullivan M, Meegan MJ, Donnelly DM (1981) Latinone, a phenanthrene-1, 4-quinone from *Dalbergia latifolia.* Phytochemistry 20(5):1089–1092
15. Devi Prasad AG (1994) *Dalbergia latifolia*: the high-valued Indian rosewood. In: NFT Highlights: a quick guide to useful nitrogen fixing trees from around the world, NFTA 94-04. Winrock International, USA, pp 69–70
16. Farooqui P, Venkatasubramanian N, Nallasamy VK (1989) Use of cuticular studies in distinguishing species of *Dalbergia.* Proc Indian Acad Sci Plant Sci 99(1):7–14
17. Gamble JS (1881) Manual of Indian timbers. Superintendent of Government Printing, Calcutta
18. Gaulier A, Pascal JP, Puyravaud JP (1995) Assessing vegetation changes in the dry deciduous Ainurmarigudi Reserve Forest, South India. Ann For Sci 52(5):449–464
19. Hiremath SC, Nagasampige MH (2004) Genome size variation and evolution in some species of *Dalbergia* Linn. f. (Fabaceae). Caryologia 57(4):367–372
20. Homfray CK (1937) Nursery and plantation notes for West Bengal 1935. Superintendent, Government Printing, Bengal Government Press, Alipore, p 210
21. Isyaka SM, Umar A, Talatu A, Abdullahi AM (2025) A review of the botanical, ethnomedicinal uses, and phytochemistry of *Dalbergia* (Fabaceae) species. Eur J Med Plants 36(1):59–71
22. Jøker D (2004) *Dalbergia latifolia* Roxb. Seed Leaflet. Forest & Landscape, Copenhagen, Denmark
23. Kadambi K (1954) The silviculture of *Dalbergia latifolia* Roxb. Monograph of Indian trees No.1. Manager of Publications, Delhi, 40 p
24. Kamala BS, Angadi VG, Rai SN (1988) Symptoms of deficiency of trace elements and the associated changes of peroxidase isoenzyme pattern in the seedlings of Rosewood, Honne, Cassia and Neem. My Forest 24(4):307–311
25. Kamala BS, Angadi VG (1986) Effect of application of trace elements on seedling growth of some timber species. Van Vigyan 24(3):109–112
26. Khare CP (2007) Indian medicinal plants. An illustrated dictionary. Springer Sciences, New York, USA, p 200
27. Kirtikar KR, Basu BD (1981) Indian medicinal plants, vol III. International book distributors, Dehradun, India
28. Krebs CJ (1972) Ecology: an experimental analysis of distribution and abundance. Harper and Row, New York
29. Krishna S, Ramaswamy S (1932) Calorific value of some Indian woods. Forest Bulletin No. 9 (Chemistry), Government of India, Central Publication Branch, Calcutta
30. Krishnankutty CN (1990) Demand and supply of wood in Kerala and their future trends. KFRI Research Report 67, Kerala Forest Research Institute, Peechi, Kerala, India, p 66
31. Kumar S, Dev I (1993) Wood preservation in India. ICFRE. 28. Indian Council of Forestry Research and Education, Dehra Dun, India, pp 65–100
32. Laurie MV (1933) Revised working plan for the Mount Stuart Forest, Coimbatore Division, Madras
33. Lemmens RHMJ (2008) *Dalbergia latifolia* Roxb. Record from PROTA4U. In: Louppe D, Oteng-Amoako AA, Brink M (eds) PROTA (Plant Resources of Tropical Africa/ Ressources végétales de l'Afrique tropicale), Wageningen, Netherlands. http://www.prota4u.org/search.asp. Accessed 21 Jan 2025
34. Luna RK (2005) Plantation trees. International Book Distributors, Dehra Dun, India, p 826

35. Mabberley DJ (2017) Mabberley's plant-book: a portable dictionary of plants, their classification and uses. Cambridge University Press, Cambridge, UK
36. Manohara TN, Balakrishna SM, Suresh HS (2025) Genomic insights of *Dalbergia latifolia* (Indian rosewood): applications in conservation and beyond. Trees 39:39. https://doi.org/10.1007/s00468-025-02617-x
37. Manoj EM (2018) Vintage rose wood fetches record price in Wayanad. The Hindu. https://www.thehindu.com/news/national/kerala/vintage-rose-wood-fetches-recordprice-in-wayanad/article25215123.ece. Accessed 15 Jan 2025
38. Nair KKN (1986) Preservation of *Dalbergia* L. f. in Kerala by establishment of a germplasm bank. KFRI research report No. 43. Kerala Forest Research Institute, Peechi, Kerala, India
39. Nazma, Ganapathy PM, Bhat KM, Sasidharan N, Gnanaharan R (1981) *Dalbergia latifolia* Roxb. A handbook on Kerala timbers. Kerala Forest Research Institute, Peechi, Thrissur, Kerala, India
40. Nijman V (2024) The illegal trade in rosewood in Indonesia. Eur J For Res 143:1047–1055
41. Orwa C, Mutua A, Kindt R, Jamnadass R, Anthony S (2009) Agroforestree database: a tree reference and selection guide version 4.0. http://www.worldagroforestry.org/sites/treedbs/treedatabases.asp
42. Pearson RS (1908) Working plan for the Ankola high forest, Block XXIV and XXV and The Kalinadi Slope, Block XXVI, North Kanara Division, Mysore
43. Pearson RS (1912) Commercial guide to the forest economic products of India. Superintendent Government Printing, Calcutta, India, p 155
44. Pearson RS, Brown HP (1932) Commercial timbers of India, vol I. Central Publication Branch, Calcutta
45. Planchais I (1993) Evolution of the dry forest of Ainurmarigudi and impact of environmental factors: study of the effect of fire on the germination of dominant species. ENGREF, Formation des Ingénieurs Forestiers, Nancy, France, pp 117–196
46. Plants of the world online (POWO) (2024) The International Plant Names Index and World Checklist of Vascular Plants (2024) https://powo.science.kew.org/taxon/urn:lsid:ipni.org:names:490295-12024. Published on the Internet at http://www.ipni.org and https://powo.science.kew.org/ © Copyright 2023 World Checklist of Vascular Plants. http://creatvecommons.org/licenses/by/3.0
47. Prain D (1904) The species of *Dalbergia* of south-eastern Asia. Bengal Secretariat Press, Calcutta
48. Rai SN (1999) Nursery and planting techniques of forest trees in tropical south Asia. Punarvasu Publications, Dharwad, India, pp 78–83
49. Rajput SS, Shukla NK (1984) A review of classification of timber for different end uses. Controller of Publications, Delhi
50. Ramesh Rao K, Purkayastha SK (1972) Indian woods, vol III. Manager of Publication, New Delhi
51. Rao RV (1989) Natural durability of rosewood against wood rot fungi. J Indian Acad Wood Sci 20(2):13–18
52. Sasidharan KR, Prakash S, Muraleekrishnan K, Kunhikannan C (2020) Population structure and regeneration of *Dalbergia latifolia* Roxb. and *D. sissoides* Wight & Ar. in Kerala and Tamil Nadu, India. Int J Adv Res Rev 5(9):51–66
53. Sasidharan KR (2015) Know your trees—rosewood *Dalbergia latifolia* Roxb. ENVIS Newslett For Genet Resour Tree Improv 1(2):5–12
54. Sekine N, Ashitani T, Murayama T, Shibutani S, Hattori S, Takahashi K (2009) Bioactivity of latifolin and its derivatives against termites and fungi. J Agric Food Chem 57(13):5707–5712
55. Sharma AN (1936) Working plan for the forests of Coimbatore South Division, Madras
56. Singh KP, Kushwaha CP (2006) Diversity of flowering and fruiting phenology of trees in a tropical deciduous forest in India. Ann Bot 97:265–276
57. Singh A, Raizada P (2010) Seed germination of selected dry deciduous tree in response to fire and smoke. J Trop For Sci 22(4):465–468
58. Sivarajan M (1969) On the export trade of Indian rosewood. Indian Forester 95(12):852–855

59. Sundararaj R, Shanbhag RR, Nagaveni HC, Vijayalakshmi G (2015) Natural durability of timbers under Indian environmental conditions: an overview. Int Biodeter Biodeg 103:196–214
60. Sutrapu S, Pal RS, Khurana N, Vancha H, Mohd S, Chinnala KM, Kumar B, Pilli G (2024) Diabetes warriors from heart wood: unveiling Dalbergin and Isoliquiritigenin from *Dalbergia latifolia* as potential antidiabetic agents in-vitro and in-vivo. Cell Biochem Biophys 82(2):1309–1324
61. Tewari DN (1995) Monograph on Rosewood (*Dalbergia latifolia* Roxb.). International Book Distributors, Dehra Dun, India p 74
62. Thothathri K (1987) Taxonomic revision of the Tribe Dalbergiae in the Indian subcontinent. BSI, Culcutta, India, p 230
63. Tripathi K, Mishra M, Yadav J, Kumar G, Shankar R, Tiwari P (2024) Cytological and phenological variability in *Dalbergia latifolia* Roxb.: a vulnerable tree species. Silvae Genetica 73:190–200
64. Trotter H (1920) The commercial timbers of India and their uses. Central Publication Branch, Government of India, Calcutta
65. Troup RS (1909) Indian woods and their uses. Superintendent government printing, India, p 273
66. Troup RS (1921) Silviculture of Indian trees. Clarendon Press, Oxford, UK
67. UNEP-WCMC (United Nations Environment Programme World Conservation Monitoring Centre) (2017) Review of selected *Dalbergia* species and *Guibourtia demeusei*. In: UNEP-WCMC technical report. UNEP-WCMC, Cambridge, UK, pp 11–17, 49–52
68. Wiemann MC (2021) Characteristics and availability of commercially important woods. In: Wood handbook—wood as an engineering material. General technical report FPL-GTR-282. U.S. Department of Agriculture, Forest Service, Forest Products Laboratory, Madison, WI, p 45
69. Yulita KS, Wardani W, Atikah TD, Pratama BA, Subiakto A, Nurjanah A, Nopiansyah F, Nugroho A, Kamal I, Arrofaha A (2022) The non-detriment findings (NDF) report for *Dalbergia latifolia* in Java and West Nusa Tenggara, Indonesia

Chapter 5
The Present State of the Indonesian Rosewood (*Dalbergia latifolia*) from a Biological Perspective

Kusumadewi S. Yulita, Tika D. Atikah, Wita Wardani, Bayu A. Pratama, Henti H. Rachmat, Titiek Setyawati, and Fifi G. Dwiyanti

1 Introduction

Dalbergia latifolia, known as Indian rosewood or sonokeling in Indonesia, is an economically important tree species, widely distributed across tropical and subtropical regions of South America, Africa, and Asia [3, 44]. The distribution range in Indonesia covering South Sumatra, Java, Bali, the Lesser Sunda islands (West and East Nusa Tenggara), and Sulawesi [35, 45]. This widespread distribution in Indonesia is attributed to land rehabilitation programmes during the 1950s and 1970s [9, 45].

Dalbergia latifolia wood is popular for high-end furniture due to its remarkable strength, exceptional durability, and captivating aesthetic qualities. Its heartwood, characterised by hues ranging from rich brown to striking dark purple, contrasts beautifully with the lighter yellowish sapwood. This wood has become the preferred choice for exquisite carvings and musical instruments [32]. The market demand for this wood has increased rapidly, particularly in China, driven by the growing middle class [13]. On the other hand, high market demand for the rosewood in International market resulted in an increased illegal logging in some countries, which led to massive exploitation in rosewood-producing countries. In light of these

K. S. Yulita (✉) · T. D. Atikah · B. A. Pratama · H. H. Rachmat · T. Setyawati
Research Centre for Ecology and Ethnobiology, National Research and Innovation Agency, Bogor, West Java, Indonesia
e-mail: yulita.kusumadewi@gmail.com

W. Wardani
Research Centre for Biosystematics and Evolution, National Research and Innovation Agency, Bogor, West Java, Indonesia

F. G. Dwiyanti
Department of Silviculture, Faculty of Forestry and Environment, IPB University, Bogor, West Java, Indonesia

R. R. Warrier et al. (eds.), *Handbook of Asian Rosewoods*, Materials Horizons: From Nature to Nanomaterials, https://doi.org/10.1007/978-981-96-9459-4_5

issues, the entire genus *Dalbergia* was included in Appendix II of CITES during the Conference of Parties (COP) meeting in Johannesburg, South Africa, in 2016.

Prior to its listing in Appendix II of CITES, the availability of information regarding *D. latifolia* in Indonesia was limited. At present, comprehensive data on this species is accessible from Java and West Nusa Tenggara, facilitated by the CITES Tree Species Programme (CTSP) activities conducted between 2020–2022. These activities are designed to support Non-Detrimental Finding (NDF) assessments for sustainable, legal, and traceable use and to provide the baseline information on DNA fingerprints [12, 35, 45–47]. The outcomes of the NDF assessment confirm that *D. latifolia* from Java and West Nusa Tenggara does not pose a detriment to the survival of the species or the integrity of the ecosystem, thus demonstrating that its harvest and trade operations are sustainable, legal, and traceable [45]. However, information on *D. latifolia* from other islands remains unavailable, and there is a notable lack of data concerning local livelihoods and their implications for trade and sustainability.

We present detailed information on *D. latifolia* in Indonesia, focusing on its main distribution areas, Java and West Nusa Tenggara, covering the species historical context, taxonomy, population trends, habitat requirements, potential threats, conservation status, and trade management strategies. The information presented in this chapter is aimed is to support sustainable use and assess potential ecosystem impacts from overharvesting, land use changes, or climate change, thereby aiding in the development of effective management policies.

2 Taxonomy, Morphology, and Phenology

Dalbergia latifolia Roxb. is one of the tree species in the tribe *Dalbergiae* of the legume family, Fabaceae [1]. It is a pantropical genus consisting of c.a. 250 species (https://www.mobot.org/mobot/research/apweb/orders/fabalesweb.htm#Dalbergieae) with many high-value trees [43]. The synonyms of this species include *D. emarginata* Roxb., described from Bengal, and *D. javanica* Miq. from Java (https://powo.science.kew.org/taxon/urn:lsid:ipni.org:names:490295-1). The latter name, *D. javanica* was first described by Miquel in 1855, based on a specimen collected between 1802 and 1804 near "Soerakarta" (Central Java) by Dr. Th. Horsfield, an American ship doctor employed by the Dutch [33]. Miquel noted that *D. javanica* grows in teak forests [1], a habitat still observed today. During early colonial times, Java was a key source of teakwood for Dutch shipbuilding. Teak harvesting began along the north coast in the mid-seventeenth century [24] and expanded inland, including Soerakarta, as northern stocks dwindled by the late eighteenth century. Teak is not native to Java; it was likely introduced between the fourteenth and sixteenth centuries, originating from Central Laos and nearby regions [39]. Consequently, as the specimen came from a planted forest, the native status of *D. javanica* also remains uncertain.

Dalbergia latifolia is considered to have a close relationship with *D. sissoo*, an Indian species that is also widely cultivated [1]. However, *D. latifolia* clustered with a

Thailand species, *D. cochinchinensis*, in the phylogeny constructed from the complete chloroplast genome [10]. *D. latifolia* is similar in size to *D. sissoo* but different in the number of leaflets and leaflet apex [1]. Integrating morphological data with genetic markers will be valuable for forensic purposes and a better understanding of the evolution of this important species.

Dalbergia latifolia can grow over 15 m in height and 100 cm in diameter. It has a single stem, with cylindrical to slightly grooved bole, and branches usually start at above 3 m above the ground. The species has a coppicing ability, allowing multiple stems to sprout from the remaining stump. It also produces root suckers, which often serve as its primary dispersal strategy, and provide rapid means of propagation [45].

This species is a slow-growing tree with annual increments ranging from 0.7 to 1.4 cm per year in Java [12]. A mature tree flowers in February and September to November [1], while on northern coast (such as Rembang, Cepu, and Bojonegoro) flowering occurs in May. In West Nusa Tenggara, flowering is expected from September to October [45]. Information gathered from the local people in Bali indicates that trees in the northern part of the island also flower during the rainy season (from December/January to March). However, only a few populations in Java and West Nusa Tenggara produce flowers and viable seeds [48]. In addition, Rika Rahmawati (2021 pers. comm) reported that the population in Central Java and its vicinity, with abundant flowers and fruit usually have lower wood quality, while those without flowers or with flowers but no fruit or viable seeds tend to have darker and higher heartwood.

3 Distribution and Habitat of *Dalbergia latifolia* in Indonesia

In Indonesia, *D. latifolia* is found in South Sumatra, Lampung, Central Java, East Java, Bali, West Nusa Tenggara, East Nusa Tenggara, East Kalimantan, and Southeast Sulawesi [1, 2, 11, 14, 15, 20, 25, 44], with its range extending from the Banten region to East Java, and most commonly found in Central to East Java. Although it is reported to grow grows wild in Java, records also confirm its presence in West Nusa Tenggara, particularly on the islands of Lombok and Sumbawa [11] (Fig. 1).

Beyond natural stands, *Dalbergia latifolia* is widely cultivated on community and private lands [4, 11, 30]. They are often planted inter-cropped with other woody species because of its nitrogen-fixing ability. This species is also found in protected areas [35]. In some areas in West Java, however, local communities perceive the species as invasive due to its rapid growth, which suppresses agricultural cops through competition, making it difficult for farmers to cultivate food plants in its vicinity.

Dalbergia latifolia can grow optimally at 20 to around 600 m asl, in habitat ranging from flat to steep [35, 45]. It tolerate a wide variety of soil conditions, from moist and nutrient-rich to dry and nutrient-poor, although the growth of is often supressed on rocky, and drought-prone substrates [27]. This species occurs under

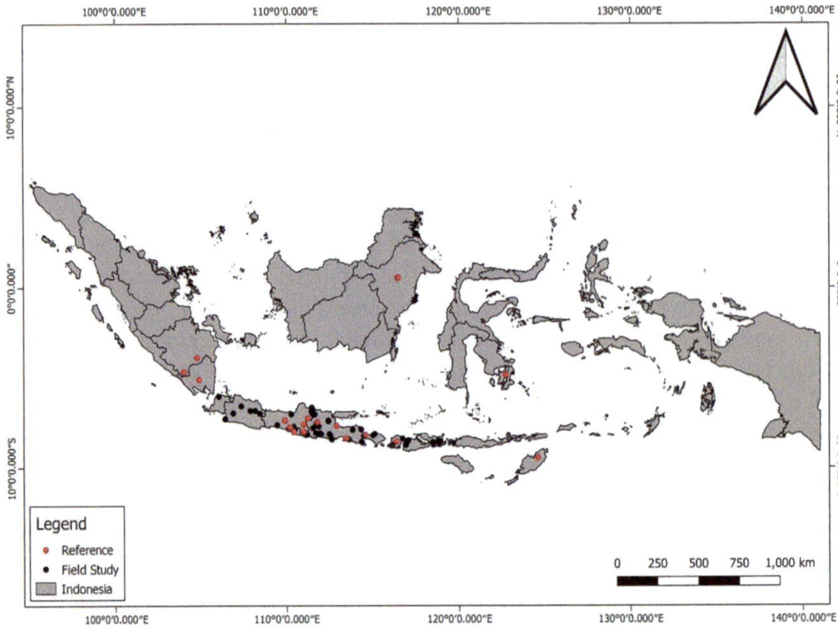

Fig. 1 Distribution of *Dalbergia latifolia* in Indonesia (*Source* [45]). Map of Indonesia highlighting various locations of distribution of *D. latifolia*. The map includes a legend indicating that red dots represent data from references, and black dots represent data from field surveys

diverse rainfall regimes, ranging from less than 1000 mm/year to over 4000 mm annually/year, with eastern Indonesia generally exhibiting drier climatic conditions than the western regions [45]. While such climatic variability influences the quality of the wood produced but it does not significantly affect the population size. Overall, evidence indicates that *D. latifolia* is widely distributed across Indonesia, occurring in nearly all major bioregions [35].

4 Propagation

Dalbergia latifolia reproduces regeneratively and vegetatively. Mature seeds showed high germination (over 75%), and saplings (around 20 cm tall) exhibited 100% growth post acclimatisation. Despite being challenging to propagate through cuttings, it is known as "difficult-to-root species" [17, 18]. Studies in Indonesia have shown that shoot tip cutting of this species exhibits a relatively high rooting ability. Root cuttings outperformed shoot cuttings, achieving a success rate of over 70%. In contrast, shoot cuttings exhibited variable success rates, ranging from 0 to 70%. Freshly cut materials demonstrated a rooting success of >70%, particularly when transported immediately to the nursery without delay [28, 38]. The origin and size of the cutting material significantly influenced the growth rate. At the same time, the propagation location—whether in a propagation tunnel or outdoor nursery beds—had no notable effect on growth outcomes [35].

Not all *D. latifolia* populations in Indonesia exhibit sexual regeneration capabilities; some heavily rely on vegetative regeneration through root suckers, leading to the formation of dense patches of new growth known as "wilding carpets". A field survey indicated varying number of new growth per hectare in different habitats, ranging from 1,000 to 200,000, with additional saplings arising from shoots from tree stumps. However, these high densities of new growth may not always be advantageous, potentially resulting in resource competition and reduced growth rates [28]. In their natural habitats, the trees demonstrate a remarkable ability to produce epicormic shoots, providing opportunities for implementing coppice harvesting techniques [29].

The root system of *D. latifolia* is characterised by rapid and extensive growth. In Cikepuh, West Java, local communities refer to it as "the soil dragging plant" due to its ability to spread widely by producing from root sprout that emerge above the soil surface [28]. This vigorous root growth is not limited to *D. latifolia* alone but is also characteristic of *D. sissoo*. These invasive roots can take over an area within a few years [3]. Moreover, both species can produce shoots at significant distances from the parent tree. While neither species is officially classified as invasive, many horticulturists consider them nuisance trees due to their aggressive root growth and sprouting tendencies [45, 48]. Nursery-scale experiments also reveal that *D. latifolia* roots can grow 4–8 m within a 6-month observation period [29]. Additionally, one-year-old seedlings produced through cutting propagation also exhibit new growth emerging from their roots. Specific treatments induced root-derived shoots in cutting-propagated seedlings by the month eight of the experiment [28].

5 Use and Trade

Dalbergia species, valued for its rich colour, density, and durability, is extensively used in high-end furniture making, musical instruments, decorative items, and architectural woodwork [7, 22, 26, 37, 49]. In China, rosewood holds deep cultural significance in luxury furniture, [19], with its use dating back to the late Ming and early Qing dynasties, when it was favoured by royal families. Later, the wood became increasingly popular among ordinary people, and the country's economy grew following the reformation. The stocks are depleting while demand continues to increase. Despite introducing non-native species for plantation purposes in the southern tropical and subtropical parts of China [42], due to its long rotation, the wood industry in China is constrained to rely on imports to meet the demand.

The rosewood price in China's domestic market increased drastically in 2006, resulting in significantly increased imports, mainly from Africa, in 2007. Subsequently, the imports escalated in 2011 as Chinese companies successfully found suppliers from African and Southeast Asian countries [42] to supply rosewood products to the US and EU markets. Despite a financial crisis in 2008 and the increasingly stringent requirement for sustainable timber in those countries, which reduced the international demand, China continues to lead as the largest rosewood importer in the world.

The international trade of rosewood initially came from Latin America and Africa, including Madagascar. In 2013, several *Dalbergia* species (including *D. nigra, D. granadillo, D. retusa* from Latin America, *D. cochinchinensis* from Indochinese countries, and all *Dalbergia* species from Madagascar) were added to Appendix II of CITES due to high trade and dwindling populations in the wild. In 2016, all species in the genus *Dalbergia* were included in Appendix II of CITES due to an increase in the number of *Dalbergia* species beyond those previously listed. Since this species is also widely distributed in Indonesia and one of the major exporters of *D. latifolia* wood in the international market, Indonesia has been applying for a CITES permit to export rosewood products since 2017.

Before the CITES listing, the trading system or mechanism of *D. latifolia* in Indonesia was similar to non-CITES commercial woods. Including this species in the CITES Appendix caused wood prices in the domestic market to soar [4, 11], accompanied by a tendency towards intensified harvesting [23]. The export level since the inclusion was 84,334 ± 3,721 m^3 annually from 2017 to 2021, and 97.6% of this number was from China [19]. In 2021, the total export of *D. latifolia*'s wood from Indonesia was 88,100.04 m^3 [45]. These volumes were not harvested from the wild. Nevertheless, there are concerns that the population in the wild would be impacted and threatened.

Timber administration in Indonesia is governed by regulations ensuring traceability and legality. Ministerial Regulation No. 8/2021 mandates transport documents verifying timber sources, while Regulation No. 20/2022 addresses domestic use of CITES-listed species like *D. latifolia.* As a CITES species, trade of such species is regulated under the Regulation of the Minister of Environment and Forestry No. 18 of 2024, formerly Decree No. 447/Kpts-II/2003, requiring permits from the Conservation of Natural Resources Regional Office (BKSDA) and the Document for Foreign Transport of Wild Plants and Animals (SATS-LN CITES). Additionally, for wood export in international trade, the company must adhere to the SVLK (Timber Legality Verification System) wood legality system, which aims to ensure the legality of legally certified wood throughout the supply chain, from the initial harvest to its market entry. The SVLK system ensures legality across the supply chain, and Regulation P.30/MenLHK/2016 mandates legality verification for all wood from state and private forests [6, 21].

6 Present Threat and Conservation Measures

The popularity of *D. latifolia* wood stems from the prohibition of exporting *Dalbergia* wood from Brazil (*D. nigra*) since 1969. As a replacement, other *Dalbergia* species, such as African blackwood (*Dalbergia melanoxylon*), Amazon rosewood (*Dalbergia spruceana*), and Indian Rosewood (*D. latifolia*—including those from Indonesia) began entering the international market. Rosewood is considered a highly traded illegal commodity [21]. Based on 67 seizure reports spanning from 2014 to 2022 [21], authorities confiscated 4,285 logs of *D. latifolia.* These seizures primarily occurred on the islands of Sumatra (16 seizures; 1,190 logs), Java (43; 1,780 logs), and Bali, Lombok, Sumbawa, and Timor (8; 1,408 logs). Although there was a rise in seizure frequency per year, no discernible seasonal patterns were observed. Nijman (2024) reported that the majority of the illegal trade is directed towards Java, from where over 95% is destined for export to China. He further mentioned that most of the rosewood trees harvested outside Java are transported to Java with major trade hubs located in cities such as Madiun, Probolinggo, Malang, and Surabaya (handling rosewood primarily from eastern Java, Bali, Lombok, Sumbawa, and Timor), as well as Tegal, Cepu, Solo, Semarang (dealing with central and eastern Java), and Bandung (serving western Java and Sumatra).

Land use change is the second threat to rosewood, particularly on community-owned land. Although export demand declined in 2021 due to the COVID-19 pandemic, which decreased purchasing power, this has not affected local harvests, particularly from community-owned lands. This often leads to illegal harvesting activities since the selling price of rosewood in the domestic market is much higher than that of teak, almost ten times that of teak and mahogany.

Another threat to the survival of rosewood is disease/pest attacks by pathogenic fungi. During field observations, two incidents of pathogenic fungi attacks were recorded in Semoyo Village, Gunung Kidul. Wilt disease in *D. latifolia* caused by *Fusarium solani* has long been reported as a common cause of tree death [36]. The spreading sucker-root reproduction of this wood species allows wider and faster transmission of fungal infection. This might destroy the clonal plants or lead to disease outbreaks that can devastate entire plantations or ecosystems. In India, the dieback of *D. latifolia* caused by *Colletotrichum capsici* has been reported [40]. Some other fungal species also cause dieback and wilt diseases in *D. sissoo* [13, 41].

Another threat to *D. latifolia* in Indonesia is land conversion, though official estimates for *D. latifolia* plantations, in private or community-owned land, are unavailable. Converting agroforestry areas previously planted with *D. latifolia* into intensive farmland for quick income generation poses a serious and alarming risk to the long-term sustainability of *D. latifolia* populations.

Conservation measures and actions to protect the species are lacking since *D. latifolia* is not protected under Indonesian regulations. Globally, the species is categorised as Vulnerable under the IUCN Red List category and criteria. Still, it is categorised as "Least Concern" in Indonesia due to its abundance, ease of cultivation, and lack of specific treatment requirements for growth, despite being in high demand. In addition, with its reputation as a slow-growing tree [35], the government must make bold and informed decisions to lead to the mitigation of those threats and prevent extinction. With threats from illegal trade and biological aspects, it is incumbent upon the Indonesian government to manage rosewood more wisely so that the trade and survival of rosewood as a species can continue sustainably.

7 Genetic Resources and Maintenance of Genetic Diversity of *Dalbergia latifolia*

Dalbergia latifolia has been widely introduced and well-adapted in different geographical regions of Indonesia [45]. To maintain the sustainability of this species, information on genetic diversity and structure is necessary as a basis for providing future conservation and management strategies for long-term survival. Genetic diversity evaluated from the major geographic distributions in Indonesia, i.e., Java Island (West Java, East Java, and Yogyakarta provinces), Lombok Island, and Sumbawa Island using the sequence-related amplified polymorphism (SRAP) markers [44, 48] revealed a relatively low overall genetic diversity ($I = 0.242$, $He = 0.156$) [48]. This is likely caused by vegetative propagation through roots, carried out intensively in the past. *D. latifolia* propagates vegetatively from roots, and only a few populations, such as Sumbawa and West Java, are known to produce viable flowers and seeds even though this species is open-pollinated [8, 31]. We identified two seed orchards in Sukabumi (West Java) and Lombok (West Nusa Tenggara), which can serve as seed stocks. When a mother tree is felled and the stump remains, new shoots will emerge as offspring from the stump and underground roots. This situation poses challenges for the trees' long-term survival, as the resulting population becomes homogeneous and is more vulnerable to internal and external stresses, including climate change, pests, and diseases [45].

Determining genetic diversity at the population or region level is also crucial for long-term management [16]. The genetic diversity of *D. latifolia* across geographical distributions varies. Central Java population has the highest genetic diversity ($I = 0.309$, $He = 0.196$) and the West Lombok population ($I = 0.202$, $He = 0.129$) has the lowest based on SRAP markers, indicating that Central Java may exhibit greater adaptability to survive and more fitness, whereas West Lombok may demonstrate less adaptation and reduced fitness compared to other populations [48]. Populations with a high level of genetic diversity have a strong positive correlation between heterozygosity and population fitness, which is vital for the long-term adaptation of populations to novel environmental conditions [34] and may be a way for populations to adapt to environmental changes [5]. In Central Java, *D. latifolia* tree stands are found in several locations, including protected forests, production forests, and farmland. Tree stands in production areas and farms may have undergone significant proliferation. Conversely, those in protected forests likely experience less human impact, which may lead to a greater diversity. This is because the regeneration of *D. latifolia* in Central Java likely comes from a combination of natural reproduction, cross-breeding, and clonal sources [48].

Different ecological conditions may influence the level of genetic diversity and structure of *D. latifolia*. For example, the West Java region has relatively higher humidity, while East Java, Lombok, and Sumbawa Island have drier climates than West and Central Java [45]. The populations were clustered according to bioregions: West–East Java, Lombok, and Sumbawa formed their respective clusters, while Central Java contains a mixture of West–East Java and Lombok populations

[48]. These patterns also indicated a complex or multiple origin of *D. latifolia* in Indonesia, which requires further studies.

Future efforts must focus on preserving the genetic diversity of *D. latifolia*. One effective strategy is to create plantations that include tree stands or populations with varied genetic makeup to enhance cross-pollination and increase genetic variability. The Central Java populations, noted for their significant genetic diversity, should be tapped as a source of germplasm for *D. latifolia* in Indonesia. Additionally, establishing clonal seed orchards utilising selected genotypes with various genetic backgrounds is recommended to conserve and produce genetically diverse seeds.

8 Conclusion

Dalbergia latifolia is widespread across Indonesia, especially on Java and the West Nusa Tenggara islands, largely due to land rehabilitation initiatives from the 1950s to the 1970s. This species is recognised for its adaptability and ease of reproduction, both sexually and asexually. It propagates effectively through root and shoot cuttings and exhibits a strong ability to coppice. Nonetheless, not every population produces viable fruits and seeds. Healthy populations can be identified as genetic resources for advancing *D. latifolia* and providing seed sources. The commercial value of *D. latifolia* increased significantly following its inclusion in Appendix II of CITES in 2017. In Indonesia, domestic trade and distribution regulated under the Ministry of Environment and Forestry's No. 18/2024, formerly Decree No. 447/Kpts-II/2003 and CITES provisions. Despite its high economic demand, the species faces serious threats from illegal logging, illegal trade, and land conversion. Conservation measures and specific protection efforts, however, remain limited, as this species is not protected under Indonesian regulations. At global level, *D. latifolia* is categorised as Vulnerable following the IUCN RedList for Threatened Species, whereas in Indonesia, it is categorised Least Concern due to its abundance, ease of cultivation, and minimal growth requirements, despite its high market demand.

Acknowledgements We thank the European Union and the CITES Secretariat for financially supporting Indonesia's CITES Tree Species Programme. We also appreciate the Indonesian Management Authority and Scientific Authority for their support in implementing the project in Indonesia. We acknowledge the help of the local forestry offices during the fieldwork and data collection. Finally, we appreciate the discussion with Dr. Vivi Yuskianti, which allowed us to complete this chapter.

References

1. Adema F, Ohashi H, Sunarno B (2016) Notes on Malesian Fabaceae (Leguminosae-Papilionoideae) 17. The genus *Dalbergia*. Blumea-Biodivers Evol Biogeogr Plants 61(3):186–206

2. Ardhana IPG, Sunardi S (2013) The technic of agroforestry at communal forest areas in Pejarakan village, district of Gerokgak, regency of Buleleng, province of Bali. J Trop Life Sci 3(1):34–42. https://jtrolis.ub.ac.id/index.php/jtrolis/article/view/66
3. Arunkumar AN, Warrier RR, Kher MM, Teixeira da Silva JA (2022) Indian rosewood (*Dalbergia latifolia* Roxb.): biology, utilisation, and conservation practices. Trees 36(3):883–898. https://doi.org/10.1007/s00468-021-02243-3
4. Atikah TD, Wardani W, Surya D (2021) Population, economic value and the trade of *Dalbergia latifolia* in Lombok, West Nusa Tenggara, Indonesia. In: IOP conference series: earth and environmental science, vol 762, no 1, p 012015. https://doi.org/10.1088/1755-1315/762/1/012015
5. Birader K (2023) Genetic diversity and the adaptation of species to changing environments. J Biodiv Endan Sp 11(3):474
6. Cetera K, Said Z, Boer FD, Qomariah IN, Suprapto E, Triyanto S (2021) How wood identification technologies help ensure timber legality in Indonesia. Working paper, WRI Indonesia, Jakarta, p 32. https://doi.org/10.46830/wriwp.18.00084
7. Chabot O, Lessard L, Dubé M (2022) A novel flax fibre composite material for stringed instrument fingerboards. J Reinf Plast Compos 41(17–18):670–678
8. Damaiyani J, Prabowo H (2019) Conservation strategy of a vulnerable species of 'Rosewood' (*Dalbergia latifolia* Roxb.) by insect pollinator identification. J Phys Conf Ser 1363(1):012005–012005
9. Danu S, Ruwendah S et al (2017) Analisis bentang alam dan bentang kehidupan industri veneer berbasis masyarakat di hutan rakyat di Kabupaten Luwu Timur. Sulawesi Community Foundation
10. Dasgupta P, Senthilkumar S, Barthwal S, Sasidharan KR, Ghosh Dasgupta M (2023) Characterisation of the chloroplast genome of *Dalbergia latifolia* (Fabaceae) and its phylogenetic relationship with other *Dalbergia* species. Nord J Bot 2023(7):e03811
11. Dwianto W, Bahanawan A, Kusumah SS, Darmawan T, Amin Y, Pramasari DA, Lestari E, Akbar F, Sudarmanto (2019) Study on the existence and characteristics of Sonokeling (*Dalbergia latifolia* Roxb) as an Appendix II CITES Wood. In: IOP conference series: earth and environmental science, vol 374, no 1, p 012063
12. Fambayun RA, Subiakto A, Rachmat HH, Atikah TD, Pratama BA (2022) Establishment of growth and yield plots of *Dalbergia latifolia* in Java and West Nusa Tenggara, Indonesia. Directorate of Biodiversity Conservation of Species and Genetic, Ministry of Environment and Forestry and the National Research and Innovation Agency, Jakarta, Indonesia
13. Fonseka DLCK, Wickramaarachchi WWUI (2018) Rosewood (*Dalbergia* spp.); A luxury timber. SSRG Int J Agric Environ Sci 5(4):25–28
14. GBIF.org (2020) GBIF Occurrence Download. https://doi.org/10.15468/dl.385fsx
15. Hani A, Suryanto P (2014) Dinamika Agroforestry Tegalan Di Perbukitan Menoreh, Kulon Progo, Daerah Istimewa Yogyakarta. Jurnal Penelitian Kehutanan Wallacea 3(2):119–128
16. Jamieson IG, Grueber CE, Waters JM, Gleeson DM (2008) Managing genetic diversity in threatened populations: a New Zealand perspective. N Z J Ecol 32(1):130–137
17. Kumar P, Patel PK, Pandro V (2023) User friendly propagation technique for quality plant production of *Dalbergia latifolia* (Roxb.) for ecological sustenance. In: Clonal forestry-principles and practices. Narendra Publishing House, Delhi, pp 140–153
18. Kumar P, Kumar PK, Sonkar MK (2022) Propagation through juvenile shoot cuttings in difficult-to-root *Dalbergia latifolia*—examining role of endogenous IAA in adventitious rooting. Plant Physiol Rep 27:242–249
19. Lyu Z (2021) Proposed strategy for decreasing the illegal logging and trade of Rosewood (*Dalbergia* spp.). Thesis submitted to University of Toronto, Canada. https://utoronto.scholaris.ca/server/api/core/bitstreams/b84e05db-a2af-46df-9478-034bc74ad153/content
20. Mando LOAS, Rosmarlinasiah, Maani A, Hasani UO (2019) Analisis Struktur dan Keanekaragaman Vegetasi Di Kawasan Hutan Lindung Jompi Kabupaten Muna. Prosiding Seminar Nasional Silvikultur VI:111–121

21. Maryudi A (2015) Choosing timber legality verification as a policy instrument to combat illegal logging in Indonesia. For Policy Econ 68:99–104. https://doi.org/10.1016/j.forpol.2015.10.010
22. NFTA (1994) *Dalbergia latifolia*: a high-valued Indian rosewood. NFTA 94-04
23. Nijman V (2024) The illegal trade in rosewood in Indonesia. Eur J For Res 143:1047–1055. https://doi.org/10.1007/s10342-024-01674-0
24. Peluso NL (1991) The history of state forest management in colonial Java. For Conserv Hist 35(2):65–75
25. Pramesthi, Haryanto KR (2010) Faktor-Faktor Yang Mempengaruhi Pemanenan Tegakan Di Hutan Rakyat (Studi Kasus Di Kelurahan Selopuro, Kecamatan Batuwarno, Kabupaten Wonogiri, Provinsi Jawa Tengah). Embryo 7(8):82–89
26. Prasad AGD, Sukandi T (1994) *Dalbergia latifolia*—the high-valued Indian rosewood. Forest, Farm, and Community Tree Network (FACT Net). https://www.winrock.org/factnet-a-lasting-impact/factsheets/dalbergia-latifolia-the-highvalued-indian-rosewood/. Accessed 05 May 2024
27. Prawirohatmodjo S, Suranto J, Martawijaya A, Den Outer RW, Sosef MSM (1993) *Dalbergia* L.f. In: Plant resources of South-East Asia No 5(1): Timber trees; Major commercial timbers. Pudoc Scientific Publishers Wageningen, Netherlands, pp 155–161. https://edepot.wur.nl/411143
28. Rachmat HH, Subiakto A, Yulita KS, Setyawati T et al (unpublished) Sustainable management of *Dalbergia latifolia* in Indonesia. CITES annual report
29. Rachmat HH, Setyawati T, Subiakto A, Yulita KS, Wardani W, Pratama BA, Rini DS, Fambayun RA, Ratnaningsih S, Purnamaningtyas NN, Ginoga KL (2022) The development prospect of coppiced *Dalbergia* plantation in Indonesia. XV World Forestry Congress: Building a Green, Healthy and Resilient Future with Forest. 2–6 May, 2022, Coex, Seoul, Republic of Korea. https://openknowledge.fao.org/handle/20.500.14283/cc1571en
30. Santoso P, Purwanto RH, Wardhana W, Andriyanti DT (2021) Potensi kayu sonokeling (*Dalbergia latifolia* Roxb.) dan jenis kayu lain di hutan rakyat Kecamatan Dlingo, Bantul Yogyakarta. J For Sci Avicennia 4(1):1–14
31. Sasidharan KR, Thangamani D, Prakash S, Muraleekrishnan K (2020) Pollination ecology and breeding system in *Dalbergia latifolia* and *D. sissoides*. Ecol Environ Conserv 26:S76–S81
32. Soerianegara I, Lemmens RHMJ (1993) Plant resources of southeast Asia. Timber trees: major commercial timbers, No. 5(1), Pudoc Scientific Publishers, Wageningen, Netherlands, pp 384–391
33. van Steenis CGGJ (1950) Flora Malesiana: Spermatophyta. (Flowering Plants), vol 1. Cyclopaedia of Collectors & Collections Noordhoff-Kolff, pp 243–244
34. Stojnić S, Avramidou EV, Fussi B, Westergren M, Orlović S, Matović B, Trudić B, Kraigher H, Aravanopoulos FA, Konnert M (2019) Assessment of genetic diversity and population genetic structure of Norway spruce (*Picea abies* (L.) Karsten) at its southern lineage in Europe. Implications for conservation of forest genetic resources. Forests 10(3):1–15
35. Subiakto A, Atikah TD, Rachmat HH, Pratama BA, Fambayun RA (2022) Distribusi, Ekologi, Pemodelan Spasial dan Regenerasi *Dalbergia latifolia* di Jawa dan Nusa Tenggara Barat, Indonesia. Direktorat Konservasi Keanekaragaman Hayati Spesies dan Genetik, Kementerian Lingkungan Hidup dan Kehutanan dan Badan Riset dan Inovasi Nasional, Indonesia
36. Suharti M, Hadi S (1974) Wilt disease of *Dalbergia latifolia* in Malang forest district, E. Java. Laporan Lembaga Penelitian Hutan, Bogor 94:9
37. Sukandi T (1994) Agroforestry-based plantations of *Dalbergia latifolia* Roxb. in Java, Indonesia. In: Westley SB, Roshelko JM (eds) Dalbergia: proceedings of an international workshop, Nitrogen Fixing Tree Research Reports (Special issue). Winrock International, Little Rock, pp 1–5
38. Sunandar R (2006) Pengaruh induksi suhu dan konsentrasi rootone-f dengan metode perendaman terhadap pembentukan akar dan tunas stek akar sonokeling (*Dalbergia latifolia* Roxb). Departemen Agronomi Dan Hortikultura Fakultas Pertanian Institut Pertanian Bogor
39. Verhaegen D, Fofana IJ, Logossa ZA, Ofori D (2010) What is the genetic origin of teak (*Tectona grandis* L.) introduced in Africa and in Indonesia? Tree Genet Genomes 6:717–733

40. Verma P, Soni KK, Verma RK, Shirin F (2016) Seedling die-back of *Dalbergia latifolia* in Central India caused by *Colletotrichum capsici*—a new record. Int J Curr Microbiol Appl Sci 5(6):350–356
41. Webb EL, Hossain SMY (2005) *Dalbergia sissoo* mortality in Bangladesh plantations: correlations with environmental and management parameters. For Ecol Manage 206(1–3):61–69
42. Wenbin H, Xiufang S (2013) Tropical hardwood flows in China: case studies of rosewood and Okoume. Forest Trends Association, Washington DC, USA
43. Winfield K, Scott M, Grayson C (2016) Global status of *Dalbergia* and *Pterocarpus* rosewood producing species in trade. In: Convention on international trade in endangered species 17th conference of Parties-Johannesburg, 24 September–5 October 2016, Johannesburg, 246p
44. Yulita KS, Atikah TD, Wardani W (2020) Unraveling genetic variations of *Dalbergia latifolia* (Fabaceae) from Yogyakarta and Lombok Island, Indonesia. Biodivers J Biol Divers 21(2):833–841
45. Yulita KS, Wardani W, Atikah TD, Pratama BA, Subiakto A, Nurjanah S, Nopiansyah F, Nugroho A, Kamal I, Arrofaha N (2022) The non-detriment findings (NDF) report for *Dalbergia latifolia* in Java and West Nusa Tenggara, Indonesia. Directorate of Biodiversity Conservation of Species and Genetic, the Ministry of Environment and Forestry, and the National Research and Innovation Agency, Indonesia. https://cites-tsp.org/sites/default/files/project_files/2023-01/NDF_Report_for_Dlatifolia.pdf
46. Yulita KS, Setyawati T, Wardani W, Atikah TD, Pratama BA, Subiakto A, Nurjanah S, Arrofaha A, Kamal I (2022) Assessment of the current management practices, the current harvest control and monitoring, and the conservation status of *Dalbergia latifolia* in Java and West Nusa Tenggara, Indonesia. Directorate of Biodiversity Conservation of Species and Genetics, the Ministry of Environment and Forestry, and the National Research and Innovation Agency, Jakarta, Indonesia
47. Yulita KS, Dwiyanti FG, Kamal I, Susila, Arrofaha N, Rachmat HH (2022) A report on the molecular characterisation of *Dalbergia latifolia* from Java and West Nusa Tenggara: DNA extraction, PCR amplification, and DNA sequencing. Directorate of Biodiversity Conservation of Species and Genetics, Ministry of Environment and Forestry, and the National Research and Innovation Agency, Indonesia
48. Yulita K, Susila, Rachmat HH, Dwiyanti FG, Atikah TD, Subiakto A, Pratama B, Setyawati T, Wardani W, Fambayun RA, Arrofaha N, Kamal I (2022) Population genetic of the Indonesian rosewood (*Dalbergia latifolia*) from Java and West Nusa Tenggara revealed using sequence-related amplified polymorphism. For Sci Technol 18(4):172–181
49. Yulita KS, Fambayun RA, Setyawati T, Subiakto A, Rini DS, Hendalastuti H, Pratama BA (2022) A review on taxonomy, biology, ecology, and population Status of *Dalbergia latifolia* from Indonesia. https://cites-tsp.org/sites/default/files/project_files/2023-01/A-review-on-Taxonomy-Biology-Ecology-and-Population-Status-08-2020_compressed.pdf

Chapter 6
MaxEnt Modelling for the Present Scenario and Predicting the Future Habitat Suitability of *Dalbergia latifolia* in India

S. Lalitha, P. Devamanikandan, D. Thangamani, R. Vasudevan, K. R. Sasidharan, T. Vamadevan, and Rekha R. Warrier

1 Introduction

Global climate change is a significant threat to biodiversity, particularly for species with narrow environmental tolerances and restricted geographic ranges, thereby increasing the risk of extinction for many of them. Projections indicate that the rapid climate change in the twenty-first century could impact biodiversity at different levels. The Intergovernmental Panel on Climate Change (IPCC) estimates that a 2–3 °C rise in global temperatures could lead to a high risk of extinction for 20–30% of species [16, 21]. In recent years, the rise of human civilisation has led to an increase in greenhouse gas emissions, resulting in global warming and subsequent alterations in the survival conditions of various species. Several models are being utilised to investigate the distribution of species with economic significance and the impact of climate change on their distribution. Considerable changes in climate can cause

S. Lalitha (✉) · P. Devamanikandan
Forest Genetic Resource Management Division, ICFRE-Institute of Forest Genetics and Tree Breeding, Forest Campus, Coimbatore, India
e-mail: lalithas@icfre.org; lolram66@gmail.com

D. Thangamani · R. Vasudevan · R. R. Warrier
Chemistry and Bioprospecting Division, ICFRE-Institute of Forest Genetics and Tree Breeding, Forest Campus, Coimbatore, India

K. R. Sasidharan
Forest Ecology and Climate Change Division (Retired), ICFRE-Institute of Forest Genetics and Tree Breeding, Forest Campus, Coimbatore, India

T. Vamadevan
EIACP Programme Centre Resource Partner on Forest Genetic Resources and Tree Improvement, ICFRE-Institute of Forest Genetics and Tree Breeding, Forest Campus, Coimbatore, India

R. R. Warrier et al. (eds.), *Handbook of Asian Rosewoods*, Materials Horizons: From Nature to Nanomaterials, https://doi.org/10.1007/978-981-96-9459-4_6

shifts in niches or reductions in habitat, which may lead to the isolation and loss of genetic diversity in these species. Prompt mitigation measures are recommended to prevent such range losses and enhance adaptation mechanisms.

Dalbergia latifolia, commonly known as Indian rosewood, is a tree native to low-elevation tropical monsoon forests of India and holds significant ecological and economic importance. It spreads across the sub-Himalayan tract eastwards to Sikkim, Bihar, Odisha, and throughout central and southern India. Additionally, it has been introduced to other regions such as Burma, Sri Lanka, Nepal, Nigeria, the island of Java in Indonesia, and Kenya. Even though *D. latifolia* is an evergreen species, it can exhibit deciduous characteristics in drier sub-populations. The species thrives across various habitat types, ranging from moist deciduous forests to tropical evergreen ecosystems, showcasing its remarkable adaptability to diverse environmental conditions. Western Ghats forests of Karnataka, Kerala, and Tamil Nadu provide the most favourable conditions for its growth. While *D. latifolia* can tolerate various soil types, it thrives in well-drained, deep, moist soils, including deep loams or clays containing lime and black cotton soils. Renowned for its high-quality timber and diverse medicinal properties, *D. latifolia* has long been a subject of ecological research and commercial utilisation in their region. Despite its resilience, *D. latifolia* faces numerous challenges, including habitat loss, overexploitation, and climate change-induced pressures, all threatening its long-term survival. Consequently, understanding the ecological requirements and distribution patterns of *D. latifolia* is very important for effective conservation and sustainable management practices. The International Union for Conservation of Nature (IUCN) has classified *Dalbergia* spp. as "Vulnerable," placing it under the "Red List" of plants. Following the 17th meeting of the Conference of the Parties (CoP17) to the Convention on International Trade in Endangered Species of Wild Fauna and Flora (CITES) held in 2016 in Johannesburg, South Africa, new trade regulations on *Dalbergia* spp. came into effect on 2 January 2017, with all species under the genus *Dalbergia* included under Appendix II of CITES. Keeping this in mind, our study aimed to identify the dominant environmental variables, status of the current potential distributions, and predict the projected climate change scenarios (RCP 2.6, 4.5, and 8.5) for *D. latifolia.* These projections will be based on the Representative Concentration Pathway (RCP) for two different periods, the 2050s and the 2070s. Through model-based predictions, the study emphasises the necessity of targeted conservation efforts to ensure the ongoing existence of this valuable species within the region's biodiversity.

2 Occurrence Data

The research area encompasses 14 states and the Andaman & Nicobar islands of India, spanning 289 geographical coordinates (234—Surveyed data; 55-GBIF portal) [5]. These coordinates range from 8° 25′ 37.2″ N to 30° 24′ 0″ N in latitude and from 71° 42′ 0″ E to 92° 50′ 9.6″ E in longitude. To eliminate spatial autocorrelation, georeferenced points within 500 m of each other were removed [8]. Finally, 203 occurrence points of *D. latifolia* were included in this study to model the current (Fig. 1) and future potential distribution of the species. The species distribution points were plotted in GIS environment using ArcGIS software 10.8.2.

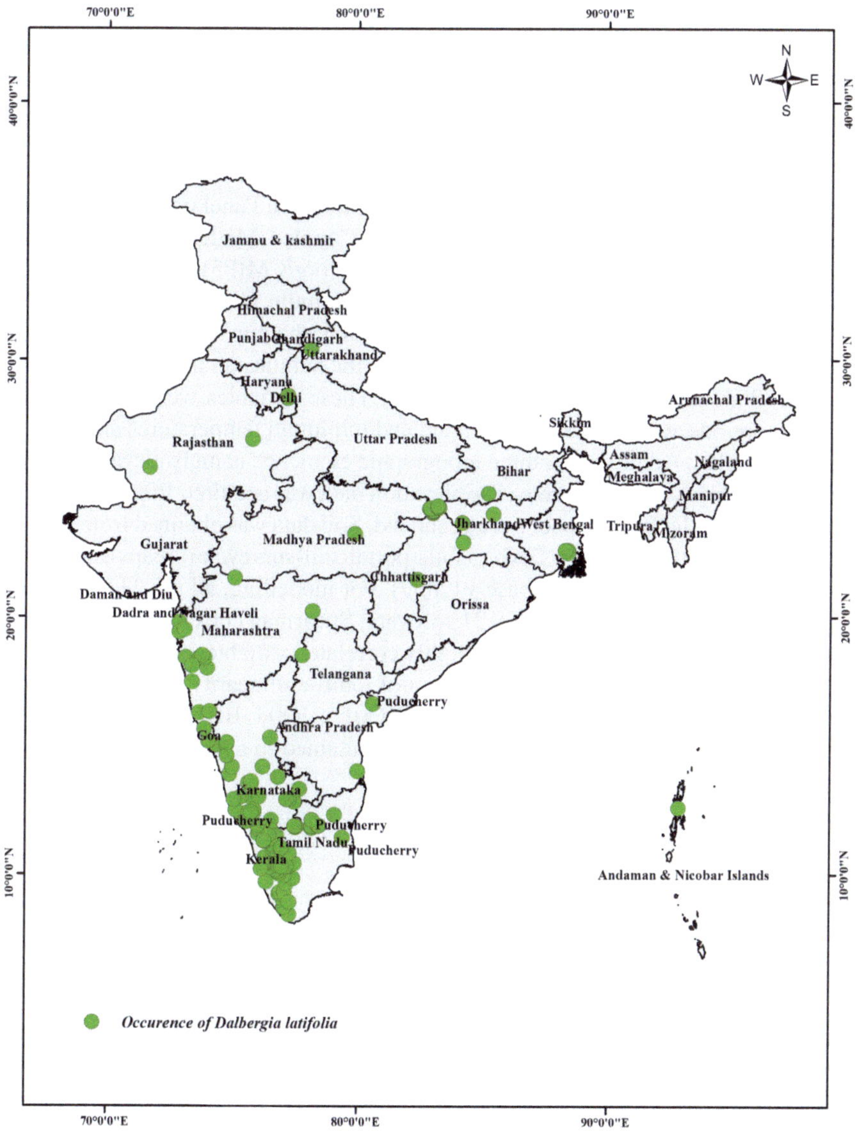

Fig.1 Distribution map of *D. latifolia* in India

3 Bioclimatic Data

The 19 bioclimatic variables in the WorldClim database (http://www.worldclim.org) were used for modelling the current potential distribution areas spanning from 1950 to 2000 [7, 23] with a spatial resolution of 30 arc seconds (~1 km). These variables were determined on various scales for duration and maximum-minimum temperature and precipitation values [6]. Datasets from the Representative Beijing Climate Center Climate System Model version 1 (BCC-CSM 1.1) for the years 2050 and 2070 were used to predict future habitat suitability under different climate change scenarios. Representative concentration pathways (RCPs including RCP2.6, RCP4.5, RCP6.0, and RCP8.5) governed by the Intergovernmental Panel on Climate Change (IPCC) based on the 5th assessment report of Coupled Model Inter-Comparison Project Phase 5 (CMIP5). (http://www.worldclim.org/CMIP5) were used. In this study, we selected RCP 2.6, representing the minimum GHG emission scenario, RCP 4.5, the medium GHG emission, and RCP 8.5, the maximum greenhouse gas emission scenario (IPCC 2014) [18] to simulate the habitat suitability of *D. latifolia* in 2050 (2041–2060) and 2070 (2061–2080). These variables were computed for different periods, encompassing maximum and minimum temperatures and precipitation values [6]. Additionally, three topographic elements, namely elevation, slope, and aspect, play a significant role. The elevation data was acquired from WorldClim, and the data for slope and aspect were generated. Soil data was obtained from the Soil Survey of India (https://www.fao.org/soils-portal/soil-survey/soil-maps-and-databases/harmonized-world-soil-database-v12/en/). For modelling, all the 23 bioclimatic variables were used. Through SPSS 21 software, Spearman correlation analysis was performed to identify and remove the highly correlated variables (coefficient value $r > 0.9$) [14] and also to minimise the impact of multi-collinearity and over-fitting of the model [19]. Ten bioclimatic variables, BIO01, BIO03, BIO09, BIO15, BIO18, BIO19, elevation, slope, aspect, and soil, were retained to simulate the current and future distribution of *D. latifolia* in India.

4 MaxEnt Analysis

The *D. latifolia* distribution data and 10 environmental variable data were processed, and the modeling operations were performed using MaxEnt (version 3.4.4). The values in the area under curve (AUC) from the receiver operating characteristic (ROC) were analysed [11] to evaluate the performance of the model. The variable importance was measured using built-in Jackknife test to determine the dominant climatic factors [9]. The Jackknife test, response curve, and logistic output format were selected, and the final output was an ASCII file. The maximum number of model iterations was set to 500, and the regularisation multiplier value was set to 0.1. Fifteen replicates were performed using cross-validation. The "write background predictions" option was selected to see if the random points were a good representation of the environmental variables to analyse, and the response curve was used to determine the threshold value for each environmental factor. Separate models were run for the current and future climatic datasets using the same settings. The final predicted model of habitat suitability was divided into four levels: unsuitable habitat (<0.2), poorly suitable habitat (0.2–0.4), moderately suitable habitat (0.4–0.6), and highly suitable habitat (>0.6). The total area of the predicted habitat of *D. latifolia* under current and future climatic conditions was also estimated by using presence grid cells multiplied by their spatial resolution.

5 Evaluation of Model Performance

The evaluation of the MaxEnt model can be verified using AUC-ROC values, True skill statistic (TSS), Kappa statistics, and Normalised Mutual Information (NMI). The AUC value is recognised as the best model predictor and an effective threshold-independent metric because it is independent of the diagnostic threshold and has a low sensitivity to species occurrence [17]. It represents the area under the curve plotted against the true positive rate (Sensitivity) and false positive rate (1—Specificity) at various threshold settings. An ideal classifier has an AUC of 1, indicating that it achieves a true positive rate of 1 and a false positive rate of 0 at all thresholds. Generally, AUC values range from 0.5 to 1.0 and can be classified into five classes. The AUC values of the models were evaluated according to the following criteria: 0.90–1.00, excellent; 0.80–0.90, good; 0.70–0.80, satisfactory; 0.60–0.70, poor; and 0.50–0.60, failed. The accuracy of the MaxEnt model for *D. latifolia* based on the AUC values was found to be >0.9, indicating the model is graded as good for our dataset for all current and future scenarios (Fig. 2). The results show that the modelling is highly reliable and can be used to analyse the impact of climate change on the distribution of *D. latifolia* in India.

The confusion matrix forms obtained from the model output raster (ASCII) files were evaluated using the true skill statistic, Kappa, and Normalised Mutual Information (NMI) analysis for understanding the model performance that evaluate their classification accuracy after selecting a threshold value and range from $+1$ to -1. The value towards $+1$ signifies an agreement between observations and prediction, and a lower value signifies agreement that is not better than random [1, 2]. TSS value >0.6, range of kappa value 0.5–0.6, and NMI value range from 0.5 to 0.6 (Table 1). Furthermore, the average AUC and TSS values of the *D. latifolia* model indicate that they are more accurate and perform better. The above metrics suggest that the species distribution and environmental variable data used in the modelling can effectively predict the potentially suitable distribution area of *D. latifolia* in India.

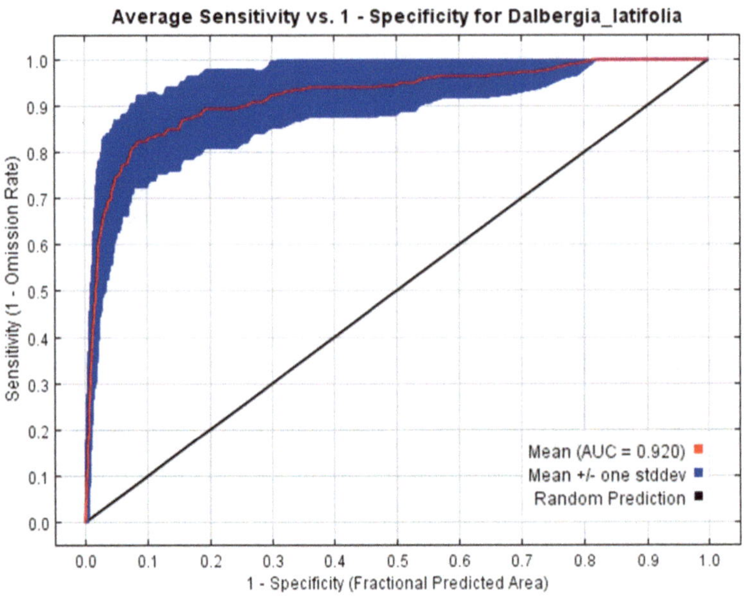

Fig. 2 The Receiver Operating Characteristic (ROC) curve for *D. latifolia*. The mean AUC value 0.920 yielded high success, inferring the predicted distribution an excellent level of accuracy

Table 1 AUC (threshold-independent), TSS, Kappa, and NMI (threshold-independent) values of habitat distribution of *D. latifolia* from different RCP scenarios (RCP 2.6, RCP4.5, and 8.5) over two future periods (2050 and 2070)

Model	RCP scenario	AUC	Accuracy		
			TSS	Kappa	NMI
Current	–	0.92	0.69	0.661	0.53
2050	RCP 2.6	0.913	0.65	0.561	0.58
	RCP 4.5	0.908	0.599	0.503	0.621
	RCP 8.5	0.914	0.6	0.518	0.62
2070	RCP 2.6	0.916	0.61	0.546	0.6
	RCP 4.5	0.917	0.587	0.531	0.615
	RCP 8.5	0.917	0.62	0.525	0.61

6 Contribution of Environmental Variables

In the present study, among the predicated variables BIO03 (Isothermality), BIO19 (Precipitation of Coldest Quarter), and Slope and Mean Temperature of Driest Quarter (BIO09) are the most important variables in all three climatic scenarios. The results showed that among 10 environmental variables, the contribution of the four variables including Isothermality (BIO03) accounted for 36.9%, Precipitation

Table 2 Contribution of variables in MaxEnt output

Code	Bioclimatic variables	Percent contribution to species suitability						
		Current	2050			2070		
			RCP 2.6	RCP 4.5	RCP 8.5	RCP 2.6	RCP 4.5	RCP 8.5
BIO03	Isothermality	36.9	48.3	55.4	55.9	55.2	51.3	51.9
BIO19	Precipitation of coldest quarter	21.3	18	10.8	8.4	9.7	12.9	15
	Slope	10.7	9.3	9.4	8.7	8.7	8.9	8
BIO09	Mean temperature of driest quarter	10.6	5.6	1.5	5.4	10.5	7.5	5.6
BIO01	Annual mean temperature	7.3	6.2	7.7	5.7	3.9	6.1	3.4
BIO18	Precipitation of warmest quarter	5.4	2.9	6.1	7.2	4.8	5.2	6
BIO15	Precipitation seasonality	4.5	2	1.9	1	1.1	2.1	1.9
	Elevation	1.7	5	4.6	5	3.9	3.8	6
	Aspect	1	2	1.7	1.5	1.7	1.7	1.8
	Soil	0.7	0.7	0.9	1	0.6	0.5	0.5

of Coldest Quarter (BIO19) for 21.3%, and Slope to 10.7% and Mean Temperature of driest quarter (BIO9) to 10.6%, collectively exceeding 79.5% of the model prediction under both current and future climatic conditions. The environmental factors that contributed less than 10% were BIO01, BIO18, BIO15, Elevation, Aspect, and Soil (Table 2). In summary, BIO03, BIO19 and slope are the dominant environmental variables predicted to limit the distribution of *D. latifolia*. These two variables provided more valuable information than the other environmental factors and significantly influenced the distribution of *D. latifolia* in India.

The contribution of each climate factor to the model was analysed and the contribution of the 10 variables calculated by the Jackknife test is shown in Fig. 3, where the gains of variable >1.2 is BIO03. Gains of variables ranging from 1 to 0.8 is BIO19, Moderate gains of 0.6–0.4 are BIO01, elevation, and slope, range of 0.4–0.2 variables are BIO09, BIO5, BIO18, and soil. Aspect had a lower gain (<0.2). The results suggest that these variables have more useful information than the others.

7 Response Curves

In the Maximum Entropy (MaxEnt) model, response curves are graphical representations that show the relationship between each predictor variable and the predicted probability of species presence. They are crucial for understanding how individual environmental variables influence the species distribution model. These curves are essential tools for interpreting and understanding the ecological niche of a species.

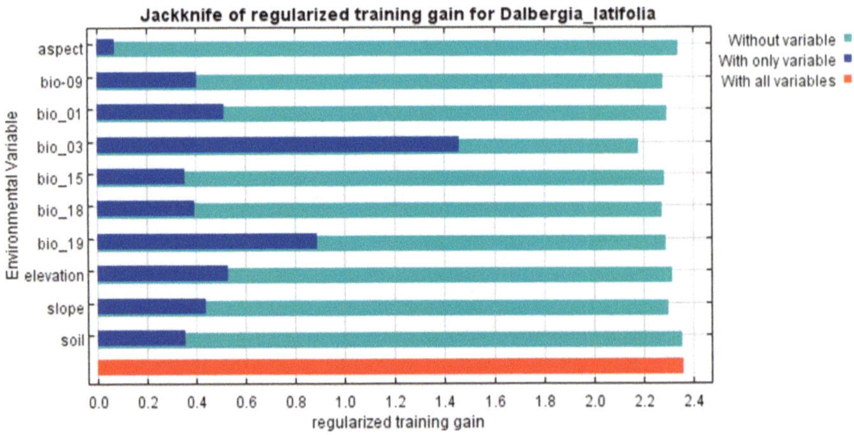

Fig. 3 The Jackknife test for evaluating the relative importance of environmental variables for *D. latifolia* MaxEnt model on habitat suitability (*Note* Aspect, "BIO09" Mean Temperature of Driest Quarter, "BIO01" Annual Mean Temperature, "BIO03" is Isothermality, "BIO15" is Precipitation Seasonality, "BIO18" is Precipitation of Warmest Quarter, "BIO19" is Precipitation of Coldest Quarter, elevation, slope, and soil)

The results of response curves of 04 variables to habitat suitability of *D. latifolia* are illustrated in Fig. 4. The suitable variable of Isothermality (BIO03) is 35.08–73.68 °C with an optimal value of 66.06 °C. Precipitation of Coldest Quarter (BIO19) is from 4.03 to 5041.05 mm with an optimum temperature of 4630.36 mm and slope is 0.07°–90.14° with an optimum value of 89.93°. The suitable Mean Temperature of Driest Quarter (BIO09) is from −7.08 to 30.42 °C with an optimal value of 948.79 °C.

8 Suitable Climate Space for *D. latifolia* Under Current Climate Scenarios

Using ArcGIS 10.8.2, the potential distribution of *D. latifolia* based on observed occurrences and current environmental conditions projected by the MaxEnt model is shown in Fig. 5. By comparing the current and future distributions under different climatic scenarios, we can evaluate the relative changes in the suitable climate space. The ASCII files were imported into ArcGIS10.8.2 and transformed into raster data before being classified into four groups using the spatial analysis tool "reclassify", namely unsuitable habitat (<0.2), low habitat suitability (0.2–0.4), medium habitat suitability (0.4–0.6), and high habitat suitability (>0.6). There are distinct patches in the region that are predicted to be highly suitable (dark red regions) as opposed to certain patches that are moderately suitable (yellow), low suitable (brown) and unsuitable (green). The total suitable habitats (including low, moderate, and

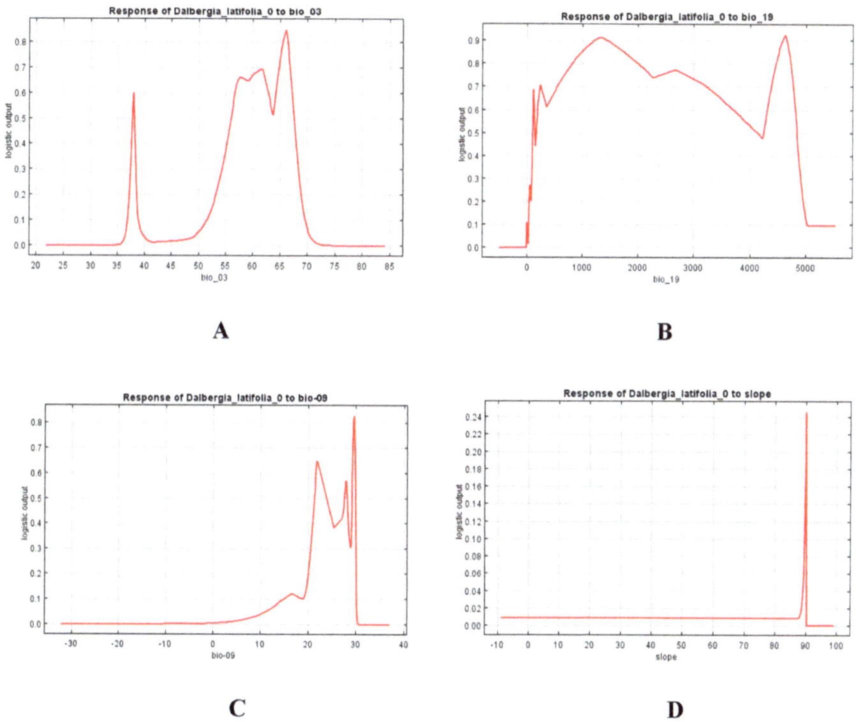

Fig. 4 Response curves illustrating the relationship of MaxEnt predicted probability of occurrence to environmental variables in *D. latifolia*. **a** Isothermality (BIO03), **b** Precipitation of coldest quarter (BIO19), **c** Mean temperature of driest quarter (BIO09), and **d** Slope

high) are primarily located in Andaman & Nicobar Islands, Andhra Pradesh, Bihar, Chandigarh, Chhattisgarh, Dadra and Nagar Haveli, Daman and Diu, Delhi, Goa, Gujarat, Haryana, Himachal Pradesh, Jammu & Kashmir, Jharkhand, Karnataka, Kerala, Lakshadweep, Madhya Pradesh, Maharashtra, Odisha, Puducherry, Punjab, Rajasthan, Tamil Nadu, Telangana, Uttar Pradesh, Uttarakhand, and West Bengal. The model predicted the unsuitable areas of *D. latifolia* in Arunachal Pradesh, Assam, Manipur, Meghalaya, Mizoram, Nagaland, Sikkim, and Tripura. The highly suitable areas are mainly distributed in the southern states, viz., Karnataka, Kerala, and Tamil Nadu. They are mainly found in the areas of Western Ghats, some parts of Eastern Ghats of Tamil Nadu, Karnataka, and some patches of Andhra Pradesh. In the central region, these areas include Bihar, Chhattisgarh, Jharkhand, Madhya Pradesh, Maharashtra, and Uttar Pradesh. The western coastal areas of Goa and Gujarat are also included, as well as the Union Territory of Andaman & Nicobar Islands, Daman & Diu, Lakshadweep, and Puducherry. Interestingly, several highly suitable habitats are also present in Punjab, Odisha, Jammu & Kashmir, Rajasthan, West Bengal, and Haryana. The current climate scenario model suggests that about 7.05% of the areas in India are suitable for *D. latifolia* growth, with the most suitable areas being

Karnataka, Kerala, and Tamil Nadu. Table 3 illustrates the total suitable area of 231,648.5 km^2 (7.05%), of which 1.35% (44,301.37 km^2) is predicted to be highly suitable, 60,767.49 km^2 area (1.85%) with moderate potential, 126,579.6 km^2 area (3.85%) showing low potential, and 3,058,471.31 km^2 as unsuitable areas for this species in India.

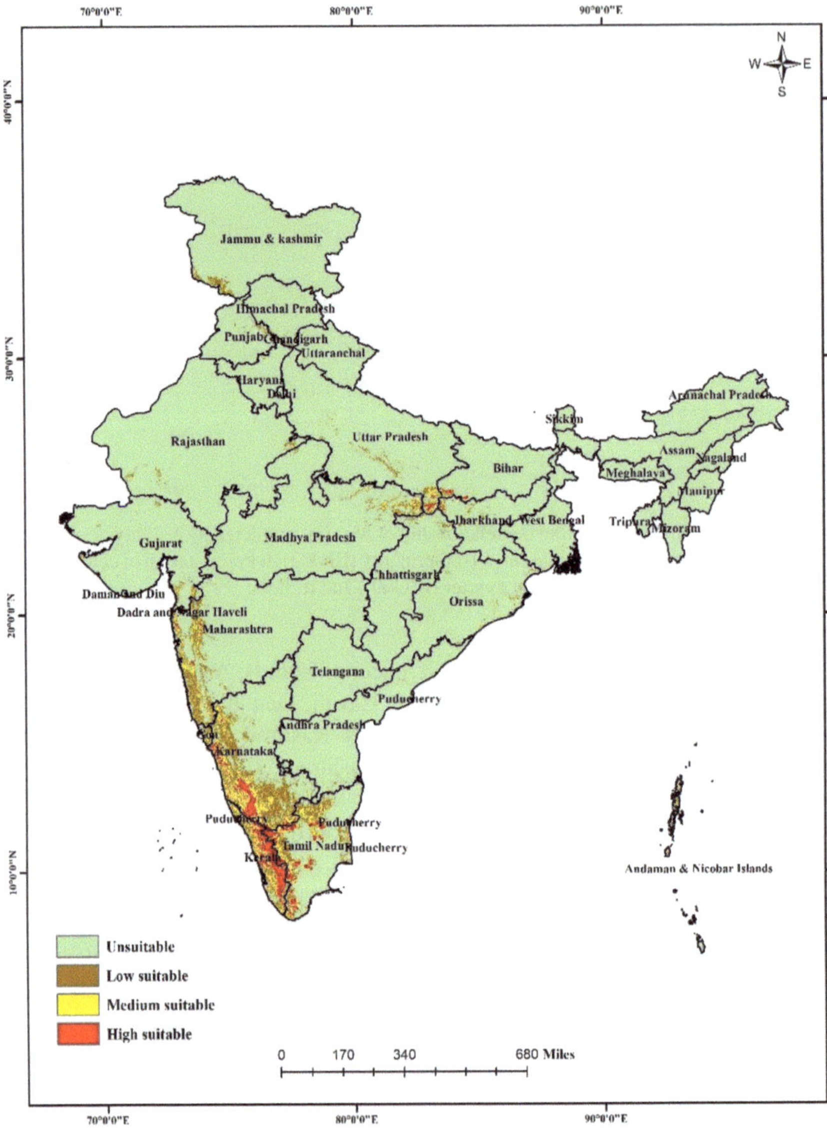

Fig. 5 Current habitat status of *D. latifolia*

Table 3 Predicted potential distribution areas for *D. latifolia* under current climatic conditions

Province	Unsuitable (km^2)	Low (km^2)	Medium (km^2)	High (km^2)	Total (km^2)	Total suitable habitat (km^2)
Jammu & Kashmir	219,716.7	3617.74	342.5482	2.900238	223,679.88	3963.183
Andaman & Nicobar Islands	2043.438	2220.89	2132.102	646.3426	7042.7766	4999.339
Andhra Pradesh	158,691.6	1204.1	22.11	0	159,917.81	1226.211
Arunachal Pradesh	84,659.68	0	0	0	84,659.68	0
Assam	81,091.13	0	0	0	81091.13	0
Bihar	92,722.82	592.551	506.7024	777.429	94,599.502	1876.682
Chandigarh	116.0461	2.4609	0	0	118.507	2.460899
Chhattisgarh	133,553	1532.69	152.4998	152.7553	135,390.94	1837.942
Dadra and Nagar Haveli	497.4157	2.09943	0	0	499.51513	2.099427
Daman and Diu	551.574	7.48283	0.654725	1.872101	561.58366	10.00966
Delhi	1505.341	1.03566	0	0	1506.3767	1.035662
Goa	901.7286	1570.6	987.723	255.2193	3715.2689	2813.541
Gujarat	184,222.4	3277.32	838.1879	166.034	188,503.94	4281.541
Haryana	43,704.2	504.982	8.577245	0.756255	44,218.516	514.3157
Himachal Pradesh	54,812.82	910.007	10.24973	0	55,733.077	920.2568
Jharkhand	77,563.78	1949.53	480.4841	433.9748	80,427.77	2863.993
Karnataka	118,024.4	40,894.6	21,052.52	12,539.04	192,510.54	74,486.14
Kerala	3259.104	9406.22	10,480.25	14,607.64	37,753.211	34,494.11
Lakshadweep	6.568031	2.50982	1.164009	0.092749	10.33461	3.766579
Madhya Pradesh	300,982.6	5844.81	1615.493	289.7761	308,732.68	7750.082
Maharashtra	276,730.6	24,082.8	6917.218	1602.813	309,333.45	32,602.85
Manipur	23,239.83	0	0	0	23,239.83	0
Meghalaya	22,951.17	0	0	0	22,951.17	0
Mizoram	21,545.94	0	0	0	21,545.94	0
Nagaland	17,319.97	0	0	0	17,319.97	0
Orissa	153,674.3	2014.87	181.0147	17.12237	155,887.31	2213.008
Puducherry	234.1884	162.015	95.51991	29.30445	521.02766	286.8393
Punjab	49,307.1	1035.31	17.19287	12.78546	50,372.39	1065.29

(continued)

Table 3 (continued)

Province	Unsuitable (km^2)	Low (km^2)	Medium (km^2)	High (km^2)	Total (km^2)	Total suitable habitat (km^2)
Rajasthan	344,331.4	1684.55	54.3931	2.347458	346,072.69	1741.291
Sikkim	7138.857	0	0	0	7138.857	0
Tamil Nadu	86,180.09	19,574.8	12,777.95	11550.71	130,083.52	43,903.43
Telangana	114,769.1	1.11458	0	0	114,770.21	1.11458
Tripura	10,561.36	0	0	0	10,561.36	0
Uttar Pradesh	233,147	4268.11	2082.693	1211.678	240,709.5	7562.48
Uttarakhand	53,503.76	70.9707	3.094357	0	53,577.825	74.07
West Bengal	85,210.29	143.489	7.143626	0.780284	85,361.703	151.41
Total	3,058,471.3	126,579.6	60,767.49	44,301.37	3,290,119.8	231,648.49

9 Distribution of *D. latifolia* in Future Climate Scenarios

Our analysis using ArcGIS 10.8.2 compared the potential future habitats for *D. latifolia* under three different climate change scenarios (RCP 2.6, RCP 4.5, and RCP 8.5) with the current distribution map. The predicted potentially suitable climatic distributions of *D. latifolia* for the years 2050 and 2070 under minimum emission (RCP 2.6), medium emission (RCP 4.5), and maximum emission (RCP 8.5) scenarios are illustrated in Fig. 6. The study found that in certain provinces of all suitable habitats of *D. latifolia*, significant changes are experienced for the period 2050 and 2070 in all climatic scenarios (RCP 2.6, RCP 4.5, and RCP 8.5). They show an expansion and contraction in these suitable areas. Overall, the total area showed a significant increase, with variation in extent between projection time periods for the same future emission scenarios. The prominent features for three climatic scenarios are as follows:

1. In the RCP 2.6 scenario for 2050 and 2070, it is predicted that there will be expansion in highly suitable areas: Andaman & Nicobar Islands, Chhattisgarh, Goa, Jharkhand, Karnataka, Kerala, Madhya Pradesh (2050), Maharashtra, Tamil Nadu, and Uttar Pradesh. Newly predicted (2050 and 2070) highly suitable regions include Andhra Pradesh and Himachal Pradesh, and medium suitable regions are Dadra and Nagar Haveli, with Delhi and Manipur holding the low-suitable area. In 2070, Manipur shifts to medium and highly suitable areas, while Arunachal Pradesh, Assam, Meghalaya, Mizoram, and Nagaland show low suitable predicted areas only for 2050 and later in 2070 these areas are predicted to be unsuitable. There will be a contraction of the highly suitable areas in 2070 in Bihar, Jammu and Kashmir and Madhya Pradesh. The regions in Lakshadweep, Daman & Diu, Haryana, Odisha, and Punjab will contract in 2050 and

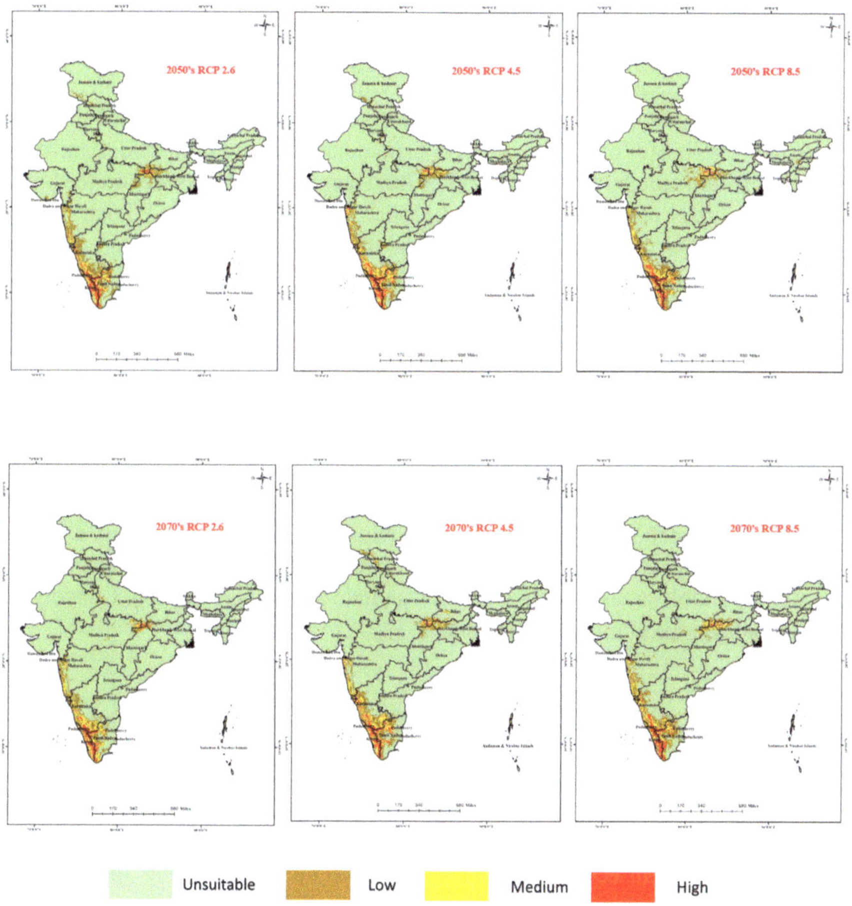

Fig. 6 The geographical distribution maps for 2050 and 2070 created according to three different climate change scenarios (minimum emission (RCP 2.6), medium emission (RCP 4.5), and maximum emission (RCP 8.5))

2070. Unsuitable regions are predicted to be Chandigarh, Sikkim, and Tripura. In 2070, Telangana is predicted to be in low suitable area.

2. In the RCP 4.5 scenario for 2050 and 2070, it is predicted that the expansion is in highly suitable areas: Andaman & Nicobar Islands, Bihar, Chhattisgarh, Goa, Jharkhand, Karnataka, Kerala, Madhya Pradesh, Maharashtra, Tamil Nadu, and Uttar Pradesh. The newly predicted highly suitable areas include Andhra Pradesh, Dadra and Nagar Haveli, Delhi, and Himachal Pradesh (2050 and 2070). In 2050, there are new predictions for Manipur in low- and medium-suitable areas. New predictions are seen in low suitable area for Meghalaya in 2050 and contraction in 2070. Uttarakhand and West Bengal are predicted to have a highly suitable area in 2070. Dadra, Nagar Haveli, and Delhi are also predicted to have a medium

and high suitable area in 2050 and 2070. The predicted regions for 2050 show low suitability for Telangana and medium suitability for Odisha. The region of Assam is predicted to have low suitability in 2050 and 2070, medium and high suitability is predicted in 2050. Mizoram and Nagaland have extended the scope of low suitability for 2050 and 2070. Daman and Diu are projected to have low and medium suitability in 2050 and 2070. There will be a contraction for highly suitable areas in Daman and Diu, Lakshadweep, Odisha, and Punjab in 2050 and 2070. On the other hand, the unsuitable regions are Arunachal Pradesh, Chandigarh, Sikkim, and Tripura.

3. In the RCP 8.5 scenario for 2050 and 2070, it is predicted that the high suitable areas Andaman & Nicobar Islands, Chhattisgarh, Goa, Jharkhand, Karnataka, Kerala, Madhya Pradesh, Maharashtra, Tamil Nadu, and Uttar Pradesh show expansion. The newly predicted high suitability areas include Himachal Pradesh (2050), Andhra Pradesh (2070), and Dadra and Nagar Haveli (2050 and 2070). In 2050, low-suitability areas include Assam, Manipur, Mizoram, and Nagaland. Medium suitable regions include Assam, Nagaland, and Tripura. Contraction is predicted to be in the highly suitable areas of Daman and Diu, Haryana, Lakshadweep, Odisha, and Punjab. Delhi and Uttarakhand regions are predicted as low suitable areas. Unsuitable areas include Arunachal Pradesh, Chandigarh, Meghalaya, Sikkim, and Telangana.

10 Analyses of the Predictions

Understanding the spatial distribution pattern of species is fundamental to effective biodiversity and conservation practices. Species Distribution Models (SDMs) are essential tools in ecology and habitat management, estimating species abundance based on environmental variables to predict habitat suitability and distribution. SDMs require species data (presence/absence records) and environmental data (climatic and topographical variables) to create habitat maps using several developed and successfully implemented models. These models include maximum entropy model (MaxEnt), ecological niche factor analysis (ENFA), generalised linear model (GLM), generalised additive model (GAM), genetic algorithm for rule-set prediction (GARP), BIOCLIM, and CLIMEX. Several studies have successfully employed MaxEnt modelling to forecast the distribution of various plant and animal species in response to climate change [3, 10, 13]. The present study aimed to use the MaxEnt model to infer species' ecological demand [4], thereby predicting the current and future potential geographical distribution of vulnerable tree species in India under changing climate conditions.

The MaxEnt model stands out as a highly reliable and widely used species distribution model (SDM) due to its accuracy and stability in predicting potential distribution ranges of species. MaxEnt uses a receiver operating characteristic curve (ROC) analysis to evaluate model performance [20]. It estimates the distribution of species using presence-only records and maximum entropy algorithm [12] and suitable even with small size of datasets [15] or rare species, and predictions may be modelled using a very few sample occurrences [22]. The importance of individual environmental variables can be evaluated using the built in Jackknife test. Compared to other models, the MaxEnt prediction results are more accurate and stable, making it one of the most widely used SDMs. MaxEnt modelling has been employed in the last few decades to predict and identify the changes in species diversity and future potential distribution changes.

According to the study, the suitability pattern of *D. latifolia* will undergo significant changes in the future due to global warming, resulting in an increase in area of suitability compared to the current scenario. The study found that environmental factors such as temperature, precipitation, and topography play an important role in the distribution. The ranking of the contribution and importance of these factors showed that Isothermality (BIO03), Precipitation of Coldest Quarter (BIO19), slope and Mean Temperature of Driest Quarter (BIO09) are found to be the key factors. These variables impact predicting the current and future climate conditions for species distribution. The response curves for *D. latifolia* show that the optimum value of the Isothermality is 66.06 °C, Precipitation of Coldest Quarter is 4630.36 mm, slope is 213.22 mm, and Mean Temperature of Driest Quarter is 948.79 °C. These response curves provide detailed information on the habitat requirements of the species, and if the predictive statistical response for distribution is close to the actual ecological response of the species, it can help to predict the species distribution accurately.

Compared to the current distribution (Table 4), in the 2050s and 2070s, the total area of highly suitable habitat would increase by 29.81 and 17.23% for the RCP2.6 scenario, 39.05 and 25.05% for the RCP 4.5 scenario, 23.88 and 26.74% for the RCP 8.5. The total area of moderately suitable regions for *D. latifolia* would increase by 20.29 and 32.81% in 2050 and 2070 under RCP2.6 scenario; 37.24 and 44.98% area would increase under scenario RCP 4.5 in 2050s and 2070s. For RCP 8.5 scenario, the regions would have increased by 29.86 and 41.50%. Low suitable regions would have increased under three scenarios in the 2050s and 2070s by 30.62 and 0.80% (RCP 2.6), 23.01 and 21.43% (RCP4.5), and 21.54 and 19.10% (RCP 8.5).

According to our predictions, the climatic distribution suitable for *D. latifolia* is expected to expand in 2050 and 2070 under future three climate scenarios compared to current scenario (2070 > 2050 > current). In summary, the habitat areas for the species are predicted to generally increase by 2050 and 2070 under the RCP 2.6, RCP 4.5, and RCP 8.5 scenarios. The prediction did not show the same tendency when compared between 2050 and 2070 under different scenarios (RCP 2.6, 4.5, and 8.5). In low-suitability habitats, there is a tendency to decrease from 2050 to 2070 for all climate scenarios. However, certain suitable areas show reduction (low suitable area 0.8% in RCP 2.6 2070s). In medium-suitable habitats, there is an increase from 2050 to 2070 for all climate scenarios; the most significant increases are observed, especially by 2070, under the RCP 4.5 scenario (44.98%). In high suitable habitat, an increase is seen only in RCP 8.5 scenario from 2050 to 2070.

Table 4 Predicted suitable areas for *D. latifolia* under current and future climatic conditions

Scenarios		Predicted area/10^4 km^2			Increase/decrease rate (%) (Compared to the current distribution)		
		Low	Medium	High	Total low suitable habitat	Total medium suitable habitat	Total highly suitable habitat
Current	–	12.6580	6.0767	4.4301	–	–	–
2050	RCP 2.6	16.5339	7.3098	5.7510	30.62	20.29	29.81
	RCP 4.5	15.5711	8.3401	6.1634	23.01	37.24	39.05
	RCP 8.5	15.3843	7.8914	5.4881	21.54	29.86	23.88
2070	RCP 2.6	12.7595	8.0704	5.1936	0.80	32.81	17.23
	RCP 4.5	15.3700	8.8093	5.5418	21.43	44.98	25.05
	RCP 8.5	15.0760	8.5985	5.6148	19.10	41.50	26.74

11 Research Limitations

Maximum Entropy (MaxEnt) modelling has become a popular tool for predicting tree species' potential distribution and suitability. However, despite its advantages, MaxEnt simulations have limitations that need to be considered when interpreting their results. Our study predicts an expansion in suitable habitat for *D. latifolia* in some areas. Firstly, current species occurrence data might not fully capture future distributions, especially considering dispersal limitations or potential shifts in competitive interactions due to climate change. Secondly, the inherent uncertainty in future climate predictions from Global Climate Models (GCMs) can translate into less precise suitability estimates in MaxEnt model. To address these limitations and strengthen the model's credibility, field surveys in areas with high predicted suitability could be helpful for validation. Despite these limitations, the MaxEnt map of potential distribution of *D. latifolia* provides a valuable tool for conservationists. It can be a guiding instrument for prioritising restoration efforts in its natural habitat by highlighting areas with the highest predicted suitability and greatest potential for successful reintroduction.

12 Conclusion

Assessing the impact of climate change on the distribution of *D. latifolia* is essential for understanding the status of this vulnerable categorised tree species. The study predicted an increase in suitable habitats for *D. latifolia* in some areas under future climate scenarios (RCP2.6, RCP 4.5, and RCP8.5 for 2050 and 2070). This expansion might be due to natural environmental changes creating more favourable conditions for the species. The study reveals a promising shift of the general fitness zone for *D. latifolia* within the Western Ghats, potentially moving closer to areas with high fitness. This signifies a valuable future resource for this species. Integrating these findings into conservation plans can ensure the long-term viability of *D. latifolia* by protecting suitable habitat while allowing for sustainable utilisation practices. Prioritising protection efforts in areas with high and moderate suitability, especially where the species is currently absent, could significantly optimise *D. latifolia* distribution. To ensure the long-term viability of this valuable resource, a balanced approach towards conservation with sustainable utilisation practices needs to be promoted.

References

1. Allouche O, Tsoar A, Kadmon R (2006) Assessing the accuracy of species distribution models: prevalence, kappa and the true skill statistic (TSS). J Appl Ecol 43(6):1223–1232. https://doi.org/10.1111/j.1365-2664.2006.01214.x
2. Bedia J, Busque J, Gutierrez JM (2011) Predicting plant species distribution across an Alpine rangeland in northern Spain: a comparison of probabilistic methods. Appl Veg Sci 14:415–432
3. Dimobe K, Ouédraogo K, Annighöfer P, Kollmann J, Bayala J, Hof C, Thiombiano A (2022) Climate change aggravates anthropogenic threats of the endangered savanna tree *Pterocarpus erinaceus* (Fabaceae) in Burkina Faso. J Nat Conserv 70:126299
4. Elith J, Phillips SJ, Hastie T, Dudík M, Chee YE, Yates CJ (2011) A statistical explanation of Maxent for ecologists. Divers Distrib 17(1):43–57
5. GBIF.org. GBIF occurrence download. https://www.gbif.org
6. Hamid M, Khuroo AA, Charles B, Ahmad R, Singh CP, Aravind NA (2019) Impact of climate change on the distribution range and niche dynamics of Himalayan birch, a typical treeline species in Himalayas. Biodivers Conserv 28:2345–2370
7. Hijmans RJ, Cameron SE, Parra JL, Jones PG, Jarvis A (2005) Very high resolution interpolated climate surfaces for global land areas. Int J Climatol 25(15):1965–1978. https://doi.org/10.1002/joc.1276
8. Istifanus AP, Abdelmutalab AGA, Pirk CWW, Yusuf AA (2023) Predicting the habitat suitability and distribution of two species of mound-building termites in Nigeria using bioclimatic and vegetation variables. Diversity 15(2):157
9. Li G, Du S, Wen Z (2016) Mapping the climatic suitable habitat of oriental arborvitae (*Platycladus orientalis*) for introduction and cultivation at a global scale. Sci Rep 6(1):30009
10. Newbold T (2018) Future effects of climate and land-use change on terrestrial vertebrate community diversity under different scenarios. Proc R Soc B 285(1881):20180792
11. Phillips SJ, Anderson RP, Schapire RE (2006) Maximum entropy modelling of species geographic distributions. Ecol Model 190(3–4):231–259. https://doi.org/10.1016/j.ecolmodel.2005.03.026
12. Phillips SJ, Anderson RP, Dudík M, Schapire RE, Blair ME (2017) Opening the black box: an open-source release of Maxent. Ecography 40:887–893

13. Puri N, Verma A, Chhetri R, Chand HB, Rijal S (2022) Species distribution modelling of two sympatric hornbills: the great hornbill and rufous-necked hornbill in Bhutan. Int J Ecol 2022(1):1475501
14. Sarikaya O, Karaceylan IB, Sen I (2018) Maximum entropy modelling (maxent) of current and future distributions of *Ips mannsfeldi* (Wachtl, 1879) (Curculionidae: Scolytinae) in Turkey. Appl Ecol Environ Res 16(3):2527–2535. https://doi.org/10.15666/aeer/1603_25272535
15. Saupe EE, Qiao H, Hendricks JR, Portell RW, Hunter SJ, Soberón J, Lieberman BS (2015) Niche breadth and geographic range size as determinants of species survival on geological time scales. Glob Ecol Biogeogr 24(10):1159–1169
16. Stocker TF, Qin D, Plattner GK, Tignor M, Allen SK, Boschung J, Nauels A, Xia Y, Bex V, Midgley PM et al (2014) Climate change 2013: the physical science basis. Working group I contribution to the IPCC Fifth assessment report. Cambridge University Press, Cambridge, 1535p
17. Swets JA (1988) Measuring the accuracy of diagnostic systems. Science 240(4857):1285–1293
18. Thakur KK, Bhat P, Kumar A, Ravikanth G, Saikia P (2022) Distribution mapping of *Bauhinia vahlii* Wight & Arn. in India using ecological niche modelling. Trop Ecol 63(2):286–299. https://doi.org/10.1007/s42965-021-00197-8
19. Wang R, Li Q, He S, Liu Y, Wang M, Jiang G (2018) Modelling and mapping the current and future distribution of *Pseudomonas syringae* pv. *actinidiae* under climate change in China. PLoS One 13(2):e0192153
20. Wang Y, Xie B, Wan F, Xiao Q, Dai L (2007) Application of ROC curve analysis in evaluating the performance of alien species' potential distribution models. Biodivers Sci 15(4):365
21. Warren R, VanDerWal J, Price J, Welbergen JA, Atkinson I, Ramirez-Villegas J et al (2013) Quantifying the benefit of early climate change mitigation in avoiding biodiversity loss. Nat Clim Change 3:678–682
22. Wisz MS, Hijmans RJ, Li J, Peterson AT, Graham CH, Guisan A (2008) NCEAS predicting species distributions working group. Effects of sample size on the performance of species distribution models. Divers Distrib 14(5):763–773
23. Worldclim – Global Climate Data. https://www.worldclim.org/

Chapter 7
Wood Structure and Radial Variation of Physico-anatomical Properties in Indian Rosewood (*Dalbergia latifolia* Roxb.)

Satish Kumar Sinha, Anil Kumar Sethy, Preeti Vats, and Nileshkumar J. Sohagiya

1 Introduction

Rosewood (*Dalbergia latifolia* Roxb.), commercially known as Indian rosewood, is a medium to large-sized deciduous or evergreen tropical hardwood species. It is one of the premier timber species of the family Fabaceae in which the world's topmost valuable timber species, the African Blackwood (*Dalbergia melanoxylon*) exists. *Dalbergia latifolia* is a durable wood and well-known for the manufacturing of high-quality wood products, primarily used in construction, railway sleepers, wagons and carriages, furniture, musical instruments like guitars, decorative carving, jewellery boxes, wall hangings, trays, and teapots, and secondarily used in mathematical instruments, agricultural equipment, sports goods, and many other luxury items [1]. It is also evidenced from the archaeological records that *Dalbergia latifolia* was renowned for its durability and preferred for coffin-making by Harappan people, which dates back to 2300 BC, during the Indus Valley Civilization [2, 3].

S. K. Sinha (✉) · P. Vats · N. J. Sohagiya
College of Forestry, Navsari Agricultural University, Navsari, Gujarat, India
e-mail: sksinha@nau.in

P. Vats
e-mail: preeti1997vats@gmail.com

N. J. Sohagiya
e-mail: njsuhagiya@gmail.com

A. K. Sethy
ICFRE-Institute of Wood Science & Technology, Bengaluru, Karnataka, India
e-mail: aksethy@icfre.in

S. K. Sinha
College of Forestry, Navsari Agricultural University, Navsari, Gujarat, India

R. R. Warrier et al. (eds.), *Handbook of Asian Rosewoods*, Materials Horizons: From Nature to Nanomaterials, https://doi.org/10.1007/978-981-96-9459-4_7

In India, rosewood is distributed throughout Peninsular India, from Uttar Pradesh, Bihar, Orissa, Madhya Pradesh, Chhattisgarh, Maharashtra, Gujarat to South India and mainly in north Kanara, south Kanara, Coorg, Malabar, and Travancore [4]. However, it has been introduced globally to Kenya, Malaysia, Mauritius, Myanmar, Nepal, Nigeria, Pakistan, the Philippines, Réunion, Singapore, Sri Lanka, Uganda, Tanzania, and Vietnam [1].

It is estimated about 300 *Dalbergia* species are in the tropical parts of the world, with the highest species diversity in Asia (119 species) followed by Africa (116 species) and Central and South America (80 species). It is recorded that India has a diversity of 25 *Dalbergia* species, of which 18 species and two subspecies occur in Western Ghats [1, 5].

In the current scenario, about 90% of wild populations of rosewood are declining and facing the peril of extinction because of overharvesting to produce luxury furniture and musical instruments that resulted in this species being listed in Appendix II of Convention on International Trade in Endangered Species of Wild Fauna and Flora (CITES) [6].

The high demand and restricted supply of *D. latifolia* has hiked the price of this species in domestic and international markets. The hiked market price of a tonne of unworked rosewood was estimated from US$ 10,000 to US$ 50,000 (~Rs. 8.0 to 40.0 lakhs) depending upon the species and trade chain location of the timber in a particular country [5, 7]. The high market price of any premier species leads to frequent adulteration with cheaper substitutes and the introduction of inferior-quality timber in the market. Hence, conceding the uniqueness of the wood quality of a particular species becomes questionable [8].

Wood quality indicates the cumulative effect of wood properties on some specified products [9]. Basic density is one of the most important wood properties for nearly all wood products, and it is influenced by the structure and variation in anatomical (vessels, parenchyma, and fibres) properties of wood. Wood property variation along radial direction of the tree is the best-known and most studied parameter. It is generally reflected as radial pattern of change in wood properties of juvenile and mature wood. Knowledge of radial variation of wood density and anatomical properties are essential to determine the uniformity of wood from pith to periphery and predict age effect on the wood quality. The greater the uniformity of wood, the greater the efficiency of producing a specific product with much improved quality of the final product [9–12]. The main objective of this chapter was to provide an insight of the wood structure and radial variation of the physical and anatomical properties of Indian rosewood to know the uniformity and maturity phase in plantation-grown timbers.

2 Sampling

Five trees were randomly selected from a 35-year-old block plantation of *D. latifolia* established at Navsari Agricultural University (20.95° N; 75.90° E), Navsari, Gujarat, India (Fig. 1a). The average mean girth and height were 85.8 cm and 17.7 m, respectively. The radial variation of physico-anatomical properties of wood was investigated using two core samples (diameter 4 mm) per tree collected by increment borer at the breast height (1.37 m above ground level) during December 2024 (Fig. 1b, d). The wood structure of rosewood was studied using two wooden blocks available at the Forestry Museum of the University and ten core samples from five trees.

Fig. 1 **a** Indian rosewood tree in block plantation, **b** Extraction of core sample from rosewood tree, **c** Wooden disc showing purplish-black heartwood (hw) and pale-yellow sapwood (sw) in the periphery, and **d** Core samples along with 1-cm long segments for studying radial variation in rosewood

3 Sample Preparation and Measurements

For the measurement of radial variation of physical properties namely basic density and moisture content in *D. latifolia,* each core sample was cut into 1-cm segment from pith to periphery, whereas for radial variation of anatomical properties such as fibre length, cell-wall thickness, vessel diameter, vessel density and tissue proportion, measurements were taken from each core sample alternatively at 1 cm interval from pith to periphery (Fig. 1d). The moisture content of each core sample and museum wood sample were determined by oven-dry method. Green volume of each 1-cm segment of core sample was calculated by measuring the diameter and length of the increment core, and volume of wooden block samples was determined by the water displacement method. The basic density was calculated by taking the ratio of oven-dry weight and green volume.

To explore the wood structure and radial variation in rosewood, anatomical features of samples were measured under the microscope equipped with image analysis system using maceration technique and sectioning 25 μm thick wood samples in transverse, radial, and tangential directions using a sliding microtome. Vessel parameters (tangential vessel diameter, and vessel frequency) were measured at 2.5x objective under Leica stereo-zoom microscope using a smoothened transverse surface of core samples and wooden blocks. However, ray parameters (ray height, width, and frequency) of wood were measured at 10x objective under Leica trinocular microscope using the slides of tangential sections. Ray compositions were recorded in radial section.

Fibre parameters such as fibre length, diameter, cell-wall thickness, and vessel parameters, namely vessel length and intervascular pit size, were determined using the tiny slivers of wood and core samples. These slivers were taken into test-tubes and macerated by Schultz's method using 50% nitric acid and few potassium chlorate crystals [13]. Fibre length was measured at 4x objective and vessel length at 10x objective, whereas fibre diameter, cell-wall thickness, and intervascular pit size were measured at 40x objective under a Leica trinocular microscope. Minimum 25 measurements were recorded for each anatomical feature to derive mean and range values as per International Association of Wood Anatomists (IAWA) committee guidelines [14].

4 General Characteristics and Structure of the Wood

The sapwood colour of rosewood is pale yellowish white often with a purplish to pink tinge and sharply distinguished from heartwood which is ranging in colour from light golden brown through shades of light purplish-pink with darker streaks to deep purple, darkening with age (Fig. 1c, d). The growth rings are usually inconspicuous but when present they are delimited by pore size, number of pores, and darker fibrous tissue in the outer portion of the ring. Wood is hard, heavy to very heavy (sp. gr.

0.83–1.22 air dry) with indistinct pleasant smell in fresh, and straight to slightly interlocked grained and medium coarse textured [15]. In the current study, similar general features and wood structure were also recorded in our rosewood samples. We reported that in fresh core samples of rosewood there was a faint pleasant smell; wood was straight to slightly interlocked grained and of medium coarse textured. The mean heartwood proportion varied from 52.07 to 92.89% with a mean of 82.92%. The basic density of plantation wood and natural wood was recorded to be 0.583 to 0.688 gcm^{-3}, and 0.75–0.88 gcm^{-3} at 12% moisture content, respectively (Fig. 1c, d).

Our results show that Indian rosewood's stem wood was diffuse-porous and often rather distant with large to medium-sized or small pores/vessels, 52–181 μm in diameter, few to moderately few, 3–18 per mm^2, mostly solitary or in short radial multiples. Growth rings were indistinct or inconspicuous and somewhere delimited by darker fibrous tissue (Fig. 1a). Vessel elements were 65–253 μm in length, intervascular pits small to medium size 6–16 μm, perforation simple, open or plugged with deposits of reddish-brown or brownish black or with white gummy deposits. Fibres were libriform and non-septate, frequently gelatinous, 475–1317 μm in length, 9.96–27.50 μm in diameter with 2.6–6.8 μm cell-wall thickness (Fig. 2d). Rays were fine to very fine, 4–9 per mm, mostly uniseriate to biseriate with a few up to 3 seriate, 4–9 per mm, homogeneous to nearly heterogeneous, composed mainly of procumbent cells, 34–125 μm in width, and 91–276 μm in height (Fig. 2b, c). Ripple marks were present, not distinct with the naked eye. Axial parenchyma was paratracheal, mostly aliform to aliform confluent and banded (Fig. 2a). Crystals were usually absent in rays and fibres but occasionally present in axial parenchyma (Table 1).

5 Radial Variation of Physical Properties in Rosewood

Basic density and moisture content are essential physical properties to determine the wood quality. It varies between trees, within a tree, and between species. It varies radially in a tree with age and distance from the pith [21].

Figure 3 represents the radial variation of basic density and moisture content. The basic density of *D. latifolia* ranged from 0.583 to 0.689 gcm^{-3} with the mean value of 0.628 gcm^{-3} and showed significant variation from pith to periphery. It gradually increased up to 6 cm distance from pith and showed almost constant value (0.598 gcm^{-3}); afterwards it sharply increased to 10 cm (0.657 gcm^{-3}) and became stabilized towards the periphery. On the contrary, the moisture content of rosewood gradually decreased up to 6 cm distance from pith and showed nearly constant value (101.5%) then it sharply decreased to 10 cm (88.0%) and levelled off towards the bark. The moisture content varied from 74.10 to 104.62% with a mean value of 94.26%. It was observed that an inverse relationship existed between the basic density and moisture content of wood (Fig. 4).

The basic density of *Casuarina equisetifolia* was divided into two groups in relation to distance from pith [22]. Similarly, in the current study, the basic density of

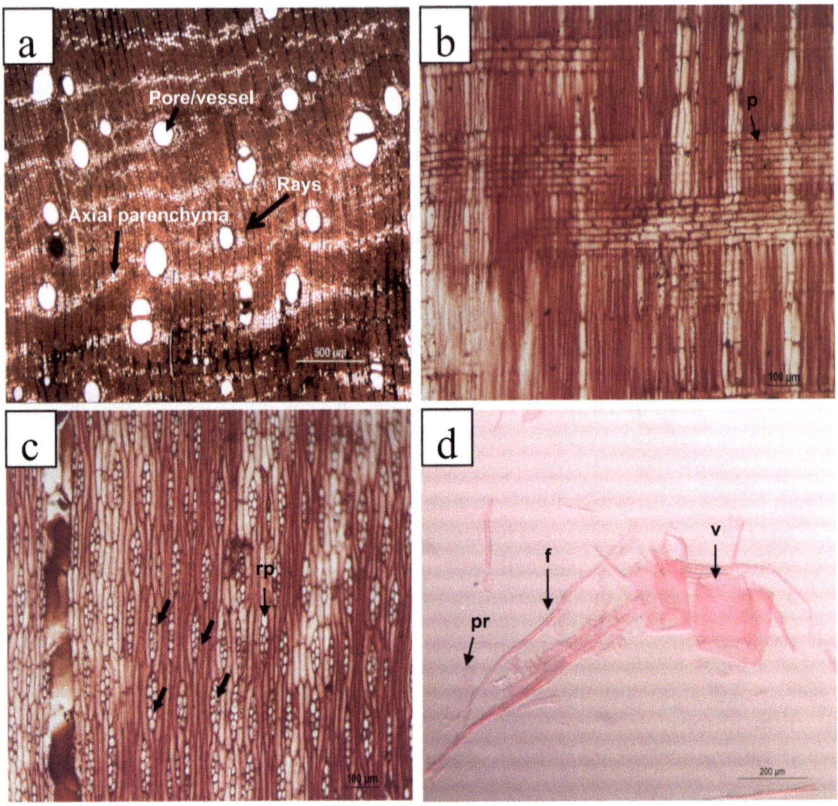

Fig. 2 **a** Diffuse-porous Indian rosewood showing aliform to confluent and banded axial parenchyma (arrows) and fine rays in transverse section, **b** Homogeneous rays composed of procumbent (p) cells in radial section, **c** Uniseriate to biseriate storied ray parenchyma (rp) (arrows) in tangential section, and **d** Maceration of wood showing vessel member (v) fibres (f) and parenchyma (pr)

Dalbergia latifolia may also be divided into two groups according to distance from pith, the boundary being at a distance of 6–7 cm. It was reported that the basic density of teak increased up to 5–6 cm from pith and then stabilized at the age of 10–12 years [23]. In contrast, in *D. sissoo* there was no significant radial variation of basic density from pith to periphery and no definite trend was observed.

6 Radial Variation of Anatomical Properties in Rosewood

The influence of various external factors such as climate is studied by examining the pattern in radial variation for different wood properties especially in tropical species [24]. The demarcation between juvenile and mature wood in diffuse-porous

Table 1 Anatomical characteristics of Indian rosewood (*Dalbergia latifolia* Roxb.) (values presented are minimum, maximum, and mean ± standard deviation in parenthesis)

Anatomical characteristics	Results from present study	Results from other studies
Vessels		
1. Type of perforation	Simple	Simple [4]
2. Length (μm)	65–253 (158.55 ± 11.63)	135–250 [4] 150–477 [15] 134.4 ± 23.30 [16]
3. Tangential diameter (μm)	52–181 (123.41 ± 3.52)	125–175 [17] 100–200 [15] 125–210 [18] 189–233 [20]
4. Frequency (per mm^2)	3–18 (8 ± 1.88)	4–8 [17] 1–7 [15] 2–12 [4, 18] 15 ± 3.03[16] 3–4 [20]
5. Intervascular pit size (μm)	6–16 (8 ± 0.533)	7–10 [4, 17] 7–12 [15]
Axial parenchyma		
1. Type	Aliform to aliform confluent	Vasicentric, aliform, and confluent [15, 17] Vasicentric, aliform, and confluent [18] Aliform [19] Aliform to confluent and banded [20]
2. Crystals		Present [17]
Rays		
1. Type	Homogenous	Mostly uniseriate, may be 2–3 seriate, homogeneous to weakly heterogeneous [15] Homogenous [19] Homogenous to nearly heterogeneous type-III [20]
2. Height(μm)	91–276 (137.24 ± 3.58)	80–150 [4] 500 [17] 200 [15] 170–315 [18] 146.5 ± 23.19 [16] 113–169 [20]
3. Width (μm)	34–125 (63.14 ± 3.47)	30–45 [4, 18] 20–56 [15] 25.3 ± 7.62 [16] 24–36 [20]

(continued)

Table 1 (continued)

Anatomical characteristics	Results from present study	Results from other studies
4. Frequency (per mm)	4–9 (7 ± 0.21)	8–12 [17] 7–16 [15] 11–16 [4, 18] 11 ± 1.71[16] 10–16 [20]
5. Oil cells	Absent	Absent [17]
6. Crystals	Absent	Absent [4]
Fibres		
1. Type	Libriform	Semi-libriform to libriform, round to angular, non-septate and storied [4, 15]
2. Length (μm)	475–1317 (835.54 ± 59.58)	315–1300 [4] 627–1353 [15] 1110–1770 [20]
3. Diameter (μm)	9.96–27.5 (17.88 ± 0.33)	25–30 [4] 16–24 [15]
4. Cell-wall thickness (μm)	2.6–6.8 (4.4 ± 0.42)	2–7 [4] 3.4 ± 0.89[16] 3.5–5.5 [15]
5. Crystals	Absent	Absent [4]

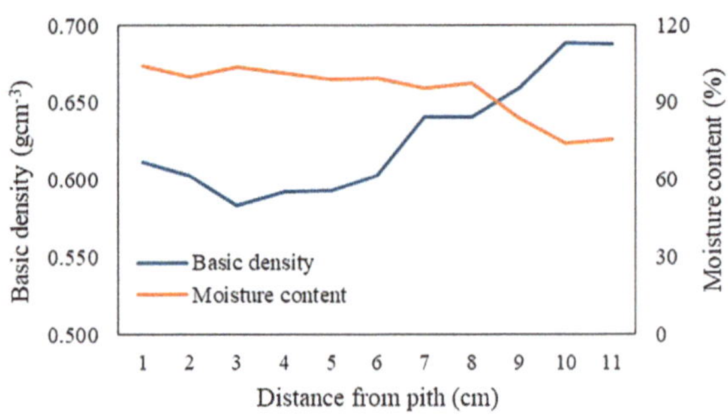

Fig. 3 Radial variation of physical properties in rosewood

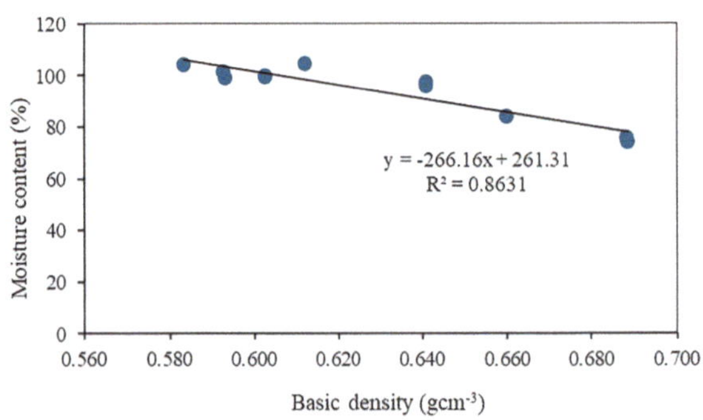

Fig. 4 Relationship between basic density and moisture content of rosewood

hardwood is also examined by investigating the radial pattern of anatomical variation at a distance from pith to periphery [25]. The radial variation of anatomical properties of *D. latifolia* is shown in Fig. 5. Considering the radial variation of anatomical properties, the fibre length, cell-wall thickness, vessel diameter, vessel density, and vessel proportion gradually increased, while the fibre and ray proportions decreased by up to 7 cm from pith and more or less stabilized towards the periphery. It was found that there was significant radial variation among all the anatomical parameters. The fibre length, cell-wall thickness, vessel diameter, vessel density, proportions of fibre, vessel, and ray varied from 927.59 to 1316.98 μm, 3.84 to 4.25 μm, 107.98 to 160.15 μm, 4 to 6 per mm^2, 23.09 to 28.91%, 15.64 to 25.09%, and 21.27 to 27.82%, respectively.

In the present study, it was observed that the radial variation of anatomical properties (fibre length, cell-wall thickness, vessel diameter, vessel density, and tissue proportion) in *D. latifolia* converged at the same distance (7 cm) from pith and relatively levelled off afterwards. Very limited studies have been carried out on the radial variation of *D. latifolia*. It was reported that among the radial variation of various anatomical properties, the fibre length, cell-wall thickness, vessel diameter, vessel

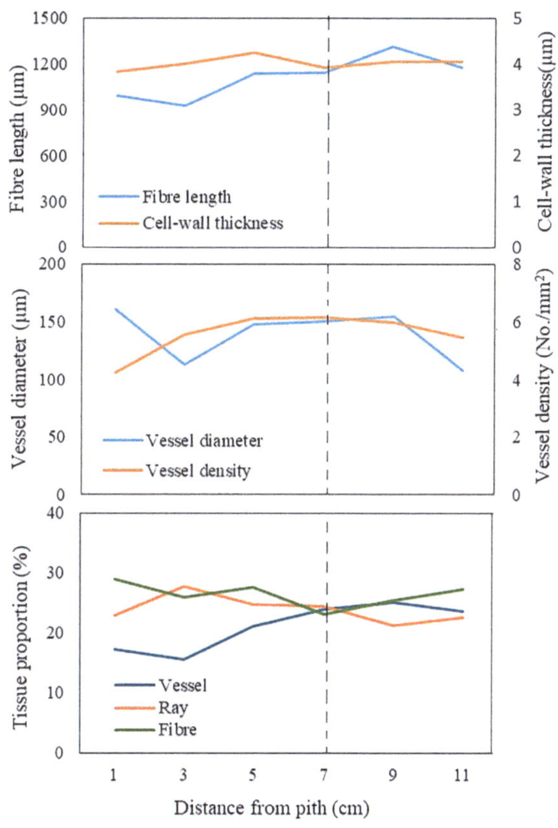

Fig. 5 Radial variation of anatomical properties in rosewood

density, and proportions of tissues such as fibre, vessel, and ray were the key parameters in deciding the demarcation between juvenile and mature wood among different *Catalpa bungei* clones [25]. Investigation on the radial variation in *Dalbergia sissoo* indicated the highest fibre length towards periphery (1273.95 μm) and the lowest fibre length towards pith (1103.57 μm) and moderately long towards the middle (1164.51 μm), and a similar pattern of variation in cell-wall thickness was also observed [26].

In the current study, the gradual increase in vessel diameter was observed up to 7 cm distance from pith and then decreased at the periphery. The increase in vessel diameter towards periphery indicated that relatively larger vessels are required in the sapwood for efficient transport of water from roots [27]. On the contrary, in 8-year-old clonal ramets of *D. sissoo*, there were no significant differences found for fibre length, cell-wall thickness, and vessel diameter in radial direction [28].

7 Conclusions

The radial variation of physical and anatomical properties of trees in a 35-year-old plantation of *Dalbergia latifolia* indicates that the boundary between juvenile and mature wood is marked at a distance of about 7 cm from the pith for effective utilization of wood. Hence, the wood adjacent to the pith with relatively low density can be utilized for non-structural purposes, and the wood with relatively high density can be utilized for structural purposes. However, for better understanding, further research is needed to be carried out on the radial variation of wood properties at different locations in plantation-grown rosewood trees.

Acknowledgements We are thankful to the Director of Research and Dean PGS, Navsari Agricultural University, Navsari, for permitting the extraction of core samples of Indian rosewood from the block plantations of the University.

References

1. Arunkumar AN, Warrier RR, Kher MM (2022) Teixeira da Silva: Indian rosewood (*Dalbergia latifolia* Roxb.): biology, utilization, and conservation practices. Trees 36(3):883–898
2. Chowdhury KA, Ghosh SS (1951) Plant remains from Harappa. Anc India Bull 7:1–19
3. Chowdhury KA (1970) Wood and its use during Pre and Proto historic time. Indian J Hist Sci 5(1):141–143
4. Pearson RS, Brown HP (1981) Commercial timbers of India their distribution, supplies, anatomical structure, physical and mechanical properties and uses, vol 1. A.J. Reprints Agency, New Delhi
5. Siriwat P, Nijman V (2023) Quantifying the illegal high-value rosewood trade and criminal trade networks in the Greater Mekong Region. Biol Conserv 277:109826. https://doi.org/10.1016/j.biocon.2022.109826

6. Riastiwi I, Witjaksono, Ratnadewi D, Siregar UJ (2022) Genetic diversity of rosewood (*Dalbergia latifolia*) in Yogyakarta, Indonesia for plus trees selection. Biodiversitas 23:2630–2639
7. Nijman V, Siriwat P, Shepherd CR (2022) Inaccuracies in the reporting of volume and monetary value of large-scale rosewood seizures. For Policy Econ 134:102626. https://doi.org/10.1016/j.forpol.2021.102626
8. Sinha SK, Vijendra Rao R (2022) Anatomy of Indian sandalwood (*Santalum album* L.). In: Arun Kumar AN, Joshi G, Warrier RR, Karaba NN (eds) Indian sandalwood. Materials horizons: from nature to nanomaterials. Springer, Singapore. https://doi.org/10.1007/978-981-16-6565-3_12
9. Zobel BJ, van Buijtenen JP (1989) Wood variation: its causes and control. Springer, Berlin Heidelberg, New York
10. Bhat KM, Bhatt KV, Dhamodaran TK (1989) Effect of tree age on heartwood proportion of *Eucalyptus grandis* grown in Kerala. Indian J For 11(2):95–97
11. Anoop EV, Jijeesh CM, Sindhumathi CR, Jayasree CE (2014) Wood physical, anatomical and mechanical properties of big leaf mahogany (*Swietenia macrophyla* Roxb.) a potential exotic for south India. Res J Agric Sci 2(8):7–13
12. Sinha SK (2015) Influence of climate on the radial variation of specific gravity and certain anatomical properties in teak (*Tectona grandis* L.f.) from Chandrapur and Thane, Maharashtra. Doctoral thesis, Forest Research Institute Deemed University, Dehradun
13. Jane FW (1956) The structure of wood. Adam and Charles Black Ltd., London
14. Wheeler EA, Baas P, Gasson PE, IAWA Committee (1989) IAWA list of microscopic features for hardwood identification. IAWA Bull 10(3):219–332
15. Chauhan L, Rao RV (2003) Wood anatomy of legumes of India: their identification, properties and uses. Bishen Singh Mahendra Pal Singh Publishers, Dehradun, India
16. Liu M, Peng L, Lyu S, Lyu J (2020) Properties of common tropical hardwoods for fretboard of string instruments. J Wood Sci 66(1):1–11
17. Richter HG, Dallwitz MJ (2000) Commercial timbers: descriptions, illustrations, identification, and information retrieval. Version: 4th May 2000. http://biodiversity.uno.edu/delta/
18. Sahu VK (2005) Wood property profile of rosewood (*Dalbergia latifolia* Roxb.) ceylon rosewood (*Albizia odoratissima* Linn. f. Benth.) and raintree (*Samanea saman* Jacq. Merr.). Doctoral Thesis, Department of Tree Physiology and Breeding, College of forestry, Vellanikkara, Kerala Agricultural University. https://kalnet.kshec.kerala.gov.in/vufind/Record/kau.27421
19. Mamoňová M (2017) Anatomic and spectrophotometric differentiation of *Dalbergia nigra* from other species of rosewood. Acta Facultatis Xylologiae Zvolen res Publica Slovaca 59(1):5–16. https://doi.org/10.17423/afx.2017.59.1.01
20. Sathish N (2020) Development of anatomical key for the identification of selected timbers of Kerala. Doctoral thesis, Department of Forest Products and Utilization, College of Forestry, Vellanikkara, Kerala Agricultural University. https://krishikosh.egranth.ac.in/items/a24f4925-f1f1-4045-ae21-6c5e72ba79c2
21. Wimmer R, Downes GM, Evans R (2002) High-resolution analysis of radial growth and wood density in *Eucalyptus nitens*, grown under different irrigation regimes. Ann For Sci 59:519–524
22. Chowdhury MQ, Ishiguri F, Iizuka K, Takashima Y, Matsumoto K, Hiraiwa T, Ishido M, Sanpe H, Yokota S, Yoshizawa N (2009) Radial variations of wood properties in *Casuarina equisetifolia* growing in Bangladesh. J Wood Sci 55:139–143
23. Okuyama T, Yamamoto H, Wahyudi I, Hadi YS, Bhat KM (2003) Some wood quality issues in planted teak. In: International conference on quality timber products of teak from sustainable forest management (abstracts), KFRI, Kerala, India, 2–5 December 2003, p 21
24. Honjo K, Furukawa I, Sahri MH (2005) Radial variation of fiber length increment in *Acacia mangium*. IAWA J 26(3):339–352
25. Liu Y, Zhou L, Zhu Y, Liu S (2020) Anatomical features and its radial variations among different *Catalpa bungei* clones. Forests 11(8):824. https://doi.org/10.3390/f11080824

26. Karimanisha K, Krishnakumar N, Parthiban KT (2020) Wood anatomical properties of *Dalbergia sissoo* Roxb. Life Sci Leaflets 119:1–12
27. Rao RV, Shashikala S, Sreevani P (2003) Variation in basic density and anatomical properties of *Eucalyptus tereticornis* clones. J Trop For Prod 9(1–2):59–67
28. Pande PK, Singh M (2005) Inter-clonal, intra-clonal, and single tree variations of wood anatomical properties and specific gravity of clonal ramets of *Dalbergia sissoo* Roxb. Wood Sci Technol 39(5):351–366. https://doi.org/10.1007/s00226-004-0273-1

Chapter 8
Properties and Working Qualities of *Dalbergia latifolia*

Anil Kumar Sethy, Kamal Mishra, and Satish Kumar Sinha

1 Introduction

Dalbergia latifolia Roxb., commonly known as Indian rosewood, is a premium hardwood highly regarded for its exceptional qualities, making it one of the most sought-after timbers in various industries. The wood is used not only within its native regions (such as India and Southeast Asia) but has gained global recognition, particularly in markets for luxury furniture and wood products. Due to its versatility and increasing demand, *Dalbergia latifolia* holds significant value in trade and commerce. For instance, in 2012, the market price for *D. latifolia* reached approximately US$ 49,656 per cubic meter for instrument blanks and US$ 16,575 per cubic meter for sawn wood [1]. The high demand for this wood can be attributed to its unique and desirable properties. This chapter presents the quality attributes of *D. latifolia* wood, focusing on its physico-mechanical properties, durability, drying behaviour, wood working qualities and utilization.

2 Gross Structure of *D. latifolia* Wood

The timber exhibits a clear distinction between the heartwood and sapwood. The sapwood is initially yellowish-white, gradually darkening as it ages. The heartwood, in contrast, is light purplish-brown to deep purple and occasionally brown to dark brown with darker streaks. Freshly cut heartwood sometimes appears bright purple, which fades overtime when it comes in contact with air (Fig. 1). When measurement

A. K. Sethy (✉) · K. Mishra
ICFRE-Institute of Wood Science and Technology, Bengaluru, India
e-mail: sethyanil@gmail.com

S. K. Sinha
Navsari Agricultural University, Navsari, Gujarat, India

R. R. Warrier et al. (eds.), *Handbook of Asian Rosewoods*, Materials Horizons: From Nature to Nanomaterials, https://doi.org/10.1007/978-981-96-9459-4_8

Fig. 1 Colour variation in the heartwood of *D. latifolia*

of surface colour parameter of heartwood samples was carried out using CIELAB colour space, the values obtained for the three colour coordinates L*, a* and b* were 32.2, 8.4 and 2.8 respectively. The L*, a* and b* values for sapwood of *D. latifolia* were 75.0, 5.0 and 20.7, respectively. L* scale represents light versus dark where 0 represents black and 100 represents white; a* scale represents red versus green where a positive value indicates red and a negative value indicates green. The values of b* scale represent yellow versus blue, where a positive number indicates yellow and a negative number indicates blue. Its grain is typically straight, though it may have some interlocking patterns, and the texture is medium to coarse. The wood is very hard and heavy to very heavy, with a specific gravity ranging from 0.70 to 1.02 in air-dried condition. Freshly cut heartwood gives a faint fragrance which fades upon drying [2].

The water extract is colourless to pale in colour without fluorescence. Colour of ethanol extract varies from yellowish brown to orange, red and purple. The froth test for the wood is positive. When splinters are burned, they turn to partial ash, with the resulting ash varying in colour from white to grey [3].

3 Physical and Mechanical Properties of *D. latifolia*

The physical properties of Rosewood (*D. latifolia)* have been presented in Table 1 [4]. The properties of Teak (*Tectona grandis*) wood, which is considered a standard timber in India, have also been provided in the table for comparison. Rosewood is comparatively heavier than Teak having a standard specific gravity of 0.754 as compared to 0.598 for Teak. Higher specific gravity suggests better mechanical properties of Rosewood. The shrinkage values of rosewood are slightly more than that of Teak. The volumetric shrinkage of rosewood is about 8.7% as compared to 6.8% in Teak. The ratio of tangential to radial shrinkage (T/R ratio) in rosewood is 2.1 as compared to 1.8 for Teak. Due to the lower shrinkage values, the wood is dimensionally stable and undergoes minimal movement during service.

Table 1 Physical properties of *D. latifolia* wood [4]

Species	Condition	SG based OD weight and		MC at test (%)	Shrinkage (%): green to air-dry condition			Shrinkage (%): green to oven-dry condition		
					R	T	V	R	T	V
		Vol. at test	Vol at O.D							
Rose wood	G	0.754	0.811	35.1	1	2.3	3.5	2.7	5.8	8.7
	AR	0.789	–	11.8	–	–	–	–	–	–
Teak	G	0.598	0.629	51.8	–	–	–	2.3	4.2	6.8
	AR	0.611		13.9	–	–	–	–	–	–

Note SG—Specific gravity, OD—Oven dry, MC—Moisture content, G—Green, AR—Air dry, R—Radial, T—Tangential, V—Volumetric

The mechanical properties of rosewood sourced from two Indian localities, such as South Chanda (presently Chandrapur), Maharashtra, dry deciduous forest, [4], and Malabar, Kerala, moist deciduous forest [5], and tested at Forest Research Institute, Dehradun, have been presented in Table 2. The properties of Teak wood have also been included in the table for comparison. The rosewood sourced from South Chanda (dry deciduous forest) exhibited superior properties compared to the rosewood from Malabar (moist deciduous forest). While, most of the mechanical properties of rosewood from Malabar, Kerala, are comparable to that of standard Teak, the rosewood from South Chanda exhibits superior static bending properties, particularly the bending strength (modulus of rupture—MoR) as compared to Teak. With a MoR value of 116.7 MPa in air-dry condition, rosewood from Maharashtra is about 21% stronger in bending than Teak. However, the flexural modulus (modulus of elasticity—MoE) of both the samples are comparable with a value of 12.2 GPa for rosewood and 12.5 GPa for Teak. The height of drop in the impact test is also significantly higher (44%) for rosewood than Teak. The maximum compressive strength (MCS) of air-dry rosewood in parallel to grain and perpendicular to grain directions are 63.6 MPa and 18.0 MPa respectively which are about 17% and 85% higher than those of Teak wood. Rosewood is a very hard timber with an average surface hardness of 15.3 kN as compared to 4.9 kN for Teak. Similarly, the shear and tensile properties of rosewood are also significantly higher than those of Teak. The better mechanical properties of rosewood can be attributed to its higher density.

The physical and mechanical properties presented in Tables 1 and 2 are comparable to the values reported by various authors. For instance, a study on the properties of 40-year-old *D. latifolia* grown in central Java reported an average basic density of 0.82, MoE of 11.5 GPa and MoR of 110 MPa [6]. Another study documented an MoE values of 13.1 GPa, MoR of 97 MPa and Janka hardness of 10.47 kN [7]. Reflecting the material's stiffness, the static and dynamic Young's modulus

Table 2 Mechanical properties of Rosewood (*D. latifolia*) and Teak (*Tectona grandis*) wood in air-dry condition

Properties	Rosewood		Standard Teak*
	South Chanda (Maharashtra)	Malabar (Kerala)	
Static bending			
• FS at EL (N/mm^2)	68.9	50.9	63.3
• MoR (N/mm^2)	116.7	92.5	95.9
• MoE (N/mm^2)	12,259	9970	12,535
Impact bending			
FS at EL (N/mm^2)	129.1	122.3	130.6
Max. drop height (cm)	99.1	111.8	68.6
MoE (N/mm^2)	17,444	16,030	16,382
Compression parallel to grain			
• Maximum Crushing Stress (N/mm^2)	63.6	47.6	55.4
• MoE (N/mm^2)	12,473	10,328	12,535
Compression perpendicular to grain			
• CS at EL (N/mm^2)	18.0	12.5	9.7
Hardness			
• Radial (N)	17.4	7.0	5.0
• Tangential (N)	15.2	8.3	5.0
• End (N)	13.2	8.4	4.7
Shear parallel to grain			
• Radial (N/mm^2)	17.1	9.7	8.4
• Tangential (N/mm^2)	11.8	15.3	9.8
Tensile stress perpendicular to grain			
• Radial (N/mm^2)	5.7	3.2	3.7
• Tangential (N/mm^2)	7.1	3.5	4.4

(modified from [4])

recorded were 13.3 GPa and 13 GPa, respectively [8]. Additionally, a Brinell hardness value of 43 MPa was observed, indicating the material's resistance to surface indentation and its overall hardness.

Wood is used in a wide range of applications due to its diverse properties. To effectively assess and compare the suitability of a wood species for specific purposes, it is beneficial to evaluate them quantitatively using specific coefficients or indices. A method for evaluating these suitability indices was developed at the Forest Research Institute, Dehradun [9]. The basic suitability indices that have been worked out to evaluate wood species include the following: Strength as a Beam, Stiffness as a Beam, Suitability as a Post, Shock Resisting Ability, Retention of Shape, Shear,

Surface Hardness, Refractoriness (Splitting coefficient), Nail Holding Power and Screw Holding Power.

To calculate these suitability indices, each relevant physical and mechanical properties are first multiplied with an "adjusting factor" and a "weightage factor". These are then added together to obtain an average value separately for green and air-dry condition. Then by taking all conditions into consideration, a final "composite figure" is obtained and compared with a similar figure of Teak, a well-known and generally acceptable species, to indicate comparative suitability indices. Table 3 shows the relevant physical and mechanical properties and the adjusting as well as weightage factors for the calculation of the basic suitability indices.

The suitability indices of *D. latifolia* wood sourced from Maharashtra and Kerala and as derived from the physical and mechanical properties [10] have been presented in Fig. 2. The weight, shock resisting ability, shear and surface hardness of *D. latifolia* are superior to Teak, while the other indices are marginally lower in comparison. The suitability indices for both locations are almost comparable.

4 Drying Behaviour

The wood tends to season more effectively when kept in log form or as rough squares, rather than being converted while still green, as it best preserves its colour while in the log form [11]. Air-drying of 1-inch-thick planks typically takes about 50–60 days. The wood is moderately refractory in nature to seasoning and can be satisfactorily kiln-dried by following schedule IV (Table 4) of Indian standard, IS-1141 [12]. The timber dries slowly, but with little degradation. One-inch-thick planks of rosewood can be seasoned within 12–15 days [12] following the schedule IV. In addition to the initial steaming, an intermediate steaming and one steaming towards the end of the drying at 55 °C and 100% relative humidity for 2–4 h should be given to enhance the drying process. At the end of the drying cycle, the dried stack should be subjected to a conditioning treatment, which includes steaming the stack at 55 °C and 70% relative humidity for 24 h to minimize the moisture gradient in the dried materials and moisture content variation between the kiln samples [12].

The schedule provided in Table 4 is meant for 1-inch-thick planks. It is important to note that thicker sections of wood will require significantly longer drying times. For wood that measures between 25 and 50 mm in thickness, the drying schedule needs to be modified by maintaining the relative humidity by 5% higher at each seasoning stage. In cases of wood with thicknesses from 50 to 100 mm, it is advisable to start with preliminary air-drying until the moisture content reaches approximately the fibre saturation point which is about 25–30%. If further drying is needed, kiln drying can follow, using the same humidity conditions as for the 50 mm sections [12]. Notably, kiln seasoning is reported to enhance the value of the timber by darkening the colour. Once the drying process is complete, the wood achieves remarkable dimensional stability during use.

Table 3 Adjusting and weightage factors for different mechanical properties for evaluation of suitability indices

Property	Adjusting factors		Weightage
	Green	Air dry	
Strength as a beam			
MoR (Static bending)	1	1	2
FS at EL (Static bending)	1.69	1.73	1
FS at EL (Impact bending)	0.69	0.81	1
Stiffness as a beam			
MoE (Static bending)	1	1	2
MoE (Impact bending)	0.70	0.72	1
Suitability as a post			
Max. crushing stress (Compression parallel to grain)	1	1	1
CS at EL (Compression parallel to grain)	1.4	1.46	1
MoE (Static bending)	3.49	4.24	1
Shock resisting ability			
Work to Max. load (Staticbending)	1	1	2
Total work (Static bending)	0.39	0.47	1
Max. height of drop (Impact bending)	0.007	0.01	1
Retention of shape			
Volumetric shrinkage	1	2	-
Radian shrinkage	3	1	-
Tangential shrinkage	1.44	1	-
Ratio = (Tangential/Radial)	3.3	2	-
Shear			
Maximum shear stress—Radial	1	1	1
Maximum shear stress—Tangential	0.89	0.92	1
Hardness			
CS at E.L. (Comp. parallel to grain)	1	1	2
Hardness—Radial	0.146	0.166	1
Hardness—Tangential	0.146	0.158	1
Hardness—End	0.143	0.144	1
Refractoriness			
Tensile stress perpendicular to grain—Radial	1	1	1
Tensile stress perpendicular to grain—Tangential	0.81	0.78	1

(modified from [9])

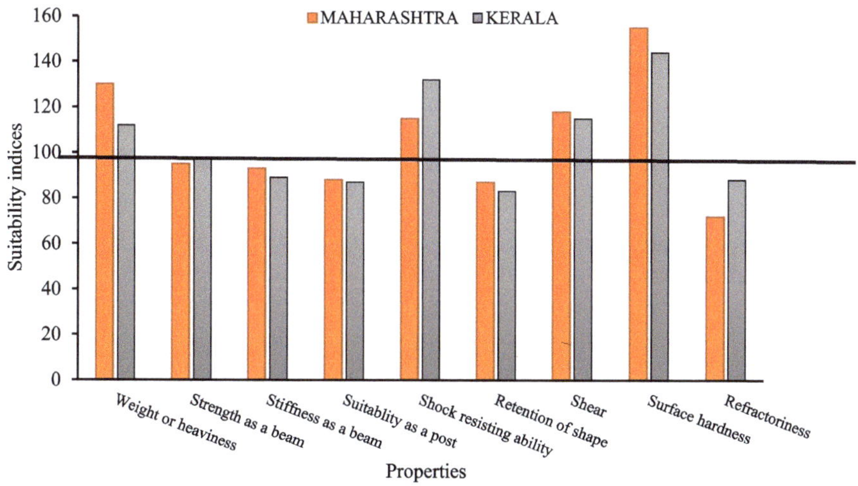

Fig. 2 Suitability indices of *Dalbergia latifolia* (in terms of *Tectona grandis* as 100) (modified from [10])

Table 4 Recommended kiln-drying schedule for 25 mm thick planks [12]

Moisture content of the wettest timber on air inlet side	Temperature		Relative humidity (%)
	Dry bulb (°C)	Wet bulb (°C)	
Green	42	38.5	80
60%	42	38	76
40%	45	40	72
35%	47	40	64
30%	49	40	56
25%	52	40	47
20%	55	40	39

5 Durability

Rational utilization of timber requires an understanding of its natural durability, which varies from species to species. While the sapwood of all species is perishable, heartwood durability differs depending on the species. Based on the resistance of the heartwood to degradation in ground contact, timbers are classified into three durability classes, such as Class I, Class II and Class III [13]. The average life of each class of the timber in ground contact is given in Table 5.

The heartwood of rosewood is highly durable, both when exposed to external conditions as well as when used in indoor locations. The heartwood is not attacked by borers and is rarely infested by white ants. Wood sourced from different locations also differs in its natural durability. Rosewood from Madhya Pradesh showed better

Table 5 Classification of timbers into durability classes [13]

Durability class	Average life	Remarks
Class I	≥120 months	Highly durable
Class II	≥60 months	Moderately durable
Class III	≤60 months	Non-durable

natural durability with an average life of 291 months as compared to rosewood sourced from Bombay, Madras and Mysore with an average life of 163 months in the graveyard [14]. Based on the average life, the heartwood has been classified as Class I timber [13] and hence does not need preservative treatment. Despite the exceptional durability of rosewood in terrestrial conditions, the wood is very susceptible to attack by marine borers. The heartwood has been reported to get rapidly destroyed by marine borers within a period of 6–12 months. The destruction is mainly caused by *Martesia striata*, which attacks the timber only for shelter [15, 16]. The sapwood is perishable and easily attacked by borers and fungi.

The natural durability of the heartwood can be attributed to its extractive content such as latifolin, 4-methoxy dalbergione and dalbergiphenol that are present in the heartwood [17]. The heartwood contains about 15% extract [17]. When Rubberwood is treated with this extract and exposed to white rot and brown rot fungi, it demonstrated significantly enhanced resistance compared to untreated control blocks [18].

6 Treatability

Unlike heartwood, the sapwood of rosewood is perishable, as it is susceptible to attack by fungi and borers, necessitating preservative treatment to prolong its service life. While preservatives can improve the durability, not all species are equally receptive to preservative treatment. The treatability of wood differs from species to species and timbers are classified into five treatability classes [13] based on the ease of heartwood treatment as depicted in Table 6.

Table 6 Classification of heartwood based on its treatability [13]

Class	Remark
a	Heartwood easily treatable
b	Heartwood treatable, but complete penetration not always obtained when the least dimension is more than 6 cm
c	Heartwood only partially treatable
d	Heartwood refractory to treatment
e	Heartwood very refractory to treatment, penetration of preservative being practically nil even from the ends

Fig. 3 Appearance of the planed surface of *D. latifolia*

The sapwood of rosewood is easily treatable; however, the heartwood is highly refractory (Class e) to preservative treatment [2, 13]. Due to its natural durability, the heartwood does not need additional preservative treatment.

7 Working Qualities

The timber is very hard and relatively difficult to saw. The best results are typically achieved using a reciprocating saw. If the logs are converted in green state, then a band saw gives very good results. Logs often have natural splits at their centre, which are sometimes filled with a chalky substance. These deposits tend to dull the saw blade quickly, which necessitates the need for boxing the heart of the log to prevent further complications during sawing [11]. The wood is rather difficult to work with hand because of its higher density and interlocked grains, but it is very easy to work with machines. The wood can be turned, glued, screwed, and polished with excellent results (Fig. 3). The logs can be peeled and sliced after softening but has a tendency to develop cracks. Rotary cut as well as sliced veneers present a highly ornamental appearance.

8 Wood utilization

Construction: For construction, timbers have been classified into three groups (Table 7), based on the modulus of elasticity (MoE) and safe permissible bending stress on extreme fibres (f_b).

Table 7 Classification of timbers for construction purposes [19]

Group	Properties	Values (N/mm^2)
A	f_b	>18.0
	MoE	>12,600
B	f_b	>12.0 and <18.0
	MoE	>9800 and <12,600
C	f_b	>8.5 and <12.0
	MoE	>5600 and <9800

Based on the MoE and fb values, rosewood has been recommended as one of the most important Group C species for construction purposes. Its safe working stresses for structural applications in comparison to Teak have been provided in Table 8.

Furniture and Cabinets: The suitability of timber for furniture is determined based on strength coefficients. Ideal furniture timber should possess adequate strength, shape retention, ease of workability, good seasoning characteristics, a smooth finish, and desirable aesthetic qualities such as colour, texture, figure and overall appearance. Quantitative evaluation of the suitability of a species for furniture and cabinet requires

Table 8 Safe working stresses of *D. latifolia* compared with *Tectona grandis*

Property	*Dalbergia latifolia*	*Tectona grandis*
Modulus of elasticity (GPa) All grades and location	9.09	10.76
Bending and tension along grain Extreme fibre stresses (MPa)		
Inside locations	16.57	16.47
Outside locations	13.83	13.73
Wet locations	11.08	10.98
Shear (MPa)		
Horizontal (All locations)	1.15	1.02
Along grain (All locations)	1.64	1.46
Compression		
MCS parallel to grain (MPa)		
Inside locations	10.2	10.2
Outside locations	9.12	9.02
Wet locations	7.45	7.35
Compressive stress perpendicular to grain (MPa)		
Inside locations	6.37	4.8
Outside locations	4.9	3.73
Wet locations	4.02	3.04

(Modified from [10])

consideration of its four suitability indices, such as strength as beam, suitability as post, retention of shape and splitting coefficient. Based on these indices, the strength coefficient is determined. Retention of shape value should be more than 60 compared to Teak as 100 [20]. Based on the strength coefficient, weight and gross features, timber for furniture and cabinets has been classified into 4 groups as given below.

Super group: Strength coefficient—75 or more.
Weight (compared to Teak as 100): 85 to 100.
Grain, texture, finish, polish and other general appearance—Excellent.

Group 1: Strength coefficient—75 or more.
Weight (compared to Teak as 100): 85 to 110.
Grain, texture, finish, polish and other general appearance—Very good.

Group II: Strength coefficient—65 or more.
Weight (compared to Teak as 100): 70 to 120.
Grain, texture, finish, polish and other general appearance—Good.

Group III: Strength coefficient—50 or more.
Weight (compared to Teak as 100): 60 to 130.
Grain, texture, finish, polish and other general appearance—Satisfactory.

Rosewood is highly valued for the creation of premium furniture and cabinets, because of its moderate weight, good working and finishing qualities, attractive colour, beautiful grain and fine texture. With a comparative strength coefficient of 115, on the basis of Teak as 100, rosewood comes under the super group for furniture and cabinet [20]. It is also regarded as an excellent timber for decorative and intricate carving.

Flooring: The suitability of timber for flooring is determined by its strength, hardness and shape retention [21]. With a hardness index of 150, retention of shape index of 85, strength as a beam index of 96 and flooring suitability figure of 120, compared to 100 for Teak, rosewood is highly suitable for heavy-duty block flooring, board flooring and parquet flooring.

Other Uses

In addition to the above-described uses, rosewood is also highly valued for a variety of other applications, including decorative plywood, manufacturing of tool handles, knobs, drawer pulls, furniture handles, agricultural implement and tool handles. Rosewood is excellent for peeling as well as slicing for producing decorative veneers with exceptional gluing qualities. Hence, the wood is highly regarded for decorative plywood [22]. With a strength coefficient of 110, compared to 100 for Teak, rosewood is very suitable for agricultural implements and tool handles [23]. The wood also finds application in the production of cutlery, artificial limbs, textile mill accessories, chess pieces, carom dots, tennis and badminton rackets [24]. The wood is also used to make calico-printing blocks, mathematical instruments, bentwood items, and handicrafts. In China, it is specifically used for making the spokes of chariot wheels [25, 26]. It is also considered as a superior material for shutters and door frames. With a suitability coefficient of 104, compared to 100 for Teak, rosewood has been classified as a super group timber for door and window shutters and Group I timber for door and window frames [10, 27]. The Ordnance department utilizes the wood in the production of ammunition and other products, including wagons, side bars, ammunition boxes, axle beds, brake blocks, and more [11].

D. latifolia is a popular material for crafting guitar fret boards, often used alongside other types of wood like ebony and maple [7]. This widespread use in musical instruments is attributed to its high stiffness and dimensional stability, alongside the beautiful colour variations which enhance the visual appeal of the instrument [8]. The wood is also used for high class bent wood furniture like superior walking sticks and umbrella handles.

References

1. Jenkins A, Bridgland N, Hembery R, Malessa U, Hewitt J, Chen HK (2012) Precious woods: exploitation of the finest timber. Tackling the trade in illegal precious woods. Chatham House, London, UK
2. Rao KR, Purkayastha SK (1972) Indian woods: their identification, properties and uses, vol 3. Manager of Publications, New Delhi, India, Government of India
3. Richter HG, Gembruch K, Koch G (2023) CITESwoodID: descriptions, illustrations, identification, and information retrieval. In English, French, German, and Spanish (2014) Version: 4th April 2023. www.delta-intkey.com
4. Limaye VD (1993) Physical and mechanical properties of woods grown in India. Third interim report on project No.1. Indian Forest Records 18(1)
5. Sekhar AC, Bhatnagar SS (1955) Physical and mechanical properties of woods tested at Forest Research Institute, Dehra Dun, Indian Forest Record (NS), TM 1(6). Manager of publications, Delhi
6. Dwianto W, Bahanawan A, Kusumah SS, Darmawan T, Amin Y, Pramasari DA, Lestari E, Akbar F (2019) Study on the existence and characteristics of Sonokeling (*Dalbergia latifolia* Roxb.) as an Appendix II CITES Wood. In: IOP conference series: earth and environmental science, 374(1)
7. Shirmohammadi M, Faircloth A, Redman A (2021) Assessment of sound quality: Australian native hardwood species for guitar fretboard production. Eur J Wood Wood Prod 79:487–497
8. Sproßmann R, Zauer M, Wagenführ A (2017) Characterization of acoustic and mechanical properties of common tropical woods used in classical guitars. Results Phys 7:1737–1742
9. Sekhar AC, Gulati AS (1972) Suitabilty indices of Indian timbers for industrial and engineering uses. Indian Forest Records (N.S.) Timber Mechanics 2(1), Manager of Publications, Delhi
10. Rajput SS, Shukla NK, Guta VK, Jain JD (1991) Timber mechanics. Strength, classification and grading of timber. Indian Council of Forestry Research and Education. New Forest, Dehradun, India
11. Chauhan L, Rao RV (2003) Wood anatomy of legumes of India: their identification, properties and uses. Bishen Singh Mahendra Pal Singh, India
12. Anonymous (1993) Code of practice for seasoning of timber. IS: 1141, Bureau of Indian Standards, New Delhi, India
13. Anonymous (2001) Code of practice for preservation of timber. IS: 401, Bureau of Indian Standards, New Delhi, India
14. Kumar S, Dev I (1993) Wood preservation in India. ICFRE Publication-28. Dehradun, India
15. Santhakumaran LN, Rao MV (1992) Natural durability of some common Indian timbers and marine plywood against bio deterioration in Kochi waters (India). International Research Group on Wood Preservation, Document No. IRG/WP/4177–92:1–16

16. Rao S, Balaji M, Srinivasan VV (1994) Resistance of 25 species of timber to marine borer attack at Visakhapatnam, East coast of India. International Research Group on Wood Preservation, Document No, Document No. IRG/WP/94–30036:1–10
17. Sekine N, Ashitani T, Murayama T, Shibutani S, Hattori S, Takahashi K (2009) Bioactivity of latifolin and its derivatives against termites and fungi. J Agric Food Chem 57(13):5707–5712
18. Sethy AK, Nagaveni HC, Mohan S, Chandrashekar KT (2005) Fungal decay resistance of Rubber wood treated with heartwood extract of Rosewood. International Research Group on Wood Preservation, Document No. IRG/WP 05–30367
19. Anonymous (1994) Design of structural timber in building- code of practic. IS: 1141, Bureau of Indian Standards, New Delhi, India
20. Anonymous (1993) Indian timbers for furnitures and cabinets-Classification. IS: 13622, Bureau of Indian Standards, New Delhi, India
21. Rajput SS, Shukla NK (1989) Classification of timber species for wooden flooring. J Timber Dev Assoc 35(3):38–44
22. Anonymous (1982) Specification for veneered decorative plywood. IS: 1328, Bureau of Indian Standards, New Delhi, India
23. Anonymous (1963) Classification of commercial timbers and their zonal distribution. IS: 399, Bureau of Indian Standards, New Delhi, India
24. Tewari DN (1995) A monograph on rosewood: *Dalbergia latifolia* Roxb. International Book Distributors
25. Anoop EV, Pasha MKS (2017) Timber identification manual: manual of timbers used by wood based handicrafts industry of Kerala, Uttar Pradesh and Rajasthan. European Union/WWF India, New Delhi
26. Orwa C (2009) Agroforestree database: a tree reference and selection guide, version 4.0. http://www.worldagroforestry.org/sites/treedbs/treedatabases.asp
27. Anonymous (1990) Indian timbers for door and window shutters and frames. IS: 12896, Bureau of Indian Standards, New Delhi, India

Chapter 9
Phytochemistry of *Dalbergia latifolia*: Bioactive Compounds and Their Therapeutic Potential

Rohit Sharma and Rakesh Kumar

1 Introduction

Dalbergia latifolia Roxb., also known as Rosewood, is a highly prized and commercially significant tree species that is sparsely distributed across its natural habitat [1, 2] and is of great economic value in terms of trade in national and international market [3]. The purplish-brown to dark purple heartwood has a specific gravity ranging from 0.83 to 1.22 and is valued for its exceptional uses in premium furniture, veneers, and especially in the guitar industry for its unique sound qualities [2, 4–7]. Its beauty and durability have made it a preferred timber for over 300 years [8, 9], but overharvesting and habitat destruction have led to its classification as "Vulnerable" by the International Union for Conservation of Nature (IUCN) and listing in Appendix II of the Convention on International Trade in Endangered Species of Wild Fauna and Flora (CITES) [1].

The wood from *Dalbergia* species contains a significant and wide variety of compounds, including several flavonoids and neoflavanoids components, along with associated compounds [10]. The chemical analysis of the bark and sapwood of *D. latifolia* showed the presence of several compounds [11]. *D. latifolia* has been the subject of numerous phytochemical investigations across its various plant parts. The methanolic extract of its leaves has confirmed the presence of alkaloids, glycosides, phenols, resins, and tannins [12]. Similarly, in another study the methanolic extracts of the leaves were found to contain flavonoids, terpenoids, sterols, glycosides, phenols, and tannins [13]. Additional studies on the methanol extract have identified resins, terpenoids, carbohydrates, tannins, and steroids [14]. The ethanol extract of its roots showed the presence of carbohydrates, glycosides, flavonoids, alkaloids, phenolic compounds, and tannins [15]. The bark extract showed the presence of steroids, phenolic compounds, tannins, flavonoids, and triterpenes [16]. The bark

R. Sharma · R. Kumar (✉)
ICFRE-Institute of Wood Science and Technology, Bengaluru, India
e-mail: colourrakesh@gmail.com

R. R. Warrier et al. (eds.), *Handbook of Asian Rosewoods*, Materials Horizons: From Nature to Nanomaterials, https://doi.org/10.1007/978-981-96-9459-4_9

also contains minerals, organic acids, flavonoids, and phenolic compounds, demonstrating antioxidant and mast cell-stabilising properties. Traditionally, polyphenols have been recognised for their therapeutic roles in promoting health and combating illness [17]. *D. latifolia* has been traditionally used to treat different illnesses. The plant is a rich source of various pharmaceutically active compounds, including carbohydrates, glycosides, flavonoids, alkaloids, phenolic compounds, and tannins [15]. These findings highlight the diverse phytochemicals in *D. latifolia* across its different extracts and plant parts.

2 Ethnomedicinal Uses

D. latifolia is a traditional resource, containing flavonoids and other phytochemicals having excellent therapeutic and medicinal value [18]. It is acknowledged as a medicinally important tree within the Indian forest flora, offering a diverse array of phytochemicals with substantial potential in Ayurveda and other traditional medicine systems [5]. Traditionally, the bark of *D. latifolia* is utilised to ease body pain, while the entire plant is used in treating conditions such as diarrhoea, dyspepsia, leprosy, and obesity. It has been widely used in traditional Indian medicine to treat conditions such as obesity, scabies, dysentery, syphilis, dyspepsia, leprosy, and ringworm [19, 20]. Extracts from various parts of *D. latifolia* have shown inhibitory effects against many microbes, highlighting its significance in regions where many communities depend on plant-based medicines [5]. Heartwood extract contains numerous flavonoids used to treat various conditions, including blood diseases, eye and nose ailments, syphilis, stomach troubles, and leprosy [21]. The ethanolic extracts from the leaves of *D. latifolia* showed peak antioxidant activity at 300 μg/ml, and the strongest anthelmintic activity was observed at 200 mg/ml [22]. Various parts of the tree possess traditional medicinal properties and are utilised as stimulants, appetisers, and remedies for dyspepsia, diarrhoea, skin infections, leprosy, obesity, and worm infections [19, 23, 24].

3 Phytochemicals

Wood of *Dalbergia* species is important and has a wide variety of compounds (Table 1). The flavonoid and neoflavanoid constituents of the *Pterocarpus* and *Dalbergia* species were reviewed together with related compounds [10]. Latinone[5], a substituted phenanthrene-1, 4-quinone was isolated and characterised from the heartwood extract [25]. Benzofuran derivatives were identified in the heartwood of *D. latifolia*, and they were named 2-[5-hydroxy-4-methoxy-2-(3-p-henyl-trans-allyloxy) benzyl]-5-hydroxy-6-methoxy-3-phenylbenzofuran, (+)-obtusafuran[7] and isoparvifuran[8]. A new

Table 1 Phytochemicals present in different parts of *D. latifolia*

Plant Part	Alkaloids	Flavonoids	Tannins	Steroids	Carbohydrates	Phenols	Reference
Leaves	+	+	+	×	+	+	[12, 13]
Bark	×	+	+	+	×	+	[29]
Roots	+	×	+	×	+	+	[15]
Wood	×	+	+	×	×	×	[27]

diaryl 1, 2-diketone, named 1-(2,5-dihydroxy-4-methoxyphenyl)-2-phenylethane-1,2-dione was present in the heartwood [26, 27]. *D. latifolia* wood have 1,7,7-trimethyl-3-phenethylidenebicyclo-[2.2.1] heptan-2-one, 4,4'-methylenebis-2,6-dimethylphenol,4,2',3',4'-tetramethoxy-5'-methyl-6-methylaminomethyl 1,1'-biphenyl, (4-methyl-sulfanylphenyl) carbamic acid 2,6-dimethoxyphenyl ester [28].

Chemical examination of the bark and sapwood of *D. latifolia* consisted several compounds [11]. Bark extracts of *D. latifolia* in aqueous and alcoholic forms were positive for flavonoids, proteins, carbohydrates, and glycosides [29]. Hydro-alcoholic bark extract [30] and sequential bark extracts [16] from *D. latifolia* showed the presence of tannins, phenolic chemicals, triterpenes, flavonoids, and steroids. HPTLC analysis showed the presence of β-sitosterol in methanolic extract of bark [20]. The methanolic extract of leaves of *D. latifolia* shows the presence of alkaloids, glycosides, phenols, resins, and tannins [12]. In another research resins, flavonoid, terpenoids, sterols, carbohydrates, and steroids were also reported from methanol extract of its leaves [13, 14]. Ethanol extract from the roots of *D. latifolia* showed the presence of carbohydrates, glycosides, phenolic compounds, alkaloids, tannins, and flavonoids [15].

D. latifolia has other metabolites, such as carotenoids and glycosylics, which were isolated from the alcoholic extract of the seeds [31]. Seeds of *D. latifolia* also contain dalbin, a new rotenoid glycoside and dalbinol[9] (a new 12a-hydroxyrotenoid), sisafolin, coumarin and a substituted 4-phenylcoumarin[17] [32, 33].

3.1 Flavonoids

The wood extract of *D. latifolia* contains high levels of flavonoid compounds [34]. Within the plant kingdom, flavonoids are a broad class of low-molecular-weight phenolic chemicals that are widely distributed and prevalent. There are more than 10,000 known structural variations of flavonoids. Their interaction with various targets in different subcellular locations results in various activities in microbes, plants, and animals due to their physical and biochemical properties. [35]. The synthesis of flavonoids occurs through either the phenylpropanoid or acetate-malonate pathway. Flavonoids are identified by a fifteen-carbon structure that includes two benzene rings (A & B) connected by a heterocyclic pyrone ring C (Fig. 1) [36].

Fig. 1 Flavone skeleton

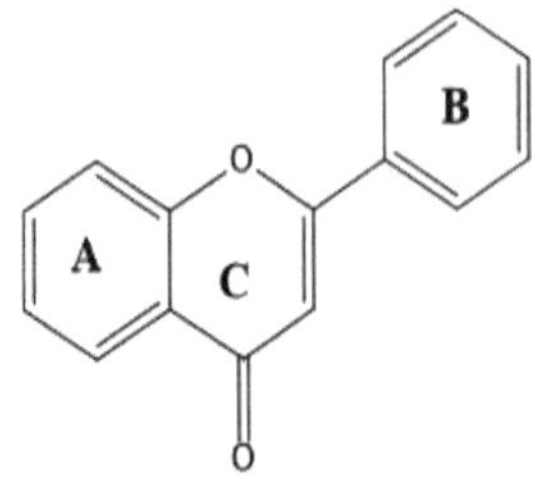

Latifolin[2], a natural flavonoid, was reported in *Dalbergia* species, including *D. latifolia* [37]. *D. latifolia* wood contains latinone[5] and dalcriodain[4] (Fig. 2) flavonoids which have excellent medicinal value [29]. Dalbergiphenol[1], 2,3-dimethoxy-2,4-dihydroxychalcone, latifolin dimethyl ether isomer, and latifolin dimethyl ether, 5-O-methyl latifolin and 2-O-methyl latifolin isomer, Latifolin[2], O-methyl dalbergin, 3-hydroxy-2,4,5-trimethoxydalbergiquinol, 2,4,5-trimethoxydalbergiquinol, and Dalbergin[3] are the major flavonoids found in *D. latifolia* wood [34].

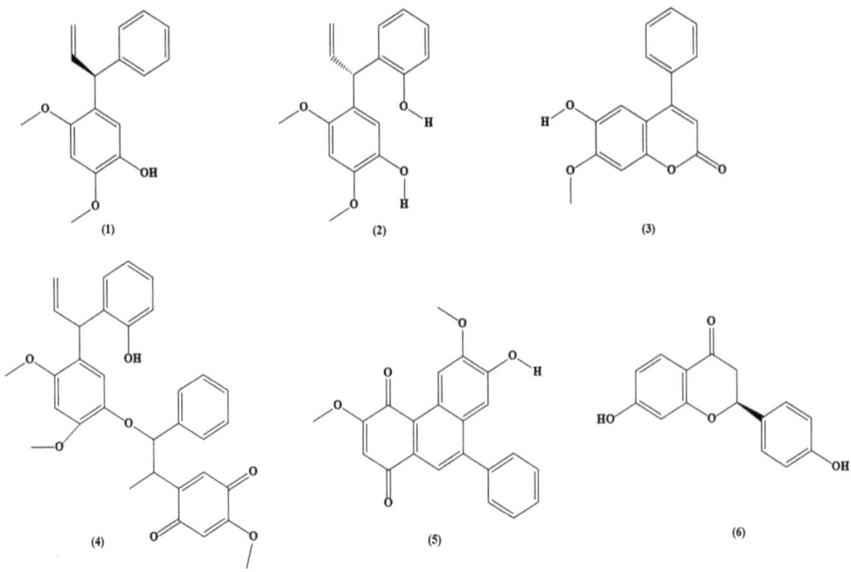

Fig. 2 Major flavonoids present in *D. latifolia*. (1) Dalbergiphenol, (2) Latifolin, (3) Dalbergin, (4) Dalcriodain, (5) Latinone, and (6) Liquiritigenin

Fig. 3 Chemical structure of Benzofuran

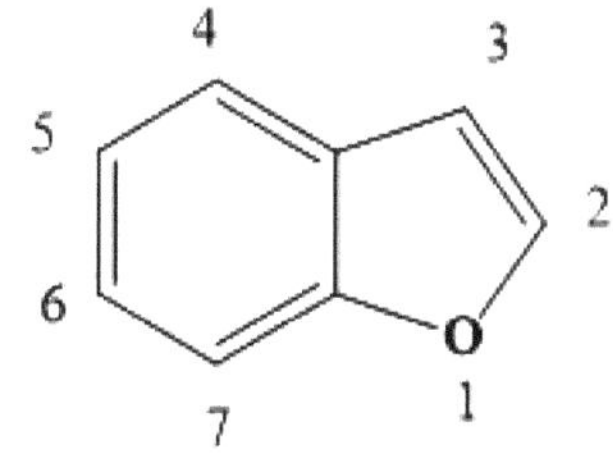

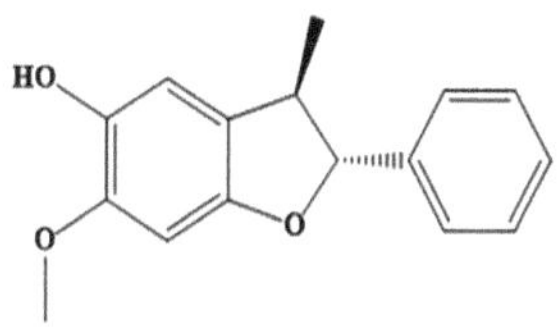

(7)

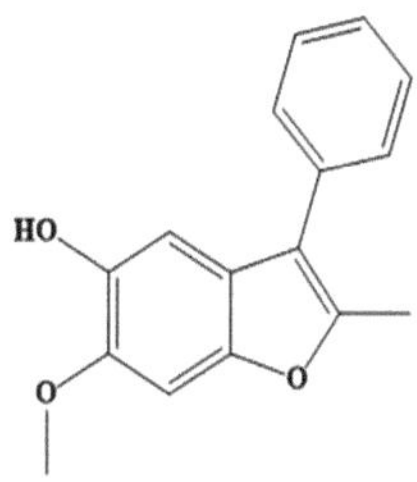

(8)

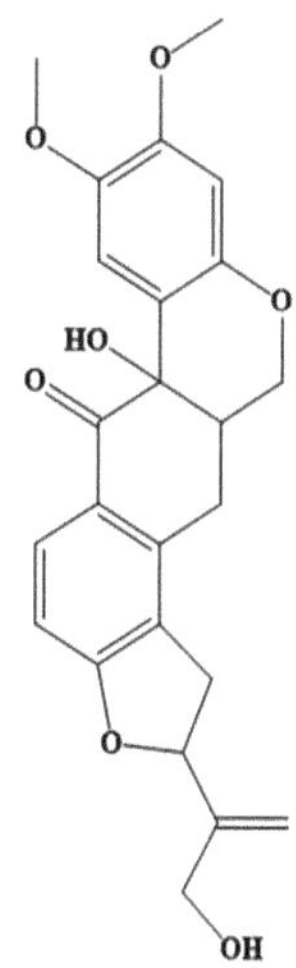

(9)

Fig. 4 Benzofurans from *D. latifolia.* (7) Obtusafuran, (8) Isoparvifuran, and (9) Dalbinol

3.2 Benzofurans

Benzofuran contains a fused benzene and furan ring, as shown in Fig. 3. Benzofuran is the basis for numerous related compounds with more intricate structures [38]. Three benzofurans from the heartwood of *D. latifolia* (Fig. 4)*;* were obtained including a new benzofuran compound, named 2-[5-hydroxy-4-methoxy2-(3-phenyl-trans-allyloxy) benzyl]-5-hydroxy-6-methoxy-3-phenylbenzofuran, together with (+)-obtusafuran[7] and isoparvifuran[8] [27]. The seeds of *D. latifolia* also contains dalbin, which is a recently discovered rotenoid glycoside, and dalbinol[9], which is a newly identified 12a-hydroxyrotenoid.

3.3 Phytosterol

Phytosterols resemble cholesterol structurally, possessing a four-ring steroid nucleus, a 3b-hydroxyl group, and frequently a 5, 6-double bond. In terms of function, they

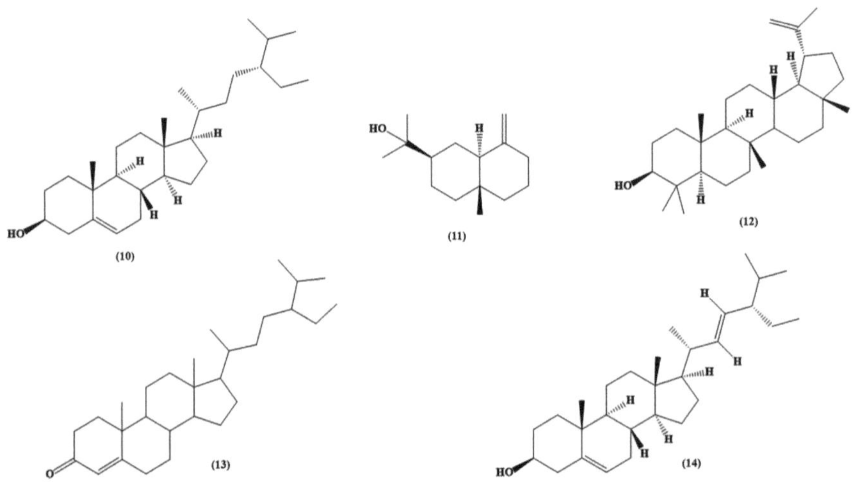

Fig. 5 Phytosterols from *D. latifolia* (10) β-sitosterols, (11) β-eudesmol, (12) lupeol, (13) β-sitostenone, and (14) stigmasterol

play a crucial role in stabilising phospholipid bilayers in cell membranes [39]. Plants usually possess more complex sterol compositions. Research shows that all sterols have the same capacity to control the membranes in a microfluid condition where they can communicate significant biological processes. This feature is influenced by evolutionary changes to molecules, which vary depending on the class of sterol [40].

A phytosterol triterpene, present in *D. latifolia,* is lupeol[12] (Fig. 5). Over the last three decades, several important pharmacological activities of lupeol[12] have been revealed by extensive research. It acts as an anti-inflammatory, antimicrobial, antiproliferative, anti-protozoal, anti-angiogenic, anti-invasive, and cholesterol lowering agent [41].

3.4 Triterpenoids

A vast class of naturally occurring compounds called triterpenoids is produced from C_{30} precursors. There are almost 200 different skeletons in the triterpenoids category of which the majority of triterpenoids consist of 6–6-6–5 tetracyclic structures, 6–6-6–6-5 pentacyclic structures, or 6–6-6–6-6 pentacyclic structures. However, triterpenoids with acyclic, monocyclic, bicyclic, tricyclic, and hexacyclic structures have also been discovered in natural sources [42, 43]. Acetyloleanolic acid[15] and β-amyrin acetate[16] were important triterpenoids present in heartwood of *D. latifolia* [44] (Fig. 6).

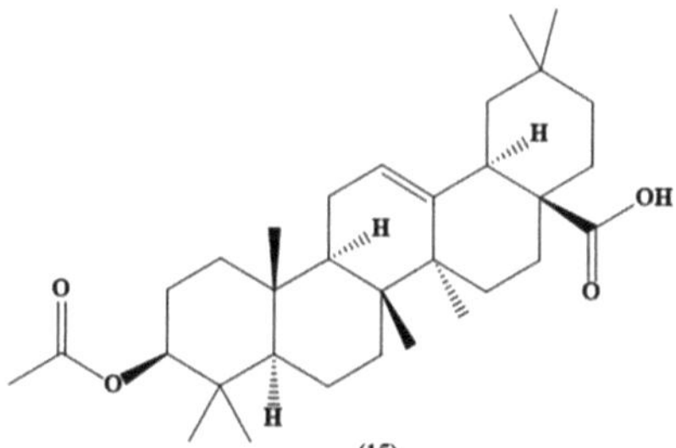

(15)

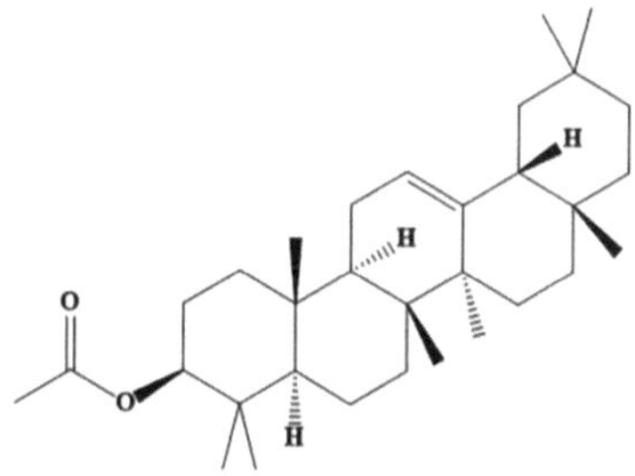

(16)

Fig. 6 Triterpenoids from *D. latifolia*. (15) acetyloleanolic acid and (16) β-amyrin acetate

3.5 Lignans

Lignans are categorised into eight distinct groups based on their structural characteristics, such as their carbon skeletons, the incorporation of oxygen into these skeletons, and their cyclisation patterns. These groups include arylnaphthalene, aryltetralin, dibenzocyclooctadiene, dibenzylbutane, dibenzylbutyrolactol, dibenzylbutyrolactone, furan, and furofuran [45]. Lignans are composed of a distinct dibenzylbutane skeleton (Fig. 7) [36].

These substances exhibit dimeric structures characterised by a β, β'-linkage between two phenyl propane units, which vary in side-chain oxidation levels and aromatic ring substitution patterns [46]. Figure 8 shows the structure of some other important compounds present in various parts of *D. latifolia.*

The bioactive constituents of this tree (Table 2) have significant potential in the healthcare sector, offering various medical benefits, including the prevention and treatment of diseases.

4 Analytical Techniques

Identifying and quantifying phytochemicals from *D. latifolia* involves a variety of analytical techniques. The most commonly employed methods include chromatographic, spectroscopic, mass spectrometry, and other advanced techniques. Column chromatography was performed with silica gel for both TLC and preparative TLC

Fig. 7 Dibenzylbutane skeleton of lignan

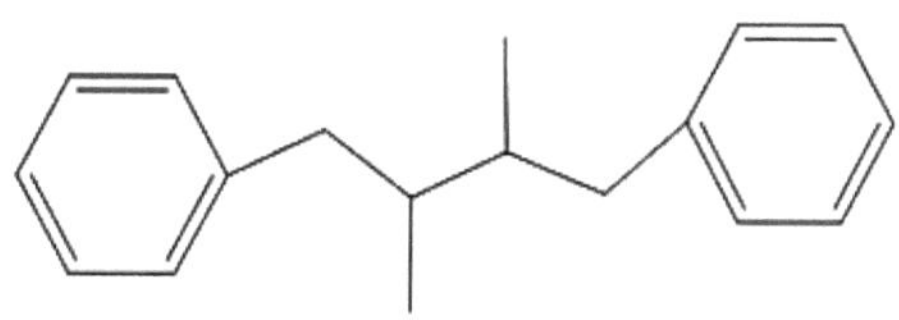

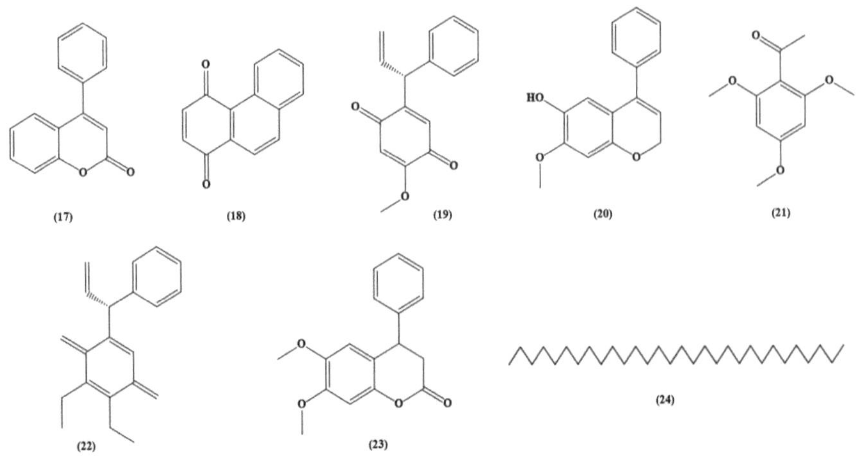

Fig. 8 Some other important compounds of *D. latifolia.* (17) phenylcoumarin, (18) phenanthrene-1, 4-quinone, (19) (R)-4-pethoxydalbergione, (20) palbergichromene, (21) 2',4',6'-trimethoxyacetophenone, (22) dimethoxydalbergione, (23) methyldalbergin, and (24) hentriacontane

to analyse the heartwood extract of *D. latifolia*. After extraction of *D. latifolia* heartwood, the fractions (I–IV) were subdivided using TLC and column chromatography. A red crystalline solid deposited from IV fraction of Methanol solution gave a compound identified as latinone**5**. The result was supported for the proposed structure from the ^{1}H NMR spectrum of latinone**5** [25]. By using the 1D and 2D NMR spectroscopy along with Infrared (IR) and FT-IR spectroscopy, a new diaryl 1,2-diketone compound, named 1-(2,5-dihydroxy-4-methoxyphenyl)-2-phenylethane-1,2-dione, was identified along with eight known compounds; phthalic acid butyl isobutyl ester, dibutyl phthalate, methyl-4-hydroxybenzoate, ethyl-4-hydroxybenzoate, p-hydroxybenzaldehyde, eucomic acid, 2-ethoxy-5-methoxy-1,4-benzoqunione, and phenyl β-D-glucopyranoside, were isolated [26].

By the comparison of NMR spectroscopic data (+)-obtusafuran**7** and isoparvifuran**8** were identified [27]. Fourier transform infrared (FTIR) spectroscopy was used to examine functional group variations in the extractive components by analysing characteristic peak positions, intensities, and shapes to infer the main chemical components. Gas chromatography-mass spectrometry (GC-MS) was employed to characterise the extractive components in detail. These analyses identified eight compounds in *D. latifolia*, with 4-methyl-2-[5-(2-thienyl)pyrazol-3-yl] phenol (12.41%) and 4,2',3',4'-tetramethoxy-5'-methyl-6-methylaminomethyl-1,1'-biphenyl (29.75%) being the most abundant [28]. Similarly, GC-MS was used to identify and characterise the latifolin**2** from *D. latifolia* wood extract [34].

High-performance thin layer chromatography (HPTLC) was used for characterisation of methanol extract of *D. latifolia* bark [20]. Researchers have used these methods to isolate and identify numerous compounds. Advanced techniques such as

Table 2 Major compounds reported in different parts of *Dalbergia latifolia* and their bioactivities

Sl. No	Compound name	Molecular weight*	Activities/other properties
1	(R)-4-methoxydalbergione	253.28	Anti-inflammatory [47], antifungal and antitermite activity [48]
2	(R)-latifolin	286.32	Antifungal and antitermite activity [49], anti-Skin Inflammatory and anti-oxidative [50], anti-inflammatory [51]
3	2',4',6'-Trimethoxyacetophenone	210.23	Antibacterial activity [52, 53], antifungal activity [54], cytotoxic, anxiolytic [55]
4	Acetyloleanolic acid	498.7	Antiviral, anti-HIV, antibacterial, antifungal, anticarcinogenic, hepatoprotective, hypolipidemic and anti-atherosclerotic activities [56, 57], antibacterial activity, anti-inflammatory, antitumor, anticancer, hypoglycaemic, antifungal [58]
5	β-Amyrin acetate	467.8	Antidiabetic, antioxidant, cytotoxicity, anti-inflammatory, analgesic, gastroprotective, anti-Parkinson effects, antifibrotic, and antihyperglycemic antimicrobial activity [59, 60]
6	β-Eudesmol	222.37	Anticancer, antianoxic, anti-ulcer, hepatoprotective, pesticide [61]
7	β-Sitostenone	412.7	Anti-inflammatory and anti-oxidative [62]
8	β-sitosterols	414.71	Anti-inflammatory activity, inducing apoptosis, angiogenic effect, and anticancer [63], benign prostatic hyperplasia [64]
9	Catechol	110.11	Antioxidant activity, anti-inflammatory [65, 66]
10	Dalbergichromene	253.28	Antihelmintic, antipyretic and analgesic [67], blood disorders, ischemia, and anti-inflammatory [68]
11	Dalbergin	267.86	Antiosteoporotic, antibacterial activity, antiulcergenic, and anti-inflammatory [69, 70]
12	Dalbergiphenol	268.32	Inhibitory activity against testosterone 5R-reductase [71], antifungal activity [48], anti-inflammatory, antioxidant, and cardiomyocyte protective effects [72]

(continued)

Table 2 (continued)

Sl. No	Compound name	Molecular weight*	Activities/other properties
13	Dalbinol	426.4	therapeutic effect against hepatocellular carcinoma (HCC), and antiproliferative [73–75]
14	Dimethoxydalbergione	284.31	Antiplasmodial activity, anti-inflammatory activity [76] cytotoxicity [77]
15	Elemicin	208.25	Antioxidant activity, antibacterial antifungal activity [78]
16	Ethyl 4-hydroxybenzoate	165.17	Anti-hyperuricemic [79], anti-melanoma agents, cytotoxic [80]
17	Formononetin	267.26	Anti-inflammatory, antioxidant and anticancer [81, 82]
18	Hentriacontane	436.8	Antitumor activity [83], anti-inflammatory [84]
19	Hydroxybenzaldehyde	122.12	Antimicrobial, antibacterial [85]
20	Isoliquiritigenin	255.25	Anti-inflammatory [86], anticancer, antibiotic [87], antioxidant activity, neuroprotective activity [88]
21	Isoparvifuran	253.28	Anticancer [89], antioxidant [90]
22	Latinone	359.4	Antioxidants [29]
23	Liquiritigenin	255.25	Anti-inflammatory [91], antidiabetic [92], antipsychostimulant, anti-Alzheimer's activities, antidepressant and antianxiety activities [93]
24	Lupeol	426.72	Anticancer, antiprotozoal, chemo preventive and anti-inflammatory [94], antimicrobial and antiproliferative [95], cytotoxic activity [96]
25	Methyldalbergin	282.29	Antiosteoporotic, antibacterial activity [69]
26	Obtusafuran	255.3	Anticarcinogenic, insect antifeedant [97, 98]
27	Phenanthrene-1,4-quinone	208.21	Cytotoxicity [99], genotoxic [100]
28	Phenylcoumarin	222.24	Cytotoxicity [101], antiplasmodial [51], antimicrobial, anti-adherence, and antibiofilm [102]
29	Phthalic acid, butyl isobutyl ester	278.34	Antioxidant, anticancer, antimicrobial activity [103]

(continued)

Table 2 (continued)

Sl. No	Compound name	Molecular weight*	Activities/other properties
30	Stigmasterol	412.7	Antiosteoarthritic, antihypercholestrolemic, cytotoxicity, antitumor, hypoglycaemic, antimutagenic, antioxidant, anti-inflammatory [104, 105]
31	Triacontane	422.8	Antibacterial, antidiabetic, and antitumor activity [106, 107], antimicrobial [108]

* National Centre for Biotechnology Information (2024) (https://pubchem.ncbi.nlm.nih.gov/)

FTIR, NMR, and GC-MS were instrumental in identifying and characterising functional groups and extractive components, highlighting the diversity and complexity of compounds present in different parts of *D. latifolia.*

5 Bioactivity

5.1 *Antibacterial Activity*

Methanol extract of *D. latifolia* leaves reported moderate activity against tested urinary tract infection causing bacteria and minimum inhibitory concentration (MIC) and minimum bactericidal concentration (MBC) were observed [14]. Several compounds extracted from different parts of *D. latifolia* have shown antibacterial properties. These include 2',4',6'-trimethoxyacetophenone**[21]** [52, 53], acetyloleanolic acid**[15]** [56–58], dalbergin**[3]** [69, 70], elemicin [78], hydroxybenzaldehyde [85], methyldalbergin**[23]** [69], and triacontane [106, 107]. The MIC ranges for all multidrug resistant (MDR) *Salmonella enterica* isolates were 3.125 to 75 mg/mL and susceptible to methanolic leaf extracts. A mixture of antibiotics and leaf extract showed a greater synergistic impact than individual [12]. *D. latifolia* leaf hydroalcoholic extract showed a moderate inhibitory effect on *B. subtilis* and *E. coli*, with MIC values of 250 μg/ml for both [109]. Compared to the other examined extracts (chloroform and methanol), the ethanol extract of *D. latifolia* roots showed the largest zone of inhibitions against *S. aureus*, *K. pneumoniae*, *E. coli*, and *B. subtilis* [15].

5.2 Anti-Termite Activity

D. latifolia extract was tested for protection against termites. Extracts of heartwood were tested against termite and mortality was studied. Sequential extraction was conducted on the heartwood using n-hexane, ethyl acetate (EtOAc), and methanol (MeOH). The bioactivity of these extracts against termites and various fungi was assessed. The n-hexane extract demonstrated the most potent activity against termites (in both mortality and antifeedant effects) and wood decay fungi. Latifolin[2], dalbergiphenol[1], and 4-methoxydalbergione[19] [47], neoflavonoids, were among the key compounds isolated and identified from the extracts [48]. Methanol, ethyle acetate, and n-hexane extract showed the mortality percentage of 6.67%, 63.33%, and 73.33% respectively. Compounds extracted from heartwood latifolin[2] showed the maximum mortality of 80%, dalbergiphenol[1] (10%) and 4-methoxydelbergione[19] 16.67%. Latifolin[2] and its derivatives showed resistance against termites [49]. *D. latifolia* heartwood n-hexane and ether soluble fractions showed substantial toxicity against *R. speratus* in the antifeedant bioassay test [110].

5.3 Antifungal Activity

The methanol, ethanol, and chloroform extracts of *D. latifolia* roots were studied for antifungal activity. The study revealed that ethanol extract showed the highest zone of inhibitions against *Candida albicans* with 12.3 ± 0.8 mm zone of inhibition followed by 9.2 ± 0.3 for methanol extract and 8.4 ± 1.0 for chloroform extract [15]. Some important compounds like 2',4',6'-trimethoxyacetophenone[21] [52, 53], acetyloleanolic acid[15] [56–58], dalbergiphenol[1] [48, 49], elemicin [78] showed antifungal properties. The antifungal activity against *Rhizopus oryzae* and *Cladosporium cladosporioides* for latifolin[2] and its derivatives was relatively low compared to white- and brown-rot fungi. Latifolin[2] showed the highest activity level against all fungi, while the activity levels of all derivatives were significantly lower than that of latifolin[2] [49]. The n-hexane extract from the heartwood of *D. latifolia* was the most effective against wood decay fungi compared to the ethyl acetate and methanol extracts. Among the compounds isolated, latifolin exhibited the most significant activity (80%) against *Fomitopsis palustris* and *C. cladosporioides*, whereas dalbergiphenol[1] and 4-methoxydalbergione[19] showed moderate activity [49].

5.4 Anti-Obesity Activity

Obesity increases the risk of other major diseases, including cardiovascular problems, making it a global concern. The bark is used to treat body pain, while the entire plant is utilised for addressing diarrhoea, dyspepsia, leprosy, and obesity [19]. In another study, extracted bark of *D. latifolia* (DLBE) was segregated into 50% hydro-ethanolic extract (yield = 16.34%) and a methanolic extract (yield = 4.32%). High-performance thin-layer chromatography (HPTLC) was used to analyse the methanolic extract, and rats were used to evaluate the DLBE for acute oral toxicity and anti-obesity effectiveness. At doses of 200 and 400 mg/kg, DLBE dramatically decreased blood urea nitrogen, liver weight, food intake, cholesterol, triglycerides, liver enzymes, serum creatinine, and body weight in rats fed a high-fat diet in anti-obesity experiments. According to histopathological examination, rats receiving treatment had normal liver and kidney architecture. These results point to the possibility of using *D. latifolia* bark extract as an obesity treatment and call for more research in this field [20]. The heartwood of *D. latifolia* has neoflavonoid dalcriodon, along with latinone[5], which is a substituted phenanthrene-1, 4-quinone[18]. The stem bark is used for treating leprosy, obesity, and worm infestations due to its tannin content [22].

5.5 *Cytotoxic Activity*

The wood extract of *D. latifolia* is a rich source of various flavonoids, including chalcones, isoflavones, neoflavones, and flavones, which are known for their potential cytotoxic activities. Mainly, isoliquiritigenin (ILTG, 2',4',4'-trihydroxychalcone), a flavonoid compound chalcone, exhibits various pharmacological characteristics, including cytotoxic activity. Methanol extract of *D. latifolia* wood showed cytotoxicity with IC50 value of 20 μg/ml against MCF10A ATCC cells. Additionally, the compounds' toxicity was evaluated on normal human LO2 cell lines, and it was shown that they were non-toxic [21]. Assessment of cytotoxicity using brine shrimp showed LC50 values of 319 and 5.86 μg/mL for acute and lethal doses, respectively, indicating high toxicity relative to the reference drug, cyclophosphamide, which had an LC50 value of 2506 μg/mL [64]. Lupeol[12] [96] and S-3-hydroxy-4,4-dimethoxydalbergione showed cytotoxicity [77]. β-sitostenone[13], a compound obtained from heartwood extract, has been shown to have considerable cytotoxic properties [44, 62, 99]. The ethyl 4-hydroxybenzoate's chemical structure contains an ester group that hinders ionisation, allowing the molecule to diffuse across the cell and contribute to cell toxicity freely. This leads to a 150-fold increase in toxicity compared to its carboxylic acid derivative [80]. Stigmasterol[14] isolated from heartwood [44] exhibited cytotoxic effects on *Spodoptera litura* cells, demonstrating stronger action compared to the plant's other active components, friedelin and rotenone obtained from different parts of *Cacalia tangutica* [104, 111]. However, it was reported that the cytotoxicity of stigmasterol[14] was relatively low. Therefore, combining stigmasterol with sorafenib may offer potential benefits in breast cancer treatment [105].

5.6 Antioxidant Efficacy

A novel benzofuran from the heartwood of *D. latifolia* displayed moderate DPPH (2,2-diphenyl-1-picrylhydrazyl) free radical scavenging activity, with an IC50 of 96.7 ± 8.9 μM. In contrast, compounds (+)-obtusafuran[7] and isoparvifuran[8] showed poor antioxidant potential, with IC50 values of 951.4 ± 10.1 μM and 743.8 ± 12.7 μM, respectively, compared to the control [27]. DPPH assay is widely used to determine the antioxidant activity of crude extracts or plant purified compounds [112]. The methanolic extract of *D. latifolia* wood showed the highest DPPH scavenging activity, with an IC50 value of 70 ± 2.55 μg/ml [21, 113]. A study revealed various antioxidant activities among different extracts of *D. latifolia*, in which the ethanolic extract of *D. latifolia* bark exhibited strong antioxidant activity [114]. On the other hand, the ethanol extract of *D. latifolia* leaves showed poor antioxidant potential in the DPPH assay, with only 2.6 ± 1.9% electron donating ability, though it performed better in the reducing power assay with a 273.2 ± 1.0% reduction of Fe + 3 [115] and exhibited dose-dependent strong antioxidant activity in the DPPH method, with a maximum free radical scavenging activity of 46.74% at a concentration of 300 μg/ml [22]. The analysis of the DPPH antioxidant assay showed that the beta-amyrin acetate fraction exhibited a higher antioxidant potential, as evidenced by an IC50 value of 158.2 μg/mL, compared to gallic acid, which had an IC50 of 201.1 μg/mL [59]. The functionalised polymer catechol show bioadhesive, anti-inflammatory, and antioxidant properties, all of which are crucial characteristics for the wound-healing process [65, 66]. Dalbergiphenol[1], dalbergiones, dalbergin[3], and benzophenones isolated from the heartwood of *Dalbergia cochinchinensis* showed a wide range of pharmacological properties, including anti-inflammatory and antioxidant properties, and these compounds are present in the heartwood of *D. latifolia* [72]. Elemicin was used to perform lipid peroxidase, catalase, and DPPH assays, and the findings indicated significant antioxidant activity [78]. Formononetin is extensively studied for its anti-inflammatory, anticancer, and antioxidant characteristics. Additionally, Formononetin can trigger the apoptotic pathway by increasing reactive oxygen species (ROS) and disrupting the intracellular antioxidant system [81, 82]. The antioxidant activities were also shown by isoliquiritigenin, isoparvifuran[8], latinone[5], phthalic acid, butyl isobutyl ester, and stigmasterol[14] [29, 88, 90, 103–105].

5.7 Anti-Inflammatory Activities

The numerous compounds present in *D. latifolia* show anti-inflammatory activities such as, (R)-4-methoxydalbergione[19], (R)-latifolin[2], acetyloleanolic acid[15], β-amyrin acetate[16], β-sitostenone[13], β-sitosterols[10], catechol, dalbergichromene[20], dalbergin[3], dalbergiphenol[1], dimethoxydalbergione[22], formononetin, hentriacontane[24], hydroxybenzaldehyde, isoliquiritigenin, liquiritigenin[6], and lupeol[12] [47, 51, 58–60, 62, 63, 65, 66, 69, 70, 72, 76, 78,

81, 82, 84, 85, 88, 91, 94, 104, 105]. These substances have been observed to exhibit multiple health-promoting effects. They possess a broad spectrum of medicinal properties, such as anti-osteoporosis, anti-inflammatory, antimicrobial, antiplasmodial, anti-androgen, anti-allergic, antioxidant, antifungal, antidiabetic, and anticancer properties [51]. The heartwood of *D. cochinchinensis* and *D. latifolia* contains dalbergiphenol[1], dalbergiones, dalbergin[3], and benzophenones, which exhibit various pharmacological properties such as anti-inflammatory and antioxidant effects [72]. A research finding suggests that 4-methoxydalbergione and 4'-hydroxy-4-methoxydalbergione demonstrate anti-inflammatory properties in BV2 microglial cells stimulated with LPS, while only 4-methoxydalbergione was found to have a cytoprotective effect in murine HT22 hippocampus cells [47].

The significance of *D. latifolia* is evident from the various bioactivities mentioned earlier. Additionally, the multitude of compounds in different components of *D. latifolia* were utilised in diverse treatments and display a broad spectrum of bioactivities. The β-Eudesmol[11] was reported to be used in the treatment of cancer, as well as for its anti-ulcer, and hepatoprotective properties [61]. Another significant compound, dalbinol[9], has demonstrated therapeutic effects against hepatocellular carcinoma (HCC) and has shown antiproliferative properties [73–75]. Liquiritigenin[6] has exhibited antidiabetic activities [92], as well as antipsychostimulant, anti-Alzheimer, antidepressant, and antianxiety activities [93]. Additionally, obtusafuran[7] has proven effective as an anticarcinogenic and insect antifeedant [97, 98], phenanthrene-1,4-quinone[18] has demonstrated cytotoxicity [99] and genotoxicity [100], and triacontane shows antimicrobial activities [108].

5.8 Compounds Contributing to the Colour of **D. latifolia** *Wood*

Wood colour variation is associated with the composition of its extractive components, with flavonoid-type compounds being noted for their contribution to the colouration of plant parts [116, 117]. Previous research has shown that certain flavonoids [116] are responsible for the colouration of wood, while lignans such as plicatic acid contribute to the colour of heartwood [118]. Masendra et al. [34] conducted a study to investigate the flavonoids in the ethanol-toluene extract of *D. latifolia* wood and their contribution to its colour. The findings revealed that insoluble polymers contributed 62.6% of wood colour. However, soluble flavonoids accounted for 34.2% of the wood's colour. Further through methylation, 12 flavonoids were identified and categorised into latifolin[2], dalbergiphenol[1], chalcone, and dalbergin[3], all of which may influence the colour of *D. latifolia* wood. In addition to frequently occurring dalbergiones, a red crystalline quinone, latinone[5], in the heartwood, which may account for its red tinge, was reported [25].

6 Conclusion

The extracts from various parts of *D. latifolia* demonstrated significant inhibitory activities against a broad spectrum of microbes. A diverse array of biomolecules has been isolated and characterised from various parts of this tree, revealing impressive antibacterial, antifungal, antioxidant, and antitermite activities. However, few studies have been reported in specific fields of medical values. Apart from antitermite, antifungal, and antibacterial properties, many other properties are yet to be explored. This new area of pharmacology can be explored in the near future. This chapter extensively highlights the phytoconstituents and pharmacological properties of various parts of *D. latifolia*, showcasing its immense potential as a cornerstone for plant-based therapeutic applications.

References

1. Arunkumar AN, Warrier RR, Kher MM, Teixeira da Silva J.A (2022) Indian rosewood (*Dalbergia latifolia* Roxb): biology, utilisation, and conservation practices. Trees 36(3):883–898. https://doi.org/10.1007/s00468-021-02243-3
2. Chauhan L, Rao RV (2003) Wood anatomy of legumes of India: their identification, properties and uses. Bishen Singh Mahendra Pal Singh
3. Atikah T, Wardani W, Surya D (2021) Population, economic value and the trade of *Dalbergia latifolia* in Lombok, West Nusa Tenggara, Indonesia. Present IOP Conf Ser: Earth Environ Sci 762(1):012015. https://doi.org/10.1088/1755-1315/762/1/012015
4. Rao KR, Purkayastha S (1972) Indian woods: their identification, properties and uses. Volume III, Forest Research Institute, Dehradun, India
5. Deshmukh VP, Lunge MS, Rajurkar AV, Dharkar NS, Raut SR, Dhoran VS (2021) Chemical characterisation and therapeutics of *Dalbergia latifolia* Roxb.: a review. J Pharmacogn Phytochem 10(4):340–345. https://doi.org/10.22271/phyto.2021.v10.i4d.14174
6. Liu M, Peng L, Lyu S, Lyu J (2020) Properties of common tropical hardwoods for fretboard of string instruments. J Wood Sci 66:1–11. https://doi.org/10.1186/s10086-020-01862-7
7. Yulita KS, Atikah TD, Wardani W (2020) Unraveling genetic variations of *Dalbergia latifolia* (Fabaceae) from Yogyakarta and Lombok Island, Indonesia. Biodiversitas J Biol Divers 21(2):833–841. https://doi.org/10.13057/biodiv/d210251
8. Orwa C (2009) Agroforestree Database: a tree reference and selection guide, version 4.0. http://www.worldagroforestry.org/sites/treedbs/treedatabases.asp
9. Sundararaj R, Shanbhag RR, Nagaveni H, Vijayalakshmi G (2015) Natural durability of timbers under Indian environmental conditions–An overview. Int Biodeterior Biodegradation 103:196–214
10. Seshadri TR (1972) Polyphenols of Pterocarpus and Dalbergia woods. Phytochemistry 11(3):881–898. https://doi.org/10.1016/S0031-9422(00)88430-7
11. Dhingra V, Mukerjee S (1971) Chemical examination of the bark and sapwood of *Dalbergia latifolia*. Phytochemistry 10(10):2551. https://doi.org/10.1016/S0031-9422(00)89912-4
12. Jemiseye O, Idowu P, Agidigbi T (2012) In vitro assessment of selected antibiotics, crude extract of *Dalbergia latifolia* leaf and their combination on MDR Salmonella enterica strain. J Complement Altern Med Res 4(3):1–9. https://doi.org/10.9734/JOCAMR/2017/39008
13. Nagarathana PKM, Naresh K, Rajendra M, Varee VJ (2018) Nephroprotective effect of leaves of *Dalbergia latifolia* on gentamicin-induced nephrotoxicity in rats. World J Pharm Res 7(8):860–876

14. Mishra MP, Padhy RN (2013) In vitro antibacterial efficacy of 21 Indian timber-yielding plants against multidrug-resistant bacteria causing urinary tract infection. Osong Public Health Res Perspect. 4:347–357
15. Prasad D, Pranavi A, Kavimani S, Suba V (2013) Antimicrobial activity of the root extracts of *Dalbergia latifolia* (Roxb). Global Trends Pharm Sci J 4(1):1039–1043
16. Nagarathna P, Yadav CK, Yadav SK (2015) Evaluation of mutagenic effect (antimutagenic) of *Dalbergia latifolia* on Swiss albino mice. Asian J Pharm Clin Res 8(3):154–158
17. Khalid M, Akhtar J, Badruddeen M, Arif M, Singh K (2015) Pharmacognostical investigation and total phenolic content of *Dalbergia latifolia* (Roxb) bark. Internat J Pharmacognosy 2(5):248–253
18. Parrotta JA (2001) Healing plants of peninsular India. CABI, Oxfordshire, UK
19. Saha S, Shilpi JA, Mondal H, Hossain F, Anisuzzman M, Hasan MM, Cordell GA (2013) Ethnomedicinal, phytochemical, and pharmacological profile of the genus *Dalbergia latifolia* (Fabaceae). Phytopharmacology 4(2):291–346
20. Khalid M, Shoaib A, Akhtar J, Alqarni MH, Badruddeen Siddiqui HH, Foudah AI (2020) Anti obesity prospective of *Dalbergia latifolia* (Roxb) hydroalcoholic bark extract in high fat diet induced obese rats. 3 Biotech. 10:1–12. https://doi.org/10.1007/s13205-020-02491-z
21. Dhandapani T, Sasidharan K, Shanmugam G, Sadanandam V, Vasudevan R, Lalitha S, Anandalakshmi R (2020) Characterization and cytotoxic activity of *Dalbergia latifolia* Wood Extract. Am J Agric For 8(5):208–213. https://doi.org/10.11648/j.ajaf.20200805.14
22. Vadivel E, Tendulkar RD, Harmalkar VV (2015) Comparative anti-oxidant and anthelmintic activity of *Dalbergia sissoo* Roxb. ex DC and *Dalbergia latifolia* Roxb. ex DC. Int J Curr Pharm Res 7(4):70–72
23. Nair K (1986) Preservation of *Dalbergia* L. f. in Kerala by establishment of a germplasm bank. KFRI Research Report No.43. Kerala Forest Research Institute, Peechi, Kerala, India
24. Khare CP (2008) Indian medicinal plants: an illustrated dictionary. Springer Science & Business Media, New York, USA, p 200
25. Criodain TO, O'Sullivan M, Meegan MJ, Donnelly DM (1981) Latinone, a phenanthrene-1, 4-quinone from *Dalbergia latifolia*. Phytochemistry 20(5):1089–1092. https://doi.org/10.1016/0031-9422(81)83033-6
26. Liu R, Wang D, Zhang P, Shao F, Chen L, Huang H, Lin S (2018) A new diaryl 1, 2-diketone from the heartwood of *Dalbergia latifolia*. Nat Prod Res 32(1):91–96. https://doi.org/10.1080/14786419.2017.1338280
27. Liu RH, Mei DY, Lin S, Wang DQ, Shao F, Chen LY, Guo SL (2019) A new benzofuran from the heartwood of *Dalbergia latifolia*. Nat Prod Res 33(1):47–52. https://doi.org/10.1080/14786419.2018.1431639
28. Yin X, Huang A, Zhang S, Liu R, Ma F (2018) Identification of three Dalbergia species based on differences in extractive components. Molecules 23(9):2163. https://doi.org/10.3390/molecules23092163
29. Khalid M, Siddiqui H, Freed S (2011) In-vitro assessment of antioxidant activity of *Dalbergia latifolia* barks extract against free radicals. Am-Eurasian J Sci Res 6(3):172–177
30. Yadav SK, Nagarathna P, Yadav CK (2015) Research article of evaluation of immunomodulatory activity of *Dalbergia latifolia* on Swiss albino mice. J Pharm Biol Sci 10:58–64. https://doi.org/10.9790/3008-10135864
31. Melo E, Ramos R, Almeida S (2016) Phytochemical study, chemicalphysical analysis and toxicological testing of Stem Bark of *Dalbergia monetaria* L. f. Br J Pharm Res 12(4):1–7. https://doi.org/10.9734/BJPR/2016/20204
32. Chibber SS, Khera U (1979) Dalbin: a 12a-hydroxy rotenoid glycoside from *Dalbergia latifolia*. Phytochemistry 18(1):188–189. https://doi.org/10.1016/S0031-9422(00)90950-6
33. Chibber SS, Khera U (1978) Dalbinol-a new 12a-hydroxyrotenoid from *Dalbergia latifolia* seeds. Phytochemistry 17(8):1442–1443. https://doi.org/10.1016/S0031-9422(00)94613-2
34. Masendra M, Purba BAV, Irawati D, Lukmandaru G (2021) Investigation of flavonoid extractives and their contribution to color of *Dalbergia latifolia* Roxb. Wood Research J 12(2):53–60. https://doi.org/10.51850/wrj.2021.12.2.53-60

35. Weston LA, Mathesius U (2013) Flavonoids: their structure, biosynthesis and role in the rhizosphere, including allelopathy. J Chem Ecol 39:283–297. https://doi.org/10.1007/s10886-013-0248-5
36. Kumar R, Tsvetkov DE, Varshney V, Nifantiev NE (2020) Chemical constituents from temperate and subtropical trees with reference to knotwood. Ind Crops Prod 145:112077. https://doi.org/10.1016/j.indcrop.2019.112077
37. Yun H-M, Park JE, Lee JY, Park K-R (2022) Latifolin, a natural flavonoid, isolated from the heartwood of *Dalbergia odorifera* induces bioactivities through apoptosis, autophagy, and necroptosis in human oral squamous cell carcinoma. Int J Mol Sci 23(21):13629. https://doi.org/10.3390/ijms232113629
38. Khanam H (2015) Bioactive benzofuran derivatives: a review. Eur J Med Chem 97:483–504. https://doi.org/10.1016/j.ejmech.2014.11.039
39. Fernandes P, Cabral JMS (2007) Phytosterols: applications and recovery methods. Biores Technol 98(12):2335–2350. https://doi.org/10.1016/j.biortech.2006.10.006
40. Dufourc EJ (2008) Sterols and membrane dynamics. J Chemical Biology. 1(1):63–77. 10.10072Fs12154-008-0010-6
41. Siddique HR, Saleem M (2011) Beneficial health effects of lupeol triterpene: a review of preclinical studies. Life Sci 88(7–8):285–293. https://doi.org/10.1016/j.lfs.2010.11.020
42. Xu R, Fazio GC, Matsuda SP (2004) On the origins of triterpenoid skeletal diversity. Phytochemistry 65(3):261–291. https://doi.org/10.1016/j.phytochem.2003.11.014
43. Ayad R, Akkal S (2019) Phytochemistry and biological activities of algerian Centaurea and related genera. Stud Nat Prod Chem 63:357–414. https://doi.org/10.1016/B978-0-12-817901-7.00012-5
44. Liu R, Wang D, Lin S, Zhang P, Shao F, Chen L (2017) Triterpenoids from heartwood of *Dalbergia latifolia.* Chin J Exp Tradit Med Formulae 23(20):42–46. https://doi.org/10.13422/j.cnki.syfjx.2017200042
45. Teponno RB, Kusari S, Spiteller M (2016) Recent advances in research on lignans and neolignans. Nat Prod Rep 33(9):1044–1092. https://doi.org/10.1039/C6NP00021E
46. Bernini R, Gualandi G, Crestini C, Barontini M, Belfiore MC, Willför S, Eklund P, Saladino R (2009) A novel and efficient synthesis of highly oxidised lignans by a methyltrioxorhenium/hydrogen peroxide catalytic system. Studies on their apoptogenic and antioxidant activity. Bioorganic and Med Chem 17(15):5676–5682. https://doi.org/10.1016/j.bmc.2009.06.010
47. Kim DC, Lee DS, Ko W, Kim KW, Kim HJ, Yoon CS, Oh H, Kim YC (2018) Heme oxygenase-1-inducing activity of 4-Methoxydalbergione and 4'-Hydroxy-4-methoxydalbergione from *Dalbergia odorifera* and their anti-inflammatory and cytoprotective effects in murine hippocampal and BV2 microglial cell line and primary rat microglial cells. Neurotox Res 33:337–352. https://doi.org/10.1007/s12640-017-9796-8
48. Safaeian S, Sekine N, Ashitani T, Murayama T, Ogiyama K, Takahashi K (2009) The bioactive extracts of heartwood of *Dalbergia latifolia* Mokuzai Gakkaishi. 55:29–36. https://doi.org/10.2488/jwrs.55.29
49. Sekine N, Ashitani T, Murayama T, Shibutani S, Hattori S, Takahashi K (2009) Bioactivity of latifolin and its derivatives against termites and fungi. J Agric Food Chem 57(13):5707–5712. https://doi.org/10.1021/jf900719p
50. Dong L, Lee H, Liu Z, Lee DS (2023) Anti-skin inflammatory and anti-oxidative effects of the neoflavonoid latifolin isolated from *Dalbergia odorifera* in HaCaT and BJ-5ta cells. Int J Mol Sci 24(8):7371. https://doi.org/10.3390/ijms24087371
51. Kumar P, Ahamad T, Mishra DP, Khan, MF (2020) Plant neoflavonoids: chemical structures and biological functions. In: Plant-derived bioactives: chemistry and mode of action. 35–57 https://doi.org/10.1007/978-981-15-2361-8_3
52. Freitas TS, de Xavier J, da C, Pereira RL, Rocha JE, Muniz DF, da Silva PT, da Hora JP, Dos Santos HS, Bandeira PN, Nogueira CE (2020) Direct antibacterial and antibiotic resistance modulatory activity of chalcones synthesised from the natural product 2-hydroxy-3, 4, 6-trimethoxyacetophenone. FEMS Microbiology Letters 367(15):fnaa124. https://doi.org/10.1093/femsle/fnaa124

53. Santiago R, Freire P, Ayala A, Teixeira A, Santos H, Bandeira P, Gonçalves F, Oliveira M, Cruz B, Sena Jr D (2018) Crystal structure, vibrational spectra and quantum chemical parameters of 2-hydroxy-3, 4, 6-trimethoxyacetophenone isolated from the *Croton anisodontus* Müll. Arg (Euphorbiaceae). J Mol Struct 1171:815–826. https://doi.org/10.1016/j.molstruc.2018.06.064
54. de Aguiar FLL, de Morais SM, Albuquerque MRJR, Bandeira PN, de Brito EHS, Marcos F, dos Santos Fontenelle RO (2016) Antifungal activity and synergistic effect of acetophenones isolated from species Croton against dermatophytes and yeasts. J Med Plants Res 10(17):216–222. https://doi.org/10.5897/JMPR2016.6048
55. Da Silva P, da Cunha Xavier J, Freitas T, Oliveira M, Coutinho H, Leal A, Barreto H, Bandeira P, Nogueira C, Sena D Jr (2021) Synthesis, spectroscopic characterisation and antibacterial evaluation by chalcones derived of acetophenone isolated from *Croton anisodontus* Müll. Arg. J Molecular Structure 1226:129403. https://doi.org/10.1016/j.molstruc.2020.129403
56. Castellano JM, Ramos-Romero S, Perona JS (2022) Oleanolic acid: extraction, characterisation and biological activity. Nutrients 14(3):623. https://doi.org/10.3390/nu14030623
57. Khwaza V, Oyedeji OO, Aderibigbe BA (2018) Antiviral activities of oleanolic acid and its analogues. Molecules 23(9):2300. https://doi.org/10.3390/molecules23092300
58. Hichri F, Jannet HB, Cheriaa J, Jegham S, Mighri Z (2003) Antibacterial activities of a few prepared derivatives of oleanolic acid and of other natural triterpenic compounds. C R Chim 6(4):473–483. https://doi.org/10.1016/S1631-0748(03)00066-3
59. Fabiyi OA, Atolani O, Adeyemi OS, Olatunji GA (2012) Antioxidant and cytotoxicity of β-amyrin acetate fraction from *Bridelia ferruginea* leaves. Asian Pac J Trop Biomed 2(2): S981–S984. https://doi.org/10.1016/S2221-1691(12)60347-5
60. Fernandes CP, Corrêa AL, Lobo JF, Caramel OP, De Almeida FB, Castro ES, Souza KF, Burth P, Amorim LM, Santos MG (2013) Triterpene esters and biological activities from edible fruits of *Manilkara subsericea* (Mart.) Dubard, Sapotaceae. BioMed Research Internat 2013(1):280810. https://doi.org/10.1155/2013/280810
61. Shanti V, Neerakkal I (2021) GC-MS analysis and in silico activity prediction of phytocompounds in the roots of *Chrysopogon zizanioides* (L.) Roberty. Plant Sci Today 8(1):218–224. https://doi.org/10.14719/pst.2021.8.1.1008
62. Tugume P, Kakudidi EK, Buyinza M, Namaalwa J, Kamatenesi M, Mucunguzi P, Kalema J (2016) Ethnobotanical survey of medicinal plant species used by communities around Mabira Central Forest Reserve. Uganda. J Ethnobiol Ethnomedicine. 12:1–28. https://doi.org/10.1186/s13002-015-0077-4
63. Saeidnia S, Manayi A, Gohari AR, Abdollahi M (2014) The story of beta-sitosterol-a review. European J Med Plants 4(5):590–609
64. Wilt M (1999) Ishani: β-sitosterol for the treatment of benign prostatic hyperplasia. BJU Int 83(9):976–983. https://doi.org/10.1046/j.1464-410x.1999.00026.x
65. Puertas-Bartolomé M, Vázquez-Lasa B, San Román J (2018) Bioactive and bioadhesive catechol conjugated polymers for tissue regeneration. Polymers 10(7):768. https://doi.org/10.3390/polym10070768
66. Puertas-Bartolomé M, Benito-Garzón L, Fung S, Kohn J, Vázquez-Lasa B, San Román J (2019) Bioadhesive functional hydrogels: Controlled release of catechol species with antioxidant and antiinflammatory behavior. Mater Sci Eng, C 105:110040. https://doi.org/10.1016/j.msec.2019.110040
67. Devi P, Singh S (2017) Phytochemical and pharmacological profiling of *Dalbergia sissoo* Roxb. Stem. J Pharmacogn Phytochem. 6(6):2483–2486
68. Li S-R, Chen L-Y, Tsai J-C, Tzeng J-Y, Tsai I-L, Wang E-C (2007) New syntheses of dalbergichromene and dalbergin from vanillin via neoflavene intermediate. Tetrahedron Lett 48(12):2139–2141. https://doi.org/10.1016/j.tetlet.2007.01.117
69. Khan E, Tandon P, Maurya R, Kumar P (2019) A theoretical study on molecular structure, chemical reactivity and molecular docking studies on dalbergin and methyldalbergin. J Mol Struct 1183:100–106. https://doi.org/10.1016/j.molstruc.2019.01.065

70. Xiao-qiang W, Chu-sheng H, Hong-xing L, Yong-qiang H (2009) Distribution, synthesis and biological activity of Dalbergin. Nat Prod Res & Dev 21(5):900
71. Shirota O, Pathak V, Sekita S, Satake M, Nagashima Y, Hirayama Y, Hakamata Y, Hayashi T (2003) Phenolic Constituents from *Dalbergia cochinchinensis*. J Nat Prod 66(8):1128–1131. https://doi.org/10.1021/np0300683
72. Ren JH, Liu Y, Shen CX, Zheng QW, Meng XW, Li JR, Chen LY, Liu RH (2023) Three novel dalbergiphenol hybrids from the heartwood of *Dalbergia cochinchinensis*. Fitoterapia 170:105663. https://doi.org/10.1016/j.fitote.2023.105663
73. Rahman M, Almalki WH, Alrobaian M, Iqbal J, Alghamdi S, Alharbi KS, Alruwaili NK, Hafeez A, Shaharyar A, Singh T (2021) Nanocarriers-loaded with natural actives as newer therapeutic interventions for treatment of hepatocellular carcinoma. Expert Opin Drug Deliv 18(4):489–513. https://doi.org/10.1080/17425247.2021.1854223
74. Zhu X, Wu X, Cheng J, Liao H, Di X, Li L, Li R, Zhou Y, Zhang X (2017) Dalbinol, a rotenoid from *Amorpha fruticosa* L., exerts anti-proliferative activity by facilitating β-catenin degradation in hepatocellular carcinoma cells. Oncotarget 8(29):47755. 10.186322Foncotarget.17766
75. Rasheed HM, Bashir K, Gurgul A, Wahid F, Che CT, Shahzadi I, Khan T (2022) UHPLC-MS/MS-GNPS based phytochemical investigation of *Dryopteris ramosa* (Hope) C. Chr. and evaluation of cytotoxicity against liver and prostate cancer cell lines. Heliyon 8(11):e11286. https://doi.org/10.1016/j.heliyon.2022.e11286
76. Shao F, Panahipour L, Omerbasic A, Tang F, Gruber R (2022) Dalbergiones lower the inflammatory response in oral cells in vitro. Clin Oral Invest 26(8):5419–5428. https://doi.org/10.1007/s00784-022-04509-7
77. Rahman MM, Khan MA (2013) Anti-cancer potential of South Asian plants. Nat Prod Bioprospecting 3:74–88. https://doi.org/10.1007/s13659-013-0027-6
78. Al-Qahtani WH, Dinakarkumar Y, Arokiyaraj S, Saravanakumar V, Rajabathar JR, Arjun K, Gayathri P, Appaturi JN (2022) Phyto-chemical and biological activity of *Myristica fragrans*, an ayurvedic medicinal plant in Southern India and its ingredient analysis. Saudi J Biol Sci 29(5):3815–3821. https://doi.org/10.1016/j.sjbs.2022.02.043
79. Lv SM, Zhang MQ, Chen JS, Liu JS, Wang MX, Sun JY, Liu C, Chen YY, Guo X, Liao ZX (2022) Study on the anti-hyperuricemic bioactivity and chemical components of *Sterculiae lychnophorae* Semen. J Funct Foods 95:105173. https://doi.org/10.1016/j.jff.2022.105173
80. Vad NM, Kandala PK, Srivastava SK, Moridani MY (2010) Structure–toxicity relationship of phenolic analogs as anti-melanoma agents: An enzyme directed prodrug approach. Chem Biol Interact 183(3):462–471. https://doi.org/10.1016/j.cbi.2009.11.020
81. Jiang D, Rasul A, Batool R, Sarfraz I, Hussain G, Mateen Tahir M, Qin T, Selamoglu Z, Ali M, Li J (2019) Potential anticancer properties and mechanisms of action of formononetin. Biomed Res Int 2019(1):5854315. https://doi.org/10.1155/2019/5854315
82. Tay KC, Tan LTH, Chan CK, Hong SL, Chan KG, Yap WH, Pusparajah P, Lee LH, Goh BH (2019) Formononetin: a review of its anticancer potentials and mechanisms. Front Pharmacol 10:820. https://doi.org/10.3389/fphar.2019.00820
83. Takahashi C, Kikuchi N, Katou N, Miki T, Yanagida F, Umeda M (1995) Possible anti-tumor-promoting activity of components in Japanese soybean fermented food, Natto: effect on gap junctional intercellular communication. Carcinogenesis 16(3):471–476. https://doi.org/10.1093/carcin/16.3.471
84. Kim S, Chung W, Kim S, Ko S, Um J (2011) Antiinflammatory effect of Oldenlandia diffusa and its constituent, hentriacontane, through suppression of caspase-1 activation in mouse peritoneal macrophages. Phytother Res 25(10):1537–1546. https://doi.org/10.1002/ptr.3443
85. Heras-Mozos R, López-Carballo G, Hernández R, Gavara R, Muñoz PH (2023) PH modulates antibacterial activity of hydroxybenzaldehyde derivatives immobilised in chitosan films via reversible Schiff bases and its application to preserve freshly-squeezed juice. Food Chem 403:134292. https://doi.org/10.1016/j.foodchem.2022.134292
86. Chen Z, Ding W, Yang X, Lu T, Liu Y (2023) Isoliquiritigenin, a potential therapeutic agent for treatment of inflammation-associated diseases. J Ethnopharmacol 318(B):117059. https://doi.org/10.1016/j.jep.2023.117059

87. Peng F, Du Q, Peng C, Wang N, Tang H, Xie X, Shen J, Chen J (2015) A review: the pharmacology of isoliquiritigenin. Phytother Res 29(7):969–977. https://doi.org/10.1002/ptr.5348
88. Shi D, Yang J, Jiang Y, Wen L, Wang Z, Yang B (2020) The antioxidant activity and neuroprotective mechanism of isoliquiritigenin. Free Radical Biol Med 152:207–215. https://doi.org/10.1016/j.freeradbiomed.2020.03.016
89. Munshi M, Zilani MNH, Islam MA, Biswas P, Das A, Afroz F, Hasan MN (2022) Novel compounds from endophytic fungi of *Ceriops decandra* inhibit breast cancer cell growth through estrogen receptor alpha in in-silico study. Inform Med Unlocked 32:101046. https://doi.org/10.1016/j.imu.2022.101046
90. Yin Z, Park R, Choi B-M (2020) Isoparvifuran isolated from *Dalbergia odorifera* attenuates H_2O_2-induced senescence of BJ cells through SIRT1 activation and AKT/mTOR pathway inhibition. Biochem Biophys Res Commun 533(4):925–931. https://doi.org/10.1016/j.bbrc.2020.09.096
91. Kim Y, Zhao R, Park S, Lee J, Cho I, Yang C, Kim S, Kim S (2008) Anti-inflammatory effects of liquiritigenin as a consequence of the inhibition of NF-κB-dependent iNOS and proinflammatory cytokines production. Br J Pharmacol 154(1):165–173. https://doi.org/10.1038/bjp.2008.79
92. Gaur R, Thakur JP, Yadav DK, Kapkoti DS, Verma RK, Gupta N, Khan F, Saikia D, Bhakuni RS (2015) Synthesis, antitubercular activity, and molecular modeling studies of analogues of isoliquiritigenin and liquiritigenin, bioactive components from *Glycyrrhiza glabra*. Med Chem Res 24:3494–3503. https://doi.org/10.1007/s00044-015-1401-1
93. Ramalingam M, Kim H, Lee Y, Lee Y-I (2018) Phytochemical and pharmacological role of liquiritigenin and isoliquiritigenin from radix glycyrrhizae in human health and disease models. Frontiers in Aging Neuroscience 10:348. https://doi.org/10.3389/fnagi.2018.00348
94. Gallo MB, Sarachine MJ (2009) Biological activities of lupeol. Int J Biomed Pharm Sci 3(1):46–66
95. Rao SD, Rao BN, Devi PU, Rao AK (2017) Isolation of lupeol, design and synthesis of lupeol derivatives and their biological activity. Orient J Chem 33(1):173–80. https://doi.org/10.13005/ojc/330119
96. Bednarczyk-Cwynar B, Więcaszek T, Ruszkowski P (2016) Cytotoxic activity of some lupeol derivatives. Nat Prod Commun 11(9):1237–1238. https://doi.org/10.1177/1934578X1601100910
97. Bertolini F, Pineschi M (2009) Recent progress in the synthesis of 2, 3-dihydrobenzofurans. Org Prep Proced Int 41(5):385–418. https://doi.org/10.1080/00304940903240836
98. Chen C, Weisel M (2013) Concise asymmetric synthesis of (+)-conocarpan and obtusafuran. Synlett 24(02):189–192. https://doi.org/10.1055/s-0032-1317704
99. Dev AA, Joseph SM (2021) Anticancer potential of *Annona* genus: A detailed review. J Indian Chem Soc 98(12):100231. https://doi.org/10.1016/j.jics.2021.100231
100. Genevois C, Pfohl-Leszkowicz A, Boillot K, Brandt H, Castegnaro M (1998) Implication of cytochrome P-450 1A isoforms and the AH receptor in the genotoxicity of coal-tar fume condensate and bitumen fume condensates. Environ Toxicol Pharmacol 5(4):283–294. https://doi.org/10.1016/S1382-6689(98)00013-1
101. Yesumarian Y, Krishnamoorthy MK, Ramasamy G, Ramachandran M (2007) Photodynamic action and antimicrobial activity of some excited metabolites of *Dalbergia sissoides* and their ability to cleave DNA. Nat Prod Commun 2(2):159–168. https://doi.org/10.1177/1934578X0700200211
102. da Cunha MG, Sardi J, de CO, Freires IA, Franchin M, Rosalen PL (2020) Antimicrobial, anti-adherence and antibiofilm activity against *Staphylococcus aureus* of a 4-phenyl coumarin derivative isolated from *Brazilian geopropolis*. Microb Pathog 139:103855. https://doi.org/10.1016/j.micpath.2019.103855
103. Shaheed K, AlGaraawi N, Alsultany A, Abbas Z, Khshayyish I, Al Khazali M (2019) Analysis of bioactive phytochemical compound of (*Cyperus iria* L.) by using gas chromatography–mass spectrometry. In: IOP conference series: earth and environmental science 38(1):012064. https://doi.org/10.1088/1755-1315/388/1/012064

104. Kaur N, Chaudhary J, Jain A, Kishore L (2011) Stigmasterol: a comprehensive review. Int J Pharm Sci Res 2(9):2259
105. Bakrim S, Benkhaira N, Bourais I, Benali T, Lee L-H, El Omari N, Sheikh RA, Goh KW, Ming LC, Bouyahya A (2022) Health benefits and pharmacological properties of stigmasterol. Antioxidants 11(10):1912. https://doi.org/10.3390/antiox11101912
106. Mallick SS, Dighe VV (2014) Detection and estimation of alpha-Amyrin, beta-Sitosterol, Lupeol, and n-Triacontane in two medicinal plants by high performance thin layer chromatography. Advances in Chemistry 2014(1):143948. https://doi.org/10.1155/2014/143948
107. Amudha P, Jayalakshmi M, Pushpabharathi N, Vanitha V (2018) Identification of bioactive components in *Enhalus acoroides* seagrass extract by gas chromatography-mass spectrometry. Asian J Pharm Clin Res 11(10):313–315. https://doi.org/10.22159/ajpcr.2018.v11i10.25577
108. Karabay-Yavasoglu NU, Sukatar A, Ozdemir G, Horzum Z (2007) Antimicrobial activity of volatile components and various extracts of the red alga *Jania rubens*. Phytother Res: Int J Devoted Pharmacol Toxicol Eval Nat Prod Deriv 21(2):153–156. https://doi.org/10.1002/ptr.2045
109. Abdulah R, Milanda T, Sugijanto M, Barliana MI, Diantini A, Supratman U, Subarnas A (2017) Antibacterial properties of selected plants consumed by primates against *Escherichia coli* and *Bacillus subtilis*. Southeast Asian J Trop Med Public Health 48(1):109–116
110. Syafii W (2000) Antitermitic compounds from the heartwood of Sonokeling wood (*Dalbergia latifolia* Roxb). Indonesian J Trop Agric (IJTA) 9(03):55–58
111. Huang JG, Zhou LJ, Xu HH, Li WO (2009) Insecticidal and cytotoxic activities of extracts of *Cacalia tangutica* and its two active ingredients against *Musca domestica* and *Aedes albopictus*. J Econ Entomol 102(04):1444–1447. https://doi.org/10.1603/029.102.0407
112. Njoya EM (2021) Medicinal plants, antioxidant potential, and cancer. In: Cancer, Academic Press, CA, USA 349–357. https://doi.org/10.1016/B978-0-12-819547-5.00031-6
113. Masendra M, Irawati D, Ridlo AS, Lukmandaru G (2020) Phenol contents and antioxidant activity of Sonokeling (*Dalbergia latifolia* Roxb.) wood. Wood Research J 11(1):27–34. https://doi.org/10.51850/wrj.2020.11.1.27-34
114. Niraimathi V, Sundaraganapathy R (2014) In vitro anti cancer activity of various fractions of hydro alcoholic extract of *D. latifolia* Roxb. Int J Phytopharm 5(4):272–83
115. Choi EM, Hwang JK (2005) Screening of Indonesian medicinal plants for inhibitor activity on nitric oxide production of RAW264. 7 cells and antioxidant activity. Fitoterapia 76(2):194–203. https://doi.org/10.1016/j.fitote.2004.11.010
116. Yazaki Y (2015) Wood colors and their coloring matters: a review. Nat Prod Commun 10(3):505–512. https://doi.org/10.1177/1934578X1501000332
117. Panche AN, Diwan AD, Chandra SR (2016) Flavonoids: an overview. J Nutr Sci 5:e47. https://doi.org/10.1017/jns.2016.41
118. Kai Y, Swan E (1990) Chemical constituents contributing to the color of western red cedar heartwood. Mokuzai Gakkaishi 36(2):218–224

Chapter 10
Diseases and Insect Pests of *Dalbergia latifolia*

K. Charishma, Deepak Kumar Mahanta, R. Athulya, G. S. Uma, and A. Muthu Kumar

1 Introduction

The Leguminosae (Fabaceae) family, known for its flowering plants, ranks as the third largest, encompassing approximately 19,300 species found across a wide range of the world's vegetation types. Several bean genera, including *Bauhinia* L., *Dalbergia* L.f., *Desmodium Desv.*, *Mucuna Adans.*, and *Vigna Savi.*, have been prioritised for biodiversity evaluations as part of the Global Legume Diversity Assessment (GLDA) [47]. The relationship between specific plant families, such as the Fabaceae, and nitrogen-fixing bacteria, enriches soil nutrients, while many species also offer food and habitat for various animals. Beyond their biological importance in their natural habitats, these taxa deliver invaluable ecosystem services by supplying high-quality wood, particularly lumber, from species belonging to the genus *Dalbergia.*

The premium timber species *D. latifolia*, commonly known as Indian Blackwood or Rosewood, is native to the low-elevation tropical monsoon forests of South East India [45]. The term "rosewood" is associated with several highly traded varieties of valuable wood, encompassing various *Dalbergia* species and other tree families that produce vividly coloured hardwoods, making them ideal for furniture and other luxury items [21, 30].

Seedlings are cultivated in nurseries for extensive plantations. The quality of these seedlings is crucial for successful plantation growth and meeting the needs of farmers. However, seedlings in nursery beds and polyethylene bags are vulnerable to various biotic stresses (Fig. 1).

The occurrence of diseases during the nursery phase can lead to considerable mortality, affecting both seedling growth and the regeneration of internal quality

K. Charishma · D. K. Mahanta
Forest Protection Division, ICFRE-FRI, Dehradun, India

R. Athulya · G. S. Uma · A. Muthu Kumar (✉)
Forest Protection Division, ICFRE-IWST, Bengaluru, India
e-mail: iwst.muthukumar@gmail.com

R. R. Warrier et al. (eds.), *Handbook of Asian Rosewoods*, Materials Horizons: From Nature to Nanomaterials, https://doi.org/10.1007/978-981-96-9459-4_10

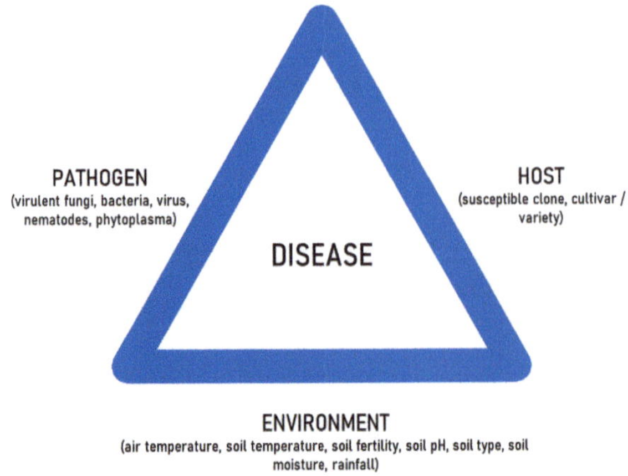

Fig. 1 Disease triangle—factors involved in causing disease

[45]. There is a scarcity of information regarding diseases affecting *D. latifolia*. The research effort uncovered the presence of diseases affecting nursery seedlings, root suckers, and trees in both plantation settings and natural habitats, linked to 13 distinct fungal species [26]. Additionally, over forty insect pests have been identified as threats to *D. latifolia*, with frequent outbreaks of these pests attributed to detrimental impacts on the species' growth and productivity [32]. This chapter compiles all the existing knowledge on the primary diseases and insect pests affecting *D. latifolia*.

2 Diseases of *D. latifolia*

2.1 Seedling Diseases of **D. latifolia**

a. **Seedling Dieback**

Causal organism: *Colletotrichum capsici*.

Symptoms: Plants that are affected start to die from the top downward. They may become completely or partially dehydrated/dried. Infections can cause significant damage to seedlings in nurseries by generating a large amount of viable conidia.

Management: Ridomil, a 0.2% fungicide application at fortnightly intervals as a soil drench, can be used to control the disease. In nursery conditions, excess moisture should be avoided in the root zone of seedlings [48].

b. **Tar Spot**

Causal organism: *Phyllachora dalbergiae*

Symptoms: The affected leaves show red, dark coal tar-like spots surrounded by a faded yellow halo, initially appearing on the upper surface. Over time, shiny, black, cushion-like ascomata develop individually or in clusters, eventually forming large patches encompassing the entire leaf. Spots rarely appear on the underside of the leaf. The damaged areas turn pale and eventually light brown. Mature leaves exhibit significant severity and may lead to premature defoliation [45].

Management: Carbendazim (0.2%) and Propiconazole (0.1%) is found effective [45].

c. **Sooty Mould**

Causal organism: *Meliola* sp. [27, 43]

Symptoms: The disease is identified by a dark grey or black coating that envelops leaves, twigs, and stems. It starts as small spots, but this discolouration can lead to large sections of leaves turning completely black over time.

d. **Leaf Spot**

Causal organism: *Cercospora dalbergiae-latifoliae, Colletotrichum gloeosporioides* [41], *Phomopsis* sp. [24, 41].

Symptoms: Small, discoloured spots may appear on leaves, ranging from grey to dark brown. These spots can enlarge over time, potentially covering a significant portion of the leaf surface.

Cercospora dalbergiae-latifoliae causes yellowish to greyish-green discolouration on the lower surface of the leaves it affects. The pustules are primarily intra-epidermal, and the stomata are brown with simple or forked conidiophores.

e. **Rust**

Causal organism: *Uredo sissoo, Maravalia achroa* and *Maravalia pterocarpi* [27, 43]

Symptoms: *Maravalia achroa* (Syd.) causes leaf rust on seedlings in nurseries. Rust is characterised by pustules on leaves.

Uredo sissoo also causes twig rust. The symptom starts as a powdery, dense mass on the lower side of the leaves that later expands, covering the entire leaf surface. The underside of the leaves develops yellowish uredinial sori. Affected plants appear weak and stunted, often displaying malformed leaves. There is an apparent reduction in growth.

Management: The disease can be adequately managed using 0.08 per cent Bayleton topically every two weeks.

f. **Rot of Seedling**

Causal organism: *Phytopthora* sp. [19]

Symptoms: The pathogen infection can cause root and collar rot, branch dieback, and defoliation, leading to seedling mortality in case of severe diseases.

2.2 Tree Diseases

Many seedling pathogens are capable of causing diseases on root suckers and trees in natural stand. Foliage diseases caused by *Phyllachora, Cercospora,* and *Uredo sissoo* are also reported on trees [26].

a. **Wilt**

Causal organism: *Fusarium solani*

Symptoms: Trees of age 15 and above are severely affected by wilt disease. Around 300 ha area was severely affected by the wilt disease in East Java, Indonesia. This disease is characterised by rolling of the young leaves first, then yellowing of the older leaves, and, lastly, red streaks on the sapwood's outer layers. The symptoms eventually lead to the death of trees [3, 42].

Management: Fungicides Topsin, Bavistin, and Dithane M45 are recommended for *F. solani* management after the preliminary in vitro studies.

b. **Root Rot**

Root rot of *D. latifolia* is considered as a minor disease in India.

Causal organism: *Phellinus gilvus, Ganoderma lucidum* and *Coriolopsis sanguinaria* [27, 43]

Symptoms: They target the bases of trees and their roots, causing zone lines and white, spongy rot in sapwood and, to a lesser degree, heartwood. The characteristic fruit bodies, such as sporophores, can be visualised near the base of the stem or collar region. The impacted trees survive for a few years until a windstorm blows them down, at which point their roots significantly deteriorate [3, 29].

3 Insect Pests of *D. latifolia*

Major insects causing significant damage and considerable economic loss to the nursery and plantations belong to orders Coleoptera, Hymenoptera, Lepidoptera, and Isoptera [39]. The most significant are cutworms, termites, and cockchafers; other main pests include certain defoliators, sapsuckers, and wood borers. Many insect pests were recorded on *D. latifolia* (Table 1), falling into three groups: sap suckers, defoliators, and wood borers. They are enumerated below, along with their host range and damage symptoms.

3.1 Major Defoliators

1. ***Chrysochus cobaltinus***

Table 1 Important insect pests of *Dalbergia latifolia*

Sl. No	Species	Order: Family	Nature of damage	References
1	*Kerria lacca*	*Kerriidae*: Hemiptera	Sap sucker	[6, 8, 15, 22]
2	*Lecanium hesperidum*	*Coccidae*: Hemiptera	Sap sucker	[6, 8, 17, 22]
4	*Oxyrachis mangiferana*	*Membracidae*: Hemiptera	Sap sucker	[6, 8, 17, 22]
5	*Oxyrachis trandus*	*Membracidae*: Hemiptera	Sap sucker	[6, 8, 17, 22]
6	*Drosicha stebbingi*	*Margarodidae*: Hemiptera	Sap sucker	[6, 8, 17, 22]
7	*Leptocorisa acuta*	*Alydidae*: Hemiptera	Sap sucker	[6, 8, 17, 22]
8	*Leptocentrus taurus*	*Membracidae*: Hemiptera	Sap sucker	[6, 8, 17, 22]
9	*Ricania speculum*	*Ricaniidae*: Hemiptera	Sap sucker	[6, 8, 17, 22]
10	*Anomala dalbergiae*	*Scarabaeidae*: Coleoptera	Defoliator	[2, 6, 8, 22]
11	*Argyroploce aprobola*	*Tortricidae*: Lepidoptera	Defoliator	[6, 8, 17, 22]
12	*Anisodes obrinaria*	*Geometridae*: Lepidoptera	Defoliator	[6, 8, 17, 22]
13	*Ectropis bhurmitra*	*Geometridae*: Lepidoptera	Defoliator	[6, 8, 17, 22]
14	*Tapena thwaitesi*	*Hesperiidae*: Lepidoptera	Defoliator	[6, 8, 17, 22]
15	*Cosmotriche laeta*	*Lasiocampidae*: Lepidoptera	Defoliator	[6, 8, 17, 22]
16	*Gastrophysa viridula*	*Chrysomelidae*: Coleoptera	Defoliator	[6, 8, 17, 22]
17	*Chrysochus cobaltinus*	*Chrysomelidae*: Coleoptera	Defoliator	[6, 8, 17, 22]
18	*Apoderus crenatus*	*Attelabidae*: Coleoptera	Defoliator	[6, 8, 17, 22]
19	*Plecoptera reflexa*	*Noctuidae*: Lepidoptera	Defoliator	[6, 8, 22, 35]
20	*Hyposidra talaca*	*Geometridae*: Lepidoptera	Defoliator	[6, 8, 17, 22]
21	*Euplatypus parallelus*	*Curculionidae:*Coleoptera	Wood borer	[20]

Host plants: *Dalbergia latifolia, Apocynum cannabinum* (Apocynaceae) and *Asclepias* spp. (Asclepiadaceae) [9, 38]; *Asclepias speciosa* and *A. eriocarpa* [9, 10, 36].

Identification: The adult beetle has an oval form and is around 7.35–7.75 mm long. It has a dark metallic iridescent cobalt-blue exoskeleton and brown legs [17].

Damage symptoms: The mature beetle engages in communal feeding behaviour, consuming leaves and tender shoots while removing leaf margins [17].

2. ***Gastrophysa viridula***

Host plants: *D. latifolia* [17], *Begonia* (*Begoniaceae*) [37], dock weed [5, 13, 45], *Rheum palmatum* [4].

Identification: The adult beetle has an oval form and is around 5.02 mm long. It is dark metallic green with reddish brown legs [17].

Damage symptoms: Beetles consume young leaves by severing the leaf from its edge, creating a circular design in the middle of the leaf [17].

3. ***Apoderus crenatus***

Host plants: *Dalbergia sissoo, D. latifolia* [17].

Identification: The weevil seems small, measuring around 4.75 mm in length. It showcases an orange-red hue and has a row of elytra. The weevil's rostrum is smooth and conical, while its prothorax is subconical. Additionally, it features a semicircular scutellum and densely hairy meso and metasternal epimera [17].

Damage symptoms: The adult stage inflicts harm on new growth by either rolling them off or cutting through the midrib, leading to defoliation [16].

4. ***Plecoptera reflexa***

Host plants: *Dalbergia sissoo, D. latifolia* [17].

Identification: The weevil appears small, measuring around 4.75 mm in length. It displays an orange-red colour and has a row of elytra. The weevil's rostrum is smooth and conical, while its prothorax is subconical. Additionally, it features a semicircular scutellum and densely hairy meso and metasternal epimera [17].

Damage symptoms: The adult stage inflicts harm on new plant growth by either rolling them off or by cutting through the midrib, resulting in defoliation [6, 16, 31].

5. ***Hyposidra talaca***

***Host* plants**: *Bombax ceiba, Cassia* sp., *Cedrela toona, Ficus glomerata, Shorea robusta, Syzygium cumini, Tectona grandis* [40].

Identification: The larvae are initially black, turning creamy brown as they mature.

Damage symptoms: The semi-looper larvae injure delicate foliage by feeding on the leaf's edge. Caterpillars harm young leaves in their fourth and fifth instars [17].

6. ***Ectropis bhurmitra***

Host plants: *D. latifolia, Aleurites montana, Artemisia vulgaris, Bombax malabaricum, Lantana aculeata, Phoebe lanceolata, Shorea robusta,* and *Tectona grandis* [1, 6].

Identification: The larva is light grey-brown, with a body that has a swollen, dark diagonal ridge that terminates in a dorsal tubercle. The head is light grey with a dark brown herringbone pattern on the lobes. A horizontal streak is also seen at the top of the front diamond on the mid-abdominal segments [17].

Damage symptoms: The larvae of this species cause harm to the delicate leaves of the plant in both nurseries and plantations.

3.2 Major Sap Sucking Insects

1. ***Ricania speculum***

Host plants: *D. latifolia* and other forest trees.

Identification: The mature individual has deep brown wings adorned with distinct undulating horizontal bands and asymmetrical translucent areas of varying dimensions. The forewings' precostal region has a high density of transverse veinlets, whereas the costal border displayed a pronounced convex shape towards the base.

Damage symptoms: The pest is very polyphagous, with females depositing eggs into the woody twigs of many host plants. The nymphs then extract sap from the plant and undergo development [23, 33, 34].

2. ***Leptocorisa acuta*** (**Alydidae: Hemiptera**)

Host plants: *D. latifolia* and other forest trees [14].

Identification: The insect appears greenish-yellow to yellow–brown. The head is longer, but the rostrum is comparatively shorter. The scutellum, a triangular plate, is located on the thorax, behind the pronotum. The fourth part of the antenna has a curved shape.

Damage symptoms: Both nymphs and adults feed foliage and young stems of the plant. Nymphs exhibit gregarious feeding behaviour on the delicate shoots, leading to the withering and desiccation of such shoots.

3. ***Leptocentrus taurus***

Host plants: *D. latifolia, D. sissoo, Zizyphus jujuba, Solanum melongena, Parthenium hysterophorus* [7, 16, 24, 46].

Identification: The insect's body is triangular, 4–7 mm long. It has yellow eyes and is winged, with three pairs of legs. Additionally, there are two curving horn-like extensions on the thorax.

Damage symptoms: The female of this insect deposits eggs in the young shoots, and both the nymphs and adults feed collectively on the tender shoots.

4. ***Drosicha stebbingi***

Host plants: *Shorea robusta, Mangifera indica, Carica papaya, Zizyphus jujuba, Prunus persica, P. domestica, Artocarpus heterophyllus, Bauhinia variegata, Helianthus annuus, Rosa indica, Althaea rosea, Citrus* sp., *Nerium odorum, Eugenia jambolana, Eriobotrya japonica, Vitis vinifera, Jasminum sambac, Albizzia lebbeck, Hibiscus* sp., *Pyrus malus, Juglans regia, and Pyrus communis, Litsaea polyantha, Butea frondosa, Holarrhena antidysenterica,* and *Mallotus philippinensis* [6, 11, 18, 29].

Identification: The mealybug has an oval white body around 1.0–1.5 cm long.

Damage symptoms: Both nymphs and adults suck the sap from the stem and excrete honeydew leading to the formation of the sooty mould.

3.3 Major Wood Borers

1. ***Euplatypus parallelus*** [12]

Host plants: Acacia spp., *Anacardium occidentale* (cashew nut), *Areca catechu* (betel nut palm), *Cocos nucifera* (coconut), *Dalbergia latifolia* (Indian rosewood), *Eucalyptus* spp., *Ficus* spp., *Hevea brasiliensis* (rubber tree), *Khaya senegalensis* (African mahogany), *Mangifera indica* (mango), *Persea americana* (avocado), *Pinus oocarpa*, and *Tectona grandis* (teak).

Identification: Adult beetles are yellowish-brown and 3.8–4.5 mm long.

Damage symptoms: Pinholes (entrance holes) on trunks and branches which are usually associated with frass and "toothpicks" (compacted frass expelled from entrance holes), accumulation of frass at the base of the trunk, wilting, fallen leaves and eventually tree mortality.

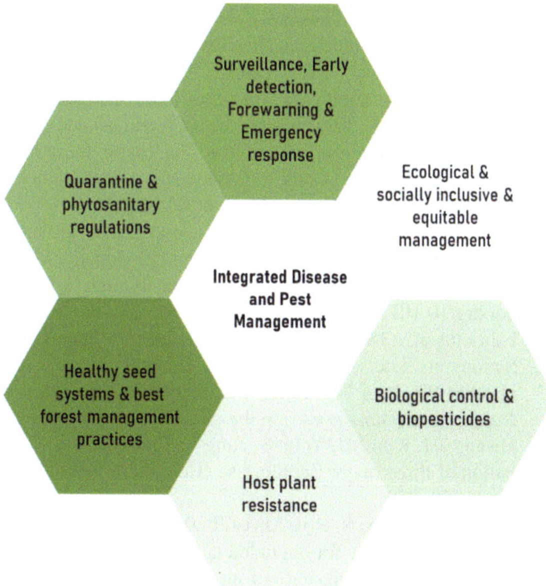

Fig. 2 Holistic approach for management of disease and insect pest [28]

4 Conclusion

Dalbergia latifolia is a species with considerable ecological and economic value; however, it is vulnerable to various pests and pathogens, including stem borers, defoliators, and fungal diseases like wilt and root rot. Addressing these challenges sustainably requires a comprehensive approach (Fig. 2) that integrates resistant germplasm, biological control agents, cultural practices, and environmentally friendly interventions.

References

1. Antony B, Sinu PA, Rehman A (2012) Looper caterpillar invasion in North East Indian tea agro-ecosystem: change of weather and habitat loss may be possible causes? J Tea Sci Res 2(1):1–5
2. Arshad M, Aslam A, Chi DF, Akbar W (2021) Insect pests of *Dalbergia sissoo*: impact and risk mitigation. In: *Dalbergia sissoo*. CRC Press, pp 89–101
3. Arunkumar AN, Warrier RR, Kher MM, Teixeira da Silva JA (2022) Indian rosewood (*Dalbergia latifolia* Roxb.): biology, utilisation, and conservation practices. Trees 36(3):883–898
4. Balachowsky A, Mensil L (1936) Insectes nuisibles aux Polygonées. In: Les Insectes nuisibles Plantes Cultivées, pp 1421–1422
5. Barbattini R, Zandigiacomo P, Parmegiani P (1986) Preliminary investigation on the pests of *Rumex obtusifolius* L. and *Rumex crispus* L. in vineyards of the Friuli region. Redia 69:131–142

6. Beeson CFC (1941) The ecology and control of forest insects in India and neighboring countries. Govt. of India Publications New Delhi, 767
7. Biswas S, Basu RC, Ghosh LK (1994) Fauna of West Bengal (INSECTA: HEMIPTERA). Zoological Survey of India, Culcatta, pp 69–90
8. Chattopadhyay S (2021) First record of broad headed bugs (Hemiptera: Heteroptera: Alydidae) on shisham (*Dalbergia sissoo*) from Jharkhand. India J Exp Zool 24(1):203–205
9. Dickinson JL (1995) Trade-offs between postcopulatory riding and mate location in the blue milkweed beetle. Behavioral Ecol 6(3):280–286
10. Dobler S, Farrell BD (1999) Host use evolution in Chrysochus milkweed beetles: evidence from behaviour, population genetics and phylogeny. Mol Ecol 8(8):1297–1307
11. Dutt GR (1925) The Giant mealy bug and its control. Bull Entomol Res 16(2):155–158. https://doi.org/10.1017/S0007485300028467
12. Fabricius JC (1801) Systema Eleutheratorum Secundum Ordines, Genera, Species adiectis Synonymis, Locis, Observationibus, Descriptionibus. Bibiliopoli Academici Novi, Kiliae, 506
13. Hatcher PE, Paul ND, Ayres PG, Whittaker JB (1997) Added soil nitrogen does not allow *Rumex obtusifolius* to escape the effects of insect-fungus interactions. J Appl Ecol 34:88–100
14. Hwang JH, Kim SH, Yoon S, Jung S, Kim DH, Lee WH (2022) Evaluation of spatial distribution of three major Leptocorisa (Hemiptera: Alydidae) pests using MaxEnt Model. Insects 13(8):750–759
15. Jaiswal AK, Roy S, Roy MM (2020) Lac-based agroforestry system for degraded lands in India. Agroforestry for degraded landscapes. Rec Adv Emerg Challeng 2:389–405
16. Kumar A (2017) The study of insect pests of *Dalbergia sissoo* Roxb. and their seasonal incidence in Jharkhand, India. Amer J Agri For 5(5):137–144. https://doi.org/10.11648/j.ajaf.20170505.11
17. Kumar A, Rajwar N (2023) New records of insect pests infesting *Dalbergia latifolia* (Roxb.) from Uttarakhand, India
18. Latif A (1949) The taxonomic status of *Drosicha stebbingi* (Green) and *Drosicha mangiferae* (Green) (Hem., Coccid.). Bull Entomol Res 40(3):351–354
19. Louppe D, Oteng-Amoako AA, Brink M (2008) Plant resources of tropical Africa 7: timbers 1. Backhuys Publishers, Leiden, the Netherlands, PROTA Foundation
20. Manohara TN, Balakrishna SM, Shetteppenavar VS (2023) Infestation of pinhole borer (*Euplatypus parallelus* (F.) on *Dalbergia latifolia* Roxb. Plant Sci Today 10(3):75–78. https://doi.org/10.14719/pst.2074
21. Mason J, Parker M, Vary L, Lowry II PP, Hassold S, Ruta G (2016) Malagasy precious hardwoods: scientific and technical assessment to meet CITES objectives. Report submitted by the World Resources Institute and the World Bank, Washington. https://www.profor.info/content/malagasy-precious-hardwoods-scientific-and-technical-assessment-meet-cites-objectives
22. Mathur RN, Singh B (1959) A list of insect pests of forest plants in India and the adjacent countries. Indian Forest Bull 171(4):5–7
23. Mazza G, Pennacchio F, Gargani E, Franceschini I, Roversi PF, Cianferoni F (2014) First report of *Ricania speculum* (Walker, 1851) in Europe (Hemiptera: Fulgoromorpha: Ricaniidae). Zootaxa 3861(3):297–300
24. McKamey S (2017) MOWD: membracoidea of the world database (version 1011, Nov 2010). In: Species 2000 & ITIS Catalogue of Life, 30th June 2017. Naturalis, Leiden, the Netherlands
25. Mohanan C, Ratheesh N, Nair LP, Rajesh KC (2005) Disease problems in root trainer forest nurseries in Kerala State and their management. In 5th meeting of IUFRO working party, pp 7–12
26. Nair R (2000) Evaluation of provenances for seedling attributes in rosewood (*Dalbergia latifolia Roxb.*) Doctoral dissertation, Department of tree physiology and breeding, College of Forestry, Vellanikkara
27. Orwa C, Mutua A, Kindt R, Jamnadass R, Anthony S (2009) *Arbutus unedo* L. agroforest tree database: a tree reference and selection guide, version, 4. http://www.worldagroforestry.org/treedb/AFTPDFS/Dalbergia_latifolia.PDF

28. Kawarazuka N, Liu Y, Mulema AA, McCutcheon S, Ibabao X (2022) Sustainable management of transboundary pests requires holistic and inclusive solutions. Food Sec 14:1449–1457. https://doi.org/10.1007/s12571-022-01301-z
29. Rahman MA (1944) Diseases and pests of *Dalbergia sissoo* (roxb.) and their control. Doctoral dissertation, Khulna University
30. Rakotonirina N, Nowak MM, Lowry PP II, Rakouth HN, Rakouth BR (2024) Will Dalbergia species survive climate change? Predicting the potential future distribution of threatened species in Madagascar. GEcoland Cons 52:e02936
31. Rawat V, Singh S (2003) Early epidemic of *Plecoptera reflexa* on *Dalbergia sissoo* an indication of Doon valley climate change. Indian Forester 129(4):539–540
32. Remadevi OK (2022) Status, issues, and challenges of biodiversity: forest insects. In: Biodiversity in India: status, issues and challenges. Springer Nature, Singapore, pp 325–362
33. Rossi E, Lucchi A (2015) The Asian planthopper *Ricania speculum* (Walker) (Homoptera: Ricaniidae) on several crops in Italy: a potential threat to the EPPO region? EPPO Bull 45(1):119–122
34. Rossi E, Stroinski A, Lucchi A (2015) Egg morphology, laying behaviour and record of the host plants of *Ricania speculum* (Walker, 1851), a new alien species for Europe (Hemiptera: Ricaniidae). Zoota 4044(1):93–104
35. Roychoudhury N, Mishra RK (2021) Shisham defoliator *Plecoptera reflexa* and its control measures. Van Sangy 8(3):30–32
36. Sady MB (1994) Survey of the blue milkweed beetle, *Chrysochus cobaltinus* LeConte (Coleoptera: Chrysomelidae), in western Nevada. Coleop Bull 8(3):299–305
37. Salisbury A, Platoni A (2013) The green dock beetle, *Gastrophysa viridula* (De Geer) (Chrysomelidae)-a pest on Begonia (Begoniaceae), a new host record. Coleop 22(3):124–125
38. Sennblad B, Bremer B (1996) The familial and subfamilial relationships of Apocynaceae and Asclepiadaceae evaluated with rbc L data. Plant Syst Evol 202:153–175
39. Sen-Sarma PK (2008) Forest insect pests and their management. A Man For Exten Edu 341–368
40. Shankar U, Lama SD, Bawa KS (1998) Ecosystem reconstruction through taungya agrosystems following commercial logging in dry mixed deciduous forest in Darjeeling Himalaya. For Ecol Manage 102:131–142
41. Shivanna MB (2005) Fungal diseases in forest nurseries in Shivmogga district, Karnataka, India. In: Diseases and insects in forest nurseries. Proceedings of the 5th meeting of IUFRO working party S703–04, pp 27–35
42. Suharti M, Hadi S (1974) Wilt disease of *Dalbergia latifolia* in Malang forest district, E. Java. Laporan Lembaga Penelitian Hutan No.94, Bogor, p 9
43. Sukhadiya M, Dholariya CA, Behera LK, Nayak D, Patel SM, Mehta AA (2020) *Dalbergia latifolia* Roxb: biography of an indigenous multipurpose tree species of India. MFP News 1(1):1–7
44. Swamy KR, Pai V, Hanumantha M, Suryanarayan V (2019) Assessment of tar spot disease in *Dalbergia latifolia* and their management. JPP 8(3):621–623
45. Swatonek F (1972) A contribution to the biology of the dock-leaf beetle (*Gastroidea viridula*). Anzeiger fur Schadlingskunde und Pflanzenschutz 45(8):116–119
46. Thangavelu K (1980) Report of *Leptocentrus taurus* fabricius (Membracidae: Homoptera) feeding on *Parthenium hysterophorus* Linn. Entomo 5(4):357–364
47. Vatanparast M, Klitgård BB, Adema FA, Pennington RT, Yahara T, Kajita T (2013) First molecular phylogeny of the pantropical genus Dalbergia: implications for infrageneric circumscription and biogeography. S Afri J Bot 89:143–149
48. Verma P, Soni KK, Verma RK, Shirin F (2016) Seedling die-back of *Dalbergia latifolia* in Central India caused by *Colletotrichum capsici* a new record. IJCMAS 5(6):350–356

Chapter 11
Propagation of Difficult-To-Root *Dalbergia latifolia* (Roxb.): Physiology of Adventitious Rooting

Pramod Kumar Tiwari

1 Introduction

The use of improved quality planting material developed through selection and vegetative propagation is the primary requirement for improving the quality and productivity of a species. Different rooting abilities of tree species are categorised as 'easy-to-root' (ER) and 'difficult-to-root' (DR) species. Commercial production of improved planting stock at a large scale for ER species like Eucalypts, Poplar, and conifers is being done from hedged shoot cuttings from either full-sib or half-sib donor families. However, large-scale production is a constraint for DR species due to their low rooting ability, and *Dalbergia latifolia* is one of the challenging species in this category. *D. latifolia* Roxb. (Family—Fabaceae), popularly known as 'Indian Rosewood' is a deciduous, commercially valuable timber species that requires moderate sunlight [68]. Though the species is distributed from the sub-Himalayan region to the southern states of India and Indonesia [32], it is overexploited and less preferred for plantations due to slow growth habit which leads to its status as vulnerable (VU) in the Red List of the International Union for Conservation of Nature and Natural Resources (IUCN) and is also included in Appendix II of the Convention on International Trade in Endangered Species (CITES). In central India, its occurrence is sporadic, and only a few plantations exist. The wood is of high density, becomes dark brown at a mature age without clear distinction in annual rings, and is recognised as a premium quality timber tree [66]. Wood is mainly used for furniture, doors, panelling, toys, and ornamental products. Tannins from bark and leaves are important in the pharmaceutical industry. N_2 fixation through root nodules makes the species suitable for rehabilitating degraded forest areas.

P. K. Tiwari (✉)
Genetics and Tree Improvement Division, ICFRE-Tropical Forest Research Institute, Jabalpur, Madhya Pradesh, India
e-mail: pramod_kt@rediffmail.com

R. R. Warrier et al. (eds.), *Handbook of Asian Rosewoods*, Materials Horizons: From Nature to Nanomaterials, https://doi.org/10.1007/978-981-96-9459-4_11

Both environmental and endogenous factors regulate adventitious root formation, but the complexity of controlling this quantitative genetic trait in trees hampers a clear understanding of their smart and efficient propagation. Adventitious rhizogenesis involves different developmental phases with their specific requirements and control of genetic and surrounding environment [20, 33, 34, 65]. Availability of endogenous growth regulators, storage and transport of required substrates to the meristematic tissues of the rooting zone, dormant stage of cells, and accumulation of inhibitory compounds at the base of the cutting and preconditioning treatments influence adventitious rhizogenesis in hardwood cuttings [1, 8, 30, 35, 44]. Of these, the availability of endogenous auxin and carbohydrate metabolism is crucial to determining the rooting ability of shoot cuttings [27, 57, 60]. Endogenous auxin concentration varies with various interdependent physiological events of the rooting process [41, 45]. Exogenous auxin application accelerates the process of adventitious root formation in plants. Commercially available auxins viz. indole-3-acetic acid (IAA), indole-3-butyric acid (IBA), and naphthalene acetic acid (NAA) are being used prominently for preconditioning treatment. However, various studies confirm relatively more effectiveness of IBA than the IAA and NAA in adventitious rooting in many tree species, e.g., *D. sissoo* [2], *Azadirachta indica* [46], and *Gmelina arborea* [6, 51].

Adventitious roots develop from non-root tissues, and the ability to form them differs significantly in forest trees. In branch cuttings of tree species, e.g., willows and poplars, roots emerge shortly after placing in the suitable substratum and environment conditions favourable to its rooting and growth [50]. Meanwhile, in cases of wound-induced adventitious rooting mechanisms, callus develops from dividing parenchyma cells, which later develop into adventitious roots with dedifferentiation and elongation of cells [28]. However, most of the research on the physiology and molecular genetics of adventitious root formation in forest trees are confined to significantly associated two factors, i.e., the auxin regulation at various growth phases and, second, the genetics of the parent plant material [4, 40].

Comparative growth performance of seedlings and cuttings exhibit different dimensions [16, 17, 21, 47]. Low functional effectiveness of adventitious roots [58] and no significant difference in seedlings and cuttings in *Eucalyptus globulus* were observed [21]. The shallow root system and absence of tap roots are the characteristics of cutting-raised plants, which thus differ from seedlings in establishment and slower initial growth [11, 15, 23, 24, 48, 59, 64]. Contradictory results lead to an unclear understanding due to the lack of genetic background of seedlings and cuttings [18, 61].

Economically viable and user-friendly propagation of *D. latifolia* [53, 54] describes the use of juvenile shoot cuttings and the development of rooted hardened plants in a short duration of three months. Propagation of superior genotypes will help conserve them and enhance productivity. However, the field performance of seedlings and cuttings of *D. latifolia* are limited.

2 Propagation Through Root Cuttings

In this propagation method [69], root cuttings measuring 30 to 40 cm in length and 15 to 20 cm in girth were collected from five selected *D. latifolia* trees by excavating the surrounding soil. These cuttings were then planted horizontally in sand beds within a shade house structure, covered with polytunnels. Sprouting occurred within 46 ± 20 days, and the resulting sprouts were utilised for further rooting. The excised sprouts were treated with IBA at concentrations of 2000 and 4000 ppm in powder and liquid forms. Treated sprouts were placed in a vermiculite medium and maintained under polytunnels within the shade house. The liquid IBA solution led to a rooting success rate of 40–70%, compared to 15–50% with the powder form. The 2000 ppm IBA liquid solution yielded 65–70% rooted sprouts. Among the tested media, vermiculite proved the most effective, producing more rooted sprouts in a shorter time. However, propagation using root cuttings has certain drawbacks, including (i) requiring over a year to develop into hardened plants suitable for field transplantation, (ii) the risk of mortality at two stages, and (iii) potential infections entering through cut root ends, which may threaten the donor tree. Additionally, this technique is costly due to the need for specialised propagation structures such as polytunnels, a substratum like vermiculite, and an extended maintenance period before field planting.

3 Propagation of Juvenile Shoots: A User-Friendly Option

Propagation of juvenile shoots obtained through annual hedging of stock plants from selected trees is a viable method for propagating difficult-to-root species, including *D. latifolia.* To standardise this technique, a multiplication garden was established in 2009 at the ICFRE-TFRI institute campus (N 23.099735, E 79.989341) using seedling progenies from 10 superior *D. latifolia* trees. Experiments were conducted in 2010, 2014, 2018, 2020, and 2023 at different ontogenic ages—1–2 years, 4–5 years, 8–9 years, 10–11 years, and 13–14 years—to evaluate rooting ability.

Juvenile shoot growth from the previous year was harvested annually in the second fortnight of April, following complete leaf fall—an optimal period for regeneration in deciduous trees. These shoots were cut into 15–20 cm segments with a 1.0–1.5 cm diameter and surface disinfected by dipping in a 1% Carbendazim 50% W.P. solution for five minutes. The basal ends of the cuttings were then treated with a rooting solution: 2.0 mM IAA for 4 h (1–2-year-old cuttings), 5.0 mM IAA for 18 h (4–5-year-old cuttings), and 5.0 mM IAA + 1.0 mM Boric Acid for 18 h (8–9, 10–11, and 13–14-year-old cuttings). Since secondary metabolites like phenols and tannins inhibit adventitious rooting, a small amount of activated charcoal was added to the treatment solution to counteract this effect.

Treated cuttings from 1–2-year-old stock plants were placed in 12 cm deep root trainers, while cuttings from older stock plants (4–5 years and beyond) were planted

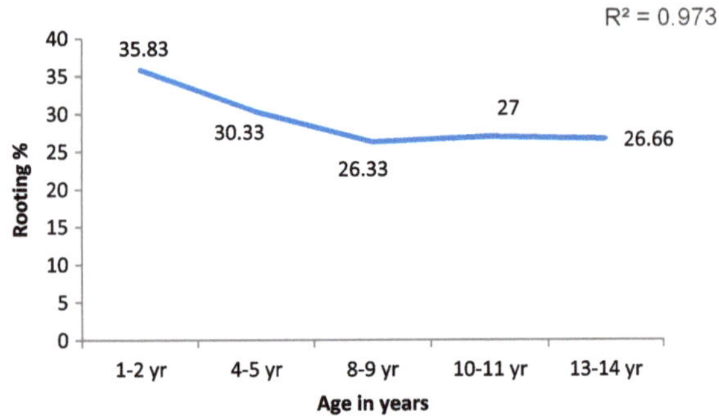

Fig. 1 Adventitious rooting trend in *Dalbergia latifolia* with increasing age of progenies

in 15 × 22 cm polybags filled with soil, sand, and farmyard manure (FYM) mixture in a 1:2:1 ratio. Propagation occurred under a shade net house, allowing 60–70% light transmission, with temperatures maintained at 35 ± 2 °C and humidity at 60–70% through regular fine mist spraying at least thrice a day. Observations were recorded for 60 days (1–2-year-old cuttings) and 75 days (older cuttings). The survival count of hardened rooted cuttings, identified by their broader leaf lamina, was used to determine final rooting success while minimising potential mortality caused by root exposure during transplantation.

The recorded mean rooting ability was 35.83% for 1–2-year-old cuttings, 30.33% for 4–5-year-olds, 26.33% for 8–9-year-olds, 27% for 10–11-year-olds, and 26.66% for 13–14-year-olds. Untreated cuttings only developed callus, with no root formation. Rooting ability declined with increasing stock plant age up to 8 years, after which it remained stable until 13–14 years. The decrease in rooting ability from 35.83 to 26.33% was still significantly higher than the <10% rooting success observed in mature trees [52].

This propagation technique is well-suited for forest nurseries, enabling the production of high-quality planting material for *D. latifolia*. The use of polybags and a partially shaded environment reduced the hardening time, allowing rooted cuttings to be field-ready within three months, making them available for plantation during the same year's rainy season. While the success rate is moderate, the method is cost-effective and ensures the rapid production of quality plants (Figs. 1 and 2).

4 Juvenile-Mature Gradient in Adventitious Rooting

Research indicates that the rooting ability of donor trees declines with increasing chronological age and as they transition from the vegetative to the reproductive phase [39, 56, 73]. In *D. latifolia*, annual flowering begins in September after an initial vegetative period of 5–6 years. A juvenile-to-mature gradient exists during this phase transition, gradually reducing the regeneration potential of stock plants.

Fig. 2 Rooting ability in *Dalbergia latifolia*: **a** hedged progenies with regenerating shoots, **b** shoot cuttings (1–2 yr) in root trainers, **c** shoot cuttings (8–9 yr) in polybags, **d**, **e**, **f** adventitious rooting in different age progenies, **g**, **h** rooted plantlets in polybags, and **i** plantation of cutting-raised plants at the age of 2 years

This shift is a key factor behind the variability in adventitious root formation observed across progenies of different ages [28].

Although annual hedging was used to arrest the reproductive growth of donor stock plants and promote reinvigoration, genetic signals triggering the reproductive phase transition at 5–6 years still led to reduced rooting ability, stabilising at 8–10 years. The phase change is associated with lower cell division activity at the base of the cuttings, contributing to the decline in rooting ability [57]. Additionally, the accumulation of phenols and tannins at the base of cuttings with age further reduces rooting efficiency by interfering with the oxidative decarboxylation of IAA, a crucial process for adventitious root formation [10, 13, 49].

Furthermore, high wood density in older trees impairs the lateral movement of endogenous IAA, limiting its transport to dividing cells and thereby affecting rooting success. While genetic lineage may play a role in determining variations in rooting ability among progenies [62], studies suggest that the lack of significant variation is primarily due to similar environmental conditions influencing the growth of both tree accessions and their progenies [72].

5 Endogenous Auxin and Adventitious Rooting

Indole-3-acetic acid (IAA) is a key endogenous plant hormone that effectively induces adventitious root formation in many woody species [9, 12]. It is primarily synthesised in the stem tips and young leaves of aerial plant parts and is actively transported to other tissues, coordinating growth and facilitating responses to environmental changes [37]. Auxin plays a crucial role in the physiological process of adventitious rooting, influencing fluctuations in endogenous auxin concentrations [8, 20, 29, 33, 41]. Several factors, including the availability and transport of endogenous growth regulators, the plant's dormant stage, carbohydrate storage, inhibitory compounds, and preconditioning treatments, impact adventitious root formation in the complex system of woody trees [30, 71]. Plant growth regulators (PGRs) other than auxins have been extensively studied for their role in signalling adventitious root formation [5]. The regulation of PGRs directly influences the rooting ability of an ortet, which declines with maturity [67]. Reduced basipetal transport of endogenous auxin generally lowers the likelihood of root initiation [71]. However, the exogenous application of synthetic auxins enhances the movement of free basal endogenous auxin, stimulating adventitious root formation [14, 22, 55]. Auxin-mediated processes regulate cell division, elongation, and differentiation across the plant [7]. In juvenile trees, branch cuttings treated with exogenous auxins promote adventitious rooting, whereas, in long-lived forest tree species, the root formation is typically not feasible without external auxin supplementation. Endogenous IAA plays a cohesive role in supporting physiological processes necessary for adventitious rhizogenesis.

The endogenous IAA content at the base of shoot cuttings was quantified using a modified spectrophotometric method [63], revealing variations among progenies from different tree accessions and ages. IAA levels ranged from 1.18 to 12.14 $\mu g.g^{-1}$ fresh tissue weight, corresponding to ages between 1–2 years and 8–9 years. However, the variation in endogenous IAA across age groups was not statistically significant. When analysing the relationship between rooting ability and endogenous IAA levels, a weak and non-significant correlation was observed across all progeny age groups (Fig. 3). This suggests an insufficient supply of endogenous IAA to the target meristematic cells. In contrast, rooting ability declined significantly with increasing ontogenic age, showing a strong correlation ($R^2 = 0.973$). While endogenous IAA initiates adventitious root development, the weak relationship with rooting ability suggests that rooting competence is indirectly regulated. Exogenously applied auxin moves passively, supporting the lateral distribution of endogenous IAA to target cells, thereby enhancing adventitious rooting, as observed in *Pongamia pinanata* [3].

In plants, IAA exists in its natural form as an undissociated molecule at $pH < 4.0$ and as an anionic form at $pH > 4.0$. Ionised IAA is transported basipetally through electrogenic carrier proteins [36], reducing its lateral distribution to target cells in the pericyclic region responsible for adventitious rooting [31, 38]. Since shoot cuttings lack continuous vascular tissue, cell dedifferentiation is necessary for adventitious root initiation, and exogenous auxin accelerates this local redistribution [3]. Undissociated IAA molecules diffuse passively across the plasma membrane

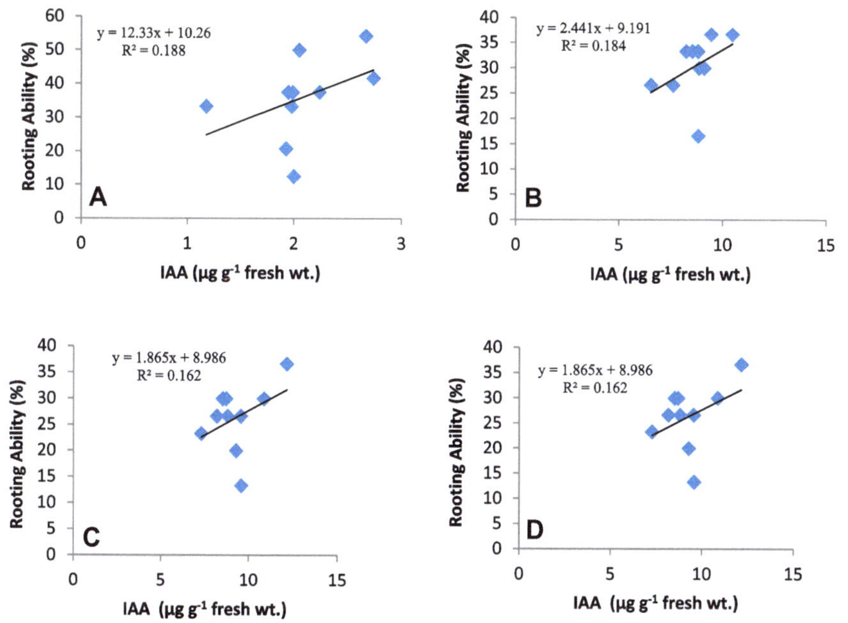

Fig. 3 Relationship of endogenous IAA with rooting ability in *D. latifolia* at progeny age **A** 1–2 Yr, **B** 4–5 Yr, **C** 8–9 Yr, and **D** 10–11 Yr

[36], while exogenous auxin application compensates for endogenous IAA oxidation at the cut base, increasing its availability [19]. Additionally, IAA conjugation plays a significant role in determining rooting capacity [43]. Rapid auxin transport in ER (easy-to-root) plants enhances the success rate of adventitious rooting [42].

6 Carbohydrate Metabolism During Adventitious Rooting

Stored carbohydrates mobilise after detachment from the mother plant, supporting the adventitious rooting process [25, 26, 70]. The carbohydrate reserves in cuttings serve as a crucial carbon source, while specific enzymes and metabolites play essential roles in various stages of root formation.

In *D. latifolia* cuttings, the pattern of soluble sugars and changes in meristematic tissues at the base were observed throughout the rooting process. New sprouts began emerging around the third or fourth day after planting while dividing cells accumulated at the cut base in the form of callus by the 12th day. Adventitious root protrusion occurred between the fourth and fifth weeks, with the first roots appearing on the 28th day after planting. Soluble sugar content was measured every four days until the 36th day post-planting. During the initial 12 days, the hydrolysis of polysaccharides led to an accumulation of soluble sugars, providing energy to the dividing meristematic

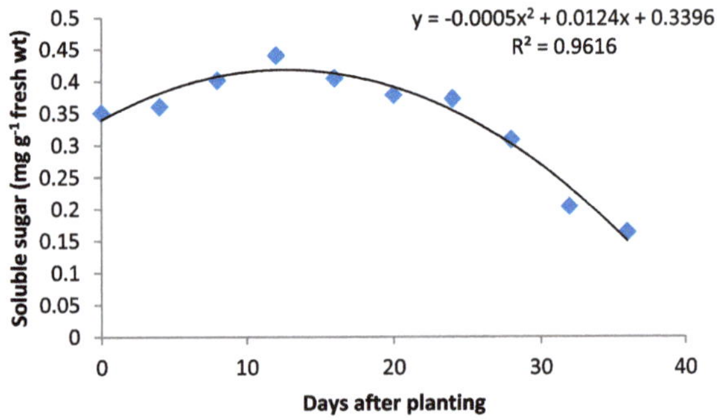

Fig. 4 Polynomial relationship between soluble sugar and adventitious rooting in *D. latifolia*

cells to form root initials and primordia. As root formation progressed, sugars were increasingly utilised for structural development, leading to a gradual decline in their levels at the cut base up to the 36th day.

The process of sugar unloading plays a critical role in adventitious rooting, with an initial apoplastic phase serving as a preparatory stage, followed by a symplastic phase for establishing and maintaining root growth. A polynomial relationship between rooting percentage and soluble sugar levels was observed, with a regression coefficient (R^2) value of 0.961 (Fig. 4), indicating a significant requirement of sugars for the initiation and development of adventitious roots.

7 Field Establishment of Cuttings-Raised Rooted Plants

Rooted cuttings were successfully established in the field, demonstrating a 100% survival rate. Their growth performance was compared to seedlings planted in a randomised block design (RBD) within the same field. Observations recorded 30 months after planting, assessing traits such as height, number of branches, and foliar diameter, showed a significant increase over seedlings from selected half-sib families (Table 1).

The enhanced growth traits during the initial years can be attributed to a two-stage selection process: first, at the donor tree stage, and, second, by selecting superior seedlings for field establishment. Despite having greater ontogenetic age than seedlings, the annual hedging of progenies suppressed reproductive growth while enhancing their morphogenetic potential.

The root architecture of cuttings differs from that of seed-derived plants. While seedlings establish quickly due to a well-developed taproot, roots in cuttings remain delicate and fleshy for an extended period. Over time, one primary adventitious root strengthens the plant, and the gradual development of lateral roots expands the root system, improving nutrient and water uptake. This broader absorption area supports metabolic processes essential for morphogenetic growth, accelerating vegetative development in *D. latifolia* during the early establishment years.

Table 1 Comparative growth traits of seedlings and cutting-raised plants after 30 months of planting

Tree/ progeny	Seedlings				Cutting-raised plants			
Accession	Height (cm)	Collar diameter (mm)	Number of branches	Foliar dia (mm)	Height (cm)	Collar diameter (mm)	Number of branches	Foliar dia (mm)
D-1	80.77	24.78	6.87	45.55	91.69 (13.52)	28.41 (14.65)	16.48 (139.88)	50.94 (11.83)
D-2	68.88	11.61	7.10	39.34	77.40 (12.36)	15.96 (37.46)	9.36 (31.83)	44.90 (14.13)
D-3	55.88	10.43	4.22	29.44	73.77 (32.01)	13.99 (34.13)	8.44 (100.00)	41.80 (42.00)
D-4	70.99	18.67	9.00	45.04	81.55 (14.87)	22.47 (20.35)	12.25 (36.11)	52.00 (15.46)
D-5	68.77	10.46	6.77	40.32	78.53 (14.19)	12.98 (24.09)	8.71 (28.65)	43.17 (7.07)
D-6	64.99	11.73	7.10	40.32	83.33 (28.21)	13.65 (16.36)	8.61 (21.26)	48.57 (20.46)
D-7	64.99	10.65	6.66	34.43	81.92 (26.05)	14.75 (38.49)	10.44 (56.75)	41.94 (21.81)
D-8	75.55	13.71	6.10	43.32	86.18 (14.07)	15.86 (15.68)	8.81 (44.42)	47.70 (10.11)
D-9	72.66	15.74	7.21	41.05	84.18 (15.85)	18.62 (18.30)	9.03 (25.24)	45.47 (10.76)
D-10	83.21	10.69	9.99	45.94	98.10 (17.89)	16.24 (51.91)	13.44 (34.53)ara>	54.85 (19.39)
Significance at $p \leq 0.05$					0.0012 (S)	0.1218 (NS)	0.0024 (S)	0.0066(S)

Figures in parenthesis are % increase over seedlings

Additionally, while the ability to form adventitious roots has a genetic basis, favourable environmental conditions and suitable site characteristics further optimise nutrient uptake, promoting faster field growth and overall plant development.

8 Concluding Remarks

The process of adventitious rooting in trees is influenced by the interaction of intrinsic and extrinsic factors, along with ontogenic age-related changes, making it highly complex. *Dalbergia latifolia*, typically resistant to adventitious rooting, enhances cellular competence for root initiation when treated with exogenous auxin. A practical and efficient propagation method involves collecting juvenile shoot cuttings from the previous year's growth through hedging, applying a basal dip treatment of IAA combined with Boric Acid, and cultivating the cuttings in a partially shaded open environment. This user-friendly approach enables the production of approximately 30% rooted, hardened plants within three months.

Using a soil mix in polybags minimises the time required for hardening and reduces mortality during transplantation, ensuring the availability of rooted cuttings for plantation during the rainy season. The combined effect of selecting both superior donor trees and progenies enhances the early growth of rooted plantlets. Therefore, this method is recommended for raising high-quality *D. latifolia* plants to promote faster growth and improved productivity.

Several key factors determine rooting ability, including cellular competence at the cutting base, moisture levels, carbohydrate mobilisation, and the availability of IAA to target cells. The wood density of tree species affects the distribution and accessibility of endogenous IAA necessary for root induction. While exogenous IAA treatment significantly enhances adventitious rooting, the weak and non-significant correlation between endogenous IAA levels and rooting ability in *D. latifolia* suggests that endogenous IAA alone may be insufficient for optimal rooting. Thus, an adequate supply of auxins is essential, with endogenous IAA supporting various physiological processes involved in adventitious rhizogenesis. Exogenously applied IAA, in turn, facilitates broader lateral distribution of auxin to meristematic target cells in the pericycle region, promoting their organisation, division, and subsequent root formation.

Acknowledgements The author is grateful to the Indian Council of Forestry Research and Education, Dehradun, for the award of projects [Project ID: 131/TFRI-2008/GEN-1(17) and 227/TFRI/2016/Gen-2(33)].

References

1. Andersen AS (1986) Stock plant conditions. In: Jackson MB (ed) New root formation in plants and cuttings. Martinus Nijhoff, Dordrecht/Boston/Lancaster, pp 223–255
2. Ansari SA, Kumar P, Mandal AK (1995) Effect of position and age of cuttings and auxins on induction and growth of roots in *Dalbergia sissoo* Roxb. Indian For 121(3):201–206
3. Ansari SA, Singh S, Rani A (2004) Inorganic salts influence IAA ionization and adventitious rhizogenesis in *Pongamia pinnata*. J Plant Physiol 161:117–120
4. Blakesley D, Weston GD, Hall JF (1991) The role of endogenous auxin in root initiation. Part I. Evidence from studies on auxin application and analysis of endogenous levels. Plant Growth Regul 10:341–354
5. Blakesley D (1994) Auxin metabolism and adventitious root initiation. In: Davis TD, Haissig BE (eds) Biology of adventitious root formation. Plenum Press, New York, pp 143–154
6. Chandra JP, Gandhi JN (1995) Rootings of cuttings of *Gmelina arborea*. Indian For 121:427–428
7. Davies PJ (2004) Plant hormones: biosynthesis, signal transduction, action! Kluwer Academic Publishers. The Netherlands, Dordrecht
8. Davis TD, Haissig BE, Sankhla N (1988) Adventitious root formation in cuttings. In: Advances in plant sciences series, vol II. Dioscorides Press, Portland, Oregon
9. De Klerk GJ, Van Der Krieken W, de Jong JC (1999) Review the formation of adventitious roots: New concepts, new possibilities. In: Vitro cellular & developmental biology-plant 35(3), pp 189–199

10. De Klerk GJ, Guan H, Huisman P, Marinova S (2011) Effects of phenolic compounds on adventitious root formation and oxidative decarboxylation of applied indole acetic acid in Malus 'Jork 9.' Plant Growth Regul 63:175–185
11. Delwaulle JC (1996) Comparative responses of cuttings and seedlings of *Carapa procera* to water stress. Tree Physiol 16:287–294
12. Diaz-Sala C, Hutchison KW, Goldfarb B, Greenwood MS (1996) Maturation-related loss in rooting competence by loblolly pine stem cuttings: the role of auxin transport, metabolism and tissue sensitivity. Physiol Plant 97(3):481–490
13. Diaz-Sala C (2020) A perspective on adventitious root formation in tree species. Plants 9:1789–1796
14. Ford YY, Bonham EC, Cameron RWF, Blake PS, Judd HL, Harrison-Murray RS (2002) Adventitious rooting examining the role of auxin in an easy and a difficult-to-root plant. Plant Growth Regul 36:149–159
15. Foster GS, Campbell RK, Adams WT (1985) Clonal selection prospects in western hemlock combining rooting traits with juvenile height growth. Can J For Res 14:628–638
16. Foster GS, Lambeth CC, Greenwood MS (1987) Growth of loblolly pine rooted cuttings compared with seedlings. Can J For Res 17:157–164
17. Frampton LJ Jr, Foster GS (1993) Field testing vegetative propagules. In: Ahuja MR, Libby WJ (eds) Clonal forestry I: genetics and biotechnology (.). Springer-Verlag, New York, pp 110–134
18. Frampton J, Li B, Goldfarb B (2000) Early field growth of loblolly pine rooted cuttings and seedlings. Southern J Appl For 24:98–105
19. Fu JH, Sun XH, Wang JD, Chu JF, Yan CY (2011) Progress in quantitative analysis of plant hormones. Chinese Sci Bull 56:355–366
20. Gasper T, Kevers C, Hausman JF (1997) Indissociable chief factors in the inductive phase of adventitious rooting. In: Altman A, Waisel M (eds) Biology of root formation and development. Plenum Press, New York, pp 55–63
21. Gasper MJ, Borralho N, Gomes AL (2005) Comparison between field performance of cuttings and seedlings of *Eucalyptus globulus*. Ann For Sci 62:837–841
22. Gatineau F, Fouche JG, Kevers C, Hausman JF, Gaspar T (1997) Quantitative variations of indolyl compounds including IAA, IAA aspartate and serotonin in walnut microcuttings during root induction. Biol Plant 39:131–137
23. Gemmel P, Orlander G, Hogberg KA (1991) Norway spruce cuttings perform better than seedlings of the same genetic origin. Silvae Genetica 40:198–202
24. Goudet J, Raymond M, deMeeus T, Rousset F (1996) Testing differentiation in diploid populations. Genetics 144:1933–1940
25. Haissig BE (1986) Metabolic processes in adventitious rooting of cuttings. In: Jackson MB (ed) New root formation in plants and cuttings. Martinus Nijhoff Publishers, Dordrecht, pp 144–189
26. Haissig BE, Davis TD, Riemenschneider DE (1992) Researching the controls of adventitious rooting. Physiol Plant 84:310–317
27. Haissig BE, Davis TD (1994) An historical evaluation of adventitious rooting research to 1993. In: Davis TD, Haissig BE (eds) Biology of root formation. Plenum Press, New York, pp 275–331
28. Hartmann HT, Kester DE, Davies FT, Geneve RL (2002) Plant propagation principles and practices. Prentice Hall, New Jersey, pp 367–374
29. Heloir MC, Kevers C, Hausman J, Gasper T (1996) Changes in the concentrations of auxins and polyamines during rooting of in vitro propagated walnut shoots. Tree Physiol 16:515–519
30. Howard BH (1994) Manipulating rooting potential in stock plants before collecting cuttings. In: Davis TD, Haissig BE (eds) Biology of root formation. Plenum Press, New York, pp 123–142
31. Jacobs M, Gilbert SF (1983) Basal localization of the presumptive auxin transport carrier in pea stems. Science 220:1297–1300
32. Kadambi K (1954) The silviculture of *Dalbergia latifolia*. Monograph of Indian Trees, No. 1. Government of India, Manager of Publications, Delhi

33. Kevers C, Hausman JF, Faivre-Rampant O, Evers D, Gasper T (1997) Hormonal control of adventitious rooting progress and questions. J Appl Bot 71:71–79
34. Klerk GJD, Krieken WVD, Jong JC (1999) The formation of adventitious roots: new concepts, new possibilities (Review). In Vitro Cell Dev Biol Plant 35:189–199
35. Legue V, Rigal A, Bhalero RP (2014) Adventitious root formation in tree species: involvement of transcription factors. Physiol Plant 151:192–198
36. Lomax TL, Muday GK, Rubery PH (1995) Auxin transport. In: Davies PJ (ed) Plant hormones: physiology, biochemistry and molecular biology. Kluwer Academic Publishers, Dordrechtpp, pp 509–530
37. Ljung K, Bhalerao RP, Sandberg G (2001) Sites and homeostatic control of auxin biosynthesis in Arabidopsis during vegetative growth. Plant J 28(4):465–474
38. Mitsuhashi-Kato M, Shibaoka H, Shimokoriyama M (1978) Anatomical and physiological aspects of developmental processes of adventitious root formation in Azukia cuttings. Plant Cell Physiol 19:393–400
39. Mitchell RG, Zwolinski J, Jones NB (2004) A review on the effects of donor maturation on rooting and field performance of conifer cuttings. Southern African For J 201:53–63
40. Mullin TJ, Morgenstern EK, Park YS, Fowler DP (1992) Genetic parameters from a clonally replicated test of black spruce (*Picea mariana*). Can J For Res 22:24–36
41. Nag S, Saha K, Choudhuri MA (2001) Role of auxin and polyamines in adventitious root formation in relation to changes in compounds involved in rooting. J Plant Growth Regul 20:182–194
42. Nakhooda M, Watt MP, Mycock D (2011) Auxin stability and accumulation during in vitro shoot morphogenesis influences subsequent root induction and development in *Eucalyptus grandis*. Plant Growth Regul 65:263–271
43. Negishi N, Nakahama K, Urata N, Kojima M, Sakakibara H, Kawaoka A (2014) Hormone level analysis on adventitious root formation in *Eucalyptus globulus*. New For 45:577–587
44. Olatunji D, Geelen D, Verstraeten I (2017) Control of endogenous auxin levels in plant root development. Int J Mol Sci 18(12):2587–2615
45. Osterc G, Stampar F (2011) Differences in endo/exogenous auxin profile in cuttings of different physiological ages. J Plant Physiol 168:2088–2092
46. Palanisamy K, Kumar P (2011) Seasonal effect on induction of adventitious rooting in stem cuttings of neem (*Azadirachta indica* A. Juss.). Indian J For 19(2):183–186
47. Pangou SV, Neela De Z, Gema L (2011) Comparison between field performance of cuttings and seedlings of *Carapa procera* D.C. (Meliaceae). Int Res J Plant Sci 2(9):281–287
48. Paul AD, Foster GS, Lester DT (1993) Field performance, C effects, and their relationship to initial rooting ability for western hemlock clones. Can J For Res 5:47–57
49. Pizarro A, Díaz-Sala C (2019) Cellular dynamics during maturation-related decline of adventitious root formation in forest tree species. Physiol Plant 165:73–80
50. Pijut PM, Wowste KE, Michler CH (2011) Promotion of adventitious root formation of difficult-to-root hardwood tree species. Hortic Rev 38:213–251
51. Kumar P, Jharia SK, Ansari SA (2011) Change in pH regimes and adventitious root induction in semi-hardwood cuttings of *Gmelina arborea* Roxb. Plant Growth Regul 65(3):531–536
52. Kumar P, Chaturved A, Vaishnav V (2022) Relationship between endogenous auxin (Indole-3-acetic acid) and adventitious rooting in *Dalbergia* species of different rooting ability. J Trop For Sci 34(2):149–158
53. Kumar P, Patel PK, Sonkar MK (2022) Propagation through juvenile shoot cuttings in difficult-to-root *Dalbergia latifolia*- role of endogenous IAA in adventitious rooting. Plant Physiol Rep 27(2):242–249. https://doi.org/10.1007/s40502-022-00664-x
54. Kumar P, Patel PK, Pandro V (2023) User friendly propagation technique for quality plant production of *Dalbergia latifolia* (Roxb.) for ecological sustenance. In: Singh S, Tomar A (eds) Clonal forestry: principles and practices. Narendra Publishing House, Delhi, pp 140–153
55. Quaddoury A, Amssa M (2004) Effect of exogenous indole butyric acid on root formation and peroxidase and indole-3-acetic acid oxidase activities and phenolic contents in date palm offshoots. Botanical Bull Acad Sinica 45:127–131

56. Ramussen A, Hunt MA (2010) Ageing delays the cellular stages of adventitious root formation in pine. Aust For 73:41–46
57. Ramussen A, Hosseini SA, Hajirezaei MR, Druege U, Geelen D (2015) Adventitious rooting declines with the vegetative to reproductive switch and involves a changed auxin homeostasis. J Exp Bot 66(5):1437–1452
58. Sasse J, Sands R (1996) Comparative responses of cuttings and seedlings of *Eucalyptus globulus* to water stress. Tree Physiol 16:287–294
59. Sasse J, Sands R (1997) Configuration and development of root systems of cuttings and seedlings of *Eucalyptus globulus*. New For 14:85–105
60. Steffens B, Ramussen A (2016) The physiology of adventitious roots. Plant Physiol 170:603–617
61. Stelzer HE, Foster GS, Shaw V, McRae JB (1998) Ten-year growth comparison between rooted cuttings and seedlings of Loblolly pine. Can J For Res 28:69–73
62. Stevens ME, Woeste K, Pijut PM (2018) Localized gene expression changes during adventitious root formation in black walnut (*Juglans nigra* L.). Tree Physiol 38:877–894
63. Stoessl A, Venis MA (1970) Determination of submicrogram levels of Indole-3- acetic acid. Anal Biochem 34:344–351
64. Struve DK, Talbert JT, McKeand SE (1984) Growth of rooted cuttings and seedlings in a 40-year-old plantation of eastern white pine. Can J For Res 14:462–464
65. Stuepp CA, Zuffellato-Ribas KC, Wendling I, Koehler HS, Bona C (2014) Vegetative propagation of mature dragon trees through epicormic shoots. Bosque Valdivia 35(3):337–345
66. Tewari DN (1995) A monograph on Rosewood: *Dalbergia latifolia* Roxb. International Book Distributors, Dehradun, p 74
67. Thorpe TA, Harry IS, Kumar PP (1991) Application of micropropagation to forestry. In: Debergh PC, Zimmerman RH (eds) Micropropagation, technology and application. Kluwer Academic, Dordrecht, pp 311–336. https://doi.org/10.1007/978-94-009-2075-0_21
68. Troup RS (1921) The silviculture of Indian trees, vol I. Clarendon Press, Oxford, pp 318–325
69. Vasudevan R, Sasidharan KR, Venkataramana KS, Thangamani D, Svanantham S (2020) Standardization of macropropagation technique for *Dalbergia latifolia* Roxb. through root cuttings, its application in production of quality planting stock and conservation of genetic resources. Int J Adv Res Rev 5(11):9–19
70. Veierskov B (1988) Relation between carbohydrates and adventitious root formation in cuttings. In: Davis TD, Haissig BE, Sankhla N (eds) Adventitious root formation in cuttings. Dioscorides Press, Portland, pp 70–78
71. Wilson PJ (1994) The concept of a limiting rooting morphogen in woody stem cuttings. J Hort Sci 69:91–600
72. Zalesny RS, Riemenschneider E, Hall RB (2005) Early rooting of dormant hardwood cuttings of *Populus*: analysis of quantitative genetics and genotype x environment interactions. Can J For Res 35(4):918–929
73. Zargar AR, Kumar D (2018) Effect of maturity stage of donor plant on propagation of *Diploknema butyracea* through branch cuttings. World J Agric Res 6(1):15–19

Chapter 12
Micropropagation Challenges in *Dalbergia latifolia*: A Comprehensive Review

Rekha R. Warrier, A. N. Arunkumar, and Geeta Joshi

1 Introduction

Micropropagation in trees is a crucial biotechnology tool enabling rapid multiplication of superior genotypes, conservation of endangered species, and production of disease-free planting material [1]. *Dalbergia latifolia* is one such commercially important timber species of India valued for its finest woods [2]. Huge international demand for its wood, driven by its exceptional wood properties such as long, straight trunks, durability, and high density, has led to significant market value [3]. Due to these factors, the IUCN has classified it as "Vulnerable" on the Red List [4] and classified as an Appendix II CITES wood [5]. The Indian Forest Act (1927) prohibits the export of rosewood logs or sawn timber [6]. This chapter collates knowledge on the micropropagation of this woody plant species, offering insights for developing a more targeted and comprehensive protocol. It examines existing literature on the choice of explants, risks of contamination and browning, media for culture initiation and multiplication, acclimatization, and field survival.

R. R. Warrier (✉)
ICFRE-Institute of Forest Genetics and Tree Breeding, Coimbatore, India
e-mail: rekha@icfre.org

A. N. Arunkumar
ICFRE-Institute of Wood Science and Technology, Bengaluru, India

G. Joshi
Indian Council of Forestry Research and Education (ICFRE), Dehradun, India

R. R. Warrier et al. (eds.), *Handbook of Asian Rosewoods*, Materials Horizons: From Nature to Nanomaterials, https://doi.org/10.1007/978-981-96-9459-4_12

1.1 Explant Selection and Preparation

1.1.1 Explants Selection

The choice of explant plays a critical role in the success of micropropagation. The most frequently used explants come from seedlings derived from in vitro cultures (Fig. 1), preferred due to their juvenile state and the absence of microbial contamination—a common issue when using explants from mature trees. Commonly used explants for rosewood micropropagation include shoot tips, nodal segments, axillary buds, and occasionally leaves (Fig. 2). Factors such as the donor plant age and the explant collection season significantly impact regeneration potential. Nodal segments have been particularly successful due to their meristematic activity, which promotes rapid shoot proliferation. Nodal explants of *D. latifolia* have been shown to produce multiple shoots faster than shoot tips [7]. Season influences explant quality; explants collected during spring and early summer tend to show higher proliferation rates, likely due to their increased metabolic activity.

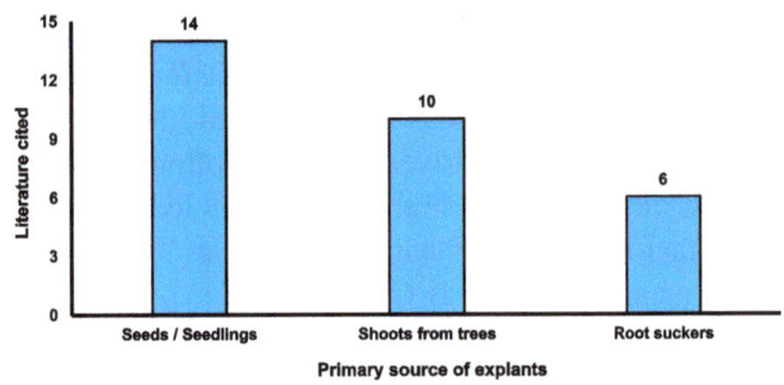

Fig. 1 Literature citing different sources to obtain primary explants for tissue culture of *Dalbergia latifolia*

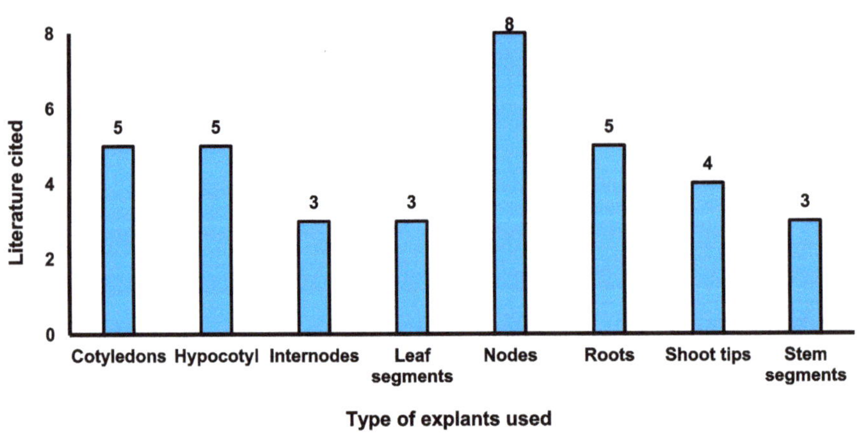

Fig. 2 Literature citing commonly used explants for tissue culture of *Dalbergia latifolia*

1.1.2 Explant Preparation

The challenges associated with explant contamination and browning have been a major obstacle in developing reliable and efficient in vitro propagation protocols for this species. A combination of ethanol, sodium hypochlorite, and rinsing with sterile water is employed. Over-sterilization can damage tissues and inhibit growth, necessitating careful balancing between sterility and tissue viability. Explants from mature trees often exhibit significant browning and contamination, a problem widely reported in the literature [3].

Trees in the 60- to 80-year-old age range have been observed to exhibit similar issues, with shoot tip necrosis and leaf abscission being common occurrences [8, 9]. These age-related physiological changes hinder the success of tissue culture propagation, making it more challenging to maintain the desired characteristics of these elite genotypes. Even explants taken from 5- to 6-year-old saplings release phenolic compounds during the establishment of in vitro culture, hindering the culture initiation's success [10].

Various substances have been utilized to overcome the issue of explant browning in *D. latifolia in vitro* cultured, such as polyvinylpyrrolidone, activated charcoal, citric acid, and ascorbic acid. Activated charcoal (AC) has proven effective [9] as it effectively adsorbs and removes the phenolic compounds released by the explants, thereby reducing the browning problem and enhancing the success of in vitro culture establishment. Juvenile explants could be inoculated without any treatment to control the phenol leach, whereas mature explants required a combination of methods to overcome polyphenol interference. Poly vinyl Pyrrolidone (PVP) was another proven chemical to reduce leaching of phenolics. Overall, a combination of antioxidants and adsorbents in the media helped to control polyphenol interference [11].

High rates of contamination (greater than 80%) were observed in shoot tips and nodes collected in all months from June to February. Survival of buds without contamination could be obtained only in the dry season [12].

1.2 Culture Media and Composition

Murashige and Skoog (MS) medium has been widely adopted in *Dalbergia* tissue culture due to its balanced composition of macronutrients and micronutrients that supports a wide range of species. Studies have shown that alterations in nitrogen and potassium levels can influence shoot and root formation in *Dalbergia* species. Lowering the nitrogen content has been reported to reduce hyperhydricity in *D. latifolia*, thus improving shoot quality [13]. In addition to inorganic nutrients, organic supplements such as coconut water, casein hydrolysate, and activated charcoal can improve growth outcomes. Activated charcoal included in the medium has been effective in adsorbing phenolic compounds released by *Dalbergia* explants, which can otherwise inhibit growth and lead to tissue browning [14]. Use of 0.2% (w/v)

Bavistin in the medium has been effective in reducing the contamination to less than 60% and in delaying the appearance of the contamination [12].

1.3 Role of Plant Growth Regulators (PGRs)

The use of plant growth regulators (PGRs) is central to *Dalbergia* micropropagation. A combination of auxins and cytokinins were commonly employed to induce shoot and root development. Benzylaminopurine (BAP), kinetin (Kn), and indole-3-butyric acid (IBA) have demonstrated promising results in inducing multiple shoot formation [8, 15] and root initiation [9]. A 2.0 mg/L BAP in combination with 0.5 mg/L Kn resulted in optimal shoot multiplication [7]. BAP has been observed to be critical for shoot induction, with optimal concentrations ranging from 0.5 to 3.0 mg/L [16, 17]. The addition of auxins like indole-3-acetic acid (IAA) and naphthalene acetic acid (NAA) supported both callus formation and rooting [18]. A rapid shoot multiplication protocol was developed for *Dalbergia latifolia* with shoot initiation after 7 days (on BAP) and 10 days (on Kn) of inoculation [19]. Though several studies have been attempted, there is a need to optimize concentration and ratio of PGRs to control hyperhydricity and vitrification in *D. latifolia.* Polyploidy was induced using oryzalin to obtain fast-growing and more vigorous rosewood seedlings for quicker harvest; however, it did not help produce faster growing seedlings [20].

1.4 Callus Induction and Somatic Embryogenesis

While organogenesis is the primary pathway for *Dalbergia* micropropagation, somatic embryogenesis presents an alternative method. Callus induction has been achieved using auxins such as 2,4-dichlorophenoxyacetic acid (2,4-D) and naphthalene acetic acid (NAA), with somatic embryos generated through different combinations of cytokinins and auxins. High cytokinin levels are shown to inhibit embryo maturation; thus reduced cytokinin concentration during later stages of callus formation has shown to improve somatic embryo development and conversion to plantlets. Somatic embryogenesis was achieved in *D. latifolia* using a two-phase method with 2,4-D, sucrose, and coconut water [21, 22], while somatic embryos were generated on MS medium with 0.5 mg/L 2,4-D and 0.1 mg/L kinetin, and successfully regenerated plantlets using 0.5 mg/L BA and abscisic acid [23].

2 Challenges in *Dalbergia* Micropropagation

Though identified as a very important species, very few literature cites are available (<25) related to micropropagation as a means to propagation of the species. Some of the major challenges which deter the efforts are detailed here.

2.1 Phenolic Exudation and Explant Browning

Phenolic exudation is a significant challenge in *Dalbergia* tissue culture, often leading to explant browning and cell death. Various strategies, including the use of antioxidants (e.g., ascorbic acid, citric acid) and adsorbents (e.g., activated charcoal, polyvinylpyrrolidone), have been shown to reduce browning [9]. Adding 0.1% activated charcoal effectively reduced phenolic oxidation in *D. latifolia*, allowing for improved shoot initiation rates, improving the overall success of cultures [11, 14].

2.2 Hyperhydricity and Vitrification

Hyperhydricity (vitrification) in *Dalbergia* micropropagation results in translucent, waterlogged tissues, caused by high cytokinin levels, low light, and high humidity. Reducing cytokinin concentrations, improving ventilation, and using semi-solid media can mitigate this disorder. Adjusting light and humidity further enhances tissue quality and promotes healthy shoot elongation [9].

2.3 Rooting and Acclimatization of Plantlets

Rooting has been effectively achieved using half-strength MS medium supplemented with IBA (Indole-3-butyric acid) at 2.0 mg/L, resulting in high rooting success [16, 18]. Direct rooting from stem segments has also been documented, indicating flexibility in propagation methods [24]. The transition from in vitro to ex vitro conditions is a delicate phase, with high mortality rates often observed. Acclimatization is challenging due to the weak root systems and high humidity requirements of in vitro grown plantlets. Methods such as gradual reduction of humidity, soil adaptation, and the use of mycorrhizal fungi have shown promise in improving survival rates during the acclimatization stage.

3 Future Line of Work

Micropropagation remains a vital tool for the conservation and sustainable production of *Dalbergia* species. Despite the numerous challenges, advances in tissue culture protocols, environmental control, and molecular techniques help enhance the efficiency of *Dalbergia* micropropagation. Continued research and development in this field will be essential for the preservation of these economically and ecologically valuable species.

1. **Refinement of Somatic Embryogenesis Protocols**: Refining protocols to enhance the conversion of somatic embryos into viable plantlets remains a promising avenue for large-scale propagation of *Dalbergia* species.
2. **Application of Genomic Tools**: Understanding the genetic basis of traits related to micropropagation success, such as rooting ability and phenolic resistance, could enable targeted breeding. Engineering genotypes, potentially through CRISPR-based gene editing, could significantly advance micropropagation efficiency.
3. **Scaling Production Through Cost-Effective Bioreactor Systems**: Developing cost-effective bioreactor systems that minimize labor and improve yield per unit area could enable viable large-scale propagation, benefiting both conservation programs and commercial forestry.
4. **Integration of Cryopreservation in Conservation Strategies**: Cryopreservation of rare *Dalbergia* genotypes could serve as a critical component of ex situ conservation efforts, especially gene banks, enabling restoration efforts if natural populations decline further.

4 Conclusion

Micropropagation presents a valuable tool for the conservation and sustainable production of *Dalbergia* species. Advances in tissue culture protocols, understanding of PGR interactions, and the application of biotechnological innovations have improved the success rates of *Dalbergia* micropropagation. However, challenges related to phenolic exudation, hyperhydricity, and acclimatization persist. Addressing these challenges and pursuing further research in molecular genetics, bioreactor technology, and cryopreservation will be instrumental in enhancing the effectiveness of *Dalbergia* micropropagation for conservation and commercial applications.

References

1. Murthy HN, Joseph KS, Paek KY, Park SY (2023) Bioreactor systems for micropropagation of plants: present scenario and future prospects. Front Plant Sci 14:1159588
2. Arunkumar AN, Warrier RR, Kher MM, Teixeira da Silva JA (2022) Indian rosewood (*Dalbergia latifolia* Roxb.): biology, utilisation, and conservation practices. Trees 36:883–898
3. Kher MM, Nataraj M, Arun Kumar AN, Sitther V, Shekhawat MS, Warrier RR, Teixeira da Silva JA (2021) Tissue culture of Indian rosewood (*Dalbergia latifolia* Roxb.). Biologia 6,3595-3604. https://doi.org/10.1007/s11756-021-00914-7
4. Lakhey P, Pathak J, Adhikari B (2020) *Dalbergia latifolia* Roxb. The IUCN Red List of Threatened Species 2020. https://doi.org/10.2305/IUCN.UK.2020-3.RLTS.T32098A67777757.en
5. Dwianto W, Bahanawan A, Kusumah SS, Darmawan T, Amin Y, Pramasari DA, Lestari E, Akbar F (2019) Study on the existence and characteristics of Sonokeling (*Dalbergia latifolia* Roxb) as an Appendix II CITES wood. In: IOP conference series: earth and environmental science, vol 374, 012063
6. Nijman V, Chavez J, Simons D, Siriwat P, Widiaswari RA, Svensson MS (2025) Quantifying illegal rosewood trade, seizures and forestry law enforcement in Indonesia. Appl Geogr 176:103509
7. Banerjee R, Shrivastava N (2008) Micropropagation of *Dalbergia latifolia* Roxb. through proliferation of axillary buds. In Vitro Cell Dev Biol Plant 44:489–494
8. Lakshmi Sita G, Raghava Swamy BV (1993) Regeneration of plantlets from leaf disc cultures of rosewood: control of leaf abscission and shoot tip necrosis. Plant Sci 88:107–112
9. Warrier KCS, Vijaykumar NK (1999) Clonal propagation of selected plus trees of Indian Rosewood (*Dalbergia latifolia* Roxb.) through tissue culture. In: Srivastava GC, Singh K, Pal M (eds) Plant physiology for sustainable agriculture. Pointer Publishers, Jaipur, India, pp 422–432
10. Sankara Rao K (1986) Plantlets from somatic callus tissue of the East Indian rosewood (*Dalbergia latifolia* Roxb.). Plant Cell Rep 5(3):199–201. https://doi.org/10.1007/BF00269118
11. Warrier KCS, Warrier RR, Vijaykumar NK (2005) Polyphenol interference in micropropagation of Indian rosewood (*Dalbergia latifolia* Roxb.). MyForest 41:561–566
12. Jaiswal M, Jha Z, Singh L, Lal J, Tirkey J (2023) Problem and its remedy of micropropagation in woody trees: a review. Environ Ecol 41:633–639
13. Rathore JS, Shekhawat NS (2009) Plant tissue culture and micropropagation of *Dalbergia* species: a review. Biotechnol Adv 27:343–350
14. Singh AK, Shukla AK (2015) Micropropagation and somatic embryogenesis of *Dalbergia* species: current status and future prospects. Tree Biotechnol Regen 33:157–165
15. Pradhan C, Pattnaik S, Chand PK (1998) Rapid in vitro propagation of East Indian rosewood (*Dalbergia latifolia* Roxb.) through high-frequency shoot proliferation from cotyledonary nodes. J Plant Biochem Biotechnol 7:61–64
16. Boga A, Ram B, Ravi G, Reddy S (2012) Effect of benzyl amino purine and gibberellic acid on in vitro shoot multiplication and elongation of *Dalbergia latifolia* Roxb.: an important multipurpose tree. Biotechnol Bioinforma Bioeng 2:597–602
17. Subasinghe S, Warakagoda PS, Abeywickrama WHT (2016) In vitro shoot proliferation of rosewood (*Dalbergia latifolia*: Fabaceae): a potent high-quality timber species. Trop Agric Res Ext 18:94–101
18. Raghava Swamy BV, Himabindu K, Sita GL (1992) In vitro micropropagation of elite rosewood (*Dalbergia latifolia* Roxb.). Plant Cell Rep 11:126–131
19. Manikandan G, Bhuvaneswari RT, Ramya MA, Ramasubbu R (2017) In vitro propagation and mass multiplication of *Dalbergia latifolia* Roxb.: a vulnerable tree species from Eastern Ghats, Tamil Nadu, India. Int J Bot Stud 2:52–57
20. Riastiwi I, Witjaksono UJS, Ratnadewi D (2024) Induction of polyploidy in *Dalbergia latifolia* Roxb. using oryzalin. Sains Malaysiana 53:1889–1900

21. Chandra S, Singh R (2006) Somatic embryogenesis in *Dalbergia sissoo* Roxb.: factors influencing embryogenic callus formation and somatic embryo development. Plant Cell Rep 25:251–259
22. Muralidhar Rao M, Lakshmi Sita G (1996) Direct somatic embryogenesis from immature embryos of rosewood (*Dalbergia latifolia* Roxb.). Plant Cell Rep 15:355–359
23. Shirin F, Bhadrawale D, Mishra JP, Sonkar MK, Maravi S (2020) Evaluation of biochemical changes during different stages of somatic embryogenesis in a vulnerable timber tree *Dalbergia latifolia* (Indian rosewood). In Vitro Cell Dev Biol Plant 56:894–902
24. Sudhadevi AM, Nataraja K (1987) In vitro regeneration and establishment of plantlets in stem cultures of *Dalbergia latifolia* Roxb. Indian For 113:501–506

Chapter 13
Distribution, Timber Quality, and Value of *Dalbergia* Species in Sri Lanka

S. M. C. U. P. Subasinghe and R. Rajathewa

1 Introduction

Sri Lanka is an island with an area of 65,610 km^2, located southwest of the Bay of Bengal in the Indian Ocean and southeast of the Arabian Sea. The northern tip of the country, Jaffna, lies about 31 km away from the southern coast of India. The country is divided into three primary climatic regions based on average rainfall: wet, dry, and intermediate zones. The wet zone receives over 2,500 mm of rainfall per year, the dry zone receives less than 1,750 mm, and the intermediate zone receives between 1,750 and 2,500 mm of rainfall annually. The wet zone receives rainfall mainly from the southwest monsoons from April to June, while the dry zone receives from the northeast monsoons from October to January [5].

Sri Lanka's geography is diverse, with a coastal plain encircling a central mountainous region. The landscape includes tropical forests, fertile plains, and numerous rivers, with the Mahaweli River being the longest. The country is also home to various ecosystems, including forests, wetlands, lagoons, and coral reefs, which support a rich biodiversity [3].

Sri Lanka's topography is divided into three zones based on elevation from mean sea level: the central highlands, the plains, and the coastal belt (Fig. 1). The central highlands, located in the south-central part of the country, feature the highest mountain, Pidurutalagala, which stands at 2,524 m above mean sea level. The largest zone in Sri Lanka is the plains, which range from 30 to 200 m above mean sea level. The coastal zone lies up to 30 m from the coastline [4].

S. M. C. U. P. Subasinghe (✉)
Department of Forestry and Environmental Science, University of Sri Jayewardenepura, Nugegoda, Sri Lanka
e-mail: upuls@sjp.ac.lk

R. Rajathewa
Bandaranayakapura, Indigolla, Gampaha, Sri Lanka

R. R. Warrier et al. (eds.), *Handbook of Asian Rosewoods*, Materials Horizons: From Nature to Nanomaterials, https://doi.org/10.1007/978-981-96-9459-4_13

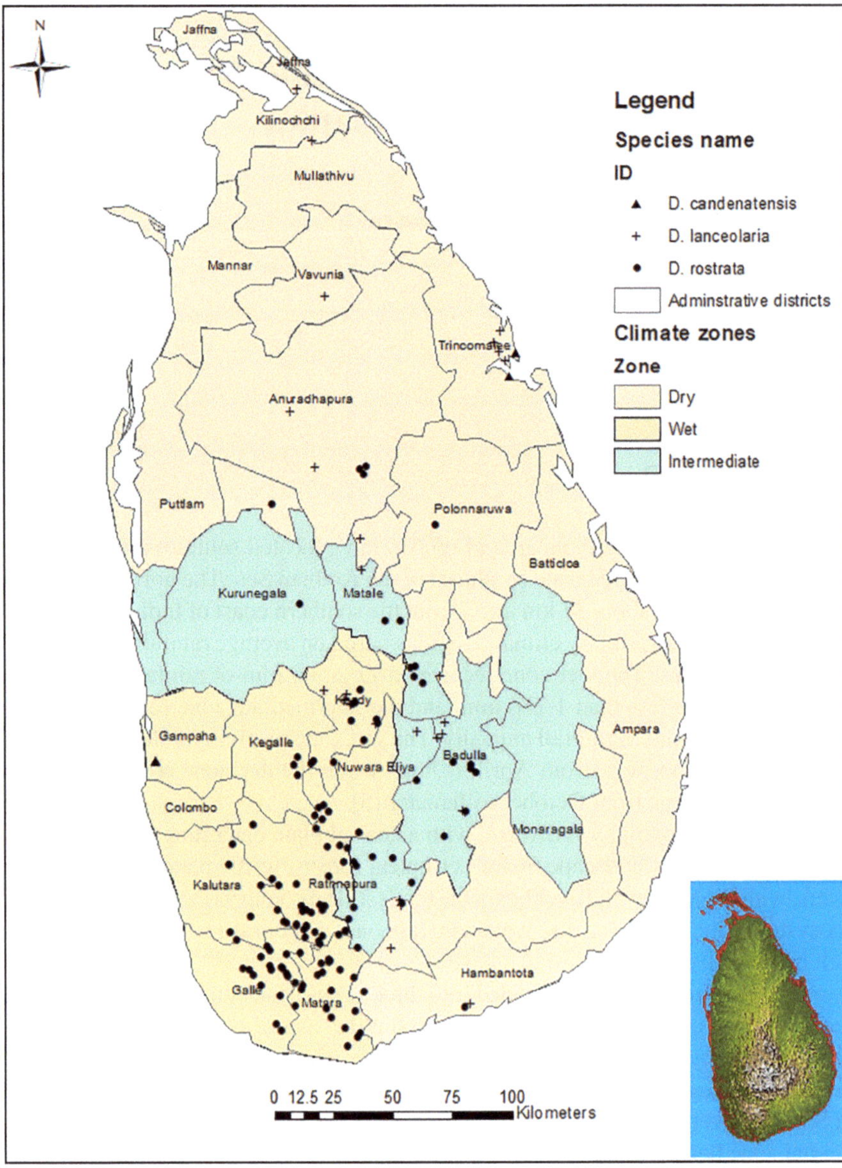

Fig. 1 Distribution of native *Dalbergia* species in Sri Lanka. Bottom right: elevation map of Sri Lanka (*Source* [11])

Table 1 Status of the four *Dalbergia* species growing in Sri Lanka

Characteristic	*D. candenatensis*	*D. lanceolaria*	*D. rostrata*	*D. latifolia*
Local name	–	Bol mara (S) Velaruvai (T)	Bambara wel	–
Habit	Shrub or liana	Tree	Woody climber	Tree
Origin	Native	Native	Native	Exotic
NCS	VU-vulnerable	VU-vulnerable	LC-least concern	
Legal status	Not protected	Not protected	Not protected	–

S = Sinhala language, T = Tamil language, NCS = National conservation status

Due to the climate and geographic variations, Sri Lanka has a wide variety of flora. Some species grow throughout the country, while many are restricted to specific climates or topographic zones [9]. Numerous introduced species, such as teak and mahogany, have adapted well to the local environment and provide valuable timber. Among approximately 250 species of *Dalbergia* growing in the tropics [7], a few have been introduced to Sri Lanka (Table 1), while three species are native to the country.

2 *Dalbergia* Species in Sri Lanka

According to the country's floristic history, there are three *Dalbergia* species: *D. candenatensis, D. lanceolaria,* and *D. rostrata* are native to Sri Lanka (Table 2; Fig. 1). Additionally, *D. armata, D. assamica, D. latifolia, D. melanoxylon,* and *D. volubilis* have been introduced, mainly at the Royal Botanic Gardens, Peradeniya. However, *D. latifolia* has been planted in locations outside the botanical gardens. Records of the remaining introduced species are lacking.

2.1 Native Dalbergia Species

2.1.1 *Dalbergia candenatensis*

This species is a scandent shrub or liana, which climbs with divaricate lateral twigs often hooked at the ends. It is distributed in the coastal areas of Sri Lanka, especially in mangrove swamps [1]. Additionally, this species has been recorded in India, Malaysia, the Philippines, Polynesia, China, and northern Australia [1]. No records are available on this species' use in Sri Lanka.

Table 2 Morphological differences of the native *Dalbergia* species in Sri Lanka

Characteristic	*D. candenatensis*	*D. lanceolaria*	*D. rostrata*
Height	Liana	Tree-20 m	Woody climber
Stem	Glabrous	Glabrous	Pubescent
Leaves	4–7-foliolate	(7–)10–15-foliolate	1–5-foliolate
Leaflets	Elliptic to obovate-elliptic, obtuse, or emarginated	Elliptic, broadly oblong, or somewhat obovate, obtuse, or retuse	Subcoriaceous, ovate, acuminate
Leaflet size	2–5 cm × 1.5 cm	(1.5–)2.5–4.5 cm × 1–2.5(–3) cm	4–10 cm × 2.5–7 cm
Petiole	Short	5 mm long	Short
Inflorescences	Short-paniculate or racemose	Axillary, short-paniculate	Axillary, paniculate
Flowers	6–10 mm long (Aug–Nov)	7–10 mm long (May–Aug)	6–7 mm long (Apr, May)
Petals	White to pinkish	Violet	White
Fruits	Glabrous, falcate lunate, acute 2.5–5 cm × 1–1.5 cm	Glabrous with stipe 1 cm long 3.5–5.0-(10) cm × 1.2–2 cm	Oblong, straight, short-stipitate, 7.5–10 cm × 1.8 cm

Source [1, 7]

2.1.2 *Dalbergia rostrata*

D. rostrata grows in moist areas at elevations of 300–1,200 m. Besides Sri Lanka, it has also been recorded in southern India, Borneo, and Java. In ancient times, this vine was used to tie felled tree stems to elephants so they could drag them to stacking places or local sawmills.

2.1.3 *Dalbergia lanceolaria*

D. lanceolaria is the only native tree species of *Dalbergia* in Sri Lanka. It is distributed up to about 900 m from the mean sea level in the country. It can grow up to 30 m in height and 30–60 cm in diameter [1, 2]. Its wood is yellowish to white in colour, soft, and not durable. Heartwood and softwood of this species cannot be distinguished [2, 8]. Due to the softness of the wood, this species is locally called "Bol mara". "Bol" in the local language means "not useful", and "mara" is another local name given to some *Albizia* species in Sri Lanka. Due to these characteristics of timber, *D. lanceolaria* is not used for construction purposes of furniture manufacturing on a wide scale. However, cottage-level carpenters occasionally use the wood of this species due to the danger of falling, damaging properties, etc., for interior structures of wardrobes and beds. According to the president of the Furniture Manufacturing Association in the Moratuwa region of Sri Lanka, where the highest furniture supply is made, the timber of *D. lanceolaria* has never been used in any product manufacturing.

Additionally, the bark, leaves, root, and seed oil of this species have ethnomedical value. Traditional medicine practitioners of Sri Lanka used bark, leaves, and roots for treating rat snake bites, orthopaedic lesions, and chronic ulcers [10]. Indian communities used the bark for dyspepsia. The leaves are used for treating leprosy and allied obstinate skin diseases, while the seed oil is used for rheumatism and cutaneous diseases [6].

2.2 *Exotic Dalbergia Species*

Among the exotic species, *D. armata, D. assamica, D. melanoxylon,* and *D. volubilis* were first recorded from the 1805–1852 period in Sri Lanka at the Royal Botanic Garden, Peradeniya. However, their recent records or descriptions are not available. Records, to a certain extent, are available only for *D. latifolia.*

2.2.1 *Dalbergia latifolia*

D. latifolia, Indian rosewood, was introduced to Sri Lanka in the eighteenth century [1]. It is a deciduous tree which grows 20–40 m with a girth of 1.5–2.0 m. Its bark is thin, grey with short cracks that exfoliate into fibrous flakes. The leaves are pinnate and alternate, with dark green leaflets on top and pale green underneath. The white flowers are 5–10 mm long and grow in axillary panicles. The tree produces oblong-lanceolate brown pods and pointed at both ends, containing 1–4 smooth brown seeds [1].

D. latifolia has been recorded outside of botanical gardens in lower elevations of the hilly areas and in the low country, especially in home gardens, where the annual rainfall is between 1,100 and 3,300 mm [8]. The heartwood timber of *D. latifolia* is highly valuable, ranging from rose to brown, with a density of 750 gcm^{-3}. However, major woodworkers have not used the wood of *D. latifolia* as construction material or for furniture manufacturing in Sri Lanka.

3 Concluding Remarks

Among the three native *Dalbergia* species, *D. candenatensis* has a highly restricted distribution in the coastal area of Trincomalee in northeast Sri Lanka. *D. lanceolaria* mainly grows in the low and high elevations of the dry and intermediate climatic zones. *D. rostrata* is primarily distributed in the low and high elevations of the wet zone, with a few areas in the dry and intermediate climatic zones. None of these three species bears any commercial value in Sri Lanka. However, there is occasional use of

D. lanceolaria timber for interior structures of wardrobes and beds by cottage-level carpenters.

References

1. Dassanayake MD (1991) A revised handbook of the flora of Ceylon, vol 7, 1st ed. Routledge, London, p 452. https://doi.org/10.1201/9781315136455
2. Dholariya CA, Sukhadiya M, Behera LK, Mehta AA, Nayak D, Patel SM (2019) *Dalbergia lanceolaria* L.f.: prospect of a lesser known tree species in India. Van Sangyan 6(4):9–11
3. Dilakshi WPN (2023) Geogrpahy of Sri Lanka. University of Kelaniya, Sri Lanka, Faculty of Social Sciences
4. Heitzman J (1990) The physical environment. In: Ross RR, Savada AMS (eds) Sri Lanka: a country study. Federal Research Division, Library of Congress, Washington, D.C., pp 61–68
5. Karunaweera ND, Galappaththy GN, Wirth DF (2014) On the road to eliminate malaria in Sri Lanka: lessons from history, challenges, gaps in knowledge and research needs. Malar J 13:1–10
6. Khare CP (2007) Indian medicinal plants: an illustrated dictionary. Springer-Verlag, New York, USA, pp 199–201
7. Mabberley DJ (2008) Mabberley's plant book–a portable dictionary of plants, their classifications and uses. Cambridge University Press, Cambridge
8. Ruwanpathirana N (2016) Sri Lankan timber trees: timber properties and uses. Mihilaka Forestry Service, Mattegoda, Sri Lanka
9. Nigamuni SS, Subasinghe SMCUP (2015) Floral diversity of six mangrove forests along the northwestern coastline of wet, intermediate and dry climate zones of Sri Lanka. Wildlanka 3(4):184–194
10. http://www.instituteofayurveda.org/plants/plants_detail.php?i=610&s=Scientific_name. Accessed 2 July 2024
11. https://www.jpl.nasa.gov/images/pia06670-sri-lanka-colored-height

Chapter 14
Dalbergia oliveri Gamble ex Prain

Chaloun Bounithiphonh and Soulivanh Lanorsavanh

1 Introduction

Dalbergia oliveri (*D. oliveri* Gamble ex Prain) (Common name—Burmese rosewood) is a species in the family Fabaceae with the vernacular names *Kham phee leung* in Lao, Tamalan or Chingchan in Burmese, Neang Nuon in Central Khmer, Mai Ching Chan in Thai, and Cam Lai in Vietnamese [1, 2]. This species is native to Laos PDR, Myanmar, Cambodia, Thailand, and Vietnam. It is widespread in Cambodia, Laos PDR, and Myanmar, with a limited range in Northeast Thailand. Its potential natural range is 997,383 km^2, while its extent of occurrence is estimated at 1,489,222 km^2 [3]. Though the species has a wide geographic range, the spread is limited to small, scattered subpopulations. Once common, the wood now faces pressure from the international timber market due to rising demand for rosewood, especially for hongmu furniture in China and Vietnam. With soaring rosewood imports by China, there was heavy sourcing of *Dalbergia oliveri* [4]. A report on the factors that influence the growth and advancement of small-scale handicraft enterprises in Myanmar revealed that *D. oliveri* was used extensively until the last decade [5]. Myanmar became the biggest log supplier to China worldwide in 2014 (Fig. 1), surpassing more traditional suppliers such as Vietnam (Fig. 2) and Laos.

C. Bounithiphonh (✉)
Forest Research Center, National Agriculture and Forestry Research Institute, Vientiane, Laos
e-mail: chalounb@yahoo.com; K.jular28@gmail.com

S. Lanorsavanh
Department of Biology, Faculty of Natural Sciences, National University of Laos, Vientiane, Lao PDR

R. R. Warrier et al. (eds.), *Handbook of Asian Rosewoods*, Materials Horizons: From Nature to Nanomaterials, https://doi.org/10.1007/978-981-96-9459-4_14

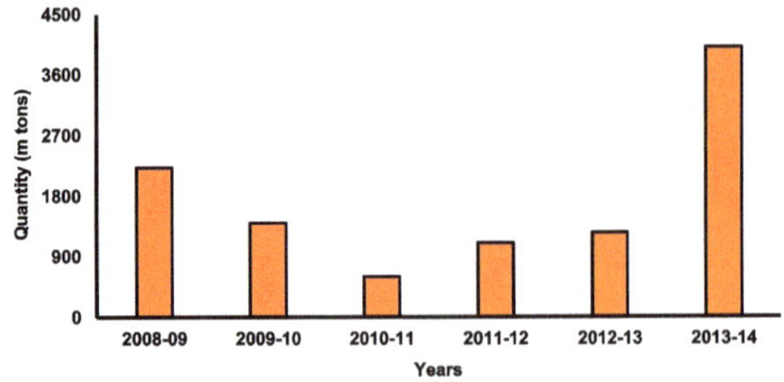

Fig. 1 Quantity of *D. oliveri* wood seized in Myanmar from 2008 to 2013

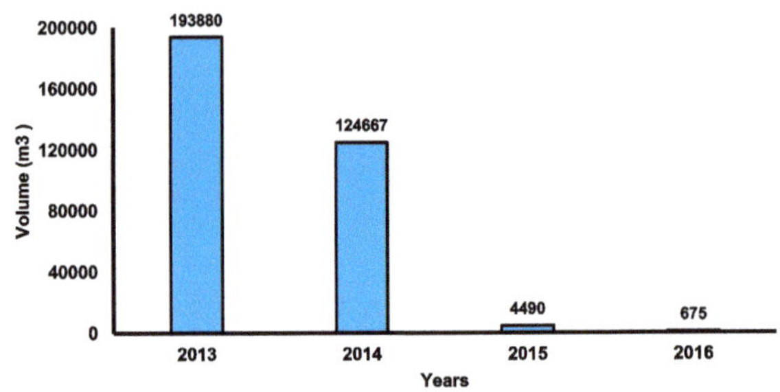

Fig. 2 Sawn wood of *D. oliveri* exported from Vietnam to China for the period 2013–2016

2 Morphology and Taxonomy

The tree grows 10–30 m tall (Fig. 3A), with a round stem and grey to dark grey scaly bark that flakes in small pieces (Fig. 3B). Young shoots are glabrous or slightly hairy, with a purplish hue. Stipules are small and shed early. Leaves are alternate, imparipinnate, with a 15–25 cm long glabrous rachis. Each leaf consists of 13–17 ovate-oblong to ovate-lanceolate leaflets, measuring 4–8 cm in length and 1.2–3 cm in width (Fig. 3C). The base is cuneate to rounded, the margin entire, and the apex obtuse to acute, sometimes with an apiculate tip. The upper surface is dark green, while the lower surface is greenish to greyish, with 9–12 pairs of secondary veins. Inflorescences are terminal or subterminal panicles, 10–15 cm long. Flowers are purple, pale purple, or white, measuring 10–12 mm long with a purple tinge. The calyx is 5-teethed, 4–5.5 mm long, purple, and hairy at the teeth. The corolla has five petals, including a rounded or orbicular standard petal and spoon-shaped wing petals. Stamens are monadelphous (5 + 5), 6–8 mm long. The ovary is hairy along the lower margin, with 2–4 ovules. The lanceolate pod is 9–14 cm long, indehiscent, and contains 1–2 seeds (Fig. 3D), which are reniform and reddish-brown to dark brown [6, 7] (Fig. 3E).

Fig. 3 **A** *D. oliveri* fruiting branch. **B** Grey scaly bark of *D. oliveri*. **C** Leaves and mature pods. **D** Pods with one and two seeds. **E** Stages of seed maturity

3 Habitat and Ecology

D. oliveri grows in evergreen, mixed deciduous, dry dipterocarp forests at 900–1,200 m elevations. It occurs individually or in groups of 5–10 trees, typically in evergreen or semi-evergreen forests dominated by *Lagerstroemia* and Dipterocarps. It thrives well near water sources or on hillsides. Young trees tolerate shade but require sunlight when mature. While *D. oliveri* produces many seeds, natural regeneration is low due to poor germination. Growth is generally slow in both natural forests and plantations [8].

4 Distribution

Cambodia: The species is sparsely found in Kratie, Preah Vihear, Kampong Thom, Ratanakiri, Stung Treng, Pursat, and Siem Reap provinces [7, 8]. A survey in 2021 revealed that the population density of *D. oliveri* in Cambodia is 0.8 plants/ha with DBH above 0.5 cm [9, 10].

Laos: There is limited population information for this species in Laos PDR. Based on a survey by National Agriculture and Forestry Research Institute (NAFRI) in 2021–22, the species has been identified in Savannakhet and Salavan-Khammouan, Savannakhet, Salavan, Champasack, Attapue, and Sekong provinces. The total number of *D. oliveri* recorded was 611 individuals. The species is scarce in plantations and outside natural forests; most trees are found in natural forests [7]. Between 2013 and 2014, rosewood exports from Lao PDR to China rose by 150% [11], leading to significant pressure and subpopulation loss and threatening species sustainability.

Myanmar: Majority of the populations are recorded in the upper part of the country [12] scattered in mixed and dry deciduous forests. About 40 lakh trees were reported in 2014 [12, 13] as being associated with teak and bamboo. Highest densities in Sagaing division (over two million trees—1,203,600 m^3), followed by Shan state (900,000 trees—354,000 m^3), and Mandalay and Kachin (141,600–212,400 m^3) reduced to one-third as a result of heavy extraction [13].

Thailand: It is known as common in the north, northeast, and eastern regions, especially along the Dang Rek mountain range that forms part of the international boundary between Cambodia and Thailand.

Vietnam: *Dalbergia oliveri* is recorded mainly in central provinces and some southeast provinces [14] with an estimated 200,000 trees in protected areas such as national parks [15].

There are reports of species occurrence in North Eastern India, Sikkim, and Siang in Arunachal Pradesh [16, 17].

5 Propagation

Propagation is mainly through seeds. Each kilogram of pod collected contains approximately 6,100 seeds. The young pods are green but turn dark brown upon ripening (Fig. 3D). To prevent insect damage, fruits should be collected as soon as they mature. The ideal seed collection period is December and January. Collectors lay a canvas around the base of a mother tree before shaking its branches to release seeds. Another method involves cutting small branches for easier collection. Seed pods are sun-dried for 2–3 days before seeds are extracted by threshing. To separate damaged seeds, they are placed in water—damaged seeds sink, while undamaged seeds float. The viable seeds are then sun-dried for a day before being stored in a cold room at 10 °C. Proper handling ensures seed viability and longevity, making collection and storage crucial steps in the propagation process [9, 10].

6 Uses

The heartwood is very hard and heavy, pink with fine grain. Sapwood is yellowish combined with dark-brown heartwood. Its major uses are as timber, high-quality furniture, luxury cabinets, art and handicrafts, decorations, handles of agricultural implements, tone wood and medicinal value [18].

7 Present Threat Status

The IUCN Red List Assessment, 1998, reported that there had been a rapid decline in the number of large trees because of overexploitation of the precious timber. Assessed as Endangered ("EN A1cd"), this was reassessed as "Critically Endangered" in 2022, as it was observed, estimated, inferred, or suspected to have a population reduction of at least 50% over three generations, due to a decline in the area of occupancy, extent of occurrence, and/or quality of habitat; and actual or potential levels of exploitation.

Cambodia: The Cambodia Tree Seed Project (2003) reported that regeneration effort for this species on a large scale have been limited and that mature and large-sized trees were rarely found in many areas of its natural range. With only 157 trees identified as seed sources in an area of 100 ha, the number of remaining individual trees was very low and were disappearing on a local level. It was considered that the species was facing extinction and that the effective protection measures implemented were insufficient.

Laos PDR: For years, the government has worked to conserve and protect endangered tree species, including two *Dalbergia* species. Despite national laws, illegal logging and land conversion for agriculture and infrastructure continue to threaten their habitats. A 2022 survey by NAFRI [19] confirmed heavy harvesting over decades for timber and land use. In 2021, Department of Forestry (DoF) [20] approved converting natural forests for development, risking habitat fragmentation and reducing regeneration potential, further endangering these species' survival.

Myanmar: In 2014, total estimated stocks of *D. oliveri* were 1.6 million cubic meters. The stocks in Myanmar rapidly declined due to trade growth rates.

Vietnam: In 1998, the species was assessed as native to central and southern Vietnam. The IUCN Red List Assessment reported that overexploitation of its timber had led to decline throughout the entire population. Initially assessed as Endangered ("EN A1cd"), presently it has been reassessed as Critically Endangered [7, 18].

a. **Legislative Prohibitions and Conservation Efforts for *Dalbergia oliveri*** [21–23]

Cambodia: The Cambodian Law on Forestry (Article 29) prohibits harvesting rare tree species and trees below the minimum diameter from Permanent Forest Reserve Areas. *D. oliveri* is listed under Cambodia's priority categories 1 and 4 for endangered or rare species. A National Forest Gene Conservation Strategy has been established, designating six stands with 263 mother trees across 186 ha. A 72.5 ha in situ conservation area exists for *D. oliveri* [21].

Laos PDR: The Decision No. 0448/MAF lists *D. oliveri* under MAF-Appendix I, strictly prohibiting wild harvesting without government approval. Since 2013, logging of natural forests for commercial purposes has been suspended (Order No. 31/PM). The NAFRI-Darwin Dalbergia project (2021) restored village sacred forests and national protected areas in Savannakhet province. A study in Daen Sa Tueng identified sufficient populations of *D. oliveri* for tree breeding. Between 1975 and 2015, a seed source for *D. oliveri* was established in Houyhuang village, Champasak province. In 2000, the Forest Research Center of NAFRI collected *D. oliveri* seeds, producing over 10 million seedlings from Saravan, Vientiane, Khammouane, Savannakhet, and Champassak provinces. The National University of Laos planted *D. oliveri* on a 2-hectare plot, demonstrating successful conservation techniques. Forest restoration efforts by the DOF include natural regeneration and enrichment planting.

Myanmar: A national log export ban since 2014 prohibits exporting unprocessed logs. *D. oliveri* is classified as a "**reserve species**", requiring MOECAF authorization for harvesting and trade. Timber extraction is prohibited in National Parks, Wildlife Sanctuaries, and Nature Conservation Areas [22].

Thailand: A nationwide ban in 1989 on logging natural forest specimens ensured protection of *D. oliveri* [23]. The In-situ genetic conservation covered for 34 ha, with *D. oliveri* considered a high priority for conservation have been established while in Tan Phu Forest (100 ha) includes enrichment plantations where *D. oliveri* seeds have been collected. Ex situ conservation stands have been established at Sakaerat Silvicultural Station, Nakhon Ratchasima, Surat Thani, and Kamphaeng Phet Silvicultural Research Stations, with 20 *D. oliveri* trees. The species is classified as "well conserved" in situ and "partly conserved" ex situ [24]. Research priorities focus on distribution, status, and conservation strategies. Occurrence has been reported in the Ban Pong Forest Sanctuary.

Vietnam: *D. oliveri* is permitted only for scientific use, breeding, and international cooperation. Protected subpopulations exist in Nam Cat Tien National Park and Yok Don National Park.

8 Conclusion

D. oliveri, a critically endangered rosewood species, faces significant threats due to overexploitation and habitat loss. While legal protections and conservation efforts are in place, their effectiveness is hindered by various challenges, including weak enforcement, insufficient resources, and lack of public awareness. A multi-faceted approach is essential to ensuring this valuable species' long-term survival. Key conservation strategies include strengthening law enforcement, protecting and restoring habitats, ex situ conservation, community engagement, scientific research, and international cooperation. By implementing these measures and addressing the underlying issues, it is possible to safeguard *D. oliveri* and preserve its genetic diversity for future generations.

References

1. Sinly S, Dany C, Sothea H, Phanit P, Sopheap L, Sobon K (2021) CITES non-detriment findings report on *Dalbergia cochinchinensis* and *Dalbergia oliveri* in the Choam Ksant district. Preah Vihear Province Forestry Administration, Phnom Penh, Cambodia, p 54
2. Center for Nature Conservation and Development (2021) Non-detriment findings for *D. cochinchinensis* and *D. oliveri* in Vietnam. CITES Management Authority of Vietnam, Ha Noi, p 28
3. Gaisberger H, Fremout T, So T, Thammavong B, Bounithiphonh C, Hoa TT, Yongqi Z, Kanchanarak T, Changtragoon S, Sreng S, Ping H (2022) Range-wide priority setting for the conservation and restoration of Asian rosewood species accounting for multiple threats and ecogeographic diversity. Biol Cons 270:109560
4. Siriwat P, Vincent N (2023) Quantifying the illegal high-value rosewood trade and criminal trade networks in the Greater Mekong Region. Biol Cons 277:109826
5. Thant MA (2004) A study on factors affecting small-scale wooden handicraft enterprises upgrading in Myanmar (case study: Bago township). Thesis submitted to Department of Economics, Yangon University of Economics, Myanmar, p 54
6. WCVP (2022) World checklist of vascular plants (WCVP), version 2.0. Facilitated by the Royal Botanic Gardens, Kew. http://wcvp.science.kew.org/. Accessed 03 Feb 2025
7. Barstow M, Boshier D, Bountithiponh C, Changtragoon S, Gaisberger H, Hartvig I, Hung H, Jalonen R, Kanchanarak T, Mackay J, Ping H, Thammavong B, Theilade I, Tran T, Win P, Zheng Y (2022) *Dalbergia oliveri*. The IUCN red list of threatened species. e.T215341339A2813403. https://doi.org/10.2305/IUCN.UK.2022-1.RLTS.T215341339A2813403.en. Accessed 03 Feb 2025
8. Dy Phon P (2000) Dictionary of plants used in Cambodia. Pauline Dy Phon and Imrimerie Olympic, Phnom Penh
9. Forestry Administration (FA) (2021) Review of the taxonomy, biology, ecology, and the status, trend, and population structure, of *D. cochinchinensis* and *D. oliveri* in Choam Ksant district, Preah Vihear province, Cambodia. Forestry Administration, Phnom Penh, Cambodia, p 15
10. Forestry Administration (FA) (2021) Systematic survey report of *Dalbergia cochinchinensis* and *Dalbergia oliveri* for piloting assessment on sustainable genetic conservation in Choam Ksant district, Preah Vihear Province. Forestry Administration, Phnom Penh, Cambodia, p 37
11. Global Witness (2015) The cost of luxury—Cambodia's illegal trade in precious wood with China. Global Witness, London

12. Nyi NK (2014) Myanmar timber trade and rosewood trade policies. Ministry of Environmental Conservation and Forestry, Myanmar. https://www.forest-trends.org/wpcontent/uploads/imported/8myanmar-ppt-pdf.pdf
13. Environmental Investigation Agency (EIA) (2014) Myanmar's rosewood crisis—why key species and forests must be protected through CITES. EIA, London, p 8
14. La QT, Bui TT, Nguyen MH, Van DTK, Nguyen HN (2021). Development of distribution maps for *Dalbergia cochinchinensis* and *Dalbergia oliveri* in Vietnam. Center for Nature Conservation and Development, Ha Noi, Vietnam
15. Dinh TKV, La QT, Ta QT, Hoang TS, Nguyen MH (2011) Report on the systematic field survey of the population distribution, abundance and stocking of *Dalbergia cochinchinensis* and *Dalbergia oliveri* in four key protected areas of Dak Uy, Bu Gia Map, Yok Don and Cat Tien in Vietnam. Center for Nature Conservation and Development, Ha Noi, Vietnam, p 53
16. Hajra PK, Verma DM, Giri GS (1996) Materials for the flora of Arunachal Pradesh, vol I. Botanical Survey of India, Calcutta, p 385
17. Sankara Rao K, Kumar D (2024) India Flora Online. http://indiafloraonline-ces.iisc.ac.in/plants.php?name=Dalbergia-oliveri. Accessed 3 Feb 2025
18. Nguyen TH, Nguyen MH, La QT (2019) Review on the taxonomy, biology, ecology, and the status, trend and population structure and dynamics of *Dalbergia oliveri* in Vietnam. Center for Nature Conservation and Development, Ha Noi, Vietnam, p 22
19. NAFRI (2022) Developing CITES non-detriment findings for rosewoods in Lao PDR. Report on the status quo of rosewood species of *D. cochinchinensis* and *D. oliveri* in Lao PDR. NAFRI, Lao PDR
20. DoF—Department of Forestry (2021) The implementation of five year report (2016–2020) and plan of forest sector in Lao PDR (2021–2025)
21. Winfield K, Scott M, Grayson C (2016) Global status of *Dalbergia* and *Pterocarpus* rosewood producing species in trade. In: Convention on international trade in endangered species 17th conference of parties, Johannesburg, South Africa, pp 1–246
22. CITES (2013) Convention on international trade in endangered species of Wild Fauna and Flora. CoP16 Prop. 60. https://cites.org/sites/default/files/eng/cop/16/prop/E-CoP16-Prop-60.pdf
23. Wiedenhoeft A (2016) Best practice guide for forensic timber identification. United Nations Office on Drugs and Crime: International Consortium on Combating Wildlife Crime. Vienna, Austria: Laboratory and Scientific Section, Global programme for combating wildlife and forest crime, New York, NY, USA pp 1–226
24. UNEP-WCMC (2014) Non-CITES timber species from South Asia (Leguminosae) potentially warranting further protection. UNEP-WCMC, Cambridge, UK

Chapter 15
Dalbergia oliveri Gamble ex Prain in China

Zheng Yongqi and Li Yongxiang

1 Introduction

Dalbergia oliveri, commonly known as Burmese rosewood, is a large tree species belonging to the *Fabaceae* family and *Dalbergia* genus. This species is primarily distributed across Southeast Asian countries such as Thailand, Myanmar, and Laos (Fig. 1). Compared to *D. cochinchinensis*, it has a wider distribution range and can grow in a broader variety of forest types, from deciduous to evergreen forests, at altitudes up to 1200 m. The heartwood with anti-inflammatory and cancer treatment properties has a long history of use in traditional Thai medicine. The isoflavonoid compounds extracted are often used as inhibitors and for mosquito prevention. Its timber is highly valued for its fine texture, high stability, and beautiful grain patterns, making it a preferred material for crafting exquisite classical furniture and other fine artworks. As demand continues to rise and natural populations decrease, *D. oliveri* has become increasingly rare and precious [1]. Due to its extremely slow growth rate (with a growth cycle of over 300 years) and the severity of human-caused destruction, it is classified as an endangered species on the IUCN Red List. Since 2017, the trade of *Dalbergia oliveri* has been strictly regulated under Appendix II of the Convention on International Trade in Endangered Species of Wild Fauna and Flora (CITES) [2].

2 Botanical Characteristics

Dalbergia oliveri is a large tree species, reaching heights of 15 to 30 m with a trunk diameter of up to 90 cm (Fig. 2a) [4]. The bark is gray, and the young branches are either glabrous or occasionally covered with fine hairs. The pinnately compound

Z. Yongqi (✉) · L. Yongxiang
Institute of Forestry, Chinese Academy of Forestry, Beijing, China
e-mail: zyq8565@126.com

R. R. Warrier et al. (eds.), *Handbook of Asian Rosewoods*, Materials Horizons: From Nature to Nanomaterials, https://doi.org/10.1007/978-981-96-9459-4_15

Fig. 1 Geographical distribution of *Dalbergia oliveri* [3]

leaves are 15–25 cm long, with petioles measuring 1–3 cm. The stipules are small and fall off very early. The leaflets, numbering 9–13, are thickly papery, ovate to ovate-lanceolate, and arranged alternately, with both surfaces being glabrous (Fig. 2b) [5]. The inflorescence is a terminal or nearly terminal panicle, with the main axis, branches, and pedicels all covered with fine pubescence. The flowers are 8–10 mm long, with pedicels measuring 2–3 cm. The calyx is bell-shaped, and the corolla is pale purple, with petals that have long claws (Fig. 2c) [5]. The species has 10 stamens arranged in two groups of 5 + 5. The ovary is stalked, with 3 to 4 ovules, a slender style, and a slightly enlarged stigma. The fruit is a broad-linguiform to oblong-lanceolate legume, with a sharply pointed to gradually pointed apex, containing 1–3 seeds that are kidney-shaped and flat, measuring approximately 6 by 12 mm (Fig. 2d) [5]. The flowering period is from March to April, and the fruiting period extends from June to November.

3 Habitat Description

Dalbergia oliveri is primarily distributed in Thailand, Myanmar, Laos, Vietnam, and Cambodia, and is cultivated in China and Malaysia. It is commonly found in riverine areas or foothills at altitudes below 900 m. The species typically occurs as individual trees or small groups of 5–10 trees in evergreen or semi-evergreen forests, often forming mixed stands with other deciduous species such as teak and ironwood. *D. oliveri* thrives in environments with abundant sunlight and high humidity, though it

Fig. 2 The partial morphological photos of *Dalbergia oliveri*

can also survive in relatively dry conditions. It is highly adaptable and can grow in various soil types, including sandy soils, loams, and clays. While it has no strict pH requirement, it prefers well-drained soils rich in organic matter.

4 Taxonomy and Identification

Dalbergia oliveri, belonging to the Fabaceae family and the *Dalbergia* genus in the Papilionoideae subfamily, plays a foundational role in both scientific research and practical applications such as ecological conservation, economic management, legal policies, and public education. The morphological differences between species within the *Dalbergia* genus are subtle, which can lead to misidentifications. Research on the wood structure of *D. oliveri* has revealed that its growth rings are not well-defined, with relatively straight grain patterns and a high presence of brown resin in the vessel pores. The predominant vessel cell type is drum-shaped, and there are numerous chambered cells containing crystals (Fig. 3) [6]. Additionally, *D. oliveri* wood contains specific volatile oils, flavonoids, and pigments, which may be absent or present in significantly different quantities in other species within the *Dalbergia* genus. These chemical components can be detected and analyzed using gas chromatography-mass spectrometry (GC-MS) technology.

As molecular markers continue to evolve, DNA barcoding has become an effective solution for the challenges posed by morphological and anatomical identification of *Dalbergia oliveri*, particularly for closely related species with similar characteristics [7]. Commonly used gene regions for this purpose include chloroplast genes such as rbcL and matK, and nuclear ribosomal ITS sequences [8]. In recent years, next-generation sequencing (NGS) technology has also been applied to the identification of *D. oliveri*, providing more detailed genetic information through whole-genome or targeted region sequencing, thus improving the accuracy of identification. To enhance identification accuracy, multidisciplinary approaches combining

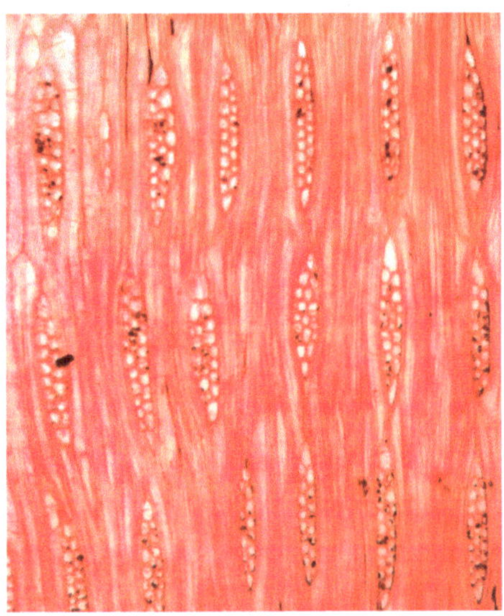

Figure 3 Microscopic image of the tangential section of *Dalbergia oliveri* wood

morphological, anatomical, chemical analysis, and molecular biology techniques are increasingly employed in the comprehensive identification of *D. oliveri.*

5 Seedling Cultivation

Dalbergia oliveri typically flowers from May to July, with fruits ripening from November to January of the following year. Each pod contains 1–3 seeds, which are brown, about 12 mm in length and 6 mm in width, with approximately 6,100 seeds per kilogram. The pods are green when immature and turn dark brown as they mature. The optimal season for seed collection is from December to January. After drying, the seeds can be stored at low temperatures for several years. Before sowing, soaking the seeds in 40 °C warm water and allowing them to cool overnight can significantly improve germination rates [9]. Once treated, the seeds can be sown in seedbeds for germination. When the seedlings produce a pair of leaves, they can be transplanted into nursery bags and cultivated in the nursery for 9–12 months. *D. oliveri* seedlings grow very slowly during the early stages, requiring careful fertilization, weeding, and protection from competitive plants until they can withstand competition from surrounding vegetation. Research shows that a 16 year-old *D. oliveri* tree can grow up to 11 m in height with a diameter at breast height (DBH) of 24 cm. In China, the domestication and acclimatization of *D. oliveri* is ongoing, with researchers focusing on the species adaptability to the subtropical climate in southern China, particularly in terms of soil, climate tolerance, and resistance to pests and diseases. Early trials in Guangdong and Guangxi have shown promising growth potential, but long-term ecological adaptability and economic viability remain areas for further investigation.

The natural distribution of *D. oliveri* is primarily in Vietnam, Thailand, Laos, and Myanmar. In these countries, research focuses on species conservation, ecological restoration, and adaptation to climate change. In Vietnam, the species is listed as endangered, and researchers conduct long-term field studies and experiments to analyze population dynamics, genetic diversity, and the impacts of habitat degradation. Thailand's research has centered on the species ecological and physiological characteristics, showing that *D. oliveri* is well-adapted to drought and high temperatures, making it valuable for afforestation and ecological restoration under changing climate conditions. Studies have also revealed that the species possesses unique water regulation mechanisms and efficient photosynthesis, providing theoretical support for its role in restoring degraded ecosystems [10].

6 Genetic Diversity

Dalbergia oliveri is primarily distributed in parts of Cambodia, Laos, Myanmar, southern Vietnam's protected areas, and northern and eastern Thailand. However, population genetic studies on this species are limited. Research using molecular markers like RAPD, ISSR, and SSR (Table 1) has demonstrated that *D. oliveri* possesses a relatively high level of genetic diversity, with Cambodian populations exhibiting particularly high diversity and distinct population differentiation [11]. In 2020, researchers conducted a study on *D. oliveri* chloroplast DNA (cpDNA) from four regions in Thailand, identifying 32 haplotypes. The frequency of these haplotypes varied across different populations. Among the 15 studied populations, the intrapopulation haplotype diversity (h) was 0.857, while the nucleotide diversity (π) was 0.00637, indicating high haplotype diversity but low nucleotide diversity [12].

Additionally, in 2020, Chinese researchers led by Yang reported the complete chloroplast genome sequence of *D. oliveri* (156,750 bp), with GenBank accession number MN823694. The specimen was preserved in the Herbarium of the Yunnan Academy of Forestry, Kunming, China, under registration number ZJFEP115 [13]. The study of the genetic diversity and genomics of *D. oliveri* plays a crucial role in developing effective conservation strategies and implementing protection measures. This research greatly aids molecular breeding and genetic improvement efforts, as well as providing the necessary data and technological support for tracking timber sources, preventing illegal logging, and promoting sustainable forest management [14].

Table 1 Genetic characteristics of *Dalbergia oliveri* populations

Country	R	N	Na	Na (rare)	Ho	He	Fis	T_2	Mm
Laos	0.87	12.25	6.18	4.74	0.80	0.73	0.27	0.31	0.32
Cambodia	0.64	20.91	7.99	5.27	0.83	0.78	–0.03	0.49	0.34
Vietnam	0.85	23.25	6.86	4.25	0.66	0.67	0.01	–0.87	0.31
Overall	0.75	20.00	7.34	4.82	0.76	0.73	–0.02		0.32

Note R = clonal richness estimate, N = number of unique genotypes used for further analyses, Na = mean number of alleles, Na (rare) = mean number of alleles corrected for sample size differences, Ho = observed heterozygosity, He = expected heterozygosity, FIS = Fixation index, T_2 = bottleneck statistic representing deviation between observed and expected heterozygosity based on number of alleles, and Mm = Garza-Williamson index representing ratio between number of alleles and allelic range

7 Threats and Risks

Despite the fact that the volume of wood traded in *D. oliveri* is lower than that of *D. cochinchinensis*, the population trend of *D. oliveri* is continuously declining within its range, with a significant reduction in the number of mature individuals. Deforestation, forest fires, and agricultural development [15] have led to habitat fragmentation, and ongoing urbanization in its native regions has caused the extinction of wild populations in some areas [16]. Additionally, genetic characteristics also impact the survival and reproductive rates of *D. oliveri*, with some individuals flowering and fruiting only once every few years, making seed collection and propagation challenging. Research in Cambodia indicates that seedlings from Thai sources have not flowered even after 10 years, while local source seedlings reach maturity in 5 years. Moreover, seedlings sprouting from stumps of trees that have been cut or damaged by fire have low seed dispersal rates, severely affecting the development of the next generation.

In Vietnam, *D. oliveri* is listed as an endangered species in the Vietnamese "Red Data Book" (2007 edition) and the IUCN Red List (version 2.3) [17]. Within its distribution area, the Ministry of Agriculture and Rural Development (MARD) and provincial forest protection departments (FPDs) have implemented measures to regulate the exploitation, trade, and conservation of the species according to legal provisions (Table 2). Cambodia has also undertaken a series of measures including sustainable forest management, authorized logging areas, trade controls, management plans, and efforts to combat illegal logging (Table 3). Despite these legal frameworks and management plans, illegal logging and timber trade issues persist. Monitoring and control measures are continuously being implemented, but ineffective enforcement has made it difficult to fully curb illegal logging and trading activities.

Table 2 Conservation actions of *D. oliveri* by the countries in their species range around early 2000s—compiled from APFORGEN reports [12]

Species	Thailand	Lao	Cambodia	Vietnam
Dalbergia oliveri	(1) Very high priority species (2) 34 ha ex situ conservation site (3) Recommended to establish more ex situ conservation areas (4) 3 genetic conservation stands, 34 ha, 30 plus trees (2002) (5) 20 plus trees (2003–2007)	Category B (vulnerable)	(1) Threat level 5 (2003) (2) 4 seed sources, 71.5 ha, 157 mother trees	(1) Priority species (2) Very high commercial value

Table 3 Wood volume (m^3) of *D. oliveri* reported as illegally harvested and/or traded in the Choam Ksant district, 2013–2019

	2013	2014	2015	2016	2017	2018	2019	Total
Sawn wood	2.36	24.956	0	0	0	0	0	27.316
Hard wood (log)	0	0	0	0	0	0	0	0
Total	2.36	24.956	0	0	0	0	0	27.316

8 Prospects

Future taxonomic studies of *D. oliveri* will increasingly rely on molecular biology techniques, such as phylogenetic research based on DNA sequence analysis. These studies will help clarify species relationships within the genus *Dalbergia* and provide scientific foundations for developing effective conservation strategies. In seedling cultivation, the current challenges for *D. oliveri* include low natural seed germination rates and slow seedling growth. Future efforts should focus on enhancing modern biotechnologies, such as tissue culture and clonal propagation, which may become crucial for improving seedling production efficiency. Additionally, mixed planting and ecological afforestation methods can enhance ecosystem stability and diversity, thus increasing the success rate of reforestation efforts.

With the genome sequence of *D. oliveri* now available, further genomic research is needed to identify superior genes and accelerate molecular breeding processes. Using resequencing technologies to understand genetic variation and population structure will aid in formulating more targeted conservation strategies. Landscape genomics can overcome traditional barriers in adaptive evolution analysis and provide scientific support for in situ and ex situ conservation efforts. The main threats to *Dalbergia oliveri* include habitat loss, over-exploitation, and illegal trade. More stringent enforcement of Convention on International Trade in Endangered Species of Wild Fauna and Flora (CITES) regulations is needed to control illegal timber trade [17]. Additionally, promoting sustainable use (Fig. 4) and cultivation techniques can reduce reliance on wild populations. Public education and community involvement are also crucial for raising local conservation awareness and encouraging participation in the protection of *Dalbergia oliveri*.

Fig. 4 Wood processing and utilization of *Dalbergia oliveri*

References

1. Nguyen TH, Nguyen MH, La QT (2019) Review on the taxonomy, biology, ecology, and the status, trend and population structure and dynamics of *Dalbergia oliveri* in Vietnam. Centre for nature conservation and development, Ha Noi, Vietnam. https://citestsp.org/wp-content/uploads/2020/11/Review-taxonomy...-of-Dalbergia-oliveri_compressed.pdf
2. Forestry Administration (2021) A review of the taxonomy, biology, ecology, and the status, trend, and population structure of *D. cochinchinensis* and *D. oliveri* in Choam Ksant district, Preah Vihear province, Cambodia
3. Plants of the World Online (POWO) (2024) The international plant names index and world checklist of vascular plants. https://powo.science.kew.org/taxon/urn:lsid:ipni.org:names:490372-1 Published on the Internet at https://www.ipni.org and https://powo.science.kew.org/ © Copyright 2023 World Checklist of Vascular Plants. http://creativecommons.org/licenses/by/3.0
4. Plant Species Information Thailand (2021). https://data.addrun.org/plant/
5. Hoang Thanh Son (2022) http://www.epharmacognosy.com/2022/04/dalbergia-oliveri.html
6. Zhongshan Redwood Furniture Laboratory, Guangzhou Analysis and Testing Center (2021) https://mp.weixin.qq.com/s/WCORBehY1F7aC6QjZhAGBA
7. Hartvig I, Czako M, Kjær ED, Nielsen LR, Theilade I (2015) The use of DNA barcoding in identification and conservation of rosewood (*Dalbergia* spp.). PLoS One 10(9):e0138231

8. Phong DT, Van Tang D, Hien VTT, Ton ND, Van Hai N (2014) Nucleotide diversity of a nuclear and four chloroplast DNA regions in rare tropical wood species of *Dalbergia* in Vietnam: a DNA barcode identifying utility. Asian J Appl Sci 2(2):116–125
9. Hung TH, Gooda R, Rizzuto G, So T, Thammavong B, Tran HT, Jalonen R, Boshier DH, MacKay JJ (2020) Physiological responses of rosewoods *Dalbergia cochinchinensis* and *D. oliveri* under drought and heat stresses. Ecol Evol 10(19):10872–10885
10. Narong N, Sobon K (2014) Review of biological and silvicultural characteristics of timber trees planted in Cambodia, p 96. https://www.irdfa.org/wp-content/uploads/2017/11/Review-of-biological-and-silvicultural-characteristics-of-timber-trees-planted-in-Cambodia.pdf
11. Phong DT, Hien VTT, Thanh TTV, Tang DV (2011) Comparison of RAPD and ISSR markers for assessment of genetic diversity among endangered rare *Dalbergia oliveri* (Fabaceae) genotypes in Vietnam. Genet Mol Res 10(4):2382–2393
12. APFORGEN (2018) Conserving rosewood genetic resources for resilient livelihoods in the Mekong—project inception workshop report
13. Zhang J, Li Y, Heng S, Wang Y (2020) The complete chloroplast genome sequence of *Dalbergia oliveri*. Mitochondrial DNA Part B 5(1):707–708
14. Hung TH (2022) Ecological genomics and adaptation of rosewoods *Dalbergia cochinchinensis* and *D. oliveri* for conservation and restoration. Doctoral thesis, University of Oxford, UK
15. Dongyang Redwood Timber (2021) https://mp.weixin.qq.com/s/t5qiqTrflCRRXCcDWM29ZQ
16. Luoma-Aho T (2004) Forest genetic resources conservation and management: proceedings of the Asia Pacific forest genetic resources programme (APFORGEN) inception workshop, Kepong, Kuala Lumpur, Malaysia, 15–18 July, 2003. Bioversity International
17. Barstow M, Boshier D, Bountithiponh C, Changtragoon S, Gaisberger H, Hartvig I, Hung H, Jalonen R, Kanchanarak T, MacKay J, Ping H (2022) *Dalbergia oliveri*. The IUCN red list of threatened species 2022 e.T215342548A2822125

Chapter 16
Taxonomy, Ecology, and Population Structure of Burmese Rosewood (*Dalbergia oliveri* Gamble ex Prain) in Myanmar

Tin Hnaung Aye and Mu Mu Aung

1 Introduction

Dalbergia oliveri Gamble ex Prain, commonly known as Burmese rosewood, is a member of the Fabaceae family and native to Southeast Asia and occurs naturally in the Indian Peninsula, Myanmar, Northern Thailand, and Northwestern Laos along the northern Thai border [1]. Forty-three species of *Dalbergia* are found natively in Myanmar's Dry Upper Mixed, Lower Upper Mixed, and Moist Upper Mixed Deciduous Forests, Deciduous Dipterocarpus or Indaing Forests and Dry Forests [2]. *Dalbergia prazeri* Prain and *Dalbergia laccifera* Laness are synonyms for *D. oliveri.* Due to its distinctive wood qualities and high market value, *D. oliveri* is a highly sought-after timber species. Despite once being more prevalent, the species has been under increasing pressure from the global timber market, primarily due to China and Vietnam's growing demand for rosewood lumber to manufacture hongmu furniture [3]. The species' population in the wild in Myanmar has significantly declined due to widespread exploitation and trafficking driven by its worth in the international market. Although complete population data for *D. oliveri* in Myanmar has yet to be available, the country has suffered the most significant loss, and estimates suggest that the species may be commercially extinct by 2023 [3]. This loss may persist in the future because enforcing laws against illicit logging is complex, and deforestation and forest degradation remain occurring. As a result, the IUCN Red List presently

T. H. Aye (✉)
Forest Department, Ministry of Natural Resources and Environmental Conservation, Nay Pyi Taw, Myanmar
e-mail: tinhnaungaye@gmail.com

M. M. Aung
Forest Research Institute of Forest Department, Ministry of Natural Resources and Environmental Conservation, Nay Pyi Taw, Myanmar

R. R. Warrier et al. (eds.), *Handbook of Asian Rosewoods*, Materials Horizons: From Nature to Nanomaterials, https://doi.org/10.1007/978-981-96-9459-4_16

lists the species as 'Critically Endangered' [4]. Although conservation efforts in Myanmar have been limited, a few studies have examined the species' biological characteristics, disease problems, and remaining wild population structure of *D. oliveri* in Myanmar [5–8].

2 Taxonomy, Habitat, and Ecology

D. oliveri is a deciduous tree species that may reach a height of 30 m and has an open, spreading crown with pinnately compound leaves and alternate leaflets (Figs. 1 and 2) [2]. Inflorescences are racemes, usually with numerous purple flowers (Fig. 3), and the fruit is a pod (Fig. 4). Trees blossom from March until May [2]. The species grows well in lowland forests or semi-shaded areas. Although the juveniles and seedlings can tolerate some shade, the mature trees require light to survive [9]. It adores well-drained soil and grows well in light (sandy), medium (loamy), and heavy (clay) soils [6]. Insects are pollinators of the species, and natural populations have low germination rates [9]. *D. oliveri* is valued for its redwood, used to make high-quality furniture, cabinets, luxury goods, artwork, and handicrafts. It is easy to polish and has a visually appealing veined structure [3]. The wood is tough, substantial, and resistant to termite damage [3]. Information about the *D. oliveri* breeding system in Myanmar is sparse; however, the seeds collected from the Sagaing, Mandalay, Bago, and Magway Regions of Myanmar had the highest germination during one to seven months of storage (Figs. 5 and 6) [6]. The seeds also showed a higher percentage of germination when partially scarred and soaked in cold water for a day [6].

Fig. 1 Natural occurrence of *D. oliveri*. *Photo source* Forest Research Institute, Myanmar

Fig. 2 *D. oliveri* tree. *Photo source* Forest Research Institute, Myanmar

Fig. 3 Flowers of *D. oliveri*. *Photo source* Forest Research Institute, Myanmar

Fig. 4 Fruits of *D. oliveri*. *Photo source* Forest Research Institute, Myanmar

Fig. 5 Fruits and seeds of *D. oliveri*. *Photo source* Forest Research Institute, Myanmar

Fig. 6 Seedlings of *D. oliveri*. *Photo source* Forest Research Institute, Myanmar

When examining the soil properties of *D. oliveri* native environment in the Sagaing Region, Mandalay Region, and Magway Region of Myanmar, it was observed to thrive in shallow soils, including small pieces of rock and gravel, and in hot and humid conditions [5]. *D. oliveri* prefers sandy-based soil types with high levels of calcium (Ca), potassium (K), and a moderate amount of organic matter (OM), along with lower levels of nitrogen (N) and phosphorus (P) [5]. Thus, *D. oliveri* may be an appropriate species to plant in degraded areas of deciduous forests. The appropriate range of soil pH is between 6.2 and 6.6; nitrogen content is between 0.03 and 0.05%, and organic matter is between 2.5 and 3% [5]. The predominant soil texture is sandy, with varying amounts of silt or clay. Humidity, elevation, and other environmental factors affect the species' geographic distribution and growth pattern [5].

3 Population Structure

According to previous studies in Myanmar, *D. oliveri* was believed to be mostly distributed in Sagaing Region and, to a lesser extent, in the Mandalay Region, Magway Region, Shan State, and Kachin State of Myanmar [5, 6, 10]. According to a Myanmar Forestry document of 2014, there were more than 2 million trees of *D. oliveri* in the Sagaing Region, slightly less than 1 million in Shan State, and less than 250,000 trees in the other states [10]. Although details on population structure and stocking of *D. oliveri* throughout the nation are lacking, the species has rapidly declined and is on the verge of extinction due to over-exploitation.

Here, we present the survey results on the population structure and stand structural characteristics of *D. oliveri* in the Zalon Taung National Park (ZNP) and its neighbouring unclassified forest (BUCF) in Banmauk Township, Sagaing Region, which covers an area of 21,616 ha [8] (Fig. 7). The elevation ranges from 100 to

1,600 msl. It has steep hills, valleys, flat plains, and flat-topped ridges. Banmauk Township in Sagaing Region is acknowledged as one of the critical locations for the wild populations of *D. oliveri* in Myanmar [8]. Thirty-four sampling plots (40 × 40 m) were randomly chosen and spaced at least 500 m apart [8] (Fig. 7). The sample areas encountered a range of human disturbances, including a local collection of medicinal plants and fuel, and legal and illicit logging. We measured the height and diameter of each tree with a diameter at breast height (dbh) of 10 cm and above. Most species in the study site are typical in tropical moist upper mixed deciduous forests.

By analysing the number of trees per hectare and diameter distribution of *D. oliveri* in the study area, we observed a disruption in the population structure due to the absence of trees with diameters ranging 21–30 cm in ZNP, showing the significant impact of historical timber extraction activities until 2011 before its declaration as a protected area (Fig. 8). In BUCF, the highest number of trees was in the 10–20 cm diameter class, with the number of trees gradually decreasing as diameters increased. Compared with BUCF, ZNP has fewer trees in the 10–20 cm and 31–35 cm diameter classes and no trees in the 21–30 cm diameter class, highlighting the impact of past logging activities on the forest's structure (Fig. 8). When comparing the stand structural characteristics (dbh, basal area and height) of five economically important tree species between BUCF and ZNP, it is depicted in Table 1 and Figs. 9, 10 and 11 that *Dipterocarpus alatus* had the highest average diameter, basal area, and height,

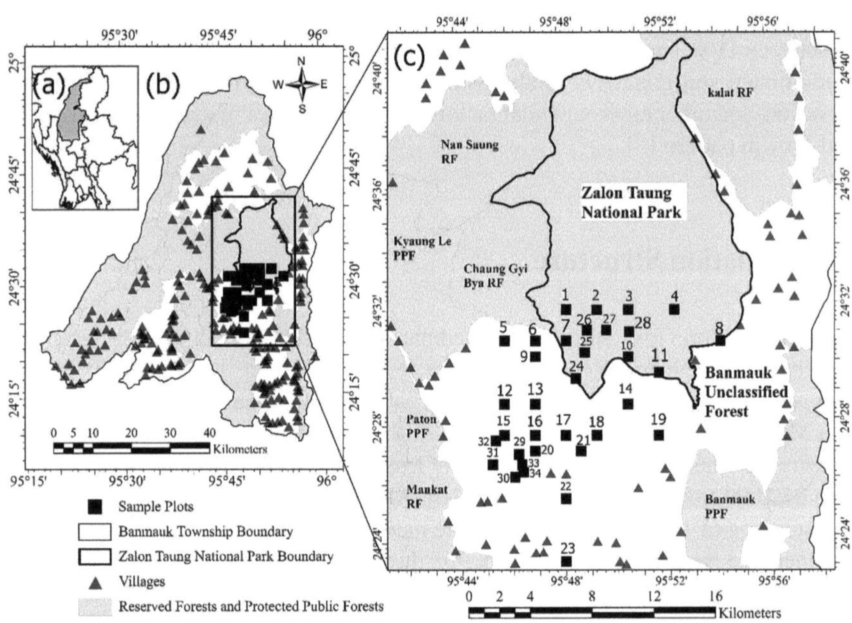

Fig. 7 Location of **a** Banmauk township, **b** Zalon Taung National Park (ZNP), and **c** 34 sample plots. *Source* [8]

Fig. 8 Population structure of *D. oliveri* in the study area

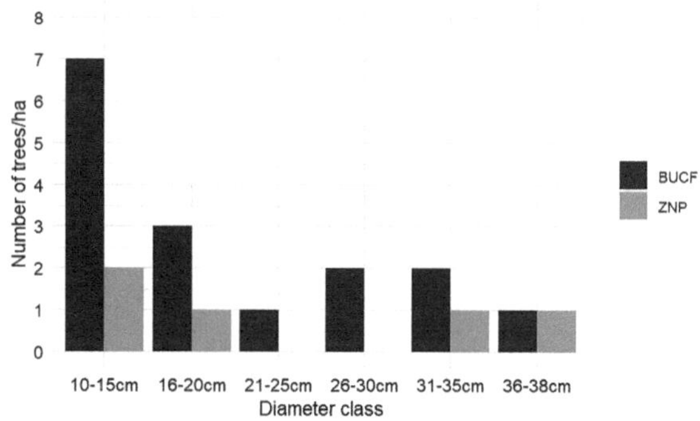

while *Tectona grandis* had the lowest [8]. The diameter at breast height and basal area of *D. oliveri* were lower than those of *D. alatus*, *Protium serratum*, and *Terminalia crenulata*, but higher than *T. grandis* (Fig. 9). Additionally, we found that the Importance Value Index (IVI) of *D. oliveri* in ZNP plots was lower than in BUCF plots, indicating its lower dominance in the tree community. We suggested that the past logging activities in ZNP before its declaration as a protected area significantly impacted the abundance and dominance of this species in the tree community. Additionally, the lower frequency and dominance of *T. grandis* and *D. oliveri* in the study area may be explained by the preference of these species for income and housing by the locals [8].

4 Trade

The over-harvesting of *D. oliveri* for its valuable lumber is causing the species' population to decline. Mature, large-sized trees are rapidly disappearing from many of their natural habitats due to a variety of threats, such as legal and illegal logging, over-exploitation of commercially viable tree species, conversion of forest land to other land uses, and the extraction of firewood and timber for local use [8]. Since *D. oliveri* is listed on CITES, much of its trade is prohibited [3]. Many countries have had logging restrictions since the early 2000s, making harvesting or trading of this endangered species illegal. However, the market value and the constrained enforcement of the law and regulations drive the continuing illegal trade of *D. oliveri* [11]. Between 2000 and 2013, US $737 million worth of rosewood (*D. oliveri* and *Pterocarpus macrocarpus*) was exported from Myanmar. In 2013, 237,000 m^3 of rosewood was traded which was six times the amount in 2010 and triple in 2012. Extra imports totalling US$52 million were made between China and Myanmar in 2014 [12]. According to the 2014 EIA report on Myanmar's Rosewood Crisis, if rosewood (*D. oliveri* and *P. macrocarpus*) extraction continues at the rate predicted, the species may go extinct in the next 13 years; however, if extraction were to continue at an exponential rate, the species would go extinct in the next three years, or before

Table 1 Mean and standard error value of structural characteristics of five species in two landscapes

	BUCF					ZNP				
	Da	Do	Ps	Tg	Tc	Da	Do	Ps	Tg	Tc
DBH (cm)	29.40 ± 1.89	19.00 ± 1.30	26.10 ± 1.41	13.80 ± 0.60	23.10 ± 1.54	36.00 ± 2.86	19.10 ± 3.17	21.40 ± 1.74	12.40 ± 1.34	27.30 ± 1.74
Basal area ($m^2\ ha^{-1}$)	0.54 ± 0.06	0.22 ± 0.03	0.43 ± 0.04	0.11 ± 0.01	0.33 ± 0.04	0.77 ± 0.16	0.23 ± 0.07	0.28 ± 0.04	0.09 ± 0.02	0.41 ± 0.04
Height (m)	14.80 ± 0.74	9.84 ± 0.50	12.80 ± 0.52	8.36 ± 0.27	11.30 ± 0.62	17.20 ± 1.12	9.99 ± 1.31	9.81 ± 0.75	8.68 ± 0.63	12.10 ± 0.80

Da—*Dipterocarpus alatus*; Do—*Dalbergia oliveri*; Ps—*Protium serratum*; Tg —*Tectona grandis*; Tc—*Terminalia crenulata*

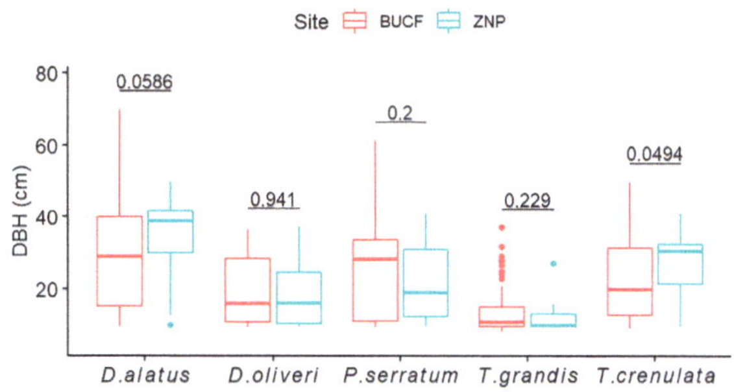

Fig. 9 Diameter at breast height (dbh) of five economically important tree species by Wilcoxon rank sum test (p-values are shown above the whisker plots). *Source* [8]

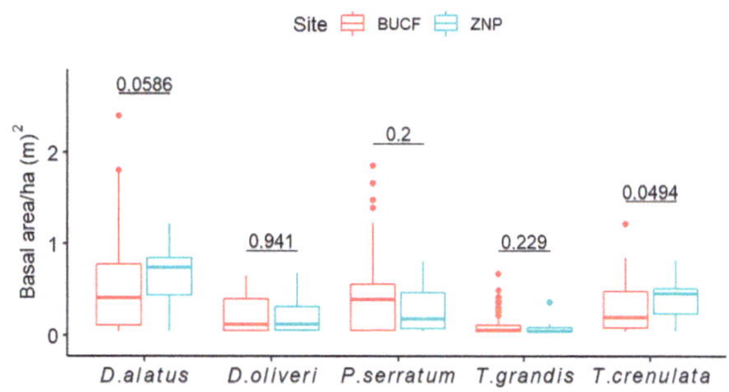

Fig. 10 Basal area of five economically important tree species by Wilcoxon rank sum test (p-values are shown above the whisker plots). *Source* [8]

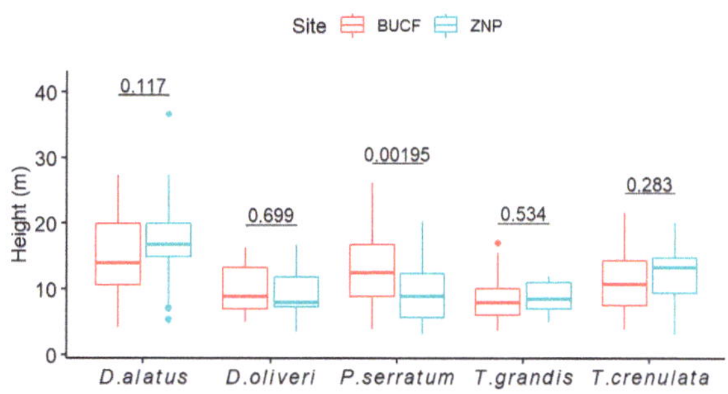

Fig. 11 Height of five economically important tree species by Wilcoxon rank sum test (p-values are shown above the whisker plots). *Source* [8]

2020 [12]. The intensive logging of *D. oliveri*, a highly traded timber species across the China-Myanmar border, has resulted in market extinction [13]. Continued over-extraction of this species, whether legal or illegal, could bring them close to extinction in their natural habitat.

5 Conservation and Management Measures

D. oliveri has been extensively harvested legally and illegally, leading to its commercial extinction in Myanmar. Without adequate conservation measures, this species has a high risk of extinction. Urgent action is necessary to conserve and restore the sustainable use of this species in the long term. This may involve implementing forest management measures such as establishing forest plantations, enrichment planting, species-wise inventory, and enforcing laws against illegal trade and logging. Thus, the local government must impose a temporary logging ban on damaged tree species, particularly *D. oliveri*, to facilitate their natural regeneration and survival in the natural forest [8]. Furthermore, law enforcement is essential to preventing illegal logging and start systematic forest land restoration in its natural range in Myanmar.

References

1. Troup RS (1921) The Silviculture of Indian trees, vol I. The Clarendon Press, Oxford
2. Kress WJ, Defilipps RA, Farr E, Kyi YY (2003) A checklist of the trees, shrubs, herbs and climbers of Myanmar. Smithsonian Institution, The United States National Herbarium, Washington, DC, USA
3. Nghia NH (1998) *Dalbergia oliveri*. The IUCN red list of threatened species 1998 (e.T32306A9693932). https://doi.org/10.2305/IUCN.UK.1998.RLTS.T32306A9693932.en
4. Barstow M, Boshier D, Bountithiponh C, Changtragoon S, Gaisberger H, Hartvig I, Hung H, Jalonen R, Kanchanarak T, Mackay J, Ping H, Thammavong B, Theilade I, Tran T, Win P, Zheng Y (2022) *Dalbergia oliveri*. The IUCN red list of threatened species (e.T215341339A2813403). https://doi.org/10.2305/IUCN.UK.2022-1.RLTS.T215341339A2813403.en. Accessed 24 Dec 2024
5. Swe PP, Tun SS, Khant HH (2018) Soil characteristics of Tamalan (*Dalbergia oliveri* Gamble ex Prain) habitat in Myanmar. Leaflet No. 8/2018. Ministry of Natural Resources and Environmental Conservation, Forest Department, Myanmar, p 23
6. Mu TT, Win SS, Aung MM (2018) Comparative study on provenances variation of Burmese rosewood (*Dalbergia oliveri* Gamble ex Prain) in Myanmar. Leaflet No. 7/2018. Ministry of Natural Resources and Environmental Conservation, Forest Department, Myanmar, p 27
7. Lin WW, Tan WW (2018) Disease problems in seed, seedling and plantation of *Dalbergia oliveri* Prain in Myanmar. Leaflet No. 9/2018. Ministry of Natural Resources and Environmental Conservation, Forest Department, Myanmar, p 19
8. Aye TH, Shibata S (2023) Vegetation structure and tree species diversity inside and outside a newly established Zalon Taung national park in northwest Myanmar. Int J For Res 2023(1):8409374
9. Winfield K, Scott M, Graysn C (2016) Global status of *Dalbergia* and *Pterocarpus* rosewood producing species in trade. In: Convention on international trade in endangered species 17th conference of parties, Johannesburg
10. Kyaw NN (2014) Myanmar timber trade and rosewood policies. In: International workshop on promoting legal and sustainable trade and investment of forest products, 25–27 March 2014, Shanghai, China. https://www.forest-trends.org/wp-content/uploads/imported/8myanmar-ppt-pdf.pdf
11. EIA (2016) Red alert: how fraudulent Siamese rosewood exports from Laos and Cambodia are undermining CITES protection. Environmental Investigation Agency (EIA), London, UK
12. EIA (2014) Myanmar's rosewood crisis. Environmental Investigation Agency (EIA), London, UK

13. Springate-Baginski O, Thein AK, Neil A, Thu WM, Doherty F (2014) Democratising timber: an assessment of Myanmar's emerging 'Forest law enforcement, governance and trade' (FLEGT) process. For Policy Econ 48:33–45

Chapter 17
Dalbergia sissoo Roxb. ex DC.

Vishwajeet Sharma, Geeta Joshi, and Praveen Verma

1 Introduction

Dalbergia sissoo Roxb. ex DC. (2n = 20), commonly known as Indian rosewood, Indian Dalbergia, Himalaya rain tree, penny leaf tree, Bombay blackwood, sisham, sissoo, North Indian rosewood, or sisu, belongs to the family Fabaceae. The genus name "*Dalbergia*" honours Swedish botanist Carl Dalberg, while the species epithet "sissoo" or "sishu" originates from its vernacular name in historical Bengal. Swiss botanist A.P. de Candolle first described this species in the second volume of *Prodromus Systematis Naturalis Regni Vegetabilis* (1825). Known for its durability and fine wood quality, the species is widely recognized for its economic and ecological significance.

2 Morphology

D. sissoo Roxb. ex DC. is an important medium to large multipurpose tree species found between 21.17°N to 32.60°N and 74.80°E to 93.43°E [1]. It is a deciduous tree reaching up to 30 m height and 8 feet girth (Fig. 1); leaves compound, imparipinnate with 3–5 alternate leaflets (Fig. 2a). It features a light crown, half-inch thick, rough, grey bark with longitudinal furrows (Fig. 2d). The tree produces 2–3 inches small, yellowish-white flowers arranged in dense panicles (Fig. 2c). Pods are linear-oblong, 1–3 inches long, containing one to four seeds (Fig. 2b), which are kidney-shaped, flattened, 8–10 × 4–5.5 mm in size, pale brown to brownish-black,

V. Sharma (✉) · G. Joshi
Indian Council of Forestry Research and Education, Dehradun, India
e-mail: vishwajeet@icfre.org

P. Verma
Forest Research Institute, Dehradun, India

R. R. Warrier et al. (eds.), *Handbook of Asian Rosewoods*, Materials Horizons: From Nature to Nanomaterials, https://doi.org/10.1007/978-981-96-9459-4_17

reniform, compressed, with papery testa [2–4]. The heartwood colour varies from dark brown to golden, while the sapwood is white to pale brown [5]. It develops long taproot early, with numerous lateral roots [6], and its bole is slightly crooked, adding to its unique appearance.

Fig. 1 Mature tree of *Dalbergia sissoo*

Fig. 2 **a** Leaves of sissoo, **b** flowering, **c** pods, **d** bark of sissoo

3 Taxonomic Classification

The taxonomic classification [7] of *Dalbergia sissoo* is as follows:

Domain	Kingdom	Division	Phylum	Class	Tribe	Genus	Species	Binomial name
Eukaryota	Plantae	Magnoliophyta	Tracheophyta	Magnoliopsida	Dalbergieae	*Dalbergia*	*Sissoo*	*Dalbergia sissoo*

4 Distribution

The species is native to Afghanistan, Bangladesh, East Himalaya, India, Iran, Myanmar, Nepal, and Pakistan and exotic to Benin, Brazil Southeast, Burkina, Cameroon, Chad, China Southeast, Cuba, Dominican Republic, Florida, Free State, Ghana, Guinea, Guinea-Bissau, Hainan, Iraq, Jawa, Kenya, Leeward Is., Malawi, Mauritania, Mozambique, Niger, Nigeria, Northern Provinces, Northern Territory, Oman, Palestine, Panamá, Puerto Rico, Queensland, Senegal, Sierra Leone, Sudan, Sumatera, Taiwan, Tanzania, Togo, Uganda, Zambia, and Zimbabwe [8]. In India, the species occurs throughout the sub-Himalayan tract and outer Himalayan valleys from the Indus to Assam, usually up to 900 m, but occasionally ascending to 1500 m [6]. In India, the species is mainly found in all states except in Andaman and Nicobar Island, Chandigarh, Lakshadweep, and Ladakh [9].

5 Silvicultural Characteristics

D. sissoo is a strong light demander. It is frost-hardy but leaves are sometimes affected by severe frost. Seedlings are susceptible to drought. It coppices vigorously up to 20 years and is highly palatable, hence browsed by herbivores. It is fairly wind-firm species and tolerates moderate winds.

5.1 Associated Species

D. sissoo either grows in pure form or in association with *Adina cordifolia*, *Albizia procera*, *Terminalia anogeissiana*, *Tamarix dioica*, *Ehretia laevis*, *Lannea coromandelica*, *Acacia catechu*, *Bombax ceiba*, *Mallotus nudiflorus*, *Holoptelea integrefolia*, *Bombax ceiba*, *Boswellia serrata*, *Kydia calycina*, and *Premna latifolia*.

5.2 Climatic Requirement

D. sissoo is well adapted to subtropical to tropical environments. The tree survives well in temperature ranges from 4 to 50 °C, annual rainfall ranging from 500 to 4500 mm, and rich well-drained alluvial soil. It can withstand periodic flooding especially in riverine areas. However, it performs poorly in waterlogged conditions. The tree adapts well in soil pH ranging from 5.0 to 7.7 and can withstand some level of salinity, though its growth is stunted on heavy clay soils. The species was not affected by snowfall and very low temperatures [6]. The species is commonly found at elevations from sea level up to 1000 m, with trees occasionally reaching as high as 1500 m in suitable environmental conditions.

6 Phenology

In its natural range, leaf shedding starts in November and becomes leafless, while in the moist and warmer areas, some leaves remain while new ones emerge by February, with complete flushing by March [10]. Flowering occurs from March to May, and fruiting starts in cold regions in December, while in some areas, the leaves persist until the end of January. The pods are prominent by April and full-sized by July. The seed ripens by December or early January and turns brown [10]. The details are elaborated in Chaps. 21 and 22 of this book.

7 Seed Collection, Storage, and Biology

Pods can be collected from November to March. Collected pods are sun-dried for 3–4 days. These can be stored for three years without much loss of viability. Pretreatment is not required; however, soaking seeds in cold water for 24 h enhances the germination. Fresh seeds with 3–5% moisture have 80–90% germination. When seeds are stored for 2 years in temperature ranges from 5 to −5 °C, they retain 86% viability [11]. In one kilogram, 40,000–50,000 seeds are present. It produces copious seeds every year [6].

8 Propagation

Shisham naturally regenerates through seeds, coppice, and root suckers. It depends on canopy density, availability of soil moisture, sunlight, soil porosity, and the absence of weeds. It can be propagated by direct sowing, entire transplanting, stump planting, and other vegetative means. In direct sowing method, seeds are sown in lines 2–3 m apart and at a depth of 1.5 cm. Stump planting is the most common practice, with 1–2 year-old seedlings having a collar diameter >1 cm and a root: shoot ratio of 10:1. Stumps are planted during the third week of March or April [12]. In the entire planting method, nine months to one-year-old seedlings, about 0.6 m tall are also planted. The spacing ranges from 1.8 × 1.8 m to 4 × 4 m. In dry areas, staggered trenches are used for planting.

9 Tree Improvement in *D. sissoo*

D. sissoo is widely used in many tree planting programmes. The species, however, has several disadvantages, such as slow growth, irregular shape, forked bole, and poor branching habits [13–15]. Various institutions in India, Pakistan, and Nepal conducted parallel tree improvement programmes. Kamaluddin and Ali [16] recommended three methods to enhance the genetics of this species viz. choosing and propagating clones with robust vertical growth for use in the field, selection of high-quality clones from specific geographical regions for plantation, and the production of hybrids to enhance desired genetic characteristics. Under the tree improvement programme, ICFRE-FRI has released first disease-resistant clone and ICFRE-AFRI released three high-yielding clones of *D. sissoo*. Assessment of variability based on biometric characteristics in 20 *D. sissoo* clones at FCRI, Mettupalayam revealed FCRIDS18 as the best clone for future tree improvement programme [17]. The details of tree improvement in *D. sissoo* are discussed in Chap. 18 of this book.

10 Biomass Accumulation in *D. sissoo*

D. sissoo grown on infertile soils at the Tropical Farm, Kom Ombo, Aswan Botanical Garden in southern Egypt at ages 2, 5, 7, 15, and 25 years exhibited consistent increase in biomass with age with mean values 60.53 kg, 203.7 kg, 591.6 kg, 1210.5 kg, and 3060.3 kg per tree respectively. Tree height, diameter growth, and heartwood to sapwood ratio increased with age [18].

11 Economics of Cultivation

Under non-irrigated conditions, the mean annual increment (MAI) for a 20 year-old plantation was 11.2 cum per ha on good sites, 8.4 cum per ha on fair sites, and 5.9 cum per ha on poor sites. The system yielded a benefit–cost ratio of 2.35 at the end of 20 years. Based on quality and distance from the source, the price of wood of *D. sissoo* in the local market varies from Rs. 800/- to 1200/- Cft (cubic feet) [19].

12 Pest and Diseases

Infestation by wood borers, leaf defoliators, leaf miners, leaf rollers, and sap suckers have been reported on *D. sissoo* [20]. Two defoliators, *Plecoptra reflexa* Guenee and *Dichomeris eridantis* Meyrick, have been reported to damage *D. sissoo*. In nurseries and young plantations, *P. reflexa* is reported to be a serious defoliator [21]. The details on wilt and die-back disease is detailed in Chap. 20 of this book.

13 Uses

13.1 Timber

Its wood has mechanical properties similar to teak and is strong, heavy, durable, and easy to work with. At 12% moisture content, the density of wood is approximately 750–800 kg/m^3 making the wood suitable for railway sleepers, wooden flooring, panelling, agricultural tools, musical instruments, plywood making, wood laminates [22].

13.2 Fuelwood

The calorific value of sapwood and heartwood of Shisham is 4,908 kcal/kg and 5,181 kcal/kg respectively [23]. The high calorific value of the wood makes the species suitable as fuelwood and charcoal [22].

13.3 Fodder

The leaves are highly palatable and consume by herbivores due to high protein and essential minerals. Young twigs and foliage form good quality fodder with a dry-matter content of 32.46% and crude protein content 2.7–24.1% [5].

13.4 Medicinal Uses

Shisham is an abortifacient, aphrodisiac, anticipaterant, antipyretic, and anthelmintic [24]. Different parts of plants are also used to treat emesis, ulcers, leucoderma, dysentery, stomach problems, skin disorders, blood diseases, syphilis, stomach problems, dysentery, nausea, eye and nose disorders, aphrodisiac, and expectorant [25–27].

13.5 Environmental Benefits

The species helps check soil erosion due to its wide-spreading root system and fixes atmospheric nitrogen in root nodules with the help of *Rhizobium*. Leaf litter decomposes in soil and improves fertility by enhancing nitrogen, potassium, iron, manganese, and organic carbon [28].

14 Conclusion

D. sissoo is an ecologically significant species due to its high-quality timber, adaptability and its contribution to the environment stability. Its ability to thrive in various climatic and edaphic conditions makes *D. sissoo* an important species for afforestation and agroforestry programmes. High-yielding clones have enhanced plantation productivity, and their nitrogen fixing ability enhances soil fertility. However, its susceptibility to pest infestations and dieback disease requires further research on disease resistance and sustainable management practices. To maximize the benefits of *D. sissoo* for future generations, it is necessary to balance its commercial use with ecological preservation, integrating technological advances with conservation initiatives.

References

1. Tewari DN (1994) A monograph on *Dalbergia sissoo* (Roxb.). International Book Distributor, Dehradun, p 316
2. Parker RN (1915) A forest flora for the Punjab with Hazara and Delhi. Superintendent Government Printing, Punjab
3. Singh SP (1989) Wasteland development. Agricole Publishing Academy, New Delhi, India, p 227
4. Luna RK (1996) Plantation trees. International Book Distributors, Dehra Dun, India, p 975
5. Orwa C, Mutua A, Kindt R, Jamnadass R, Anthony S (2009) Agroforestry database: a tree reference and selection guide version 4.0. World Agroforestry Centre, Kenya
6. Troup RS (1921) The Silviculture of Indian trees 1. Oxford Press, Oxford, UK
7. Vikrant A, Arya ML (2011) A review on anti-inflammatory plant barks. Int J Pharm Tech Res 3(2):899–908
8. POWO (2024) Plants of the World Online. Facilitated by the Royal Botanical Gardens, Kew. http://www.plantsoftheworldonline.org/. Accessed 3012 2024
9. Bhattacharjee A, Krishna G, Kumar A, Sengupta S, Chakraborty S, Dhanavate R, Sarkar S, Sahu RK, Soni AK, Chakraborty O, Mallick B, Prasad K (2018) Report on non-detriment findings (NDFs) of *Dalbergia latifolia* Roxb. in India. Botanical Survey of India, Kolkata, p 70
10. Haq R (1985) Monograph on *Dalbergia sissoo* Roxb. (Shisham). MSc thesis, Forest Education Division, Pakistan Forest Institute, Peshawar
11. Joshi G (2000) Physiological investigation on germination and viability of *Dalbergia sissoo* Roxb. Seed for ex-situ germplasm conservation. PhD thesis, Forest Research Institute (Deemed University), Dehradun

12. Bari MA, Ferdaus KMKB, Hossain MJ (2008) Callus induction and plantlet regeneration from in vivo nodal and internodal segments and shoot tip of *Dalbergia sissoo* Roxb. J Bio Sci 16:41–48
13. Vidakovic M, Siddiqui KM (1968) Heritability of height and diameter growth in shisham (*Dalbergia sissoo*) using one parent progeny test. Pak J For 18:5–94
14. Sidhu DS (1997) Selection of plus trees for genetic improvement of stem form in shisham (*Dalbergia sissoo*). Indian J For 20:20–25
15. Nautiyal S, Pal M, Sagta HC, Rawat RS, Negi DS (2003) Clonal variation in growth, flowering and seed production of *Dalbergia sissoo* in a clonal seed orchard. J Trop For Sci 15(1):97–100
16. Kamaluddin M, Ali M (1995) Genetic improvement and propagation strategy for *Dalbergia sissoo* in Bangladesh. Bangladesh J For Sci 24(2):54–61
17. Rajendran P, Arunmaharaja B, Krishnakumar N, Parthiban KT (2019) Clonal variation of *Dalbergia sissoo* genetic resources. Pharma Innov J 8(11):271–277
18. Sayed RMM, Ebeid AFA (2011) Variability in biomass production and some wood properties of *Dalbergia sissoo* trees grown in Aswan (Southern Egypt). J Plant Prod, Mansoura Univ 2(10):1331–1340
19. Kumar S, Joshi G, Warrier RR, Arunkumar AN (2023) FAQ's on Agroforestry. Indian Council of Forestry Research and Education, Dehradun, pp 55–56
20. Mathur RN, Singh B (1960) A list of insect pests of forest plants in India and the adjacent countries. Indian For Bull 171:1–130
21. Sharma MK, Singaland RM, Pokhriyal TC (2000) *Dalbergia sissoo* in India. In: Appanah S, Allard G Amatya SM (2000) Proceedings of the sub-regional seminar on dieback of sissoo (*Dalbergia sissoo*). Field Document No. 18. https://www.fao.org/4/ae910e/ae910e02.htm
22. Tyagi VC, Kamini DN, Prabhu G (2020) *Dalbergia sissoo*: an important tree with fodder value. Van Sangyan 7(10):1–8
23. Anon (1952) The wealth of India. Raw materials, vol III. D-E. CSM, New Delhi
24. Yasmeen S, Gupta P (2016) In vitro demonstration of *Dalbergia sissoo* (Indian rosewood) methanolic extracts as potential agents for sunscreening and DNA nick prevention. Int J Pharm Pharm Sci 8(6):175–181
25. Nikum RD, Nehete JY (2021) Review on traditionally medicinal importance, morphology, phyto-chemistry and pharmacological activities of *Dalbergia siss*oo. Int J Pharmacogn 8(8):329–337
26. Brijesh S, Daswani PG, Tetali G, Antia NH, Birdi TJ (2006) Studies on *Dalbergia sissoo* (Roxb.) leaves: possible mechanism(s) of action in infectious diarrhoea. Indian J Pharmacol 38(2):120–124
27. Sen A, Kaur A, Johar V, Singh V (2023) Indian rosewood: a versatile tree for economic prosperity. Int J Phytol Res 3(4):16–21
28. Bhattacharya M, Singh A, Ramrakhyani C (2014) *Dalbergia sissoo*—variability in morphology. J Med Plants Stud 2(3):8–13

Chapter 18
Variability in *Dalbergia sissoo* and Tree Improvement

Geeta Joshi, A. N. Arunkumar, Rekha R. Warrier, and Vishwajeet Sharma

1 Introduction

Genetic variability within a species is fundamental to developing effective tree improvement strategies. Conducting morphological and genetic diversity studies is essential for identifying superior genetic material for breeding and enhancement programmes. The potential for species improvement largely depends on the variations observed within and among populations [1, 2].

Dalbergia sissoo exhibits significant morphological variation, particularly in pod and seed traits, seedling characteristics, and vegetative propagation. Pod size and shape vary widely, with some trees producing long pods while others bear short, broad ones. Seed size and weight also differ, influencing germination rates and seedling vigour. Seedling variability is evident in growth rates, leaf morphology, and branching patterns, which may result from genetic potential, environmental influences, or their interaction.

Vegetative propagation methods such as cuttings and grafting are adopted to address seed variability and maintain desirable traits. However, the success of these techniques largely depends on the genotype and propagation method used.

G. Joshi (✉) · V. Sharma
Indian Council of Forestry Research and Education (ICFRE), Dehradun, India
e-mail: geeta@icfre.org; geejos@gmail.com

A. N. Arunkumar
ICFRE-Institute of Wood Science and Technology, Bengaluru, India

R. R. Warrier
ICFRE-Institute of Forest Genetics and Tree Breeding, Coimbatore, India

R. R. Warrier et al. (eds.), *Handbook of Asian Rosewoods*, Materials Horizons: From Nature to Nanomaterials, https://doi.org/10.1007/978-981-96-9459-4_18

2 Variability in Pod and Seed Characters

Forest tree seeds exhibit morphological and physiological variations due to their broad geographical distribution. Within a species, differences in seed size, germination percentage, and seedling growth rate are primarily influenced by strong genetic influence [3]. Environmental conditions also play a crucial role in shaping seed traits.

In *D. sissoo*, pod and seed trait variations across different seed sources can be attributed to its wide distribution across diverse agroclimatic zones with varying rainfall, temperature, and soil conditions. The influence of habitat on these variations has been confirmed [4, 5]. Significant positive correlations ($P < 0.01$) have been observed between pod length and breadth, while significant negative correlations ($P < 0.05$) exist between latitude and seed weight, length, and thickness [6]. Similarly negative correlations were reported between seed size and latitude [3]. Seed quality in *D. sissoo* is linked to seed size [7], with seed weight is influenced by seed length [8]. Furthermore, pod and seed size—measured by length, breadth, thickness, and weight—were highest in four-seeded pods and lowest in one-seeded pods [9].

Significant pod and seed characteristic variations have been observed among six seed sources in Uttar Pradesh, India, with Gonda yielding the best quality seeds [10]. Notable differences in pod and seed traits were observed among seeds collected from superior trees in 15 locations across Punjab, Haryana, and Rajasthan [3, 10]. In a similar study, collections from Kanpur, Uttar Pradesh, and Rewa, Madhya Pradesh, were identified as the best seed sources [11]. A study for seeds collected from 58 Candidate Plus Trees (CPTs) across seven seed sources in Uttar Pradesh reported significant variation in length, breadth, thickness, shape index (length-to-breadth ratio), seed weight, viability, and germination percentage. Mathura seeds exhibited the highest quality, while Mau seeds were of the lowest quality [12]. Variability in seed characteristics among progenies and clones in a *D. sissoo* seed orchard was also reported [13].

Extensive studies covering 60 seed sources across the natural distribution range of *D. sissoo* in India and Nepal have documented significant variation in pod and seed traits, with the highest variation observed in the number of seeds per pod and 1,000-seed weight. However, this variation was discontinuous or non-clinal [14]. High genotypic and phenotypic coefficients of variation, genetic advance, and heritability for 100-seed weight were reported [8]. Similarly, moderate to high heritability and genetic advancement were recorded for pod weight, seed size, and root-shoot ratio among 30 genotypes from different climatic zones of Haryana [15].

Significant positive correlations ($P < 0.05$) were observed between seed width and seed weight, seed length and seed weight, seed thickness and seed weight, and seed weight with germination value for 19 progenies of *D. sissoo* from the Seedling Seed Orchard (SSO) at Bithmara, Haryana [16]. Heritability values for most seed and pod traits exceeded 0.50, with seed weight exhibiting the highest heritability and genetic gain. Positive correlations were observed between various morphometric and seed traits from different sources [4, 8]. Variation in sugar content in *D. sissoo* seed sources was observed [17].

Seeds collected from 19 different sources at altitudes ranging from 120 to 1130 m in the Central Himalaya and Siwalik foothills, India, exhibit significant variation in seed traits and pod characteristics. Provenances from Rampur Mandi, Badapur, Najibabad, and Kalagarh had larger pods, more seeds per pod, larger and heavier seeds, and higher germination [6].

Significant variations was observed for germination, for progenies from different locations however for the seeds from same location the differences in germination was non-significant [18]. Seed germination percent was negatively correlated to altitude and the seed from populations in higher elevation had low germination which could be attributed to higher dormancy existing in these seed lots [19]. There was a strong positive correlation between germination percentage and seed weight, while seed thickness negatively correlated with seed length [10]. Among the seeds from fifteen seed sources, the Kanpur source had the highest germination (89.44%) and minimum germination of 83.65% was recorded for the Allahabad source [11].

3 Variability in Seedling Characters

The growth performances of *D. sissoo* seedlings are dependent on seasonal variations; height and diameter growth was more in wet to semi-dry season [9]. Growth of seedlings was better in soil mixed with cow dung as compared to clear soil, and seedlings from seed without coated showed less growth as compared to seeds with coat [9]. It was found that seedlings had higher height and collar diameter growth when raised in soil-cow-dung mixtures than control or inorganic fertiliser [20].

While studying variability for 15 seed sources from Punjab, Haryana, and Rajasthan, the differences for germination percentage, germination value, and growth traits like shoot and root length and shoot and root dry biomass were significant [21]. Germination percentage, germination value, and root length were maximum from Patiala and Amargarh in Punjab, while the highest shoot and root biomass was recorded for Muktsar. Significant genetic variation among ten indigenous seed sources was reported for rate of germination, shoot length, and seedling dry weight [3]. Similarly, among 19 different sources in Central Himalaya and Siwalik foothills of India, taller seedlings were observed for provenances from Rampur-mandi, Badapur, Najibabad, and Kalagarh [6]. Significant positive correlations were observed between seedling height, collar diameter, and various other growth traits for seeds from 20 seed sources. Poor growth was observed in seedlings from Jorhat, Tilyar Zoo, and Farinda [22]. Among seed sources significant variation in seedling growth traits has also been reported [23, 24].

In pot cultures, variability was observed in nine clones of *D. sissoo* for height, collar diameter, root length, number of leaves, and number of branches. Among all clones, best growth was observed in clone no. 9093 (1) of Galla Mandi Behrampur of Uttar Pradesh [25]. Similarly, substantial variations were observed for growth traits and positive correlation was observed between seed weight and germination percentage in seedlings, among 20 clones from north India [10].

Studies have also been carried out on the nodulation, nitrogen fixation potential, and growth characteristics of *D. sissoo* in India and other places [10, 26–28]. Significant difference in nodulation behaviour, soil nitrogen accretion, and nitrogenase activity from different seed sources has been reported [29, 30]. However, among 20 seed sources, non-significant variations were observed in collar diameter and nodule diameter, and seedlings from Koshitappu, Bahraich, and Pratapgarh had higher number of nodules in comparison to others seed sources [22].

In provenance studies, seeds were collected from 59 provenances from India and Nepal (latitudes 24° 42″ N to 32° 36″ N, longitudes 74° 30″ E to 94° 36″ E, altitudes 76 m to 460 m asl), and seedlings were raised in four nurseries. The best average growth among four nurseries was observed in Mainpuri, Shahbad, Laxmipur, and Pilibhit, from Uttar Pradesh and Simblewala from Jammu and Kashmir provenance. The minimum growth was observed in Firozabad, Sultanpur, and Kunda provenances from Uttar Pradesh and Rohtak provenance from Haryana. Correlation studies exhibited that latitude, longitude, altitude, and rainfall did not have any effect on the provenance variation [31]. However, a significant positive relation ($P < 0.05$) between plant height and longitude of seed source, and pod size and plant height increased towards eastern locations, whereas seed size and weight decreased [6]. Similar study with six year-old plants from different provenances showed significant variation for growth characters [26].

In tree improvement programmes, heritability plays an important role as it provides an index of the relative strength of heredity versus environment. High heritability for germination percentage (0.89) and low heritability for root length (0.11) and shoot length (0.39) were observed, and it was stated that for six seed source biochemical traits had strong and highly significant correlation with seedling biomass [21].

4 Vegetative Propagation

D. sissoo is predominantly propagated through seeds. However, for clonal forestry, vegetative propagation using cuttings and tissue culture is essential. This section reviews research on vegetative propagation through cuttings, while propagation via tissue culture is discussed in Chapter 19 in this book.

Rooting capability in *D. sissoo* depends on endogenous factors such as plant age, maturation state, cutting type, season of collection, and physiological condition. Exogenous factors influencing rooting include rooting medium, temperature, humidity, hormones, photoperiod, and light intensity. *D. sissoo* has an inherent ability to root, with even green sticks showing rooting when planted in the soil [32]. It can be propagated vegetatively through root suckers, stem cuttings, or root cuttings [33]. Genetic factors significantly impact rooting potential [34]. The response varies with the thickness of the cuttings, auxin treatment, and tree age [35]. Cuttings treated with IBA from four-year-old trees showed 50% rooting, whereas 30-year-old trees had only 30% rooting [36]. Lower crown cuttings treated with growth hormones had up to 60% rooting success [37].

Softwood nodal cuttings from the middle portion of shoots exhibited the best rooting response [38]. Rooting percentages varied depending on the kind of cuttings; for lower shoot cuttings it was 72.5, 52.5% for upper shoot cuttings, 72% for leafless branches, and 35% for leafy cuttings [39]. Cuttings from young trees had a rooting success of 50–90%, with the best results occurring in August [40]; however, maximum rooting in February was also reported [36]. Seasonal effects on rooting were also noted [41]. Rooting was best in young, thin cuttings and seedlings rather than mature trees [35]. On average, 30 cm long cuttings showed 86% rooting compared to 20 and 25 cm cuttings [39]. Another study found that 15 cm cuttings with a 2.0 cm diameter had 78.8% survival, 75.81% rooting, 16.1 root numbers, and 12.01 cm root length [36].

Altitude also affects rooting, cuttings from higher altitudes exhibited higher rooting percentages and better growth. In a study with cuttings from three sources in Uttar Pradesh, an average rooting rate of 25% was recorded, with a maximum of 45% [34]. Through air layering, 68% adventitious rhizogenesis was reported [42].

Growth hormones significantly enhance rooting in *D. sissoo*. IAA at 6,000–10,000 ppm was most effective under mist conditions [43]. Increasing IBA concentration up to 2,000 ppm improved sprouting and root growth [38]. Cuttings from 10 year-old trees treated with IAA and IBA at 100–200 ppm showed 80–100% rooting [40, 44]. IAA induced root formation, while NAA promoted adventitious roots [35]. NAA and IBA at 100 mg/L were most effective for root differentiation in mature hardwood and softwood cuttings [45]. Cuttings treated with 4,000 ppm IBA achieved 86.6% survival and rooting [36].

Clonal variations influence the sprouting and rooting of softwood nodal cuttings in *D. sissoo* [38]. Rooting percentage, root length, and the number of adventitious roots varied significantly among 41 clones, with rooting percentages ranging from 21.67 to 93.18% [46]. Survival ranged 80–100%, with Clone C14 from Haridwar showing maximum survival and C86 from Rajasthan excelling in rooting [47]. In clones from India and Nepal, cuttings from Tulsipur and Laxmipur (Gonda, Uttar Pradesh) had the highest rooting response at 2,000 ppm IBA. Clonal variation in leafy shoot cuttings was also observed [48].

Variations among 96 *D. sissoo* clones in hedge gardens were observed when hedged to 30 cm height in a VMG at Birplasi, Nalagarh (Himachal Pradesh). The traits included shoot emergence days, collar diameter, shoot number, shoot length, and basal diameter. Clones 86 and 91 from Rajasthan showed the fastest shoot emergence within five days. Clone 107 from Rajasthan had the highest shoot number (67), followed by Clone 42 from Gonda, U.P. (57 shoots). Clone 9 from Pathri, Haridwar, Uttarakhand, had the longest shoot length (133 cm) and largest diameter (11 mm) [49].

5 Tree Improvement

Within a tree species, variations arise from both heredity and environmental influences. It is well established that genotypes exhibit variation due to environmental differences. The growth performance and adaptability of a species differ across sites. Environmental factors can affect the growth behaviour of identical genotypes, even under uniform site and climatic conditions. In stable environments, differences between genotypes are primarily hereditary, whereas in variable environments, phenotypic differences become more pronounced [25].

D. sissoo is widely used in tree planting programmes. The drawback of this species is its slow growth, crooked form, forked bole, and poor branching traits [27, 50–54]. Improvement in the species can focus on screening and selecting from promising provenances for better stem form and vigorous growth and disease resistance. Three strategies for the genetic improvement of *D. sissoo* were proposed: a) selecting and propagating clones with strong apical dominance for field planting, b) identifying superior clones from provenance resource stands for plantation, and c) developing inter-provenance hybrids to enhance genetic traits [55].

Selections from Natural Populations: Natural populations of *D. sissoo* exhibit variations in growth traits [50, 51]. Heritability studies have indicated a significant genetic influence on stem straightness [27, 51] with high heritability for stem crookedness (42–65%) and a strong environmental effect on height and diameter [56]. Superior trees demonstrate higher apical dominance, total height, and clear bole height with fewer crooks per meter [27]. Superior trees demonstrated higher apical dominance, total height, fork diameter ratio, and clear bole height with lesser number of crooks per metre [27, 57], with Gonda in Uttar Pradesh having natural trees with the best growth, form, timber quality, and biochemical constituents [54].

The National Research Centre for Agroforestry (NRCAF), Jhansi, identified 32 plus trees in the Bundelkhand region [58]. On the basis of fast growth and straightness, PT-6 and PT-2 of Mahurani were found to be superior with mean annual increment of 15.53 m^3/ha/year and 12.73 m^3/ha/year, respectively at NRCAF [58]. Similarly, CCS Haryana Agricultural University (CCS HAU) selected plus trees from Rajasthan, Haryana, Punjab, and Nepal based on growth, clear bole, stem straightness, compact crown, apical dominance, branch angle, and pest and disease resistance [59].

Tree Improvement Initiatives: The Indian Council for Forestry Research and Education (ICFRE) has undertaken systematic tree improvement since the 1990s, focusing on plus tree selection, provenance trials, clonal multiplication, and genetic evaluations. Between 1994 and 2001, ICFRE implemented the Planting Stock Improvement Programme (PSIP) under the World Bank-assisted Forestry Research Education and Extension (FREE) Project. Selection criteria included height, collar diameter, stem straightness, and disease resistance, leading to the identification of 36 superior genotypes across multiple states (Uttar Pradesh (23), Uttarakhand (04), Haryana (02), Punjab (05), and Nepal (02)).

ICFRE established a clone bank with 350 clones at Forest Research Institute (FRI) and additional seedling seed orchards in Himachal Pradesh, Haryana, Punjab, Uttar Pradesh, and Uttarakhand. Multi-locational trials initiated in 2006 across Punjab aimed to identify superior genotypes with higher biomass yield and disease resistance.

Parallel tree improvement programmes were carried out by various institutions [3, 5, 31, 60–63] in India [26, 64], in Pakistan, and [56, 65–67] in Nepal. Further advancements include the selection of plus trees [27, 57], progeny trials, and clonal development [68–71].

Progeny Trial Evaluations and Performance of Provenances: Multi-locational trials were established by FRI at three locations in Punjab to identify superior and stable genotypes with higher marketable biomass having resistance to disease and insect-pests [70]. In Nepal, nine provenances (seven from Nepal and two from Pakistan) were evaluated and found that survival rates at 3.5 years ranged from 85 to 97%. The Pakistan provenances exhibited the poorest growth, while the Nepal provenances performed similarly, with the Makwanpur provenance showing superior growth [65]. Six provenances of *D. sissoo* from Nepal was assessed for survival, form, growth, and vigour; the Kailali (Seti River) provenance was identified as the best for volume production and was recommended for planting [56]. While tracking height and diameter growth over six years for *D. sissoo* planted at two sites in the Tarai belt of Uttar Pradesh, it was found that growth followed a nearly linear trend at both sites, with better height growth up to four years at site 2 and superior diameter growth at the same location [72].

Genetic variability, heritability, and genetic advance in total height, main stem height, basal diameter, and straightness in three-year-old progeny of 43 plus trees was analysed [18, 73]. They observed highly significant variation among progenies, with the phenotypic coefficient of variation ranging from 11.3% for straightness to 37.2% for main stem height. High heritability estimates were recorded, varying from 0.78 for total height to 0.85 for main stem height, indicating strong genetic control over these traits. Ten clones of *D. sissoo* were assessed for survival, height, girth, branching pattern, and volume yield after 12 years [74]. In a provenance trial consisting of 59 provenances covering a wide geographic range from India and Nepal, assessment after 1.5 years revealed that provenances from eastern Uttar Pradesh performed well across all sites. Shahbad, Pilibhit, Mainpuri, and Laxmipur (Uttar Pradesh), and Simbalwala (Jammu & Kashmir) provenances showed superior growth, while Firozabad, Sultanpur, Kunda, and Rohtak exhibited the least growth [31].

A progeny trial established in CCS Haryana Agriculture University (HAU) comprising twenty diverse provenances showed high phenotypic variation (47.41%) for stem form, with high heritability estimates (83.63% broad sense, 74.50% narrow sense) and genetic advance (84.61%) [23], indicating high degree of genetic control. Progeny trials at CCS HAU identified five superior selections (PT28 and PT12 from Haldwani, PT-283 from Gonda, PT-9 from Tanakpur, PT-84 from Nepal) based on total height, unforked height, and straightness. For straightness and unforked height broad and narrow sense heritability was high [23]. Six-year-old *D. sissoo* plantations from three geographical sources in Pakistan plants were assessed for growth, and it was revealed that seed sources from Chichawatni had the greatest diameter as compared to Changa Manga and Mardan [26].

As part of ICFRE's National Provenance Trials (1994–1995), seeds from 40 sources across India and Nepal were evaluated, with the Mirzapur (Nepal) and Pratapgarh (UP) sources exhibiting superior survival, height, and volume [24]. Among the tested ten provenances/seed sources, best performance was observed for provenances from Oachghat, Solan, H.P., Kukrail, Lucknow, and Orai, U.P. [19]. The seed sources from Lucknow and Muzaffarnagar displayed good performance and were recommended for planting programmes in this region [75].

Vigorous traits can be identified as early as 1–2 years, while stem form is assessed within five years [56]. Studies on biochemical markers indicated variation with respect to free proline, polyphenols, total protein, total carbohydrate, and starch [24]. Strong genetic control was observed over height, protein, chlorophyll, and sugar, whereas girth at breast height (GBH) and photosynthetic efficiency were environmentally influenced [76]. Maximum growth and Fv/Fm (chlorophyll fluorescence) was recorded for clones from Gonda and Nepal. Chlorophyll content and fluorescence traits have been identified as reliable selection criteria for improving *D. sissoo* [76].

In Pakistan trials, stem straightness of *D. sissoo* was shown to be strongly influenced by genetic factors. However, genetic control over diameter and height growth appeared less evident, likely due to field management issues. In Nepal trials, vigour, reflected by volume production, was found to be genetically controlled [56].

Clonal Evaluation: Presently, clones are being exploited for immediate genetic gains, taking full advantage of the superior genetic qualities of field-tested clones in many countries, and bypass long breeding cycles of tree improvement programmes.

Inter-clonal variation for growth traits has been observed in *D. sissoo* [5, 26, 31, 54, 77]. Longitude and rainfall of the area from where clones were collected and clear bole height of mother tree had a significant effect on the growth of clones [54]. A clonal bank comprising 350 clones was established at ICFRE-FRI, and several improvement trials such as clonal/seedling seed orchards, gene banks, and clone banks have been established in Himachal Pradesh, Haryana, Punjab, Uttar Pradesh, and Uttarakhand. At ICFRE-AFRI, Jodhpur, vegetative multiplication garden (VMG) comprising 50 CPTs was established in 5 ha area.

The first disease-resistant clone of *D. sissoo* FRI-DS-14 was released by ICFRE-FRI in 201. This clone had a productivity of ~18 m^3/ha/year [70, 71]. Based on results from multi-locational clonal trials established at Deesa, Kheralu, Gandhinagar, and Rajpipla in Gujarat, ICFRE-AFRI released three Clones AFRI DS-1, AFRI DS-2, and AFRI DS-4 for semi-arid and sub-humid regions of Gujarat, and three clones—AFRI DS-1, AFRI DS-2, and AFRI DS-4—for semi-arid and sub-humid regions, with wood yield ranging from 8.91 to 12.69 CuM/ha/year [71].

Genetic trials with 20 clones established at Forest College and Research Institute, Mettupalayam, Tamil Nadu, indicated high heritability for height and volume index

(98%) and significant genetic gains (98.55%) at six months after planting [78, 79]. Five superior clones namely FCRIDSC 3, FCRIDSC 4, FCRIDSC 7, FCRIDSC 8, and FCRIDSC 16 performed well at the end of six months after planting [78]. However, at the end of 32 months after planting, FCRIDS18 was found to be superior [79].

Similarly for height, crown width and length, broad sense heritability was higher [15, 61, 73, 76, 81]. For total chlorophyll, chlorophyll b, protein and sugar 99% estimates of heritability were observed [76]. In chlorophyll fluorescence genotypic variations have also been reported [103, 104]. For 20 clones low heritability for collar diameter and high heritability for height and number of branches was reported by [4]. Flowering and growth in thirty-eight clones assessed for four years revealed best growth and form in Clone C045 of Trilokpur, Gonda, U.P. and C018 of Khanpur, Shahmansoorpur, U.P., and highest seed production in C099 and C087 from the dry zone of Rajasthan [54].

In an evaluation trial comprising 23 clones of *D. sissoo* at Pindori Mindo Mind, Hoshiarpur, Punjab, the best growth was observed in the Clones 66, 192, 204, and 232 from Gonda, U.P., and 12 from Pathri, Haridwar, Uttaranchal. Clones from Yamunanagar, Haryana, and Hanumanghar, Rajasthan, had poor growth [80]. Excellent performance at all sites was observed for Clone C192 (Tulsipur, North Gonda, UP), while Clones C066 (Chhichrauli, HR) at Pandhori Mindo Mind (Hoshiarpur) and Lachhiwala (Dehra Dun) and C034 (Tulsipur, North Gonda, UP) at Bithmera (Hisar) and Lalkuan (Haldwani) were site-specific. In Clones C083, C087 (Hanumangarh, RJ), and C034 (Tulsipur, North Gonda, UP), early flowering, and pod and seed setting were observed while it was late in Clones C066 (Chhichrauli, HR), C019 (Haridwar, UA), and C192 (Tulsipur, North Gonda, UP) [81]. For the physiological parameters, clones from states of Uttar Pradesh (C034 and C192), Uttarakhand (C019), and Haryana (C066) had more photosynthetic rate, transpiration rate, and stomatal conductance compared to the Clones C083 and C087 from drier regions of Rajasthan [82].

In another study in pathologically tested nine clones of *D. sissoo*, the best performance was observed for clones from Uttar Pradesh, namely 9058 (1) Dhani Kakarhawo (Sidharthnagar); 9058 (2) Dhani Kakarhawo (Sidharthnagar); 9064 (2) Tilkhana Village (Maharajganj); 9093 (1) Galla Mandi (Behrampur) which could be due to adaptability to similar eco-climatic conditions as compared to clones from other places (Chaukiyal, et al. 2015) [25].

Significant genetic variability was observed in eleven clones of *D. sissoo* for seed production potential and seed quality parameters in clonal trials at Dehradun and Hisar. Arid region of Hisar was found to be more suitable for seed production while seed production potential was highest in clones from Kanpur, U.P. Based on seed quality parameters and stability parameters, Clone no. 103 from Suratgarg and Clone nos. 83 and 85 from Hanumangaeh were most promising [83].

Eight clones of *D. sissoo* were evaluated at 54 months for growth traits, adaptability, and stem quality, including clear bole height and stem straightness. Significant variation was observed among clones for growth traits and in biochemical leaf composition. Tree height was highest in Clone PS-52 (10.86 m), while, DBH (12.33 cm) and volume index (0.159 m^3/tree) was maximum in PS-54. The highest clear bole height (4.01 m) was found in PS-20, PS-90, and PS-52, while straightness was superior in PS-20, PS-52, and PS-54. PS-52 had the highest survival rate (97.60%). Disease incidence and pest infestation varied, with PAU-2 showing the least symptoms. High heritability was observed for stem quality traits. Based on index scores, PS-52 and PS-20 were identified as the most promising clones [84].

6 Diversity Assessment Using Molecular Markers

Molecular markers have become an increasingly important tool for studying genetic diversity, identifying individual genotypes, and selecting desirable traits [85]. Though widely cultivated and with elaborate programmes to support the identification of genotypes with superior growth, genetic diversity assessment using molecular markers in *D. sissoo* is very sparse. Only twenty studies have been reported, of which 75% used RAPD markers.

The conventional DNA extraction protocol was modified for *D. sissoo*, which significantly improved DNA yield (100–400 μg per milligram of leaf tissue) and ensuring its suitability for RAPD and ISSR analyses [86]. RAPD and ISSR markers were compared in genetic diversity studies, it was concluded that RAPD markers were more efficient, displaying 12.5–15% higher polymorphism [87, 88]. RAPD markers were utilized to assess 22 accessions from different regions in Pakistan [89], while over 300 superior trees from both natural forests and plantations in India were analysed for demonstrating the RAPD effectiveness in characterizing genetic relationships [90]. Genetic diversity and differentiation among natural populations in India and Nepal were investigated using RAPD markers; a relatively high level of genetic diversity with minimal differentiation indicating stable genetic structure was reported [91]. In Nepal, five *D. sissoo* populations were studied by using cpDNA and the Gdh-A isozyme gene. These markers helped to establish that the trees in the plantations had originated from distant sources. Gdh-A variations indicated differing adaptive potential [92].

Genetic diversity in *D. sissoo* appears to be influenced by mating system and geographic distribution. However, low genetic variation among *D. sissoo* individuals was observed [89], which was in consonance to the observations made while studying genetic relationships in *Dalbergia* species using molecular markers [93]. Genetic diversity was low in cultivated clones of sissoo found from diverse sissoo-growing regions of India and Nepal, and it was recommended to have more distant genotypes (clones), for future breeding and plantation programmes [94, 95]. RAPD markers were used to quantify genetic diversity among clones of *D. sissoo* and reported that it is a powerful tool for the species' future conservation and improvement programme [96, 97].

The ISSR studies by several researchers reveal that these markers could help tag and shape the scattered shisham populations or steer tree improvement towards identifying dieback-resistant shisham germplasm [87, 88, 98–100]. To differentiate and discriminate the shisham genotypes at intraspecific level in cultivated clones in Pakistan, rbcL was used as a DNA barcoding [101]. Fourteen SSR markers were identified through cross species amplification, and six SSR markers were used to validate the breeding behaviour of sissoo with other Dalbergias. This approach is of use in the conservation and development of superior hybrids for sustainable development and production of commercial populations [102].

7 Conclusion

D. sissoo exhibits considerable variability in pod and seed traits, seedling performance, and vegetative propagation, all of which are influenced by genetic and environmental factors. This variability is crucial for tree improvement programmes, which aim to enhance growth, adaptability, and disease resistance. Advances in molecular markers have strengthened breeding efforts, enabling precise genotype selection, and accelerating genetic improvement. The integration of clonal propagation techniques and genetic diversity assessments offers significant potential for improving productivity and resilience across ecological zones. Continued research and application of emerging biotechnological tools will be key to optimizing the genetic potential of *D. sissoo* for sustainable forestry and conservation.

References

1. Barner H, Ditlevsen B, Olesen K (1991) Introduction to tree improvement. Humblback (Denmark) DANIDA Forest Seed Centre, pp 1–16
2. Bangarwa KS, Thomson W, White KJ, Hansen IN, Rehman S, Zabala NQ, Shivkumar P, Ladipo DO, Read M (1994) Seed production and tree improvement. *Dalbergia sissoo* production and use—a field manual. NFTA, Morrilton (AR) USA, pp 34–38
3. Vakshasya RK, Rajora OP, Rawat MS (1992) Seed and seedling traits of *Dalbergia sissoo* Roxb. provenance variation studies among ten source in India. For Ecol Manag 48:265–275
4. Gera M, Gera N, Gupta BN (2000) Preliminary observations on genetic variability and character associations in *Dalbergia sissoo* Roxb. Indian For 126:608–615
5. Devagiri GM (1997) Evaluation of seed source variation in seed and seedling traits in *Dalbergia sissoo* Roxb. PhD thesis, F.R.I. (Deemed University), Dehradun, p 189
6. Singh B, Bhatt BP (2008) Provenance variation in pod, seed and seedling traits of *Dalbergia sissoo* Roxb., Central Himalaya, India. Trop Agric Res Ext 11:39–44
7. Leishman MR, Wright IJ, Moles AT, Westoby M (2000) The evolutionary ecology of seed size. In: seeds: the ecology of regeneration in plant communities. CABI Publishing, Wallingford, New York, pp 31–57
8. Hooda MS, Bahadur R (1993) Variability correlation and path coefficients analysis for some seed traits in Subabul. Seed Res 21:49–51
9. Sheikh AH, Matin MA, Uddin SN (2006) Seedling growth performance of *Dalbergia sissoo* (Roxb.) during nursery stage in Bangladesh. J Appl Sci Res 2(11):1047–1052
10. Singh N, Pokhriyal TC (2001) Variations in pod and seed traits in six different *Dalbergia sissoo* seed sources. J Trop For Sci 13:162–170

11. Dubey S, Tripathi S (2018) Studies on fifteen seed source variation in seed traits of *Dalbergia sissoo* (Roxb.). Int J Sci Res 7(9):2319–7064
12. Srivastava SK, Verma N, Mishra CM, Dubey P, Burfal BS (2001) Evaluation of seed source variation on different seed parameters of *Dalbergia sissoo* Roxb. Indian For 127(03):358–362
13. Sofi AH, Singh O (2011) Variation in seed germination and nursery growth among families of *Dalbergia sissoo* Roxb. Ann For 19:69–74
14. Devagiri GM, Dhiman RC, Thapliyal RC, Nautiyal S (1998) Seed source variation in pod and seed traits of *Dalbergia sissoo* Roxb. Ann For 6(2):148–155
15. Dhillon RS, Bisla SS, Dhanda SK, Singh VP (1995) Genetic variation and heritability of some growth traits in shisham (*Dalbergia sissoo* Roxb.). Ann Biol 11(1):107–110
16. Sofi AH, Singh O (2013) Variability, heritability and genetic divergence in pod and seed traits of *Dalbergia sissoo* Roxb. in a seedling seed orchard. Indian For 139(5):408–414
17. Singh N (1998) Physiological and biochemical parameters vis-a-vis. genetic improvements in *Dalbergia sissoo* Roxb. PhD thesis, F.R.I. (Deemed University), Dehradun
18. Singh J, Bangarwa KS (1995) Progeny testing for growth parameters in *Dalbergia sissoo* Roxb. Ann Biol 11(1–2):111–140
19. Gera M, Gera N, Ginwal HS (1999) Seed source variation in germination and early growth among ten indigenous populations of *Dalbergia sissoo* Roxb. Indian For 125(12):1190–1197
20. Hossain MK (1995) Growth and development of *Dalbergia sissoo* seedlings grown wider different combinations of porting media. Chittagong Univ Stud-Sci 19:259–264
21. Singh G, Dhillon GPS, Dhakad AK, Kaur N (2020) Variability among seed sources of shisham (*Dalbergia sissoo* Roxb.) for morphometric and biochemical traits. Indian J Agrofor 22(2):37–43
22. Devagiri GM, Dhiman RC, Kumar PN, Patil CSP (2007) Seed source variation in seedling and nodulation characters in *Dalbergia sissoo* Roxb. Silvae Genetica 56:88–91
23. Bangarwa KS (2002) Extent of genetic variability and selection of superior progenies in *Dalbergia sissoo*. Natl J Plant Improv 4(1):61–64
24. Gera M, Awadhiya A, Gera N (2016) Provenance trial of *Dalbergia sissoo* Roxb. Indian For 142(3):213–220
25. Chaukiyal SP, Bhatia P, Kumar M, Dobhal P, Kumar S (2015) Performance of *Dalbergia sissoo* clones in relation to morphological traits under nursery conditions. J Exp Zool India 18(1):43–48
26. Rehman S, Hussain A (1986) Growth and heritability estimates among six years old three geographical sources of shisham (*Dalbergia sissoo* Roxb.) in Pakistan. Pak J For 26(2):67–72
27. Sidhu DS (1997) Selection of plus trees for genetic improvement of stem form in shisham (*Dalbergia sissoo*). Indian J For 20:20–25
28. Singh N, Pokhriyal TC (2000) Biomass distribution pattern in relation to seed source variation in *Dalbergia sissoo* seedlings. Ann For 8(2):238–249
29. Singh N, Pokhriyal TC (2002) Nitrogen fixation and nodulation behaviour in relation to seed source variations in *Dalbergia sissoo* seedlings. J Trop For Sci 14:198–206
30. Pandey A (2002) Seed source variation vis-a-vis Rhizobium arbuscular mycorrhiza interaction in *Dalbergia sissoo* Roxb. PhD thesis, F.R.I. (Deemed University), Dehradun, p 177
31. Sagta HC, Nautiyal S (2001) Growth performance and genetic divergence of various provenances of *Dalbergia sissoo* Roxb. at nursery stage. Silvae Genetica 50(3–4):93–98
32. Anonymous (1983) Troup's Silviculture of Indian trees. Controller of Publications, Delhi
33. Nanda KK, Anand VK, Kumar P (1970) Some investigations of auxin effects on rooting of stem cuttings of forest plants. Indian For 96(3):171–187
34. Singh B, Bhatt BP (2009) Sprouting and rooting response of *Dalbergia sissoo* stem cutting from different altitude. Indian For 135(3):342–346
35. Ansari SA, Kumar P, Mandal AK (1995) Effect of position and age of cuttings and auxins on induction and growth of roots in *Dalbergia sissoo* Roxb. Indian For 121:201–206
36. Kiruba M, Devanand PS, Sivakumar B, Nelson NR, Vijayan R, Hemaprabha K, Radha P, Tilak M, Utharasu S, Sivakumar K, Mathivanan M (2023) Clonal propagation of *Dalbergia sissoo* through stem cuttings. Pharma Innov J 12(4):357–360

37. Nautiyal S, Rawat MS (1994) Macro propagation of teak (*Tectona grandis* L. f.). Indian For 120(2):46–151
38. Husen A (2004) Clonal propagation of *Dalbergia sissoo* Roxb. by softwood nodal cuttings: effects of genotypes, application of IBA and position of cuttings on shoots. Silvae Genetica 53(2):50–55
39. Singh B, Yadav R, Bhatt BP (2012) Vegetative propagation of *Dalbergia sissoo*: effect of growth regulators, length, position of shoot and type of cuttings on rooting potential in stem cuttings. For Stud China 14:187–192. https://doi.org/10.1007/s11632-012-0304-1
40. Gupta BD, Kumar A, Negi DS (1993) Vegetative propagation through branch cuttings in *Dalbergia sissoo* Roxb. Indian For 119(5):381–387
41. Rana U, Gairola M, Nautiyal AR (1987) Seasonal variation in rooting of *Dalbergia sissoo* and auxin effect on it. Indian J For 10(3):220–222
42. Prasad R, Dhuria SS, Tiwari SK (1992) Status of research on vegetative propagation of forest trees in Madhya Pradesh. In: Reddy KK (ed) Vegetative propagation and biotechnologies for tree improvement. Natraj Publishers, Dehradun, pp 44–52
43. Dhuria SS (1992) Vegetative propagation of *Dalbergia sissoo* through cuttings under mist conditions. In: Vegetative propagation and biotechnologies for tree improvement. Natraj Publishers, Dehradun, pp 25–26
44. Singh B, Yadav R, Bhatt BP (2011) Effects of mother tree ages, different rooting mediums, light conditions and auxin treatments on rooting behaviour of *Dalbergia sissoo* branch cuttings. J For Res 22:53–57
45. Puri S, Verma RC (1996) Vegetative propagation of *Dalbergia sissoo* Roxb. using softwood and hardwood stem cuttings. J Arid Environ 34(2):235–245
46. Singh O (2006) Clonal variation in rooting ability of branch cuttings of *Dalbergia sissoo* Roxb. Indian For 132(6):746
47. Chauhan A, Bakshi M, Bhandari M (2009) Rooting response of *Dalbergia sissoo* as affected by auxins and genotype. Indian J For 32(4):545–550
48. Dhiman RC, Nautiyal S, Singh J, Gupta PK, Singh J (1997) Clonal variation in macro-propagation through leafy shoot cuttings in *Dalbergia sissoo* Roxb. Ind J Soil Conserv 25(2):170–171
49. Bakshi M, Thakur KS, Kumar S (2007) Clonal variation in regeneration of shoots after hedging in vegetative multiplication garden of *Dalbergia sissoo* (shisham) Roxb. Indian For 133(1):101–107
50. Vidakovic M, Siddiqui KM (1968) Heritability of height and diameter growth in shisham (*Dalbergia sissoo*) using one parent progeny test. Pak J For 18:75–94
51. Vidakovic M, Ahsan J (1970) The inheritance of crooked bole in shisham (*Dalbergia sissoo* Roxb.). Silvae Genetica 19:94–98
52. Vidakovic M, Ahsan J (1970) The improvement of stem form in shisham (*Dalbergia sissoo* Roxb.). Pak J For 20:115–119
53. Bangarwa KS, Hooda MS, Puri S (1990) An ideal plus tree in *Dalbergia sissoo* Roxb. In: MPTs for agroforestry system. Range Management Society of India. IGFRI, Jhansi, India, pp 111–114
54. Nautiyal S, Pal M, Sagta HC, Rawat RS, Negi DS (2003) Clonal variation in growth, flowering and seed production of *Dalbergia sissoo* in a clonal seed orchard. J Trop For Sci 15(1):97–108
55. Kamaluddin M, Ali M (1995) Genetic improvement and propagation strategy for *Dalbergia sissoo* in Bangladesh. Bangladesh J For Sci 24(2):54–61
56. White KJ (1990) An improvement plan for *Dalbergia sissoo* Roxb. in Nepal. Bank Janakari 2(4):363–381
57. Monga R (2016) Selection of plus trees of shisham (*Dalbergia sissoo* Roxb.), macro propagation studies and ex-situ conservation. MSc thesis, Punjab Agricultural University, Ludhiana, Punjab
58. Gupta VK, Kumar RV, Ahalawat SP, Datta A (2003) Performance of plus tree progenies of shisham (*Dalbergia sissoo* Roxb.). In: National symposium on agroforestry in 21st century held from February 11–14, 2003 at Ludhiana, India, p 68

59. Bangarwa KS (2002) Plus tree selection and progeny testing for establishment of first generation seed orchard in *Dalbergia sissoo* Roxb. Indian J Agrofor 4(2):122–131
60. Vakshasya RK (1988) Growth and foliar variation among provenances of *Eucalyptus camaldulensis* Dehn. grown at Dehradun. Proc Indian Acad (Plant Sci) 98(6):507–513
61. Bangarwa KS (1993) Study on provenance variation, progeny testing and reproductive biology in *Dalbergia sissoo* Roxb. PhD thesis, CCS Haryana Agricultural University, Hisar, India
62. Shiv Kumar P, Rai AM, Dwivedi AK, Banerjee AC (1994) Genetic improvement of *Dalbergia sissoo*. In: Proceeding of an international workshop. NFTA, Morrilton (Ar) USA, pp 183–186
63. Tewari DN, Sharma KK (1994) Production of *Dalbergia sissoo* in India. In: Proceeding of an international workshop. NFTA, Morrilton (Ar) USA, pp 190–199
64. Rehman SA, Gunasenal JPM, Joshi RB (1994) Comparative growth of *Dalbergia sissoo* Roxb. provenance. In: Proceeding of an international workshop. NFTA, Morrilton (Ar) USA, pp 187–189
65. Neil PE (1990) *Dalbergia sissoo* provenance testing in Nepal. Nitrogen Fixing Tree Res Rep 8:130–132
66. Hawkins T (1986) *Dalbergia sissoo* provenance trial. Forest research project Adabhar trial site research result 1980 to 1986. Forest Research and Survey Office, Department of Forests, Kathmandu, Nepal
67. White KJ, Gautam I, Dahl M (1994) *Dalbergia sissoo* Roxb. provenance trial at Sagarnath in Nepal. In: Proceeding of an international workshop. NFTA, Morrilton (Ar) USA, p 200
68. Sidhu DS, Dhillon GPS, Singh A (2006) Genetic variation among progenies of *Dalbergia sissoo* Roxb. at nursery stage. Ann Biol (Hisar) 22:165–167
69. Singh A, Dhillon GPS (2014) Evaluation of *Dalbergia sissoo* Roxb. progenies at field stage under semi-arid regions of Punjab. In: Compendium of abstracts. World congress on agroforestry, Delhi, 10–13 February, p 232
70. Kumar A, Rathore TS, Palanisamy K, Viswanath S (2016) Advances in tree improvement and forest genetic resources conservation and management. Greenfields Publishers, Dehradun
71. Hegde MT (2024) Timber tree improvement in India—an overview. Wood Good 4(4):24–36
72. Tewari DN (1994) A monograph on *Dalbergia sissoo* Roxb. International Book Distributors, Dehradun, India, p 316
73. Singh J, Bangarwa KS, Singh J (1995) Heritable variation and genetic gain estimates in half sib progenies of *Dalbergia sissoo* Roxb. Crop Res Hisar 10:96–98
74. Yadav SS, Bangarwa KS (1997) Performance of *Dalbergia sissoo* Roxb. in a 12 year provenance trial in Haryana. Indian J For 21(1):13–15
75. Gera M, Gera N, Aggarwal R, Gupta BN (2002) Genetic variation in biochemical contents in foliage of twenty seed sources of *Dalbergia sissoo* Roxb. Indian For 128(7):726–737
76. Sharma A, Bakshi M (2014) Variability in growth, physiological, and biochemical characteristics among various clones of *Dalbergia sissoo* in a clonal seed orchard. Int J For Res 2014(1):ID 829368. https://doi.org/10.1155/2014/82936
77. Singh N (1998) Physiological and biochemical parameters vis-a-vis genetic improvements in *Dalbergia sissoo* Roxb. PhD thesis, F.R.I. (Deemed University), Dehradun, India
78. Dhixya Deve KR, Parthiban KT (2014) Growth performance and variability of *Dalbergia sissoo* Roxb. clones. Electron J Plant Breed 5(2):317–321
79. Rajendran P, Arunmaharaja B, Krishnakumar N, Parthiban KT (2019) Clonal variation of *Dalbergia sissoo* genetic resources. Pharma Innov J 8(11):271–277
80. Dogra AS, Nautiyal S, Nautiyal DP, Singh G (2006) Field performance of twenty three clones of *Dalbergia sissoo* in clonal seed orchard. Indian For 38(3):273–280
81. Rawat RS, Nautiyal S (2007) Inter clonal variation in *Dalbergia sissoo* Roxb. with respect to growth performance, survival and phenological characteristics in different climatic zones. Indian For 133(2):215–228
82. Rawat RS, Nautiyal S (2008) Inter-clonal variation in *Dalbergia sissoo* Roxb. with respect to photosynthetic rate, transpiration rate and stomatal conductance in different climatic zones. Indian For 134(5):657–666

83. Bamel JS (2005) Studies on genetic variability for seed production and its quality parameters in selected clones of *Dalbergia sissoo* Roxb. at two different agroclimatic sites. PhD thesis, FRI (Deemed University), Dehradun, India
84. Singh RI (2022) Studying the variation among clones of Dalbergia sissoo Roxb. for growth, stem quality and physiological trait. Unpublished MSc thesis, Punjab Agricultural University, Ludhiana, Punjab, India
85. Sahu PK, Sao R, Khute IK, Baghel S, Patel RR, Thada A, Parte D, Devi YL, Nair S, Kumar V, Mondal S (2023) Plant genetic resources: conservation, evaluation and utilization in plant breeding. In: Advanced crop improvement, volume 2: case studies of economically important crops, 2023 September 8. Springer International Publishing, pp 1–45. https://doi.org/10.1007/978-3-031-26669-0_1
86. Ginwal HS, Maurya SS (2010) Evaluation and optimization of DNA extraction method for *Dalbergia sissoo* leaf. Indian J Biotechnol 9:69–73
87. Arif M, Zaidi NW, Singh YP, Haq QMR, Singh US (2009) A comparative analysis of ISSR and RAPD markers for study of genetic diversity in shisham (*Dalbergia sissoo*). Plant Mol Biol Rep 27:488–495. https://doi.org/10.1007/s11105-009-0097-0ISI:000270780900009
88. Bal P, Panda PC (2018) Molecular characterization and phylogenetic relationships of Dalbergia species of eastern India based on RAPD and ISSR analysis. Int J Innov Sci Technol 3(7):417–425
89. Ashraf M, Abdul SM, Rabia R, Sobia T (2010) A molecular study of genetic diversity in shisham (*Dalbergia sissoo*) plantation of NWFP, Pakistan. Pak J Bot 42(1):79–88
90. Kumar A, Dobhal S, Sharma S (2011) Assessment of genetic diversity in different clones of *Dalbergia sissoo* Roxb. by RAPD markers. Afr J Biotechnol 10(35):6686–6694
91. Wang BY, Shi L, Ruan ZY, Deng J (2011) Genetic diversity and differentiation in *Dalbergia sissoo* (Fabaceae) as revealed by RAPD. Genet Mol Res 10(1):114–120
92. Pandey M, Gailing O, Leinemann L, Finkeldey R (2004) Molecular markers provide evidence for long-distance planting material transfer during plantation establishment of *Dalbergia sissoo* Roxb. in Nepal. Ann For Sci 61(6):603–606
93. Rout GR, Bhatacharya D, Nanda RM, Nayak S, Das P (2003) Evaluation of genetic relationships in *Dalbergia* species using RAPD markers. Biodivers Conserv 12:197–206
94. Ginwal HS, Maurya SS, Chauhan P (2011) Genetic diversity and relationship between cultivated clones of *Dalbergia sissoo* of wide geographical origin using RAPD markers. J For Res 22:507–517
95. Bakshi M, Sharma A (2011) Assessment of genetic diversity in *Dalbergia sissoo* clones through RAPD profiling. J For Res 393–397. https://doi.org/10.1007/s11676-011-0183-7
96. Dobhal S, Sharma S, Ahmed N, Kumar A (2019) Genetic polymorphism in *Dalbergia sissoo* Roxb. using RAPD markers. Indian J Biotechnol 18:164–173
97. Tewari SK, Tewari L, Dubey A, Kumar A, Kumar N, Kaushal R (2022) Use of the RAPD marker to determine the genetic diversity of various *Dalbergia sissoo* Roxb. (shisham) genotypes and their evolutionary relationship in Indian subcontinents. Vegetos 35(3):850–857
98. Ijaz S, Razzaq HA, Babar M, Haq I (2018) Assessment of population genetics of shisham (*Dalbergia sissoo*) based on genetic structure and diversity analysis. Int J Biosci 13(3):209–222
99. Ijaz S, Haq IU, Razzaq HA, Nasir B, Babar M (2019) ISSR based population genetics study for tagging a diverse population of shisham (*Dalbergia sissoo*) in Pakistan. Appl Ecol Environ Res 17:5851–5861
100. Javaid A, Akram W, Shoaib A, Haider MS, Ahmad A (2014) ISSR analysis of genetic diversity in *Dalbergia sissoo* in Punjab, Pakistan. Pak J Bot 46(5):1573–1576
101. Ijaz S, Haq IU, Razzaq HA, Nasir B, Babar M, Abbas M, Sakhawat MA (2022) Intraspecific discrimination power of plastid coding region rbcL as DNA barcoding marker in *Dalbergia sissoo* population of Pakistan. Phytopathogenomics Dis Control 1:25–29
102. Muhammad AJ, Haq IU, Ghafoor N, Rehman FU, Ali A, Muhammad ZB (2024) Cross-specific amplification of SSR markers in *Dalbergia latifolia* and *Dalbergia sissoo*—a case for hybrid seed production. J For Environ Sci 40(3):259–264. https://doi.org/10.7747/JFES.2024.40.3.259

103. Husen A (2009) Growth, chlorophyll fluorescence and biochemical markers in clonal ramets of shisham (*Dalbergia sissoo* Roxb.) at nursery stage. New For 38(2):117–129
104. Gautam S (2006) Study the variation in *D. sissoo* through morpho-physiological and molecular markers. PhD thesis, FRI (Deemed University), Dehradun, India

Chapter 19
Tissue Culture in *Dalbergia sissoo*: A Review

R. Kalaiselvi, Geeta Joshi, A. N. Arunkumar, and Rekha R. Warrier

1 Introduction

Tissue culture technology plays a transformative role in the propagation and conservation of tree species [37], addressing challenges such as slow natural growth, limited seed germination, and disease susceptibility [5]. This method, involving the aseptic cultivation of plant cells, tissues, or organs on nutrient media, enables rapid and large-scale production of genetically uniform plants [14]. Tissue culture also facilitates the year-round production of plants, independent of seasonal limitations. In addition, tissue culture accelerates the development of hybrid varieties with superior traits, such as faster growth, pest resistance, and enhanced environmental adaptability [22]. It also supports reforestation efforts in degraded ecosystems, making it an indispensable tool in sustainable forestry [6]. Traditional propagation methods, such as seed germination and cuttings, are constrained by low success rates and the limited availability of high-quality planting materials in *Dalbergia sissoo* [20]. Also, susceptibility to pests and slow natural regeneration are other bottlenecks observed in the species. A thorough literature review revealed that studies on micropropagation in *D. sissoo* have been attempted in India, Pakistan, Bangladesh, and Nepal, where this species is widespread in agroforestry.

R. Kalaiselvi · R. R. Warrier (✉)
ICFRE-Institute of Forest Genetics and Tree Breeding, Coimbatore, India
e-mail: rekha@icfre.org

G. Joshi
Indian Council of Forestry Research and Education (ICFRE), Dehradun, India

A. N. Arunkumar
ICFRE-Institute of Wood Science and Technology, Bengaluru, India

R. R. Warrier et al. (eds.), *Handbook of Asian Rosewoods*, Materials Horizons: From Nature to Nanomaterials, https://doi.org/10.1007/978-981-96-9459-4_19

2 Explant Sources

The choice of explant significantly influences the success of tissue culture, as different explants have varying capacities for regeneration, contamination rates, and growth responses. Below is a comparison of commonly used explants for *D. sissoo* tissue culture with applications.

Explant type	Advantages	Challenges	Applications
Nodal segments	• High regeneration potential • Easily obtained from mature plants • Low phenolic exudation	• Prone to contamination during initial stages • Limited by seasonal availability	• Micropropagation • Clonal propagation
Shoot tips	• High meristematic activity • Lower risk of somaclonal variation • Produces genetically uniform plants	• Delicate and susceptible to damage • Requires precise handling during excision	• Rapid propagation of disease-free plants • Genetic transformation
Leaf explants	• Abundant source • Can induce callus and somatic embryos	• High phenolic exudation • Difficult to induce direct organogenesis	• Callus culture • Somatic embryogenesis
Cotyledons	• High regeneration potential in young seedlings • Suitable for somatic embryogenesis	• Limited availability and viability • Requires sterilised seeds for initial culture	• Somatic embryogenesis • Production of synthetic seeds
Hypocotyls	• Suitable for inducing callus • High juvenile capacity for regeneration	• Limited to seedling sources • Variable response to plant growth regulators	• Callus-based regeneration • Research on stress responses
Root segments	• Can produce adventitious shoots under optimal conditions • Abundant in mature plants	• Difficult to sterilise • Low regeneration efficiency compared to other explants	• Rooting studies • Indirect organogenesis

Explant source, season, type, size, and disinfection are crucial for establishing aseptic *D. sissoo* tissue cultures (Fig. 1). Close examination of the tissue culture literature reveals that most of the studies lack information about one or more of these essential parameters (Tables 1 and 2), which may reduce the reproducibility of such studies, thereby hampering the large-scale production of the species. Nodes and axillary buds containing predetermined shoot meristems were preferred for micropropagation over other explants [1, 3, 4, 11, 12, 15, 24, 30, 36]. Embryos [7, 10,

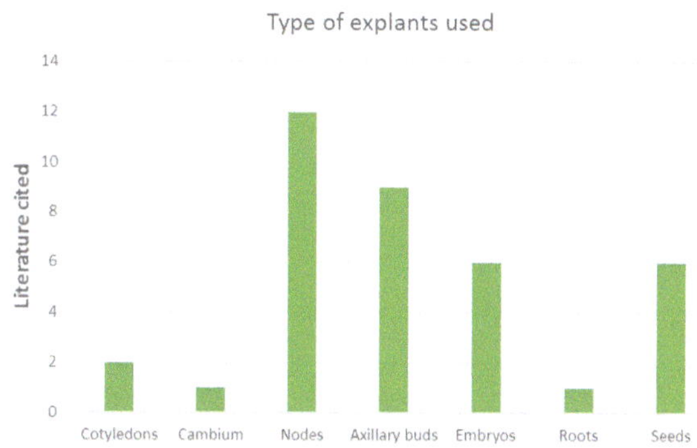

Fig. 1 A literature review of the different explants reported for use in tissue culture of *D. sissoo*

25, 33] and seeds [21, 23, 26, 29, 34, 35] were also used due to their juvenile, contamination-free nature, unlike mature trees [8, 11], which often show browning and microbial contamination.

3 Surface Sterilisation and Control of Browning

Effective surface disinfection and control of endophytic microbes are essential for successful culture establishment. All the studies adopted frequent washing with water (running tap water or distilled water) for surface cleaning and leaching of phenolics. Mercuric chloride was the commonly used sterilant to control microbial contamination. All the authors reported its use except [1, 2, 25]; four reports were data-deficient. In addition, Tween 20 [1, 7, 25, 30–33, 39], Teepol [8, 10, 15–17, 20, 26, 34, 35], and 70% ethanol [7, 13, 17, 24, 25, 30–33, 39] were used as surfactants, while Savlon [7, 30–33] and Dettol [25] were used as disinfectants. The time for sterilisation ranged from 10 min (for tender tissues such as embryos) to over 80 min (for mature tissues such as wood blocks) [20].

4 Culture Medium and Growth Regulators

The Murashige and Skoog (MS) medium was the commonly used one—either in half or full strength, except a few who used Woody Plant Medium and B5 [2, 3, 16, 36, 38]. Among the different growth regulators tested, Benzylaminopurine (BAP) at 0.5–5.0 mg/L was the most commonly used cytokinin for shoot proliferation (Table 2). In 75% of the cases, Kn (Kinetin) in concentrations ranging 0.5–12.5 mg/L, either singly or in combination with BAP, has also been used for shoot initiation. Only two studies reported the use of TDZ (Thidiazuron) [23, 38]. The auxins indole-3-acetic acid (IAA), indole-3-butyric acid (IBA), and Naphthalene Acetic acid (NAA) mainly targeted root induction and the concentrations ranged 0.05–2.0 mg/L 2,4-D

Table 1 Age, source, and sterilisation procedures adopted in processing different explants of *Dalbergia sissoo*

S.No	Type of explant (length)	Age details of explant	Study locations	Surface sterilisation process	Author (s) and years
1	Axillary buds (0.2–0.4 cm)	30–50-year-old tree	Data-deficient	Buds (soaked in 200 mg/L ascorbic acid) 95% EtOH (2–3 secs) 0.1% $HgCl_2$ (5–7 min) rinsed (3 × SDW)	Dawra et al. [13]
2	Seed	Seven days old seedling (epicotyl, hypocotyl, and cotyledon)	60–70-year-old elite tree	RTW (15 min) 5% liquid detergent Laboline (5 min) 0.1% $HgCl_2$ (8 min) rinsed (5 × SDW)	Pradhan et al. [23}
3	Seed	Data-deficient	University of Rajasthan Campus	1% $HgCl_2$ (10 min) rinsed (3 × SDW)	Sharma et al. [29]
4	Seeds	Data-deficient	Afforestation Division, Hattisar, Kathmandu	RTW (60 min) Teepol 0.1% $HgCl_2$ (20 min) rinsed (5 × SDW)	Suwal et al. [35]
5	Wood blocks from Bark (Cambial cells)	Data-deficient	Data-deficient	RTW (60 min) Teepol 0.1% $HgCl_2$ (8–10 min) rinsed (5 × SDW)	Kumar et al. [20]
6	Axillary buds/ Nodes (1–1.5 cm)	30–40 years old mature tree	M S University of Baroda	RTW (60 min) Teepol 0.05% 0.1%$HgCl_2$ (2 min) rinsed (4 × SDW)	Chauhan et al. [8]
7	Nodal segments (1 cm) from coppice shoot of mature tree	20–25-year-old tree	Forest department nursery, Rohtak	RTW 1% Teepol (20 min) 0.1% $HgCl_2$ (20 min) rinsed (5 × SDW)	Gulati and Jaiwal [15]
8	Semi-mature Zygotic embryo from green pods	40 days old	Regional Plant Resource Center, Bhubaneswar	2% Teepol (10 min) RTW (15–20 min) 0.1% $HgCl_2$ (20 min) rinsed (5 × SDW)	Das et al. [10]
9	Seed	Seven-day-old seedling (epicotyl, hypocotyl, and cotyledon)	Data-deficient	RTW (60 min) 0.1% $HgCl_2$ (20 min) rinsed (5 × SDW)	Pattnaik et al. [21]
10	Young Shoots (clone no 36)	60-year-old superior tree	CSO at New forest, Dehradun	RTW Teepol (20 min) 0.1% $HgCl_2$ (15 min) rinsed (5 × SDW)	Joshi et al. [16]
11	Mature and Immature Embryo	30-year-old Tree	Data-deficient	RTW (30 min) Tween 20 (15 min) Savlon (15 min) 70% EtOH (1 min) 0.1%$HgCl_2$ (20 min) rinsed (4 × SDW)	Singh et al. [33]

(continued)

Table 1 (continued)

S.No	Type of explant (length)	Age details of explant	Study locations	Surface sterilisation process	Author (s) and years
12	Juvenile Shoots (10 mm) from basal sprouts of mature tree. (Encapsulation of nodal segments)	Data-deficient	Data-deficient	RTW (30 min) Tween 20 (5 min) Savlon (5 min) 70% EtOH (30 Sec) 0.1%$HgCl_2$ (20 min) rinsed (4 × SDW)	Singh et al. [30]
13	1. Coppicing ability of clones from different locations 2. Nodes	Data-deficient	Clones collected from 1. Pathri, Uttaranchal 2. Gonda, Uttar Pradesh 3. Chichrauli, Haryana 4. Hanumangarh, Rajasthan	RTW Teepol (10–15 min) 0.1% $HgCl_2$ (20 min) 90% EtOH (5 min) rinsed (5 × SDW)	Kalia et al. [17]
14	Zygotic embryo from green pods	Pods used 40–50 days after anthesis	University campus, Khandwa Road, Indore	RTW (30 min) Tween 20 (15 min) Savlon (15 min) 70% EtOH (1 min) 0.1%$HgCl_2$ (20 min) rinsed (4 × SDW)	Chand et al. [7]
15	Cotyledon from Seed	Data-deficient	Data-deficient	RTW (30 min) Tween 20 (15 min) Savlon (15 min) 70% EtOH (1 min) 0.1%$HgCl_2$ (20 min) rinsed (4 × SDW)	Singh and Cchand [31]
16	Node, Internode, and Shoot tip	Data-deficient	Rajshahi University Campus, Rajshahi, Bangladesh	1% $HgCl_2$ (10 min) rinsed (3 × SDW)	Bonga et al. [4]
17	Cotyledon from Seed	Data-deficient	Data-deficient	RTW (30 min) Tween 20 (15 min) Savlon (15 min) 70% EtOH (1 min) 0.1%$HgCl_2$ (20 min) rinsed (4 × SDW)	Singh et al. [32]
18	Nodes from epicormic shoots	Data-Deficient	Data-deficient	RTW (60 min) 0.1% $HgCl_2$ (4 min) rinsed (5 × SDW)	Thirunavoukkarasu et al. [36]
19	Node (1.5–2.0 cm)	2 weeks and 6 months old nodes	University of Sargodha, Sargodha, Pakistan	RTW (5–6 min) 0.05% Tween 20 (10 min) 2% sucrose + 1.5% ascorbic acid + 1.5% citric acid (10 min) 1% NaOCl (15 min) rinsed (4 × SDW)	Ali et al. [1]

(continued)

Table 1 (continued)

S.No	Type of explant (length)	Age details of explant	Study locations	Surface sterilisation process	Author (s) and years
20	Mature and immature embryo from seed	Data-deficient	Data-deficient	RTW (25 min) Tween 20 (5 min) 3% Dettol (15 min) 70% EtOH (1–2 min) 0.01 M HCl (4 min) rinsed (3 × SDW)	Rehman et al. [25]
21	Nodal explants from cuttings of CPTs	Data-deficient	Clone 09Pathri, Uttarakhand Clone 59Chichrauli, Haryana Clone79-Hanumangarh, Rajasthan	0.1% $HgCl_2$ (15 min) rinsed (4 × SDW)	Arya et al. [3]
22	Seeds	Data-deficient	Afforestation Division, Hattisar, Kathmandu	RTW (60 min) Teepol 0.05% 0.1%$HgCl_2$ (20 min) rinsed (5 × SDW)	Suwal et al. [34]
23	Seed	6–9 year old tree	Data-deficient	RTW (15 min) Teepol (15 min) 0.1% $HgCl_2$ (8 min) 3xSDW	Sahu et al. [26]
24	Rejuvenated shoots (4–5 cm)	20–25-year-old tree	Data-deficient	0.1% Bavisitin (15 min) 0.1% $HgCl_2$ (3–5 min) 0.1% ascorbic acid (15 min) rinsed (3 × SDW)	Vibha et al. [38]
25	Axillary and epicormic shoots (1.5–2.5 cm)	Data-deficient	Data-deficient	Tween 20 0.7% chloroxylenol 70% EtOH (10 Secs) 0.1% $HgCl_2$ (5 min)	Zaidi et al. [39]
26	Axillary shoots (3.0–4.0 cm)	Data-deficient	Department of Botany, B R Ambedkar Bihar University	RTW (40 min) Detergent (10 min) 0.7% ascorbic acid + 0.7% citric acid (10 min) 0.1% $HgCl_2$ (5 min) rinsed (3 × SDW)	Anjali et al. [2]
27	Nodes from branch cuttings	Data-deficient	VMG in FRI, Dehradun	Wipe with 70% EtOH aqueous solution of Citramide (15 min) 0.2% Fungicide (15 min) 0.1% HgCl2 ()8–10 min	Raturi et al. [24]

(2,4-Dichlorophenoxyacetic acid) for callus induction. Additives such as calcium chloride [30], 3% or 2% sucrose [10], ascorbic acid [11], amino acids [12], activated charcoal [13, 15], silver nitrate/adenine sulphate [24], calcium nitrate, ammonium sulphate, potassium chloride [38], and casein hydrolysate [34] were also used for

Table 2 Media composition, hormones with concentrations, growth conditions for shoot and root initiation, and multiplication of *Dalbergia sissoo*

S.No	Media composition and concentrations	Purpose	Growth conditions	Result(s)	References
1	MS contai ns B5 amino acids + 5 mg/L ascorbic acid with Auxin (IBA, IAA, IPA and NAA) or Cytokinin (BAP, Kn) or GA	Plant regeneration	Temp 26 ± 4 °C, pp 16 h L Li 2000–3000 Lux	• MS contain B5 amino acids + 5 mg/L ascorbic acid + 2% Sucrose + 0.9% agar + 0.5 mg/L IAA, and 1.0 mg/L Kn shows 100% rooted shoots • Two auxins showed 10% and GA showed 5% rooted shoots	Datta et al. [12]
2	MS contain B5 vitamins + 5 mg/L ascorbic acid with IAA (0.1, 0.5, 1.0, 1.5) mg/L or NAA (0.5, 1.0,2.0) mg/L or IPA (0.5,1.0,2.0) mg/L or 2,4D (0.5 and 1.0) mg/L alone and in combinations	Effect of auxin in shoot development	Temp 28 ± 2 °C, pp 16 h L Li 1600 Lux	• Callus induced @ MS + B5 vitamins + 5 mg/L ascorbic acid + 2% Sucrose + 0.9% agar + 0.5 mg/L NAA alone and 0.5 mg/L IAA • Callus induced @ MS + B5 vitamins + 5 mg/L ascorbic acid + 1.0 mg/L IAA was transferred to MS + 1.0 mg/L IAA + 0.5 NAA for plant regeneration	Datta et al. [11]
3	MS + 1 mg/L thiamine HCl + 100 mg/L inositol + 3% sucrose + 1% activated charcoal with PGR (IAA alone and in combination with BAP and Kn)	Plant regeneration	Temp 27 ± 1 °C, Li 2500 Lux	• 55.4% Buds regenerate into rooted plantlet @ 0.5 mg/L NAA + 1.0 mg/L BAP	Dawra et al. [13]
4	MS + BAP (0.5–5.0) mg/L or Kn (0.5–5.0) mg/L or 2 iP (0.5–5.0) mg/L or TDZ (0.1–1.0) mg/L	Shoot proliferation	Temp 25 ± 1 °C, Li 45 μ m-2 S-1 60%RH	• Shoot proliferation noticed 4–5 days @ MS + 2.0 mg/L BAP (7.9 shoots/explant) • Rooting @ half-strength MS + IAA (1.0 mg/L) + IBA (1.0 mg/L) + IPA (1.5 mg/L)	Pradhan et al. [23}
5	MS + 2,4 D or BAP or IAA or NAA or IBA	Callus induction	Temp 26 ± 2 °C, Li 2000–3000 Lux	• Hypocotyl produced callus @ MS + 3 mg/L BAP + 1 mg/L IAA • Callus produced shoots @ MS + 5 mg/L BAP + 1 mg/L IAA • Rooting observed @ MS + 3 mg/L IBA + 3 mg/L NAA	Sharma et al. [29]

(continued)

Table 2 (continued)

S.No	Media composition and concentrations	Purpose	Growth conditions	Result(s)	References
6	MS + BAP (0.01 to 1.0) mg/L or Kn (2–4) mg/L or NAA (0.01–1.0)mg/L alone or in combination	Shoot induction	Temp 25 ± 4 °C, pp 16 h L Li3000 Lux	• Shoot initiation @ MS + 1 mg/L BAP + 0.1 mg/L NAA + 1 g/L Casein hydrolysate • Shoot multiplication @ MS + 0.25 mg/L BAP produced 10–15 shoots per explant	Suwal et al. [35]
7	MS + 2,4 D (0.5–2.0)mg/L + BAP (0.1–2.0) mg/L alone or in combination	Callus induction	Temp 25 ± 2 °C, pp 12 h, Li 45 E m-2 S-1 60–70%RH	• Callus induction @ MS + 2.0 mg/L 2,4 D + 0.1 mg/L BAP, 20 days old callus transferred to suspension • Single cells from suspension transferred to MS + 1 mg/LBAP for shoot induction • Rooting @ ½ MS + 1.0 mg/L IBA	Kumar et al. [20]
8	Ms + BAP, Kn, and NAA (0.25, 0.5, 1.0, 1.5, 2.0, and 2.5) mg/L	Shoot initiation and multiplication	Temp 26 to 30 ± 1 °C, pp 16 h L 8 h D, Li 2500 Lux (Optimum Temp 26 ± 1 °C)	• Shoot initiation on the 10.8^{th} day and grew till 25 days @ MS + 1 mg/L BAP + 0.25 mg/L NAA • Multiplication @ 1.5 mg/L BAP + 0.25 mg/L Kn. (Duration 4 weeks) • Root induction @ 1 mg/L IBA after 24 days of inoculation. (3.4 roots were observed at the length of 1.8 cm)	Ali et al. [1]
	MS + IAA, IBA and NAA (0.25, 0.5, 1.0, 1.5, and 2.0) mg/L	Root induction			

(continued)

Table 2 (continued)

S.No	Media composition and concentrations	Purpose	Growth conditions	Result(s)	References
9	MS + BAP and Kn (0.5, 1.0, 1.5, 2.0, and 2.5) mg/L	Shoot initiation	Temp 26 ± 1 °C, Li 3000 Lux, Moisture 68–70%	• Shoot bud initiation (88.48%) on the 14.68th day @ MS + 1.5 mg/L BAP (1.75 cm Shoot length) • Shoot bud initiation observed (78.62%) on the 16.56th day @ MS + 1.5 mg/L Kn (1.54 cm Shoot length) • Shoot multiplication observed (92.75%) on the 12th day @ MS + 1.5 mg/L BAP + 1 mg/L NAA (2.18 cm Shoot length) • Shoot multiplication observed (88.72%) on the 13.58th day @ MS + 1.5 mg/L Kn + 1 mg/L NAA (1.88 cm Shoot length) • Root induction @ 1 mg/L IBA (4.88 roots were observed at the length of 2.2 cm)	Anjali et al. [2]
	MS + BAP, Kn (1.5 mg/L) and NAA at (0.25, 0.5, 1.0, 1.5, and 2.0) mg/L	Shoot multiplication			
	MS + IBA and NAA (0.1, 0.25, 0.5, 0.75, 1.0, and 1.5) mg/L	Root induction			
10	MS + BAP (0, 2.5, 5.0, 7.5, 10.0, and 12.5) mg/L or Kn (0, 2.5, 5.0, 7.5, 10.0, and 12.5) mg/L	Shoot initiation	Data-Deficient	• March to May was reordered as the best season for bud induction and shoot proliferation • Shoot bud initiation observed (Clone 992.35%) @ MS + 2.5 mg/L BAP, (Clone 5989.31%) @ MS + 10.0 mg/L BAP, (Clone 7991.63%) @ MS + 5.0 mg/L BAP • Shoot multiplication rate observed in Clone 9 (2.41) @ 1X MS + 2.5 mg/L BAP, Clone 59-(2.35) @ 1X MS + 5.0 mg/L BAP, Clone 79 (2.37) @ 1X MS + 5.0 mg/L BAP • Root induction was observed in Clone 9 (65.12%) @0.5X MS + 5.0 mg/L IBA, Clone 59 (60.50%) @0.5X MS + 7.5 mg/L IBA, Clone 79 (62.74%) @ 0.5X MS + 5.0 mg/L IBA	Arya et al. [3]
	MS/B5/WPM (0.5X, 1X, 2X conc) + BAP (0, 2.5, 5.0, 7.5, 10.0, and 12.5) mg/L or Kn (0, 2.5, 5.0, 7.5, 10.0, and 12.5) mg/L	Shoot multiplication			
	MS (0.5X, 1X, 2X conc) with IBA (2.5, 5.0, 7.5, and 10) mg/L or IAA (2.5, 5.0, 7.5, and 10) mg/L or NAA (2.5, 5.0, 7.5, and 10) mg/L	Root induction			

(continued)

Table 2 (continued)

S.No	Media composition and concentrations	Purpose	Growth conditions	Result(s)	References
11	Ms + BAP (0.5, 1.0, 1.5, 2.0, and 2.5) mg/L alone, Kn (0.5, 1.0, 1.5, 2.0, and 2.5) mg/L alone and BAP + (NAA, IBA, IAA) combination, Kn + (NAA, IBA, IAA)	Callus induction Indirect organogenesis (Shoot induction)	Temp 27 ± 2°, PP 16 h, Li 3000Lux	• Callus induction from node/internode after 6 days @ 2 mg/L BAP + 0.5 mg/L NAA • Shoot induction from intermodal callus (2.4 shoots) @ 1.5 mg/L BAP + 0.5 mg/L IAA • Shoot induction from nodal callus (1.8 shoots) @ 1.5 mg/L BAP + 0.1 mg/L NAA • Root induction was observed after 12.6 days in MS + 1 mg/L IBA + 2 mg/L NAA (5.1 roots/shoot)	Bonga et al. [4]
	MS/½ MS + IBA, NAA, IAA (1.0, 2.0, 3.0, and 4.0) alone and with combination of IBA + NAA/IAA	Root induction			
12	MS + 0.05–3.0 mg/L 2,4D with 0.1/0.25 mg/L Kn	Callus induction	Temp 25 ± 2 °C, pp 16 h L 8 h D, 40 µM m^{-2} S^{-1} (40W Fluorescent tubes)	• Callus induction from zygotic embryo after 6 weeks (78.3%) @ MS + 2.0 mg/L 2,4 D + 0.25 mg/L Kn • Shoot induction @ MS + 2.0 mg/L BAP + 0.25 mg/LNAA (7.5 shoots/callus clump) observed after 10 weeks • Root induction @ ½ MS + 0.25 mg/L IBA was observed after 15 days	Chand et al. [7]
	MS + 0.5–3.0 mg/LBAP with 0.25 mg/L NAA	Shoot induction			
	MS/½ MS + 0.25–1.0 mg/L IBA	Root induction			

(continued)

Table 2 (continued)

S.No	Media composition and concentrations	Purpose	Growth conditions	Result(s)	References
13	½ MS + 1 mg/L IBA	Root primordial formation	Temp 25 ± 2 °C, pp 16 h L 8 h D, 50 μM m^{-2} S^{-1}	• 1–2 mm Root primordial was formed after 10 days	Singh et al. [30]
14	Sodium alginate (1.5, 3.0, 4.5, 5.0%), Calcium chloride (25, 50, 75, 100 mmol/L) prepared in ½ MS or ½ MSO or MSO or DH_2O	Encapsulation of nodal segments and root primordial with nodes		• Nodal and root primordial was encapsulated using 3% sodium alginate + 75 mmol/L $CaCl_2$ ($CaCl_2$ dissolved in ½ MSO) stored at 4 °C for 30 days	
15	MS or ½ MS or ½ MSO + NAA + BAP	Plantlet regeneration from Encapsulated materials		• 85% of Regeneration was observed @ ½ MS + 0.05 mg/L NAA + 0.5 mg/L BAP, and 98.3% was observed @ ½ MS0 after 20 days	
16	MS/modified MS with BAP, Kn, IBA, NAA with supplements	Shoot initiation Shoot multiplication Rooting	Temp 24 ± 1 °C, pp 16 h L	• Bud initiation @ MS + 2.0 mg/L Kn + 0.5 mg/L IBA + 3.0 mg/L BAP + 42.0 mg/L ascorbic acid, after 6 days • Multiplication observed @ MS (1/2 micro and macro) + 0.5 mg/L Kn + 1.0 mg/L Adenine sulphate, after 15 days (5–8 shoots/explant) • Rooting @ ½ MS + 0.1 mg/L NAA + 2.0 mg/L IBA after 3–5 days	Chauhan et al. [8]
17	MS, BAP, IBA	Shoot initiation	Data-Deficient	• Soot Initiation was (30%) observed @ MS + 1 mg/L BAP • Shoot multiplication was @ MS + 0.15 mg/L BAP (3 shoots/explant) • Rooting was observed (40%) @ MS + 1 mg/L IBA (4 roots/shoot)	Chinnaraj et al. [9]
18	MS (0.01–10) mg/L BAP alone or Kn (2–4) mg/L alone or in combination with NAA (0.01–0.1) mg/L	Seed germination Shoot initiation Shoot multiplication	Temp 25 ± 4 °C, pp 16 h L Li 3000 Lux	• Seed germination observed after 4 days (Solidified medium) • Shoot induction @ MS + 1 mg/L BAP + 0.1 mg/L NAA + 1 g/L Casein hydrolysate (2.5 shoots/explant after 7 days) • Multiplication @ 0.25 mg/L BAP (10–15 shoots after 8 weeks)	Suwal et al. [34]

(continued)

Table 2 (continued)

S.No	Media composition and concentrations	Purpose	Growth conditions	Result(s)	References
19	MS + Kn (0.1–0.25) mg/L + 2, 4 D (1.5–3.0) mg/L with 3% or 2% sucrose	Somatic embryos induction	Data-Deficient	• Callus induced @ MS + 0.1 mg/L Kn + 3.0 mg/L 2, 4 D + 3% sucrose, after 4 weeks • Proliferation of somatic embryos @ 0.5 mg/L Kn + 1.5 mg/L 2,4 D + 2% sucrose • Zygotic embryos formed @ ½ MS + 0.5 mg/L ABA + 2% sucrose • Developmental stages of embryos were studied using light and scanning microscope	Das et al. [10]
20	Data-deficient	Encapsulation of plant parts	Data-deficient	• Shoot tip, nodal segments, and root primordial were encapsulated with 3% sodium alginate and 75 mmol/L $CaCl_2.2H_2O$ • Synthetic seeds can be stored for 8 weeks at 4 °C • Used for long-term conservation and utilisation of germplasm	Sharma et al. [28]
21	Ms + auxin (NAA, IAA, IBA) or cytokinin (BAP, Kn) alone or in combination	Shoot induction	Temp 25 ± 2 °C, pp 16 h, Li 80 µM m^{-2} S^{-1}	• MS + 1 mg/L BAP alone produce maximum shoots (1.8/explant, 78%) • MS + 0.5 mg/L NAA induced two shoots (90% response) • MS + 1 mg/L BAP + 0.5 mg/L NAA produced 3–4 shoots with 100% response. After 60 days transferred to rooting media • Root induction @ MS + 1 mg/L IBA (7 days), transferred to ½ MS + 1% activated charcoal (14 days) finally to ½ MS without vitamins and PGR + 1% activated charcoal (7 days) • Rooted plantlets transferred to hardening	Gulati and Jaiwal [15]
	MS + IBA or NAA or NAA alone IBA with 1% activated charcoal	Root formation			

(continued)

Table 2 (continued)

S.No	Media composition and concentrations	Purpose	Growth conditions	Result(s)	References
22	Ms/B5 BAP (0.1 to 1) mg/L with IAA or NAA (0.1 to 0.5)mg/L combination	Bud breaking Shoot multiplication	Data-Deficient	• 100% bud breaking @ MS + 0.25 mg/L BAP + 0.25 mg/L NAA • Maximum 8 shoots were produced @ MS + 1 mg/L BAP + 0.25 mg/L NAA • Rooting @ ½ MS + 1 mg/L IBA after 18 days	Joshi et al. [16]
23	½ MS + NAA or IAA or IBA (0.1 to 1.5 mg/L) alone	Rooting			
24	Coppicing ability of shoot hedges MS + BAP (2.5–12.5) mg/L alone or with combination of NAA (0.5 or 1.25) mg/L	Shoot proliferation	Data-Deficient	• Coppicing ability of shoot hedges high in Uttar Pradesh clone (13.81 shoots formation), Haryana clones—9.29 shoots	Kalia et al. [17]
				• Nodes from trees of Uttar Pradesh and Uttaranchal showed shoot proliferation @ MS + 2.5 mg/L BAP, whereas those from Haryana and Rajasthan showed shoot proliferation @ MS + 2.5 mg/L BAP + 0.5 mg/L NAA	
	MS + IBA (2.5–10 mg/L)	Rooting		• Rooting @ MS + 2.5 mg/L IBA was optimised	

(continued)

Table 2 (continued)

S.No	Media composition and concentrations	Purpose	Growth conditions	Result(s)	References
25	MS + NAA (0–2.0) mg/L; 2,4D (0–2.0) mg/L alone or in combination with BAP (0–3.0)mg/L	Callus induction	Temp 25 ± 1 °C, Li 45 μ m^{-2} S^{-1} 60%RH	• Hypocotyl produced Green friable callus @ MS + 2 mg/L BAP + 0.5 mg/L BAP • Induced callus transfer to suspension with liquid media (MS + 2 mg/L BAP + 0.5 mg/L BAP) • Shoot induction at @ MS + 3 mg/LBAP + 0.5 mg/LNAA (5.9 shoots/explant) • Rooting @ ½ MS + 1 mg/L IAA + 1 mg/L IPA + 1 mg/L IBA (4.5 roots formed in 30–35 days)	Pattnaik et al. [21]
	MS + IAA + IPA + NAA	Rooting			
26	MS + BAP or Kn with silver nitrate/Adenine sulphate	Bud breaking, Bud induction, multiplication	Temp 26 °C Li 50 μ m^{-2} S^{-1} PP 16 h L/8 h D	• Bud breaking @ MS + 1.4 mg/L BAP, initiation @ MS + 1.5 mg/L BAP + 10 mg/L $AgNO_3$ (2.6 nodes/explant) • Multiplication @ MS + 0.5 mg/L BAP + 2 mg/L $AgNO_3$ + 50 mg/L Adenine sulphate (10.7 shoots/plant)	Raturi et al. [24]
27	MS + 2,4 D, Kn, BAP or NAA with a combination of auxin and cytokinin	Callus induction	Temp 26 ± 4 °C, pp 16 h L Li 2500 Lux	• Mature embryos induced callus @ 0.6 mg/L 2,4 D + 0.2 mg/L Kn • Mature embryo callus transferred to media MS + 0.3 mg/L NAA + 2.0 mg/L BAP • Rooting @ MS + 1.5 mg/L IBA	Rehman et al. [25]
	MS + BAP or Kn or NAA with combinations	Shoot proliferation			
	MS + IBA	Rooting			
28	MS + 2,4 D (0.5–4) mg/L MS + BAP (0.5–4.0)mg/L + Kn (1.0–5.0)mg/L	Callus induction Shoot proliferation	Temp 25 ± 2 °C Pp 16 h Li 2000–3000 Lux	• Callus induction from cotyledonary leaf @ MS + 3 mg/L 2,4 D • 82% Shoot proliferation from callus @ 0.5 mg/L NAA + 0.5 mg/L BAP	Sahu et al. [26]
29	MS + 2,4 D + Kn with sucrose 2% or 10% or glutamine	Somatic embryogenesis and plantlet development	Temp 26 ± 2 °C Li 50 μ m^{-2} S^{-1} PP 16 h L/8 h D	• Callus induced from cotyledons @ MS + 2.0 mg/L 2,4D + 0.1 mg/L Kn + 3% sucrose • Somatic embryos formed @ MS + 0.1 mg/L Glutamine + 3% sucrose • Plants regenerated @ ½ MS + 10% or 2% Sucrose	Singh and Chand [31

(continued)

Table 2 (continued)

S.No	Media composition and concentrations	Purpose	Growth conditions	Result(s)	References
30	MS + 2,4 D + Kn with sucrose 2% or 10%	Somatic embryos and encapsulation	Temp 25 ± 2 °C Li 50 μ m^{-2} S^{-1} PP 16 h L/8 h D	• Callus induced from cotyledons @ MS + 2.0 mg/L 2,4D + 0.1 mg/L Kn + 3% sucrose • Somatic embryos @ ½ MS + 10% or 2% Sucrose were selected and encapsulated with 2.5% sodium alginate with 75 mM $CaCl_2$	Singh et al. [32]
31	MS + BAP (1.0, 2.0, 3.0, and 5.0) mg/L + NAA (0.05 mg/L)	Induction of Somatic embryos and plant regeneration	Temp 25 ± 2 °C Li 50 μ m^{-2} S^{-1} PP 16 h L/8 h D	• Somatic embryo from mature cotyledons @ MS + 2.0 mg/L 2,4D + 0.1 mg/L Kn • Shoot bud formation @ MS + 5.0 mg/L BA • Rooting of mature and Immature embryos @ MS + 0.3 mg/L and 1.1 mg/L BAP respectively	Singh et al. [33]
32	MS/WPM + BAP (0.5–2.0 mg/L) and IAA (0.1–0.2 mg/L)	Shoot induction	Temp 25 ± 2 °C Li 45 μ m^{-2} S^{-1} PP 16 h L/8 h D	• 84.6% response recorded @ MS + 1.5 mg/L BA + 0.2 mg/L IAA (8.38 shoots/explant) • Rooting @ ½ MS 1.5 mg/L IBA	Thirunavoukkarasu et al. [36]
	MS + IAA (0.5–2.0 mg/L) or IBA (0.5–2.0 mg/L)	Rooting			
33	MS/WPM + BAP or Kn or TDZ alone or in combinations + calcium nitrate, ammonium sulphate, potassium chloride	Shoot and root formation	Temp 25 ± 1 °C, Li 45 μ m^{-2} S^{-1} 60%RH	• Initiation @ on MS medium + 2.0 mg/L of BAP • Shoot multiplication and elongation @ 0.5 mg/L BAP + 0.001 mg/L (20 shoots/plant), MS + 0.5 mg/L BAP	Vibha et al. [38]
34	MS + BAP or IAA or IBA alone or in combination	Shoot and root formation	Temp 26 ± 2 °C PP 16/8	• Shoot formation MS + 1.5 mg/L BAP + 0.3 mg/L IAA and MS + 2.5 mg/L BAP + 0.3 mg/L IAA (2.78 shoots/explant) • Root induction @ MS + 1.0 mg/L IBA + 1.0 mg/L IAA	Zaidi et al. [39]

better results. Most researchers have used a combination of auxin and cytokinins to produce either shoot regeneration in axillary buds or callus induction.

5 Shoot Initiation and Multiplication

Most in vitro protocols involve direct or callus-mediated regeneration (Table 2). Shoot initiation occurred within 10–25 days at optimal PGR concentrations. MS + 1–2.5 mg/L BAP + 0.25 mg/L NAA resulted in the highest shoot proliferation [34, 35]. The inclusion of additives in the medium enhanced shoot proliferation [24, 35, 38].

Callus induction and somatic embryo generation were the most successful routes of shoot induction and proliferation [4, 7, 10, 11, 18–21, 25, 26, 29, 31–33]. The callus could be induced from hypocotyls, cotyledons, cambium, shoot tips, and stem segments from mature trees, suggesting this route is the most appropriate for the propagation of *D. sissoo*. Responses from axillary bud proliferation were met with poor responses, which could be attributed to the recalcitrant nature of the species or the hindrance due to phenolics. Synthetic seed technology, a helpful storage method for encapsulated in vitro regenerants or explants [28], has only seen limited application in *D. sissoo*, and the only report available has provisional results [27] (Table 3).

6 Root Induction

In vitro rooting was the only method adopted for root induction. The most optimal rooting was observed in ½ MS + 1 mg/L IBA. Rooting occurred 12–35 days after inoculation. On average, four roots per shoot were obtained [2, 4, 9, 21]. About fifteen reports were available on the rooting efficiency of the shoots, while other reports only mentioned attempts made with rooting experiments. Successful rooting is a crucial stage in the mass production of a plant species. Though most authors have reported this aspect, difficulties have been suggested in the identification/adoption of a proper combination of auxins and cytokinins for root induction.

7 Acclimatisation

Successful acclimatisation ensures high survival rates during field transfer. Precise details on primary and secondary hardening, including potting mixture characteristics (pH and water conductivity), temperature, humidity, duration of each hardening stage, and the application of nutrient solutions, are crucial parameters to be considered. Only eleven studies reported the survival of the rooted plants under greenhouse conditions. The commonly used medium for transfer was sand and soil, including

Table 3 Acclimatisation, hardening protocols, and field performance of *Dalbergia sissoo*

S. No	Hardening process	Remarks	References
1	In vitro rooted ex agar plants planted in Soil:Sand: FYM (1:1:1) Kept in moist chamber (RH—80–90%, Temp 30 ± 2 °C), 4 to 5 weeks transfer to pots or polybags	20–25% survived	Arya et al. [3]
2	In vitro rooted ex agar plants Sand:soil: Compost (1:1:1) Kept in polytunnels for 30–35 days, transfer to natural conditions	60% survived	Bari et al. [4]
3	In vitro rooted ex agar plants Sterilised peat moss: soil: compost (1:1:1) for 3 weeks transfer to pots	53% survived	Chand and Singh [7]
4	Regenerated plantlets Sterilised peat moss: soil: compost (1:1:1)	56% survived	Singh and Chand [30]
5	Microshoots Sand bed (34 °C/ 15 °C, RH 70–80%) after 4 weeks transferred the pots to natural condition	35% plants developed roots	Suwal et al. [34]
6	Rooted plantlets sterile mixture of Vermiculite: sand: soil (1:1:1) for 15 days transferred to pots with Garden soil	100% survival	Gulati and Jaiwal [15]
7	Rooted plantlets liquid ¼ MS + 2% sucrose for 7 days sand:soil: Manure (1:1:1) kept in polytunnels	70% survival	Joshi et al. [16]
8	Rooted shoots Soil:Sand (1:1) in glass house Pots	45% survived	Kumar et al. [20]
9	Exagar plants vermicompost soil	60–70% survived	Pattnaik et al.[21]
10	Exagar plants vermicompost soil	Survival rate 68%	Pradhan et al. [23]
11	Encapsulated embryos germinated under ex vivo condition (Peatmoss moist with ½ MS)	22.2% germination rate	Singh and Chand [32]

peat moss [7], using vermiculite [15] and including vermicompost [21, 23] in the soil. The survival rates ranged from 20 to 100%, though no reports are available on the transfer of the plants and their performance in the field.

8 Future Directions

Tissue culture techniques have significantly supported the propagation and conservation of *D. sissoo*. However, challenges such as contamination, recalcitrance, and rooting necessitate continued research. Research on refining rooting media composition, hormone treatments, and environmental factors to improve root development and uniformity is the need of the hour. There are challenges in defining primary and secondary hardening conditions, including potting mix composition, humidity control, and acclimatisation duration, which decide survival rates. Systematic studies are required on the transfer of plants to field conditions to assess their adaptability and survival.

Future integration of molecular tools and cryopreservation will further enhance the efficiency and scalability of *D. sissoo* tissue culture. These advancements hold promise for sustainable forestry and the conservation of this economically and ecologically important species.

References

1. Ali A, Rizwan M, Majid A, Saleem A, Naveed NH (2012) Effect of media type and explants source on micropropagation of *Dalbergia sissoo*: A tree of medicinal importance. J. Med. Plants Res. 6(9):1742–1751
2. Anjali K, Kumari R (2020) Development of protocol for efficient micropropagation of a multipurpose local cultivar of *Dalbergia sissoo* a threatened tree species. Indian J. Sci. Res. 11(1):117–125
3. Arya ID, Nautiyal S, Arya S (2013) Tissue culture studies on clonal variations in micropropagation of *Dalbergia sissoo*. Int. J. Biotechnol. Res. 1(4):58–70
4. Bari MA, Ferdaus KMKB, Hossain MJ (2008) Callus induction and plantlet regeneration from In vivo nodal and internodal segments and shoot tip of *Dalbergia sissoo* Roxb. J. Bio-Sci. 16:41–48
5. Bonga JM, Aderkas P (1992) In vitro culture of trees. Kluwer Academic Publishers, Dordrecht
6. Bozzano M, Jalonen R, Thomas E, Boshier D, Gallo L, Cavers S, Bordács S, Smith P, Loo J (2014) Genetic considerations in ecosystem restoration using native tree species. State of the world's forest genetic resources–thematic study. Food and Agriculture Organization of the United Nations, Rome
7. Chand S, Singh AK (2005) Plant regeneration from semi-mature zygotic embryos of *Dalbergia sissoo* Roxb. Int J Biotechnol 4:78–81
8. Chauhan VA, Josekutty PC, Jasrai YT, Prathapasenan G (1996) Rapid in vitro multiplication of *Dalbergia sissoo* Roxb. J Plant Biochem Biotechnol 5:117–118
9. Chinnaraj S, Malimuthu C (2011) Development of micro-propagation and mini cutting protocol for fast growing Melia, Dalbergia and Eucalyptus clones for pulpwood and bio-energy plantations. BMC Proc 5(7):131

10. Das P, Samantaray S, Rout GR, Roberts AV (1997) In vitro somatic embryogenesis of *Dalbergia sissoo* Roxb. —a multipurpose timber-yielding tree. Plant Cell Rep 16, 578–582
11. Datta SK, Datta K (1984) Auxin induced regeneration of forest tree *Dalbergia sissoo* Roxb. through tissue culture. Curr Sci 52(9): 434–436
12. Datta SK, Datta K, Pramanik T (1982) *In vitro* clonal multiplication of mature trees of *Dalbergia sissoo* Roxb. Plant Cell, Tissue Organ Cult 2:15–20
13. Dawra S, Sharma DR, Chowdhury JB (1984) Clonal propagation of *Dalbergia sissoo* Roxb. through tissue culture. Curr Sci 53(15):807–809
14. George EF, Hall MA, De Klerk GJ (2007) Plant propagation by tissue culture (volume 1), Springer, Science and Business Media
15. Gulati A, Jaiwal PK (1996) Micropropagation of *Dalbergia sissoo* from nodal explants of mature trees. Biol Plant 38:169–175
16. Joshi I, Bisht P, Sharma VK, Uniyal DP (2003) Studies on effect of nutrient media for clonal propagation of superior phenotypes of *Dalbergia sissoo* Roxb. through tissue culture. Silvae Genetica 52(3–4):143–147
17. Kalia S, Kalia RK, Sharma SK (2004) Evaluation of clonal variability in shoot coppicing ability and *in vitro* responses of *Dalbergia sissoo* Roxb. Silvae Genetica 53(1–6):212–220
18. Kapoor R, Rao M (2019) Micropropagation and somatic embryogenesis in *Dalbergia sissoo*: a review. Plant Biotechnol. Rep. 13(4):345–360
19. Kaur H, Kapoor S (2018) Direct organogenesis from nodal segments of *Dalbergia sisso*o. Indian J Biotechnol 17(2):120–128
20. Kumar A, Tandon P, Sharma A (1991) Morphogenetic responses of cultured cells of cambial origin of a mature tree-*Dalbergia sissoo* Roxb. Plant Cell Rep 9:703–706
21. Pattnaik S, Pradhan C, Naik SK, Chand PK (2000) Shoot organogenesis and plantlet regeneration from hypocotyl-derived cell suspensions of a tree legume, *Dalbergia sissoo* Roxb. In Vitro Cell and Dev Biol Plant 36(5):407–411
22. Potter KM, Jetton RM, Bower A, Jacobs DF, Man G, Hipkins VD, Westwood M (2017) Banking on the future: progress, challenges and opportunities for the genetic conservation of forest trees. New Forest 48:153–180
23. Pradhan C, Kar S, Pattnaik S, Chand PK (1998) Propagation of *Dalbergia sissoo* Roxb. through in vitro shoot proliferation from cotyledonary nodes. Plant Cell Reports 18:122–126
24. Raturi MK, Thakur A (2021) Silver nitrate and silver-thiosulphate mitigates callus and leaf abscission during Shisham clonal micro-propagation. J. Plant Biotechnol. 48(3):173–178
25. Rehman HM, Rana IA, Mustafa G, Joyia FA, Khan IA, Pijut PM (2012) In vitro regeneration of *Dalbergia sissoo* Roxb. and the potential for genetic transformation. Notulae Botanicae Horti Agrobotanici Cluj-Napoca 40(2):140–147
26. Sahu J, Khan S, Sahu RK, Roy A (2014) Micropropagation of *Dalbergia sissoo* Roxb. through tissue culture technique. Pak J Biol Sci 17(4):597–600
27. Sharma N, Mehta R, Kumar S (2017) Somatic embryogenesis in *Dalbergia sissoo*: a tool for synthetic seed production. Plant Tissue Cult. Biotechnol. 27(1):101–109
28. Sharma P, Roy M, Roy B (2021) Encapsulation of plant propagules of *Dalbergia sissoo* Roxb. (Rosewood) for conservation and utilisation of germplasm. AgriAllis 3(9):16–19
29. Sharma S, Chandra N (1988) Organogenesis and plantlet formation *in vitro* in *Dalbergia sissoo* Roxb. J Plant Physiol 132(2):145–147
30. Singh AK, Chand S (2004) Somatic embryogenesis and plant regeneration from cotyledon explants of a timber-yielding leguminous tree, *Dalbergia sissoo* Roxb. J Plant Physiol 160(4):415–421
31. Singh AK, Chand S (2005) Protocol of somatic embryogenesis: *Dalbergia sissoo* Roxb.(Sissoo). In: Protocol for somatic embryogenesis in woody plants, pp. 361–368, Springer Netherlands, Dordrecht
32. Singh, A. K., Chand, S.: Plant regeneration from alginate-encapsulated somatic embryos of *Dalbergia sissoo* Roxb. Indian J. Biotechnol. **9** 319–324 (2010)
33. Singh AK, Chand S, Pattnaik S, Chand PK (2003) Adventitious shoot organogenesis and plant regeneration from cotyledons of *Dalbergia sissoo* Roxb. a timber yielding tree legume. Plant Cell Tiss Org Cult 68:203–209

34. Suwal B, Karki A, Rajbhandary SB (2014) The in vitro proliferation of forest trees: *Dalbergia sissoo* Roxb. ex Dc. In: Compilation (Micropropagation research papers 1988–2012), Bulletin No.37, Department of Plant Resources (DPR), Thapathali, Kathmandu, Nepal. pp. 1–4
35. Suwal B, Karki A, Rajbhandary SB (1988) The *in-vitro* proliferation of forest trees. 1. *Dalbergia sissoo* Roxb. ex DC. Silvae Genetica 37(1): 26–28
36. Thirunavoukkarasu M, Panda PK, Nayak P, Behera PR, Satpathy GB (2010) Effect of media type and explant source on micropropagation of *Dalbergia sissoo* Roxb.-an important multipurpose forest tree. Int Res J Plant Sci 1:155–162
37. Thorpe TA (2007) History of plant tissue culture. Mol Biotechnol 37:169–180
38. Vibha JB, Shekhawat NS, Mehandru P, Dinesh R (2014) Rapid multiplication of *Dalbergia sissoo* Roxb.: a timber yielding tree legume through axillary shoot proliferation and ex vitro rooting. Physiol Mol Biol Plants 20: 81–87
39. Zaidi S, Khan AI, Rana IA (2016) Development of source independent micropropagation system in *Dalbergia sissoo* Roxb. as a basis for germplasm conservation and disease free plants production. Mol Plant Breed 7:1-12

Chapter 20
Wilt and Dieback of *Dalbergia sissoo*: A Systematic Review on Contributing Factors and Management Strategies

Ramkrishna, Nitika Negi, Shristi Manikpuri, Nitika Bansal, Siya Sharma, Sabyasachi Banerjee, Sushant Sharma, Amit Pandey, Ajay Thakur, Rajendra Kumar Meena, Shambhavi Yadav, Maneesh S. Bhandari, and Shailesh Pandey

1 Introduction

Dalbergia sissoo Roxb., commonly known as shisham, is one of the most significant timber-yielding species of international importance. This nitrogen-fixing tree, belonging to the family Fabaceae, forms root nodules through symbiosis with rhizobia, comparable to the nitrogen-fixing mechanisms observed in crops [1]. Shisham thrives in a range of sub-tropical to tropical climates and is widely distributed across India, including the northeastern, northern, western, eastern, and central regions, as well as in neighboring countries. It is native to the sub-Himalayan and Bhabar regions and has been intentionally introduced to other areas [2]. Shisham is commonly found alongside *Acacia catechu* (Khair) trees as a prominent seral species [3]. It typically grows in alluvial soils with a pH range of 5.0–7.7 [4] and serves as a pioneer species in the riverine succession of the Gangetic alluvial plains in India [5]. This species plays a critical role in the development and management of ecosystems, particularly in stabilizing riverbanks. Its deep roots minimize soil erosion in areas prone to flooding, thereby strengthening riverbanks and preserving the soil. Owing to

The first four authors contributed equally to this work.

Ramkrishna · N. Negi · S. Manikpuri · N. Bansal · S. Sharma · S. Banerjee · S. Pandey (✉)
Forest Pathology Discipline, Forest Protection Division, ICFRE-Forest Research Institute, Dehradun, India
e-mail: shailesh31712@yahoo.co.in; pandeysh@icfre.org

A. Pandey
ICFRE-Institute of Forest Productivity, Ranchi, Jharkhand, India

S. Sharma · A. Thakur · R. K. Meena · S. Yadav · M. S. Bhandari
Genetics and Tree Improvement Division, ICFRE-Forest Research Institute, Dehradun, India

R. R. Warrier et al. (eds.), *Handbook of Asian Rosewoods*, Materials Horizons: From Nature to Nanomaterials, https://doi.org/10.1007/978-981-96-9459-4_20

its high economic and ecological value, shisham is often cultivated in plantations and agroforestry systems, ensuring a sustainable supply of this valuable wood resource across the Indian subcontinent.

Unfortunately, shisham is vulnerable to fungal diseases that seriously harm the trees and even cause their death. In a pioneering study, *Fomes* sp. was found to be the leading cause, swiftly killing standing shisham trees by attacking their roots [6]. Subsequently, *Fomes lucidus* (currently classified as *Ganoderma lucidum*) was recognized as a lethal fungus responsible for widespread shisham mortality in plantations and other areas [2]. Wilt and dieback were commonly observed in small clusters of shisham trees growing in marshy and waterlogged conditions, near highways where soil is regularly excavated, and in orchards and tea gardens where soil is ploughed, hoed, fertilized, and irrigated [7]. Trees of all ages were found to be susceptible to wilting, but those between 6 and 10 years old were particularly vulnerable. Interestingly, while the disease was not confined to specific soil types, trees in clay soils with water stagnation exhibit more severe symptoms [7]. In some cases, mortality in shisham trees aged 15–25 years was attributed to wilt caused by *Fusarium solani* (current name: *Neocosmospora solani*), while younger trees in the seedling, sapling, and pole stages did not show similar symptoms [8]. Follow-up studies revealed that the fungus survives saprophytically in soil as mycelium and spores [9]. Laboratory experiments further demonstrated that sterile loam soils with up to 20% moisture content supported the survival of wilt pathogen, while higher moisture levels were detrimental [10]. Additionally, shisham trees in irrigated plantations showed no signs of root rot, suggesting irrigation as an effective preventive measure against wilt [10]. Artificial inoculation experiments later confirmed the pathogenicity of *F. solani* as a wilt-inducing pathogen, reproducing natural wilt symptoms on inoculated trees [11]. More recently, significant mortality was reported in experimental trials of 49 clones and ramets in northern India [12]. Research has established that *F. solani* and *G. lucidum* can coexist synergistically, presenting a major challenge for effective disease management [13]. To manage and prevent fungal attacks, several strategies have been recommended, including careful site selection to avoid poorly drained soils, use of chemical fungicides, proper disposal of infected plant material, application of beneficial microbes for biological control, cultivating resistant germplasm, and implementation of best cultural practices. Despite these efforts, shisham mortality remains a persistent issue, causing substantial economic and ecological losses.

2 Losses Due to Shisham Mortality

Over the past three decades, widespread mortality of shisham trees has been reported across India, Pakistan, Nepal, Bhutan, Myanmar, and Bangladesh [14–24]. In Pakistan's Punjab province, up to 80% of shisham population along canal banks and 40% along highways and roadsides were destroyed [20]. Similarly, in Bangladesh, a significant number of shisham trees began dying in the 1990s, resulting in the loss of 2 million trees within seven years and a mortality rate of 55% [17, 19]. In

India, disease surveys indicated that more than 400,000 mature trees were destroyed between 1998 and 2001, causing an estimated economic loss of USD 200 million [18]. In eastern India, a follow-up survey recorded a staggering 70% mortality rate in Purnea, Bihar, with 350,000 dead trees observed across an area of 5,428 hectares [21]. In the Mainpuri district of Uttar Pradesh, wilt disease caused 31–57% mortality among roadside, riverbank, and other major shisham-growing localities [25]. More recent surveys in northern India have shown that plantation trials at Pinjore, Haryana, and Arcadia Tea Estate, Dehradun, are under severe threat [23]. This alarming decline has also raised concerns in Myanmar, Bhutan, Bangladesh, Sri Lanka, and Nepal, further underscoring the regional scale of the issue [24]. Overall, this widespread mortality has led to a significant reduction in shisham populations, with billions of trees lost and a noticeable decrease in tree densities. The impact of shisham mortality is far-reaching, severely affecting the economic objectives of the state forest departments and imposing heavy financial losses on farmers. Addressing this critical issue is essential to safeguarding both the ecological and economic benefits provided by this valuable species.

3 Symptoms of Wilt and Dieback

Tree decline is characterized by a slow but widespread loss of vigor, ultimately leading to the mortality of one or more tree species [26]. However, the term "decline" does not always denote a specific disease or its underlying cause. When researchers refer to conditions such as maple, oak, and ash decline, they are describing illnesses marked by gradual debilitation, regardless of their etiology [27]. Similarly, shisham decline is a broad term used to describe a gradual loss of growth and vitality in trees. The most commonly observed symptoms associated with declining shisham trees are wilt and dieback, both of which have been clearly defined [7]. Wilt refers to cases where isolated trees or patches of trees within an affected area die suddenly due to root infections caused by fungi. Symptoms of wilt include flagging of leaves, pods, and tender twigs; the formation of a thin crown; and yellowing and reduction of leaves (Fig. 1A). Dieback, on the other hand, presents more specific symptoms and is characterized by the progressive death of twigs and branches, typically beginning at the tips. This is followed by thinning of leaves, crown reduction, and the development of a "table-topped" appearance, where the dead, leafless limbs resemble a stag's head (Fig. 1B). Notably, early-stage dieback symptoms, such as crown thinning and leaf yellowing, overlap with those of wilt. As the condition progresses, small dry twigs shed continuously, leaving the tree appearing as a blunt stub with thick branches [28]. Eventually, the tree dies (Fig. 1C). Fruiting bodies of *Ganoderma* fungi are often observed at the base of the trunk (Fig. 1D). Additionally, dying trees frequently ooze sap (Fig. 1E), and different species of *Fusarium* and *Neocosmospora* have been consistently isolated from such trees (Fig. 2A–I; our unpublished data). This clear distinction between wilt and dieback, along with their overlapping symptoms in early stages, emphasizes the complexity of diagnosing and managing shisham decline.

Fig. 1 Wilt (**A**) and dieback symptoms (**B**); Dead shisham trees (**C**); Sporophores of *Ganoderma lucidum* (**D**); Ooze exuding from the tree trunk (**E**)

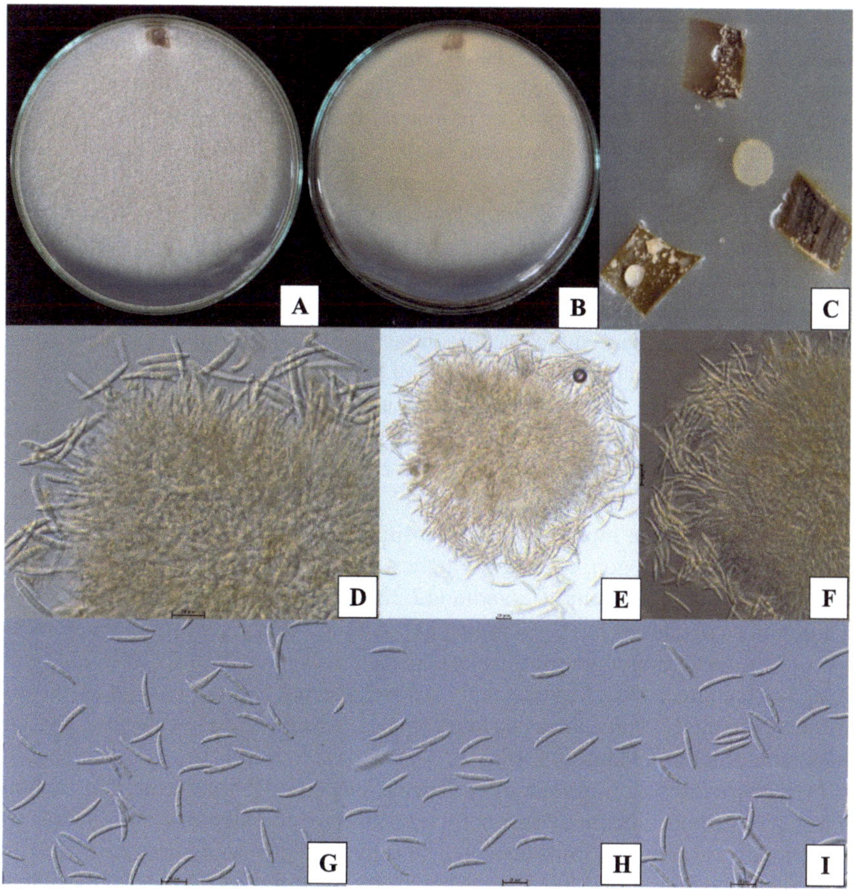

Fig. 2 Colony Front **(A)** and rear view **(B)** of *Fusarium solani* (current name: *Neocosmospora solani*) on PDA; Sporodochia produced on carnation leaf agar **(C)**; Microscopic view of sporodochia showing conidia attached to phailades **(D–F)**; Boat-shaped conidia **(G–I)**, Scale bar: 20 μm **(D–I)**

4 Major Pathological Factors Responsible for Shisham Mortality

Researchers have identified several potential factors contributing to the high mortality rate of shisham. These include edaphic (soil-related), entomological (insect-related), pathological (pathogen-related), silvicultural (forest management-related), and age-related factors. Among these, pathological factors have received the most attention, particularly following the studies by [7] and [8], which identified *F. solani* (current name: *N. solani* [29]) as the primary cause of shisham mortality in India. Further studies on fungal pathogens have reinforced the hypothesis that fungi are among the primary factors responsible for widespread shisham mortality, manifesting as wilt and dieback symptoms. However, there remains a lack of consensus among researchers regarding the specific fungal species responsible for this issue. For example, in Pakistan, *Phytophthora cinnamomi* was identified as a significant cause of shisham mortality [30]. Subsequent surveys across 11 districts in Punjab, where mortality rates ranged from 25–30%, revealed the involvement of *Botryodiplodia theobromae*, *F. solani*, and *Colletotrichum* sp. in diseased trees [31]. Pathogenicity tests indicated that *B. theobromae*, either alone or in combination with *F. solani* and *Colletotrichum* sp., caused browning of vascular bundles, gummosis, and the death of branches and leaves. However, neither *F. solani* nor *Colletotrichum* sp. individually produced disease symptoms. In contrast, pathogenicity assays demonstrated that *F. solani* alone could induce yellowing of leaves and internal browning of stems and roots. Meanwhile, the less frequently isolated *Rhizoctonia solani* and *Curvularia lunata* were unable to produce severe disease symptoms on their own [32]. Subsequently, *Ceratocystis fimbriata*, either alone or in combination with *F. solani*, was reported to induce disease symptoms, with *C. fimbriata* suspected to be the primary factor driving shisham mortality [33]. In a similar study, *F. solani* was identified as the major fungal pathogen, causing the highest disease incidence (31.39%), followed by *B. theobromae* (19%) and *C. lunata* (12.22%) in stem inoculations. Notably, no symptoms were caused by *G. lucidum* [34]. Another study confirmed the association of *Ceratocystis manginecans* (now considered a synonym of *C. fimbriata*) with dying shisham trees [35]. More recently, a novel species, *Ceratocystis dalbergicans*, distinct from the *C. fimbriata* sensu lato complex, was identified as a causal agent of shisham mortality in Pakistan [36].

In northern India, during a survey conducted in 2002 in Bhitmera, mortality among 5-year-old shisham clones due to root rot was observed in a clonal seed orchard and a seed production area [37]. Subsequently, 38 fungal isolates were recovered from the roots, stems, and ooze of dying trees. These were identified as *F. solani* (24 isolates) and *Fusarium* sp. (14 isolates) based on morphological characteristics and sequence analysis using specific primers targeting the internal transcribed spacer (ITS) region. Shisham seedlings inoculated with 18 *F. solani* isolates through soil and roots developed characteristic disease symptoms. Re-isolation successfully yielded *Fusarium* isolates identical to those used for inoculation, thus fulfilling Koch's postulates. This study strongly supported the hypothesis that *F. solani* is the primary cause of shisham mortality in India [38].

5 Abiotic Factors and Shisham Mortality

Research on abiotic factors has provided detailed insights into the complex interplay of environmental stressors that affect the health of shisham. The foundation for understanding the sensitivity of shisham trees to environmental conditions was established in the 1950s. Poor soil texture, particularly clay soils, was identified as a factor linked to high mortality rates [10]. Careful observations revealed that trees in light-textured soils exhibited better health, whereas areas with unfavorable soil composition, such as Saharanpur and Dehradun, experienced high mortality rates [10]. A comprehensive study on various abiotic stressors affecting shisham revealed waterlogging as the most detrimental factor, with a disease incidence of 73.1% [9]. Flooding and semi-flooding conditions also significantly affected tree health, with 27.4% and 52.1% mortality incidences, respectively. Other identified factors included degraded soil conditions (27.6% incidence), high irrigation practices (13.8% incidence), and high-density and moisture conditions (41.1% incidence). A study carried out in the Karnal Forest Division further emphasized the importance of water availability, demonstrating that when irrigation was stopped in 9 out of 11 coupes, disease-related mortality was observed in all non-irrigated areas after three years, while adequately watered coupes remained healthy [39]. This highlights the susceptibility of shisham to water stress and its potential to increase disease vulnerability.

During the 1980s and 1990s, researchers began investigating the impact of water-table depth on shisham tree health. A persistently high water table, particularly at a depth of 2–3 m, has been shown to cause up to 100% mortality in plantations. This effect is attributed to increased susceptibility to root infections, particularly by pathogens like *F. solani*, which thrive in saturated conditions. The age of the trees also plays a crucial role in their tolerance to high water tables, with younger trees (5–6-years-old) experiencing higher mortality at shallow water-table depths. As research progressed, more comprehensive ecological surveys provided a broader perspective on shisham mortality. A survey conducted across six northern Indian states identified water stress and nutrient deficiencies as significant contributors to mortality. The study also highlighted the regional variability in abiotic stresses impacting the species [40]. Another study in Punjab, Pakistan, focused on the role of soil moisture in disease progression and recorded mortality rates of 75–80% along canal banks, establishing a strong correlation between high soil moisture levels and the severity of wilt and dieback diseases [20]. These findings were further corroborated by research showing that waterlogged conditions exacerbate the incidence of dieback [41, 42]. Acid rain, which leaches essential soil nutrients and mobilizes toxic heavy metal ions, emerged as another significant abiotic factor impacting shisham health. This environmental pollution weakens the trees by reducing their resistance to diseases and extreme weather conditions, further exacerbating their vulnerability [43].

Climate change has emerged as a significant factor contributing to tree mortality on a global scale. Alterations in climate patterns, including global warming and unpredictable rainfall, are likely among the underlying causes of shisham decline. Research has provided quantitative evidence of the relationship between climatic factors and dieback, showing that a 1 °C increase in temperature could result in a 5.02% rise in disease incidence and a 6.30% increase in disease severity [44]. Drought has also been identified as a primary driver of shisham tree decline, underscoring the critical role of both waterlogging and drought stress in tree health [41]. A link between increased tree mortality and drought has been reported, with mortality rates rising from 1% in previous years to 4% by 2010. A linear regression model demonstrated a strong association between tree mortality and drought conditions, with a coefficient of determination (R^2) of 0.61, particularly highlighting the impact in Punjab, Pakistan [45]. Further studies expanded on the role of climatic factors, identifying unusual conditions such as extreme fog and hot summers as additional stressors for shisham trees. These conditions were found to disrupt normal growth patterns and photosynthesis, with cumulative effects leading to increased mortality. Increased solar radiation resulting from canopy openings was observed to elevate evapotranspiration rates, causing significant stress to trees during critical growth periods [46]. Collectively, these findings emphasize the complex interplay between climatic factors and shisham mortality, underscoring the need for adaptive strategies to mitigate the impacts of climate change on this valuable species.

The most recent comprehensive study revealed that dieback disease is widespread across Pakistan's shisham-growing regions, with varying levels of incidence and severity [47]. Interestingly, no significant differences in symptoms were observed among different shisham age classes, indicating that the disease can affect trees at any growth stage, except juvenile plants. The current understanding of abiotic factors influencing shisham mortality underscores the intricate interplay between soil conditions, water availability, and climatic variables. While earlier research predominantly focused on individual factors, more recent studies emphasize the cumulative and interactive effects of multiple stressors, particularly those exacerbated by climate change. These findings highlight the need for a holistic approach to addressing the complex challenges faced by this valuable tree species.

6 Management of Wilt

6.1 Botanicals

Plant-based fungitoxic compounds, including neem seed and camphor oil, as well as leaf extracts from *Argemone*, *Ricinus*, *Calotropis*, and *Aloe*, have demonstrated antifungal activity against the wilt pathogen in pot assays. However, these compounds failed to achieve significant disease control, though they did delay the onset of wilt symptoms [25]. Focused investigations into neem-based products for managing pathogen infections in shisham seedlings evaluated various formulations, including neem oil, neem seed decoction, neem seeds with and without their coats, and neem seed coat and neem leaf extract at 5% and 15% concentrations. All neem products reduced disease intensity when applied as a foliar spray, soil amendment, or root zone injection [48]. In a separate study, *Putranjiva roxburghii* oil completely inhibited the mycelial growth of the pathogen at 500 ppm, in vitro. Subsequent field trials at the nursery level yielded promising results, with *P. roxburghii*-treated shisham seedlings showing no signs of wilting, in contrast to untreated control plants [49]. These findings collectively highlight the potential of plant-derived compounds in managing wilt. However, further research is needed to optimize their efficacy and develop practical and sustainable disease management strategies.

6.2 Biological Control

Extensive research has highlighted the significant potential of biological control agents, particularly *Trichoderma* species, in effectively managing shisham wilt. In vitro studies demonstrated notable antagonistic activity of various *Trichoderma* species against the wilt pathogen, with *T. viride* showing the highest suppression of colony growth (52.4%), followed by *T. harzianum* (24%) [20]. A comparative study [50] examined the efficacy of chemical and biological control agents against wilt and found that the systemic fungicide carbendazim (Bavistin 50 WP, 0.2%) and the biological control agent *T. piluliferum* were equally effective in protecting nursery seedlings from the wilt pathogen. Extending these findings to field conditions, the efficacy of *T. viride* and *T. harzianum* in suppressing the pathogen was further confirmed [51]. Notably, in field trials involving 7-month-old seedlings planted in inoculated soils, *T. viride* demonstrated superior healing efficacy compared to *T. harzianum.* Further investigations [52] explored the antifungal activity of *Trichoderma* species against *F. oxysporum* under in vitro conditions. The results revealed that *T. viride* and *T. harzianum* inhibited *F. oxysporum* growth by 46.4% and 58%, respectively, highlighting their effectiveness against different *Fusarium* species. A comprehensive study [13] evaluated the efficacy of various fungal biocontrol agents and reported a range of growth inhibitions of the target pathogen, with *T. virens* DSRF33 showing the highest inhibition rate (88.04 $\pm$ 0.55%), followed by *T. koningii* DSRF15 (84.44 $\pm$ 4.19%). Other effective agents included *Aspergillus niger* DSRF19, *Penicillium frequentans* DSRF11 and *Penicillium decumbens* DSRF7, with inhibition rates ranging from 79.45% to 81.11%. Recent research has expanded to include bacterial biocontrol agents, such as *Serratia marcescens* and *Pseudomonas azotoformans.* These bacteria have demonstrated the ability to completely suppress wilt symptoms in 12-month-old shisham plants, underscoring their potential in integrated biological control strategies [23].

6.3 Chemical Methods

Fungicides, such as Captan, Bavistin, Topsin-M, and Pursue, have been found to be highly effective against the wilt pathogen [25, 53]. Notably, a benomyl-derived fungitoxic compound, MBC (Methyl-2-benzimidazole carbamate), has proven to be particularly effective against wilting. This stable fungicide is suitable for injection into trees and offers a targeted approach to disease management [54]. Benomyl demonstrated significant efficacy in reducing the mycelial development of the wilt pathogen, even at low doses of 10 ppm. In contrast, Ridomil was also effective but required higher doses to achieve similar results [55]. Other fungicides, such as Dithane M-45 and Vitavax, also showed substantial suppressive effects on pathogen development [20, 55, 56]. Field applications of benomyl further highlighted its potential; for example, the use of 50 L of 200 ppm benomyl successfully recovered 6–8-year-old wilting shisham trees [56]. More recently, field trials demonstrated that Bavistin and *T. viride* were effective in managing wilt-affected trees. Trees were drenched 14 times with Bavistin at a concentration of 0.1% and three times with *T. viride* (10^6 spores/ml), resulting in notable disease suppression [57]. The latest research has further identified chlorothalonil and procymidone as highly effective fungicides for inhibiting the mycelial growth of the pathogen. In comparison, sulfur was found to be the least effective [58]. These findings emphasize the importance of selecting appropriate fungicides and application methods for effective management of shisham wilt.

6.4 Resistant Varieties

Resistant varieties play a crucial role in disease management by effectively reducing pathogen load and disease pressure. Although certain unspecified shisham varieties (1–12) were reported to be disease-free, the sample size was too small to conclusively determine their resistance potential [59]. Studies involving *F. solani* isolated from dying trees, combined with pathogenicity evaluations on seedlings from 107 healthy shisham trees, revealed significant variation in resistance levels. Seedlings from Amritsar (Punjab) were found to be highly resistant, whereas those from Dehra (Kangra), Himachal Pradesh, were highly susceptible [21]. Among the isolates tested, the strain from Nihal Gate (Haldwani), Uttarakhand, was identified as the most virulent [21]. In a more recent study, the responses of 10 host plant genotypes to infection by two strains of wilt pathogen were examined in vitro. Genotypic evaluation using a dual culture setup revealed notable variations in resistance. Two genotypes, identified as 14 and 66, demonstrated resistance against both strains [60]. These findings underscore the importance of identifying and promoting resistant genotypes as a sustainable strategy for managing shisham wilt.

7 Management of *Ganoderma lucidum*

Forest diseases, particularly those caused by *G. lucidum*, often receive less attention compared to pathogens affecting cash crops and plantations. This lack of focus can result in significant reductions in timber yield, leading to substantial economic losses at the national level [61]. Current control strategies for *G. lucidum* in forest ecosystems primarily aim to delay the onset of infection, but these approaches often lack practicality and economic viability [62]. Effective management of *G. lucidum* infections requires an integrated approach that combines cultural, chemical, and biological control methods, similar to strategies employed against other phytopathogens. However, research on the effective control of root rot in shisham caused by *G. lucidum* remains limited. To address this gap, findings on the management of *Ganoderma* pathogens affecting other crops are reviewed and discussed in the following section, offering potential insights and strategies for controlling shisham root rot.

7.1 Cultural Control

Cultural practices play a vital role in managing root rot disease caused by *Ganoderma* in forest ecosystems. Prevention serves as the foundation of these strategies, starting with the selection of seeds from healthy, disease-free trees for new plantings and avoiding areas with a history of *Ganoderma* infestation. In forests or plantations where infections are localized, mechanical trenches can be excavated around infected trees to prevent root-to-root pathogen transmission [63]. Avoiding monoculture practices, particularly in areas previously affected by *G. lucidum*, is critical, as it can exacerbate tree mortality, especially in shisham plantations [62]. Sanitation measures, such as the removal of stumps and residual roots from infected trees, are essential to eliminate potential sources of inoculum. Furthermore, maintaining appropriate planting densities and avoiding practices that stress trees or create favorable conditions for pathogen proliferation—such as flooded irrigation and repeated ploughing—are crucial [24]. By implementing these cultural control measures, forest managers can enhance the resilience of tree populations to root rot, mitigate economic losses, and support the overall health of forest ecosystems.

7.2 Biological Control

Biocontrol methods for *Ganoderma* pathogens have primarily been investigated under laboratory conditions, with limited field studies that focus on basal stem rot (BSR) in coconut and Indian mesquite [64]. *Trichoderma* species, particularly *T. harzianum*, have emerged as potent biocontrol agents against *G. lucidum*. Their effectiveness arises from their ability to compete for space and nutrients, engage in mycoparasitism, stimulate plant growth, and activate plant defense mechanisms [65–70]. In vitro studies have demonstrated that *T. harzianum*, when combined with organic amendments such as neem cake (5 kg/tree), green leaves (50 kg/tree), farmyard manure (50 kg/tree), 1% Bordeaux mixture (40 L/tree), and phosphobacteria (200 g in 10 kg farmyard manure/tree/year), has been shown to significantly reduce disease severity and enhance nut yields in coconut palms [71]. *T. harzianum* has shown the most significant inhibition of *G. lucidum* radial growth (68.5%) compared to *T. pseudokoningii* and *T. viride* [24]. Other fungal antagonists, including species of *Aspergillus* and *Penicillium*, have also demonstrated efficacy in suppressing *G. lucidum* growth under in vitro conditions [72]. Additionally, bacterial biocontrol agents such as *Bacillus* and *Pseudomonas* species have shown promise against pathogenic *Ganoderma* species [73–75]. However, it is important to note that *Trichoderma* species appear to be most effective during the initial stages of infection and have a limited impact on heavily infected trees [76]. To enhance disease management, seed treatment with biocontrol agents before sowing—a technique proven effective in many crops—could be applied to tree species susceptible to pathogens [77–79]. Recent advancements in nanotechnology for plant disease management [80] have led to the development of nano-bio-fungicides, offering new potential for the effective management of *G. lucidum* and other phytopathogens [81, 82]. These nano-formulations represent a promising frontier in the biological control of *Ganoderma* diseases. While biocontrol methods show promise, their practical application in forest ecosystems requires further field validation to ensure efficacy as well as ecological safety.

7.3 Chemical Control

In vitro studies on *G. lucidum* have demonstrated 100% efficacy of fungicides, such as Bavistin, Mancozeb, and Propiconazole. Bordeaux mixture, when combined with Bavistin and Tridemorph-containing fungicides, has successfully inhibited *G. lucidum* growth [83, 84]. Chemical fungicides have also been effective in managing *Ganoderma* infections in other economically significant crops. For instance, in the case of *Ganoderma*-induced BSR in coconut palms, soil amendments with neem cake or farmyard manure have shown promise [85]. Tridemorph, applied at a concentration of 2 mL/100 mL quarterly for one year, in combination with annual applications of 5 kg neem cake per palm, has effectively suppressed the disease [71]. However, despite their effectiveness, the use of chemical fungicides raises environmental concerns, potentially impacting wildlife and non-target microorganisms. This highlights the need for a careful balance between disease control efficacy and ecological considerations in forest pathology management strategies.

8 Management Strategies for Abiotic Factors

Effective management of shisham mortality requires a multifaceted approach. Improving soil drainage and quality is crucial, and this can be achieved through the use of organic amendments [86]. Controlled irrigation practices help prevent waterlogging, while crop rotation promotes soil health and reduces the accumulation of pathogens. Regular monitoring of soil moisture and tree health enables early intervention [87]. Therefore, cultivating shisham in well-drained, light-textured soils is recommended. Adequate irrigation is particularly important in drought-prone regions. Vigilant monitoring of disease symptoms and environmental conditions is essential. In severely affected areas, harvesting diseased trees may help prevent further spread. Removal of infected plants and disinfection of seeds before sowing is required [88]. Additionally, developing predictive models based on environmental data can aid in forecasting and managing disease outbreaks in shisham plantations [89]. These integrated approaches aim to enhance tree resilience and mitigate the abiotic impacts on shisham cultivation.

9 Discussion

The decline of shisham in India and neighboring countries can be largely attributed to fungal pathogens responsible for wilt and dieback. Numerous studies across Asia have documented this decline, yet only a few have thoroughly investigated the issue. Several key research areas require further exploration to enhance understanding and develop strategies to mitigate shisham mortality. One crucial aspect is the need for more in-depth taxonomic studies of the pathogens associated with shisham populations. Current literature indicates that various fungal species, particularly those from the genera *Fusarium* and *Ganoderma*, are implicated in shisham mortality [84, 90]. However, the precise identification and classification of these pathogens remain challenging, as they are often placed in species complexes. Importantly, recent changes in the International Code of Nomenclature for algae, fungi, and plants have mandated the use of a single scientific name for each fungal species, eliminating dual nomenclature for fungi such as *Fusarium* [91–95]. While the reassignment of members of the *F. solani* species complex to *Neocosmospora* has been proposed [29], some have advocated for retaining the name *Fusarium* due to its deep entrenchment in scientific literature [92]. Moreover, the absence of standardized protocols and the subjective nature of certain taxonomic decisions have sparked disputes between "lumpers," who adopt a broad species concept, and "splitters," who favor more narrowly defined species [96]. This lack of consensus complicates the delineation of species boundaries within the genus. Nevertheless, the renaming of the members of the *F. solani* species complex to *Neocosmospora*, based on gene sequencing and morphological data [29], has been adopted in many subsequent publications [97–99].

Historically, fungal taxonomy has relied heavily on morphological traits, such as the spore shape and size. However, environmental conditions can influence these characteristics, making them unreliable indicators of genetic relationships. As a result, molecular techniques, including multi-locus sequence typing (MLST), have become the preferred method for species identification [93, 100]. For instance, a study [36] employed MLST to identify *Ceratocystis dalbergicans*, a recently described species linked to shisham mortality in Pakistan. Accurate pathogen identification is critical for implementing effective disease management strategies, as it enables targeted interventions that are essential for mitigating disease impacts. Therefore, we recommend conducting multi-gene phylogenetic analyses and pathogenicity testing of various *Fusarium* and *Ganoderma* isolates from shisham in India. This would provide a robust framework for their precise identification, thereby aiding in the development of effective management strategies. One promising approach involves the use of resistant germplasm, as research suggests that genetically resistant varieties can significantly reduce the impact of dieback, particularly in new plantations [101, 102].

Overall, cultural practices, such as proper site selection, optimal soil conditions, and effective irrigation management can improve tree health and enhance resilience against pathogens. Fungicide applications and biological control agents have been shown to have beneficial effects on shisham growth [56, 103, 104]; however, further field studies are needed to confirm their long-term efficacy. Sanitation practices, including the removal and destruction of infected plant material, are essential for preventing the spread of pathogens within and between plantations. Regular monitoring and early detection of disease symptoms enable timely interventions, ultimately reducing the severity of outbreaks. By combining these strategies—identification of shisham planting areas, resistant germplasm, cultural practices, chemical treatments, biological control, and sanitation—foresters and land managers can develop a comprehensive management plan that addresses the immediate threats posed by pathogens and ensures the long-term sustainability of shisham populations across their range.

Acknowledgements The authors acknowledge the support of the Compensatory Afforestation Fund Management and Planning Authority (CAMPA), Ministry of Environment, Forest and Climate Change (MoEF&CC), Government of India, New Delhi, for funding the All India Coordinated Research Project (AICRP) on *Dalbergia sissoo* (vide No. 75/2019/ICFRE(R)/RP/SFRESPE(CAMPA)/AICRP-11/Main File/33), under the scheme Strengthening Forestry Research for Ecological Sustainability and Productivity Enhancement. Thanks are due to the Director, ICFRE-Forest Research Institute, Dehradun, India, for providing the necessary research facilities.

References

1. Dhiman VK, Rana N, Dhiman VK, Pandey H, Verma P, Singh D (2022) Effect of rhizobial isolates and nitrogen fertilizers on nursery performance, nodulation behavior and nitrogenase activity of *Dalbergia sissoo* Roxb. seedlings. Plant Stress 4:100080
2. Troup RS (1921) The silviculture of Indian trees, Review Edition, vol 1, Clarendon Press, Oxford
3. Champion HG, Seth SK (1968) A revised survey of forest types of India. Government of India Press, Delhi
4. Ijaz S, Haq IU, Babar M (2021) Reproductive biology, botany, and taxonomical description of *Dalbergia sissoo*. In: *Dalbergia sissoo,* CRC Press, 3–12(2021)
5. Lodhiyal N, Lodhiyal LS, Pangtey YPS (2002) Structure and function of Shisham forests in Central Himalaya, India: dry matter dynamics. Ann Bot 89(1):41–54
6. Parker RN (1918) A Forest Flora for the Punjab with Hazara and Delhi. Lahore Government Printing Press, West Pakistan
7. Bagachee KD (1945) Wilt and dieback of Shisham, Babul and Khair in the artificial regeneration under agriculture-cum-forestry management. Indian For 71:20–24
8. Bakshi BK (1954) Wilt of shisham (*Dalbergia sissoo* Roxb.) due to *Fusarium solani* sensu Snyder and Hansen. Nature 174(4423):278–279
9. Bakshi BK (1955) Wilt disease of Shisham (*Dalbergia sissoo* Roxb.). Indian For 81(4):276–281

10. Bakshi BK (1957) Wilt disease of shisham (*Dalbergia sissoo* Roxb.). IV. The effect of soil moisture on the growth and survival of *Fusarium solani* in laboratory. Indian For 83(8):505–512
11. Bakshi BK, Singh S (1959) Root disease of shisham (*Dalbergia sissoo* Roxb.). VIII. Inoculation studies on wilt. Indian For 85:415–421
12. Bhandari MS, Rama Kant, Ahmed N, Dobhal S, Luna RK, Nautiyal S, Kumar VK, Kumar A (2014) Shisham mortality in Hoshiarpur, Punjab: causes and remedy. Indian For 140(2):147–153
13. Bhatia P, Harsh NSK, Dubey RC, Kaushik PK (2016) Antagonistic activity of rhizosphere fungi against *Fusarium solani* causing wilt in *Dalbergia sissoo.* J Sci Res Adv 3:352–357
14. Baksha MW, Basak AC (2000) Mortality of Shisham (*Dalbergia sissoo* Roxb.) in Bangladesh. In: Appanah S, Allard G, Amatya SM (eds) Proceedings of the sub-regional seminar on die-back of sissoo (*Dalbergia sissoo*), Kathmandu, Nepal, 1–4. Food and Agriculture Organization of the United Nations, Bangkok
15. Khan MM, Khan MH (2000) Die-back of *Dalbergia sissoo* in Pakistan. Field Document No. 18. In: Appanah S, Allard G, Amatya SM (eds) Proceedings of the sub-regional seminar on dieback of sissoo (*Dalbergia sissoo*), Kathmandu, Nepal, April 25–28, 2000, pp 51–56. Forestry Research Support Programme for Asia and Pacific (FORSPA), FAO Regional Office for Asia and Pacific, Bangkok
16. Sharma MK, Singhal RM, Pokhriyal TC (2000) *Dalbergia sissoo* in India. Field Document No. 18. In: Appanah S, Allard G, Amatya S (eds) Proceeding sub regional seminar on die-back of sissoo (*Dalbergia sissoo*), Kathmandu, Nepal, 5–16. Bangkok. Forestry Research Support Programme for Asia and Pacific (FORSPA), p 65
17. Anon (2000) Death of *Dalbergia sissoo* in Shirajgong district by sweet hole borer, Bangladesh. The Daily Prothom Alo
18. Shukla AN (2002) Mortality of *Dalbergia sissoo* in India. Indian For 128(11):1209–1215
19. Webb EL, Hossain SMY (2005) *Dalbergia sissoo* mortality in Bangladesh plantations: correlations with environmental and management parameters. Ecol Manag 206(1–3):61–69
20. Bajwa R, Javaid A (2007) Integrated disease management to control Shisham (*Dalbergia sissoo* Roxb.) decline in Pakistan. Pak J Bot 39(7):2651–2656
21. Shukla AN (2008) Resistance of *Dalbergia sissoo* to *Fusarium solani* f. sp. *dalbergiae.* For Pathol 38(6):410–418
22. Ahmad I, Hanan A, Gul S (2015) Frequency of mycoflora associated with Shisham (*Dalbergia sissoo*) decline in district Faisalabad, Pakistan. FUUAST J Biol 5(2):225–229
23. Banerjee S, Singh S, Pandey S, Bhandari MS, Pandey A, Giri K (2020) Biocontrol potential of *Pseudomonas azotoformans*, *Serratia marcescens* and *Trichoderma virens* against Fusarium wilt of *Dalbergia sissoo.* For Pathol 50(2):12581
24. Shah KK, Tiwari I, Modi B, Pandey HP, Subedi S, Shrestha J (2021) Shisham (*Dalbergia sissoo*) decline by dieback disease, root pathogens and their management: a review. J Agric Nat Resour 4(2):255–272
25. Singh KP, Kumari P, Bhadauria S (2008) Mortality of *Dalbergia sissoo* incited by *Fusarium solani* in western Uttar-Pradesh and its management. J Mycol Plant Pathol 38(3):621–624
26. Manion PD, Lachance D (1992) Forest decline concepts: an overview. In Forest Decline Concepts. APS Press, St. Paul USA
27. Sinclair WA, Hudler GW (1988) Tree declines: four concepts of causality. Arboric. J. 14(2):29–35
28. Khan MH (2000) Shisham die-back in Pakistan and remedial measures. In: Proceedings of the sub-regional seminar—Die-Back of Sissoo (*Dalbergia sissoo*), Kathmandu, Nepal. https://www.fao.org/4/ae910e/ae910e07.htm
29. Crous PW, Lombard L, Sandoval-Denis M, Seifert KA, Schroers HJ, Chaverri P, Gené J, Guarro J, Hirooka Y, Bensch K, Kema GH (2021) *Fusarium*: more than a node or a foot-shaped basal cell. Stud Mycol 98:100116
30. Gill MA, Ahmed IA, Khan AU, Aslam M, Ali S, Rafique RM, Khan M (2002) *Phytophthora cinnamomi*? A cause of shisham decline in Punjab, Pakistan. In: Proceedings of 3rd national

conference of plant pathology, NARC, Islamabad, 1–3 Oct. 2001. Pakistan Phytopathology Society, pp 33–37
31. Khan SH, Idrees M, Muhammad F, Mahmood A, Zaidi SH (2004) Incidence of shisham (*Dalbergia sissoo* Roxb.) decline and in vitro response of isolated fungus spp. to various fungicides. Int J Agric Biol 6:611–614
32. Rajput NA, Pathan MA, Jiskani MM, Rajput AQ, Arain RR (2008) Pathogenicity and host range of *Fusarium solani* (Mart.) Sacc. causing dieback of shisham (*Dalbergia sissoo* Roxb.) Pak. J Bot 40(6):2631–2639
33. Poussio GB, Kazmi MR, Akem C, Fateh FS (2010) First record of *Ceratocystis fimbriata* associated with shisham (*Dalbergia sissoo*) decline in Pakistan. Australas Plant Dis Notes 5(1):63–65
34. Ahmad BI, Khan RA, Siddiqui MT (2013) Incidence of dieback disease following fungal inoculations of sexually and asexually propagated shisham (*Dalbergia sissoo*). For Pathol 43(1):77–82
35. Al Adawi AO, Barnes I, Khan IA, Al Subhi AM, Al Jahwari AA, Deadman ML, Wingfield BD, Wingfield MJ (2013) *Ceratocystis manginecans* associated with a serious wilt disease of two native legume trees in Oman and Pakistan. Australas Plant Path 42:179–193
36. Haq IU, Ijaz S, Latif MZ, Khan IA, Ali HM, Kaur S (2023) Phylogenomics resolves the etiology of dieback disease and deciphers *Ceratocystis dalbergicans* sp. nov. causal agent of *Dalbergia sissoo* decline. Front Gen 14:1136688
37. Harsh NSK, Chandra S, Uniyal K (2011) Screening resistance of *Dalbergia sissoo* clones against *Ganoderma lucidum* root rot disease in field conditions. For Pathol 41(3):221–226
38. Arif M, Zaidi NW, Haq QMR, Singh YP, Khan S, Singh US (2013) Molecular phylogeny and pathotyping of *Fusarium solani*: a causal agent of *Dalbergia sissoo* decline. For Pathol 43(6):478–487
39. Bakshi BK, Reddy MA, Singh S (1976) Ganoderma root rot mortality in Khair (*Acacia catechu* Willd.) in reforested stands. Euro J For Pathol 6(1):30–38
40. Negi JDS, Dangwal R, Chauhan PS (2003) An ecological analysis of shisham *Dalbergia sissoo* (Roxb.) mortality in northern India. Ann For 2(1):20–26
41. Keerio GR (2001) Report on the incidence of shisham decline in Sindh. In: Proceedings of the National Seminar on Shisham Dieback, Punjab Forestry Research Institute, Faisalabad, Pakistan, pp 107–109
42. Sharma R, Singh S, Bakshi M (2000) Studies on the wilt disease of shisham (*Dalbergia sissoo*) in India. Indian Phytopathol. 53(1):1–5
43. Javaid A, Bajwa R, Anjum T (2004) Tree dieback in Punjab. Pakistan. Mycopath. 2(1):1–5
44. Afzal MZ, Khan SSH, Mahmood AUA, Akhtar AS (2006) Global warming and its impact on shisham decline in Pakistan. In: Proceedings of the 3rd National Seminar on Shisham Dieback. Punjab Forestry Research Institute, Faisalabad, Pakistan, pp 29–36
45. Allen CD, Breshears DD, McDowell NG (2015) On underestimation of global vulnerability to tree mortality and forest die-off from hotter drought in the Anthropocene. Ecosphere. 6(8):1–55
46. Dangwal R, Negi JDS, Dobriyal MJ (2014) Survey and analysis of Shisham (*Dalbergia sissoo* Roxb.) mortality in doon valley, Uttarakhand. J Tree Sci 33(2):38–47
47. Latif MZ, Ijaz S, Habib A, Khan IA (2021) Assessment of the distribution, incidence, and severity of Shisham (*Dalbergia sissoo*) dieback disease in Pakistan. Pak J Agric Sci 58(6):1819–1826
48. Rajput NA, Pathan MA, Lodhi AM, Dou D, Rajput S (2011) Effect of neem (*Azadirachta indica*) products on seedling growth of shisham dieback. Afr J Microbiol Res 5(27):4937–4945
49. Kumar N (2014) Search of a natural remedy for control of Fusariam wilting of Sisham (*Dalbergia sissoo* Roxb.). Int J Eng Res Sci Technol 3:34–44
50. Joshi K, Harsh NSK (2009) Biological control of Fusarium wilt of *Dalbergia sissoo* seedlings in nurseries using *Trichoderma* species. J Mycol Pl Pathol 39(3):439–444
51. Basak AC, Basak SR (2011) Biological control of *Fusarium solani* sp. *dalbergiae*, the wilt pathogen of *Dalbergia sissoo*, by *Trichoderma viride* and *T. harzianum*. J Trop For Sci 460–466

52. Shrivastava N, Kumar M, Kumar S, Varma A, Kumar V (2015) Diversity and biocontrol aspects of rhizospheric fungal flora of healthy and diseased plants of *Dalbergia sissoo* in Central East India. Biocontrol Sci 4(7):61–64
53. Khan AH, Asghar AG, Ghulam RC, Hamid A (1965) Observation on the mortality of shisham (*Dalbergia sissoo* Roxb.) and other trees in Khanewal plantation. Part I, II and III, Pak J For 6:109–120, 203–220, 289–301 (1965)
54. McWain P, Gregory GF (1973) A benomyl-derived fungitoxicant for tree wilt disease control. USDA Forest Service Research Note, 3
55. Ahmad M, Khan MA, Ahmad F, Khan SM (1996) Effectiveness of some fungicides on the colony growth of *Fusarium oxysporum* f. sp. *tuberosi* and *Fusarium solani* associated with potato wilt. Pak J Phytopathol 8(2):159–61
56. Bajwa R, Javaid A, Mirza JH, Akhtar N (2003) Chemical control of wilt in Shisham (*Dalbergia sissoo* Roxb.). Mycopathol 1(2):111–113
57. Kumar M, Sharma S, Chaturvedi OP (2020) Management of Shisham Wilt (*Fusarium solani* f. sp. *dalbergiae).* Ind J Agrofor 6(2)
58. Ghazali HM, Akram S, Fatima I, Hussain M, Hameed A, Arif M, Ahmed MA, Al-Ghamdi AA, Elshikh MS, Alrashidi BO (2022) Fungi species causing dieback and wilt diseases in shisham [*Dalbergia sissoo* (Roxb.)] and impact of various fungicides on their management. J King Saud Univ Sci 34(4):101970
59. Javaid A, Bajwa R, Anjum T (2004b) Dieback resistant and susceptible varieties of shisham. In: Proceedings of 4th National Conference of Plant Pathology, 14–16 October, 2003 Arid Agriculture University, Rawalpindi. 17–27
60. Chauhan M, Thakur A, Singh Y, Rather MM, Harsh NS (2021) Gnotobiotic evaluation of *Dalbergia sissoo* genotypes for resistance against *Fusarium solani* via dual culture set up. J Genet Eng Biotechnol 19:1–9
61. Ramesha MN, Dwivedi VK, Singh C, Kumar V (2017) Incidence of *Ganoderma lucidum* L. (Reishi) on *Delonix regia* and *Peltophorum pterocarpum* in semi-arid region. Indian For 143(2):139–142
62. Sankaran K, Bridge P, Gokulapalan C (2005) *Ganoderma* diseases of perennial crops in India – an overview. Mycopatholog 159:143–152
63. Mawar R, Ranganathan S (2023) *Ganoderma* root rot mortality: managing the growing threat to the tree behind traditional agroforestry in Indian arid region. Ann Arid Zone 62(3):185–195
64. Mawar R, Sharma D, Ram L (2021) Potential of biocontrol agents against *Ganoderma lucidum* causing basal stem rot in mesquite (*Prosopis cineraria*) in arid regions of India. J For Res 32:1269–1279
65. Dennis C, Webster J (1971) Antagonistic properties of species-groups of *Trichoderma.* Trans Br Mycol Soc 57(3):363
66. Srilakshmi P, Thakur RP, Prasad KS, Rao VP (2001) Identification of *Trichoderma* species and their antagonistic potential against *Aspergillus flavus* in groundnut. Int Arachis Newsl 21:40–43
67. Meena M, Swapnil P, Zehra A, Dubey MK, Upadhyay RS (2017) Antagonistic assessment of *Trichoderma* spp. by producing volatile and non-volatile compounds against different fungal pathogens. Arch Phytopathol Plant Prot 50(13–14):629–648
68. Palanna K, Narendrappa T, Basavaraj S, Shreenivasa K (2017) Efficacy of fungal and bacterial bio-control agents on *Ganoderma* spp. causing foot rot of arecanut. Int J Agric Innov Res 6(2):299–304
69. Naher L, Yusuf UK, Habib SH, Ky H, Siddiquee S (2018) Mycoparasitism activity of *Trichoderma harzianum* associated with chitinase expression against *Ganoderma boninense.* Pak J Bot 50(3):1241–1245
70. Tapwal A, Kumari A, Tiwari B (2023) Mass multiplication of *Trichoderma asperellum* on conifer needles and weed foliage. Arch. Phytopathol. Pflanzenschutz 56(15):1166–1179
71. Bhaskaran R (1993) Integrated management of basal stem rot disease of coconut. Indian Coconut J 24:5–8

72. Elkhateeb WA, Daba GM (2022) The precious *Ganoderma* mushroom and plant diseases. J Microbiol Biotechnol 7:2836–2322
73. Lodha S, Harsh LN (2009) Combined effects of biocontrol agents and residues on root rot mortality in Indian mesquite (*Prosopis cineraria*). Acta Hort 883:317–322
74. Susanto A, Sudharto PS, Purba RY (2005) Enhancing biological control of basal stem rot disease (*Ganoderma boninense*) in oil palm plantations. Mycopathol 159(1):153–157
75. Darlis D, Jalloh MB, Chin CFS, Basri NKM, Besar NA, Ahmad K, Rakib MRM (2023) Exploring the potential of Bornean polypore fungi as biological control agents against pathogenic *Ganoderma boninense* causing basal stem rot in oil palm. Sci Rep 13(1):10316
76. Badalyan SM, Garibyan NG, Innocenti G (2004) Interactions between xylotrophic mushrooms and mycoparasitic fungi in dual-culture experiments. Phytopathol Mediterr 1000–1005
77. Mancini V, Romanazzi G (2014) Seed treatments to control seedborne fungal pathogens of vegetable crops. Pest manag sci 70(6):860–868
78. Haider MR, Alam MS, Shutrodha AR (2016) Effect of pre-sowing treatment on seed germination and seedlings growth of *Sapindus mukorossi* Gaertn. -an important medicinal plant in Bangladesh. J Biosci Agri Res 6(2):570–577
79. Moumni M, Brodal G, Romanazzi G (2023) Recent innovative seed treatment methods in the management of seedborne pathogens. Food Sec 15(5):1365–1382
80. Pandey S, Giri K, Kumar R, Mishra G, Raja, Rishi R (2018) Nanopesticides: opportunities in crop protection and associated environmental risks. In: Proceedings of the National Academy of Sciences, India Section B, Biological Sci 88:1287–308
81. Chauhan P, Tapwal A (2023) Green synthesis of nanoparticles using botanicals and their application in management of fungal phytopathogens: a review. Arch Microbiol 205(3):94
82. Kumari A, Tapwal A (2023) *Trichoderma*: a beneficial antimycotic agent and insights on its utilization in nanoparticle synthesis. Bionanosci 13(4):1399–1417
83. Nirwan B, Choudhary S, Sharma K, Singh S (2016) In vitro studies on management of root rot disease caused by *Ganoderma lucidum* in *Prosopis cineraria*. Curr. Life. Sci. 2(4):118–126
84. Timilsina S, Bhattarai R, Miya MS, Gautam D (2020) Sissoo, its Pathogenic Constraints and their Management in Nepal: A review. Grassroots. J. Natural. Resources. 3(4):1–17
85. Sariah M, Zakaria H (2000) The use of soil amendments for the control of basal stem rot of oil-palm seedlings. In *Ganoderma* diseases of perennial crops Wallingford UK: CABI 89–99
86. Dayaram MK, Sharma S, Chaturvedi OP (2003) Shisham mortality in Bihar extent and causes. Indian. Phytopathol. 56(4):384–387
87. Dangwal LR (2014) Study on the mortality of Shisham trees in northern India. J. Tree. Sci. 33(2):38–47
88. Zulfiqar MA (2019) Factors leading towards *Dalbergia sissoo*: A critical review. Pak. J. Agri. Res. 32(2):302–316
89. Ahmad I, Atiq M, Nawaz MF, Ahmed S, Asif M, Gull S, Tanvir MA, Abdullah M, Azhar MF, Rajput NA (2019) Prediction of dieback disease of *Dalbergia sissoo* (Shisham) based upon environmental factors and tree age. Appl Ecol Environ Res 17(3):6483–6495
90. Kumar N, Khurana SP (2016) Bio-management of wilting of a valuable timber and medicinal plant of Shisham (*Dalbergia sissoo* Roxb.)–A review. Int J Cur Microbial App Sci 5(1):32–54
91. Geiser DM, Aoki T, Bacon CW, Baker SE, Bhattacharyya MK, Brandt ME, Brown DW, Burgess LW, Chulze S, Coleman JJ, Correll JC (2013) One fungus, one name: defining the genus *Fusarium* in a scientifically robust way that preserves longstanding use. Phytopathology 103(5):400–408
92. O'Donnell K, Al-Hatmi AM, Aoki T, Brankovics B, Cano-Lira JF, Coleman JJ, De Hoog GS, Di Pietro A, Frandsen RJ, Geiser DM, Gibas CF (2020) No to *Neocosmospora*: phylogenomic and practical reasons for continued inclusion of the *Fusarium solani* species complex in the genus *Fusarium.* Msphere 5(5):10–1128
93. Van D, Al-Hatmi AD, Brankovics AM, De B, Hoog GS (2014) Taxonomy and clinical spectra of *Fusarium* species: where do we stand in 2014? Curr Clin Microbiol 1:10–18
94. Lombard L, Van der Merwe NA, Groenewald JZ, Crous PW (2015) Generic concepts in Nectriaceae. Stud Mycol 80(1):189–245

95. Summerell BA (2019) Resolving *Fusarium*: current status of the genus. Annu Rev Phytopathol 57(1):323–339
96. Moretti AN (2009) Taxonomy of *Fusarium* genus: A continuous fight between lumpers and splitters. Zbornik. Matice. Srpske. Za. Prirodnenauke 117:7–13
97. Veloso JS, Câmara MS, Souza RM (2020) Guava decline: updating its etiology from '*Fusarium solani*' to *Neocosmospora falciformis*. Eur J Plant Pathol 159(2):455–460
98. Sandoval-Denis M, Crous PW (2018) Removing chaos from confusion: assigning names to common human and animal pathogens in *Neocosmospora*. Persoonia—Mol Phylo Evol Fungi 41(1):109–129
99. Sandoval-Denis M, Lombard L, Crous PW (2019) Back to the roots: a reappraisal of *Neocosmospora*. Persoonia—Mol Phylo Evol Fungi 43(1):90–185
100. Ekwomadu TI, Mwanza M (2023) *Fusarium* fungi pathogens, identification, adverse effects, disease management, and global food security: a review of the latest research. Agric 13(9):1810
101. Minocha R, Jain SM (2000) Tissue culture of woody plants and its relevance to molecular biology. In Mol Biol Woody Plants 1:315–339. Dordrecht: Springer Netherlands
102. Ijaz S, Haq IU, Khan IA, Ali HM, Kaur S, Razzaq HA (2022) Identification of resistance gene analogs of the NBS-LRR family through transcriptome probing and in silico prediction of the expressome of *Dalbergia sissoo* under dieback disease stress. Front Gen 13:1036029
103. Kumari A, Tapwal A, Thakur N (2024) *Ganoderma lucidum*: insights on host range, diagnosis, and management strategies. J Basic Microbiol 202300769
104. Bose R, Saini A, Bansal N, Bhandari MS, Pandey A, Joshi P, Pandey S (2022) Biological control of forest pathogens: success stories and challenges. In: Trends of applied microbiology for sustainable economy. Academic Press, pp 155–184

Chapter 21
Reproductive Biology and Breeding Strategies for the Indian Rosewood: *Dalbergia sissoo* Roxb. ex DC

Prashant Shaw, Payal Dhyani, Rajendra Kumar Meena, Maneesh S. Bhandari, and Rajesh Tandon

1 Introduction

Dalbergias or Rosewoods occupy an important place among the hardwoods owing to their multifarious uses for a range of purposes. *Dalbergia sissoo* Roxb. ex DC., popularly known as Shisham, is a medium to large, semi-deciduous leguminous tree species from the tropics and subtropical regions of the old world, primarily the Indian subcontinent. One of the 282 accepted species of *Dalbergia* [1], the Shisham trees usually have a light canopy and appear lofty when laden with strap-shaped pods, and not lopped for fuel (Fig. 1a).

The native populations of *D. sissoo* are known to occur in the sub-Himalayan stretch of the Indian subcontinent between the altitudinal limits of 200–1400 m [2, 3]. Its distribution extends to Oman, Afghanistan, Pakistan, Nepal, Bhutan, and Myanmar as well. It has been accredited as the State tree of Punjab, India, and is commonly referred to by its vernacular name, 'Tahli' [4], which is one of the two varieties known and characterized by a rectangular pattern of deep fissures in the bark. The other variety, namely 'Tahla', has deep vertical fissures that run along the bole. The species is known to exhibit wider genetic variability [5, 6].

Shisham is usually not a dominant forest-dwelling species, and well-grown trees are generally found in plantations near the riverine areas. It is a hardy species and thrives well in a wide range of habitats with a salubrious combination of sandy, loam, or alluvial porous soils that essentially support aeration and drain well [4, 7].

P. Shaw · R. Tandon (✉)
Department of Botany, University of Delhi, Delhi, India
e-mail: rajeshtandon@botany.du.ac.in

P. Dhyani · R. K. Meena · M. S. Bhandari
Division of Genetics and Tree Improvement, ICFRE-Forest Research Institute, Dehradun, Uttarakhand, India

R. R. Warrier et al. (eds.), *Handbook of Asian Rosewoods*, Materials Horizons: From Nature to Nanomaterials, https://doi.org/10.1007/978-981-96-9459-4_21

Fig. 1 *Dalbergia sissoo.* **a** The habit of the tree. **b** In inflorescence at the peak-time of blooming. **c** An inflorescence with developing fruits. **d** An infructescence with fully developed pods

Shisham is a pioneer species of primary succession in riparian zones, open woodlands, and natural riverine forest of the river Indus, the Ganges, the Yamuna, and the Brahmaputra along with their tributaries. The species shows a preference for temperature ranging 20–30 °C. It can tolerate higher temperatures but may experience reduced growth rates and increased susceptibility to pests and diseases under extreme environmental conditions. Invariably, the species requires full overhead sunlight for successful regeneration and establishment. Silviculture applications have highlighted that seedlings of *D. sissoo* are drought sensitive, but after the sapling stage, the tree becomes drought-tolerant [8].

The Shisham trees are among the most sought-after species by foresters in afforestation programmes for being a multipurpose leguminous tree and can be converted into revenue for its timber. It is perhaps one of the most widely used hardwood timber species in the tropics. With a specific gravity of nearly 0.85, Shisham wood finds a place among the heavy timber species due to the appealing grain pattern, rich colour of the heartwood, durability, resilience, decay-resistance, and endurance [9]. The Shisham wood primarily finds its use in crafting a range of furniture items and sometimes into musical instruments like drums from its empty bole [10–12]. Besides timber, there are a number of nutraceutical and medicinal uses [13, 14]. Importantly, *D. sissoo* plays a key role in maintaining ecological balance and is widely employed for restoration purposes [15, 16]. With an increasing popularity in timber trade and other benefits, there have been deliberate introductions in other parts of the world ranging from Northeast Asian, African, North American, South American, and even Middle Eastern countries [17, 18] (Fig. 2).

Considering its multipurpose uses, its propagation methods are well-known in the country. The trees are known to propagate both by sexual and vegetative means. However, there are very few empirical studies that demonstrate its reproductive performance in different regions. Thus, the main focus of this chapter is to provide insights into the current knowledge on the reproductive strategy and natural recruitment of *D. sissoo*. These attributes are important to ascertain reproductive performance, especially when the climatic pattern is shifting the distribution boundaries to more unfavourable growth conditions of many tropical plant species.

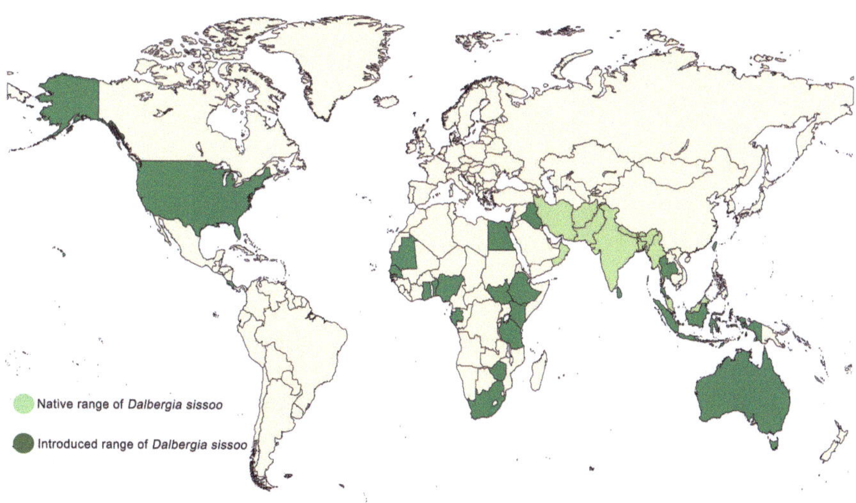

Fig. 2 Distribution of *Dalbergia sissoo* in the native and the introduced regions of the world

2 Phenology

Dalbergia sissoo shows a seasonal phenological pattern, and the pheno-events are well segregated over the year. The pheno-events are usually synchronized at the population level. The trend as seen in many parts of the country, especially the northern region, is presented here (Table 1). Major changes in the canopy begin to appear with the onset of winter (November–December), when the trees start shedding leaves, and by January the canopies become completely defoliated [19, 22]. As the temperature gradually rises by mid-February, the trees acquire new leaves [19, 20], though they appear in March in the sub-temperate region of Himachal Pradesh [21]. A burst of fresh leaves can be seen forming a canopy by March. Flowering ensues once the crown of foliage is in full bloom (last week of March). The floral buds become conspicuous nearly 15 days after the emergence of new flush of leaves, and rise in temperature seems to speed up the flowering [22]. Flowering in Shisham trees is known to persist for 4–6 weeks (March–April) at the population level and around 5 weeks for an individual tree. The peak period of blooming resides nearly around mid-April. The short flowering duration in *D. sissoo* is believed to impart mass floral display and resource availability for increased intraspecific pollinator activity. Considering the outbreeding nature of the species, sufficient and simultaneous availability of mates in bloom engenders better pollen flow among genets [23].

Table 1 Phenological events of *D. sissoo* as recorded in various parts of India

Sl. No	Pheno-event	Months	Place	Reference(s)
1	Leaf senescence	December/January	Chandigarh, Punjab	[19, 22]
2	New flush of leaves	February	Chandigarh, Dehradun	[19, 20]
		March	Himachal Pradesh	[21]
3	Floral bud initiation	March	Punjab, Dehradun	[20, 22]
4	Blooming	March	Dehradun	[20]
5	Culmination of flowering	Last week of April	Punjab	[22]
6	Fruit emergence	April	Dehradun	[20]
		Last week of June	Himachal Pradesh	[21]
7	Fruit maturation	September	Chandigarh	[19]
		October	Dehradun	[20]
		Mid December–Mid-January	Himachal Pradesh	[21]
8	Diaspore dispersal	Last week of January	Himachal Pradesh	[21]

The blooming period coincides with intense pollination activity, especially involving social bees. Interestingly pollination induces change in flower colour to orange-yellow, while the unpollinated flowers remain creamy-white that eventually wilt away. Fruit-set becomes conspicuous 15–20 days after pollination (DAP) and usually around the third/fourth week of April (Fig. 1c). The pods can be seen developing throughout April to August (Fig. 1d), and the mature pods, containing fully developed seeds, are formed by late September [20]. At maturity, the pod turns brown and usually contains one to four seeds in a pod [17, 19, 24]. The mature fruits are retained on the canopy until the onset of monsoon. Overall, the reproductive phase spanning initiation of flower buds to pod formation completes in a span of 7–8 months. In addition, it has been observed that buds and flowers appear simultaneously in various growth phases and developmental stages [22]. It also revealed that bud initiation to flowering-blooming phase is comparatively shorter than the pod to long-lasting fruit maturation phase. The mature fruits are retained until the onset of monsoon. Such correlated timing of the reproductive pheno-events and availability of water enhances effective dispersal [25].

There is credible evidence that a global rise in temperature has influenced the phenological pattern of many plant species through combined direct and indirect impacts on plant growth, development patterns, species interactions, and ecosystem dynamics [26–29]. Like many other plant species, *D. sissoo* also has been recorded to show shifts in its phenophases due to rapid changes in climatic factors. An advancement in leaf emergence and flowering initiation phase by 40 and 35 days, respectively, was reported [30]. Another study conducted to assess decadal change in phenological events of eleven multipurpose trees including *D. sissoo* reported an advancement by 15 days for leaf emergence and 5 days for flowering initiation [21]. These studies also looked into climatic parameter analysis during the period between 1999 and 2017. Trend analysis of meteorological factors during 1999–2017 using Mann–Kendall Test and Sen's slope (–0.033) shows a significant trend for the average annual temperature. Thus, it can be inferred that the phenology of *D. sissoo* was significantly impacted by climate change from 1999–2017, which extended the growth period by 3 months, specifically in the last two decades [21, 30]. However, the impacts of these phenological shifts on the reproductive fitness of the tree species are yet to be investigated.

The typical papilionaceous flag-blossoms of *D. sissoo* are arranged in axillary panicles of short racemes that open acropetally (Fig. 1b). There is synchronized behaviour in floral events within a tree and at the population level. The time of anthesis, anther dehiscence, duration of the stigma receptivity, and associated insect visitation period shows synergistic diurnal association. These floral pheno-events happen to occur in a bimodal manner during the day. Anthesis (0930–1030 h and 1500–1600 h), anther dehiscence (1000 h–1100 h and 1530 h–1630 h), and the stigma receptivity (1000 h and 1600 h) are well-coordinated in forenoon and afternoon periods to complement the efficiency of pollination process. Anther dehiscence and stigma receptivity may occur in the bud stage itself, just prior to anthesis, and the duration of the stigma receptivity may span over 4–5 h after the opening of flowers. The flowers which becomes receptive during the forenoon loses functionality by afternoon, and the ones that become receptive during the afternoon period are not available by the next day [31]. The stigma of *D. sissoo* is wet-papillate type and can be categorized as Group-III type [25, 32]. Like many other legumes, success in pollen germination on the stigma in *D. sissoo* depends on the stigmatic exudates, which can be released only after the thick cuticle is ruptured by anthesis [33].

Thus, the fresh flowers become available in different batches, maximizing reproductive success and avoiding competition for pollination service. The extent of fruit-set during the two batches of flowering is, however, yet to be established. It is important as the quality reward, especially nectar, changes with rise in temperature due to evaporation. Also, there is no information on the nectar production dynamics and the composition.

The freshly opened flowers are cream-white and mildly fragrant. Flowers offer nectar and the starchy pollen as pollination rewards, that serve as primary attractant, whereas visual and the olfactory cues are believed to act as secondary attractants.The legitimate floral foraging guild of *D. sissoo* comprises honeybees, butterflies, beetles, thrips, and some other insects. Among these honeybees, especially belonging to genus *Apis* (*A. dorsata*, *A. mellifera*, and *A. cerana*) are known to be most effective, frequent, and are the primary pollinators in the tree species [19, 25, 31, 34]. Pollen transfer happens sternotribically from the bodies of the honeybees. Birds have also been recorded visiting the trees in bloom, but their role in pollination is considered doubtful [22]. Therefore, in *D. sissoo*, the pollination mechanism is mainly entomophilous and attuned for melittophily [35].

3 Breeding Behaviour

The terms breeding system and mating system have been used interchangeably in the literature, but they essentially refer to different aspects. The breeding system refers to the compatibility status of the species among the conspecifics in a population, while the mating system indicates the extent of outcrossing [36–38]. Reproductive studies, so far, have attempted to explain the former, and information on the latter, which essentially requires the use of suitable biochemical or molecular markers, are lacking. On the contrary to the earlier notion that Shisham is largely a self-compatible species, studies suggest that *D. sissoo* is a self-incompatible and a preferential outbreeding tree species. Protandrous dichogamy is shown to be an outbreeding floral contrivance and means to prevent spontaneous autogamy in the species [25, 39].

Besides the bagging experiment, there are several indices that help in predicting the possible breeding system [40, 41]. Among these, the one based on pollen/ovule (P/O) ratio [42] is frequently used in reproductive biological studies. A couple of investigations have shown that pollen production in a flower of Shisham tree may range ~ 3500–5000 (with ~ 60% viability) and ~ 6 ovules in an ovary. The P/O ratio, estimated ~ 710–940, invariably suggests that *D. sissoo* is a facultative xenogamous species [19, 25]. Earlier studies demonstrated the absence of naturally occurring cross-pollination in *D. sissoo*, as no pod setting were observed when buds were emasculated; and left open for the pollinators to transfer pollen upon them [43, 44]. Controlled pollination experiments have clearly established a significant increment in percentage pod formation in cross-pollinated flowers (44%) than self-pollinated (6%) or open-pollinated (15%) flowers [19, 45]. Marginally higher percentage pod formation (48%) was observed when pollen mixture obtained from five different trees were used to pollinate the flowers of a tree other than the mother tree. These outcomes suggest that biparental inbreeding may have adverse effects in nature, which can be only validated through parent-progeny analysis. That the species is a preferential outbreeder with inbreeding depression effects is further validated as seed-set per pod was also significantly low upon autogamy (0.5) than that obtained from xenogamy (2.4), open (1.3), and mixed-crossing (2.6) [19, 45].

4 Recruitment

A pistil of *D. sissoo* can have up to six ovules [24], but all the ovules do not become seeds, as many abort during post-fertilization stages. By the time fruits reach maturity, the pods may have seeds between one and four (Fig. 3). In general, the extent of fruit and seed-set among flowering plants does not match with flower and ovule production, respectively; as they may abort due to a variety of reasons [46]. Young fruits are often destroyed by parakeets [47]. In Shisham, seed abortion is argued to be due to sibling rivalry rather than maternal regulation of resources or pollen limitation [48, 49]; the first fertilized seed seems to induce abortion by releasing compounds of indole nature.

Natural recruitment in Shisham occurs predominantly by seeds and vegetatively through stump planting and suckers [7]. The success of recruitment heavily depends on the intrinsic and extrinsic factors during the post-dispersal phase. The fruits are anemochorous, with fruits serving as the dispersal units; the larger pods (>2-seeded) often break into smaller units with seeds encased inside. Single-seeded pods have better flight dynamics, resulting in greater horizontal dispersal distance, reduced wing-loading, and lower vertical settling velocity [49]. Additionally, the fruits may also disperse by water during rains, depending on the location of the parent tree [50]. Like many legumes, the seeds exhibit seed-coat dormancy and require scarification. Nursery methods often employ clipping off a small margin of the coat followed by soaking them to break dormancy and promote seed germination. However, it is also recommended that for regeneration, two-seeded pods should be used as they tend to give better germination than those from one-seeded pods [51].

The recruit happens to be better in the riparian areas. Better seed germination is observed near riverbeds, where moisture levels are conducive to germination. In riverine regions, co-occurring grasses provide protection from scorching sunlight without being competitive, as they appear in separate clumps. In non-riverine areas, seeds prefer environments free of weeds with abundant sunlight [50]. Some species show substantial germination with seed burial compared to surface sowing [52]. It was observed that earthworms (*Lumbricus rubellus* and *Eudrilus eugeniae*) positively

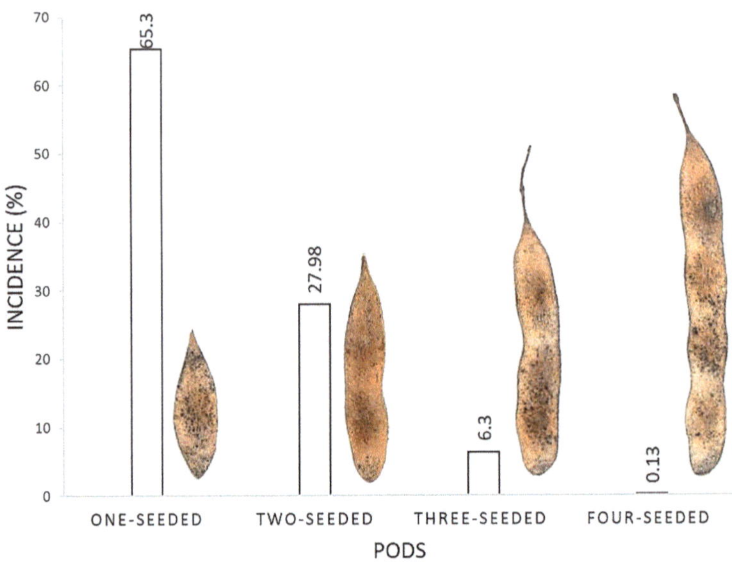

Fig. 3 Incidence of 1, 2, 3, and 4-seeded fruits expressed as percentage (n = 120 infructescences, 10 trees)

affect seed germination in *D. sissoo* by aiding seed burial and altering seed properties via their gut passage [53]. A similar positive effect on germination may occur when aquatic foragers consume seeds, although this phenomenon remains unexplored. Establishment and survival of seedlings are influenced by adequate moisture and protection against fire and grazing. Seedlings are prone to prolonged drought, especially when lying uncovered on the soil surface, while consistent shaded, moist conditions can lead to frequent rotting. Lack of soil moisture critically affects young seedlings, but once established, they develop natural adaptability to survive dry conditions.

5 Conclusions

Shisham, which is considered second only to Teak, is an important agroforestry tree species, and it has been substantially addressing an ever-increasing demand for timber in several countries. Investigations focusing on reproductive attributes, so far, indicate that there are no reproductive constraints in the tree species. There is an enormous production of fruits and there are well-established nursery practices to multiply the saplings by vegetative and sexual means. However, there is no information from different provenances on the two varieties known. Considering its natural distribution in the Indo-Gangetic plain and some parts of the sub-montane Himalayan region, the species is likely to get affected in reproductive performance due to indirect consequences of shifts in climatic patterns. As the natural recruitment of the species is seen better along the riparian zones, climate-change-associated shifts in precipitation, soil-moisture patterns, and flowering phenology would adversely affect the fecundity and distribution. Regrettably, knowledge on the mating pattern of the species is lacking and thus, any inference on the genetic diversity would be misleading. Studies carried out on the genetic diversity in different provenances have provided contrasting results, which reflect variation in siring pattern, gene flow, and genetic structure. These aspects need consideration in future investigations to sustain the quality of produce, as it depends on the purity of the germplasm and its perpetuation.

References

1. POWO (2024) Plants of the World Online. Facilitated by the Royal Botanic Gardens, Kew. http://www.plantsoftheworldonline.org/
2. Orwa C, Mutua A, Kindt R, Jamnadass R, Anthony S (2009) Agroforestree Database: a tree reference and selection guide version 4.0. World Agroforestry Centre, Kenya
3. Lakhey P, Pathak J, Adhikari B (2020) *Dalbergia sissoo*. The IUCN Red List of Threatened Species 2020: e.T62022617A62022619
4. Singh A, Jerath N, Ladhar SS, Singh G, Luna RK (2014) Tree Directory of Punjab. Punjab Agricultural University, Ludhiana and Punjab State Council for Science & Technology, Chandigarh, pp 184
5. Arif M, Zaidi NW, Singh YP, Rizwanul Haq QM, Singh US (2009) A comparative analysis of ISSR and RAPD markers for study of genetic diversity in Shisham (*Dalbergia sissoo*). Plant Mol Biol Rep 27:488–495
6. Bakshi M, Sharma A (2011) Assessment of genetic diversity in *Dalbergia sissoo* clones through RAPD profiling. J For Res 22:393–397
7. Shaltout KH, Keshta AE (2011) The biology of Egyptian woody perennials 4. *Dalbergia sissoo* Roxb. ex DC. Assiut Univ Bull Environ Res 14(2):131–152
8. Luna RK (1996) Plantation trees. International Book Distributors, Dehra Dun, India pp 975
9. Mark J, Newton A, Oldfield S, Rivers M (2014) A Working list of commercial timber tree species. Botanic Gardens Conservation International, Richmond, UK
10. Nair MNB (1998) Wood anatomy and major uses of wood. Universiti Putra Malaysia, Faculty of Forestry
11. Kumari V, Jain S (2019) Woods used for musical instruments found in the region of Sikar district. Indian J. Environ. Sci. 23(2):53–55
12. Arunkumar AN, Warrier RR, Kher MM, Teixeira da Silva JA (2022) Indian rosewood (*Dalbergia latifolia* Roxb): biology, utilisation, and conservation practices. Trees 36(3):883–898
13. Saste G, Balasubramaniam AK, Singh R, Patel S, Jagtap S, Patel S, Maurya R, Hingorani L (2023) Comprehensive standardization and rapid quality evaluation of isoflavones and isoflavone glycosides from *Dalbergia sissoo* by ultra-high-performance liquid and thin layer chromatography analysis with tandem mass spectrometry. Separation Science Plus 6(1):2200079
14. Sen A, Kaur A, Johar V, Singh V (2023) Indian rosewood: a versatile tree for economic prosperity. Int J Phytol Res 3(4):16–21
15. Tyagi VC, Dikshit N, Gautam K, Govindasamy P (2020) *Dalbergia sissoo*: An important tree with fodder value. Van Sangyan 7(10):1–8
16. Srivastava R, Bharti V (2023). Importance of *Dalbergia sissoo* for restoration of degraded land: A case study of Indian dry tropical Vindhyan regions. Ecol Eng 194:107021

17. ICRAF (2007) Agroforestree database. In: Agroforestree database, Nairobi, Kenya: World Agroforestry Centre
18. ILDIS (2007) International Legume Database and Information Service. In: International Legume Database and Information Service, Reading, UK: School of Plant Sciences, University of Reading
19. Vasudeva SP (2008) Thesis- Reproductive Biology of some heterostylous species and *Dalbergia sissoo* Roxb. Doctoral Thesis, Panjab University, Chandigarh, India. https://shodhganga.inflibnet.ac.in/handle/10603/83129
20. Chhetri R, Kumar R, Megha (2020) Phenological trend of tree species at Forest Research Institute, Dehradun. India. Res J Agricult Forestry Sci 8(2):1–8
21. Panda S, Bhardwaj DR, Sharma P, Handa AK, Kumar D (2021) Impact of climatic patterns on phenophase and growth of multi-purpose trees of north-western mid-Himalayan ecosystem. Trees Forest People 6:100143
22. Sharma R, Chauhan SK, Khajuria HN (2009) Reproductive biology and variability studies in Dalbergia sissoo (Roxb.). J Tree Sci. 3:86-88.
23. Gentry AH (1978) Antipollinators for mass-flowering plants? Biotropica 10:68–69
24. Ijaz S, Haq IU, Babar M (2021) Reproductive biology, botany, and taxonomical description of *Dalbergia sissoo*. In: *Dalbergia sissoo*: biology, ecology and sustainable agroforestry, CRC Press, Boca Raton, USA, pp 3–12
25. Menon S (1991) Thesis- Reproductive Biology of *Dalbergia sissoo*—An important Indian tree legume. Doctoral Thesis, Delhi University. https://indcat.inflibnet.ac.in/index.php/thesesrequest?id=th_6961
26. Menzel A, Fabian P (1999) Growing season extended in Europe. Nature 397(6721):659
27. Parmesan C, Yohe G (2003) A globally coherent fingerprint of climate change impacts across natural systems. Nature 421(6918):37–42
28. Parmesan C (2006) Ecological and evolutionary responses to recent climate change. Annu Rev Ecol Evol Syst 37:637–669
29. Parmesan C (2007) Influences of species, latitudes and methodologies on estimates of phenological response to global warming. Global Change Biol 13(9):1860–1872
30. Thakur PS, Dutt V, Thakur A (2008) Impact of inter-annual climate variability on the phenology of eleven multipurpose tree species. Curr Sci 1053–1058
31. Vasudeva SP, Sareen TS (2009) Pollination biology in Dalbergia sissoo Roxb. (Papilionodeae; Leguminosae). Indian For 135(9):1165–1168
32. Heslop-Harrison Y, Shivanna KR (1977) The receptive surface of the angiosperm stigma. Ann Bot 41(6):1233–1258
33. Shivanna K, Owens S (1989) Pollen-pistil interactions (Papilionoideae). Advances in legume biology: Monograph of Systematic Botany Missouri, Missouri Botanical Garden 29:157–182
34. Sasidharan KR, Thangamani D, Prakash S, Muraleekrishnan K (2020) Pollination ecology and breeding system in *Dalbergia latifolia* and *D. sissoides*. Ecol Environ Conserv 26:S76-S81
35. Frankie GW (1976) Pollination of widely dispersed trees by animals in Central America, with emphasis on wild bee pollination systems. Linn Soc Symp Ser 2:151–159
36. Shivanna KR, Tandon R (2014) Reproductive ecology of flowering plants: a manual (No. 14769), Springer, New Delhi, India
37. Sorokhaibam SS, Chandra A, Baishya R, Barik SK, Goel S, Tandon R (2024) Contradistinctive floral attributes, pollination guilds and their consequence on the outcrossing rate in two elevational morphs of *Rhododendron arboreum* Sm. Front Plant Sci 15:1355680
38. Barman C, Singh VK, Das S, Tandon R (2018) Floral contrivances and specialised pollination mechanism strongly influence mixed mating in *Wrightia tomentosa* (Apocynaceae). Plant Biol 20(3):546–554
39. Menon S, Shivanna KR, Mohan Ram HY (1993) Pollination ecology of *Dalbergia sissoo*–an important Indian tree legume. In: Pollination in tropics, International Union of the Study of Social Insects (IUSSI), Indian Chapter, Bangalore, pp 148–149
40. Zapata TR, Arroyo MTK (1978) Plant reproductive ecology of a secondary deciduous tropical forest in Venezuela. Biotropica 221–230

41. Larson BM, Barrett SC (2000) A comparative analysis of pollen limitation in flowering plants. Biol J Linn Soc 69(4):503–520
42. Cruden RW (1977) Pollen-ovule ratios: a conservative indicator of breeding systems in flowering plants. Evolution 31:32–46
43. Tewari SK, Shubhanjana S, Pandey SBS (2002) Reproductive behaviour of *Dalbergia sissoo* (Shisham). Indian For. 128(3):336–340
44. Preksha (2005) Germplasm evaluation reproductive biology and polyploidy induction in shisham (*Dalbergia sissoo* Roxb.) Doctoral Thesis. G.B. Pant University of Agriculture and Technology, Pantnagar, India
45. Vasudeva S, Sareen T (2011) Reproductive Biology of *Dalbergia sissoo* Roxb. Indian For 137(4):411–425
46. Stephenson AG (1981) Flower and fruit abortion: proximate causes and ultimate functions. Ann Rev Ecol Syst 12:253–279
47. Sahni KC (1998) The Book of Indian Trees. Oxford University Press: Delhi, pp 230
48. Ganeshaiah KN, Shaanker RU (1988) Seed abortion in wind-dispersed pods of *Dalbergia sissoo*: maternal regulation or sibling rivalry? Oecologia 77:135–139
49. Raju BM, Uma Shaanker R, Ganeshaiah KN (1996) Intra-fruit seed abortion in a wind dispersed tree, *Dalbergia sissoo* Roxb: proximate mechanisms. Sex Plant Reprod 9:273–278
50. Troup RS, Joshi HB (1983) Troup's The Silviculture of Indian Trees. Controller of Publications, India
51. The Wealth of India (2000) National Institute of Science Communication, Council of Scientific and Industrial Research, India, vol 3, D-I
52. Satchell J (1983) Earthworm ecology: from Darwin to vermiculture. Chapman and Hall, London
53. Ghatnekar SD, Kavian MF, Ghalsasi DS (2009) Activities of *Lumbricus rubellus* and *Eudrilus euginiae* and fate of *Acacia auriculiformis, Casurina equisetifolia* and *Dalbergia sissoo* seeds. Vermitechnology I. Dyn. Soil Dyn. Plant. 3:82–86

Chapter 22
Pollination Ecology and Breeding System in *Dalbergia sissoo*

Kulvir Singh Bangarwa

1 Introduction

Tree breeding depends on the knowledge of genetics and reproductive biology of a species. The focus of reproductive biology is to understand the flowering phenology, floral and fruit biology, pollination, and seed and pod characteristics. Studies on reproductive biology and breeding systems therefore play a key role in defining the success of an improvement programme [1]. The breeding systems, through the mechanism of reproduction, the method of pollination, and the degree of compatibility, regulate the amount of recombination. They help in the understanding of genetic makeup of natural variations, and of open pollinated seed origins, both of which are linked with genecology of the species.

1.1 Botanical Description of the Flowers, Fruits, and Seeds of Dalbergia sissoo

The axillary panicle inflorescence is 3.5–7.5 cm in length. The papilionaceous sessile white to yellowish white small flowers (7–9 mm long) are hermaphrodite and have pervasive fragrance [2]. The standard petal in the form of a low claw is narrow at the base, and the wing and keel petals are oblong. The bracts are small, pubescent, and caducous. The calyx is 5 mm long, serrated, unequal, ciliated, and shorter than the tube. The anthers are uniform, stamens are nine, monodelphous with slit tube on the upper side. The hairy style has capitate stigma, ovary is pubescent, 2–4 ovulate [3]. The indehiscent, smooth, and shiny pods are 1–4 seeded and are linear-oblong (4.5–10 × 0.7–1.5 cm), stipitate with acute apex, and are reticulated against the seeds.

K. S. Bangarwa (✉)
Department of Forestry, CCS Haryana Agricultural University, Hisar, India
e-mail: kulvirsb@gmail.com; kulvirsb@yahoo.com

R. R. Warrier et al. (eds.), *Handbook of Asian Rosewoods*, Materials Horizons: From Nature to Nanomaterials, https://doi.org/10.1007/978-981-96-9459-4_22

The flat kidney-shaped seeds are variable (8–10 × 4–5.5 mm), pale brown, brown to brownish-black, reniform, compressed, with papery testa [4, 5].

1.2 Floral Structure

The papilionaceous flowers of *Dalbergia sissoo* are arranged in diverse auxiliary panicles with racemose corymbose branches. An average inflorescence has 45 flowers ranging 20–75 flowers. The cream-coloured flowers have a length of 0.8 to 1.0 cm. The companulate calyx is five-toothed, 4–6 mm long, with its upper two sepals united and the lowest the longest. The creamish corolla is 7–8 mm long. The five petals are united only at the tips. The standard is broad with long claws, and the keel is obtuse with wings below. The stamens are 9 in number monadelphous with tube splitting along the upper side, anthers minute, basifixed; dehiscence occurred by terminal slits. Pollen grains were psilate and sticky. A typical flower produces an average of 4000 pollen grains. Gynoecium is a monocarpellary, superior ovary, 4–8 mm long with marginal placentation housing 3–8 ovules, average number being 5.0. The style is short, thick, and glabrous, ending into a papillate wet stigma. The pollen ovule ratio is 710 [6].

1.3 Inflorescence Characteristics

Dalbergia sissoo has small (7–9 mm long), pea-shaped, bisexual, and inconspicuously white to yellow flowers having a pervasive fragrance borne on small axillary branches. The inflorescence 3.5–8 cm long bears 7–14 sessile to sub-sessile papilionaceous flowers. Flowers contain campanulate calyx and five petals, each with an average length of 1.0 cm; standard petals are narrower at the base forming a claw for keel (1) and wings (2), which are oblong [7].

The androecium consists of 9 monadelphous stamens, grouped together on a broad stalk, with 5 larger stamens and 4 smaller ones. The average length of the stamens is 0.79 ± 0.2 cm, which is slightly shorter than the hairy pistil, measuring 0.85 ± 0.06 cm. The pistil, which is stalked and hairy, has a narrow ovary containing 3 to 8 ovules, a short style, and a dot-like stigma. An examination of ovaries from flowers collected from 10 selected trees showed that the ovary contained 3, 4, 5, or 6 ovules in 5, 18, 48, and 27% of the samples, respectively. The mature brown pod is flat, measuring 5 to 8 cm in length, and contains 1 to 4 seeds. In the bud stage, the ovary is bent and surrounded by anthers [8].

1.4 Pollen Structure and Viability

The spherical pollens are thin-walled, with three furrows (zonocolpate) and an average size of 10.2 $\pm$ 0.11 µm. A pollen viability test conducted at various stages of floral development using acetocarmine staining solution revealed that viability is 100% during the bud stage but decreases to 56% after the flowers open. This test also suggests that the role of pollens in pollination and fertilization patterns significantly diminishes after anthesis [9].

1.5 Stigma Receptivity

The gynoecium comprises a single hairy pistil with a narrow ovary containing 3–8 ovules, a short style, and a dot-like stigma. The stamens are slightly shorter than the pistil. Anther dehiscence occurs early in the morning, just before the flowers open. The stigma, which is shiny and sticky—indicating its receptivity—becomes active shortly before the flower opens and remains receptive for a few hours afterwards [7]. Data on anthesis indicate that most flowers open between 10:00 and 14:00, with peak activity occurring from 11:30 to 13:30 [5, 10].

2 Flowering and Fruiting Phenology

The flowering starts at the beginning of March and continues until the end of April. The flowering is at its peak in the first two weeks of April when almost all the trees in the observed population flower profusely; high temperatures can hasten this process. Young flower buds appear alongside new leaves and open in March–May [8] and fruits in June–November [6]. The flowering period of marked trees was noticed to be five weeks for an individual tree and about 5–7 weeks for the whole population. The early to late flowering period and light to dense intensity (light to dense) may be attributed to different phonological factors at different sites [8].

The emergence of floral buds commences after about 15 days of emergence of new leaves indicating that stimulus for flowering is the new flush. An average inflorescence has 45 flowers ranging from 20 to 75. A medium-sized tree on an average has 14,450 inflorescences (3200–20,000) indicating an average of about 6.5 lakh flowers (ranging from 1.44 lakh to 9 lakh flowers) produced. In an inflorescence 2–7 flowers open per day depending upon the total number of buds present in it. It takes 3–5 days for a bud, with just emerged petals to develop into a flower. The mean period of anthesis is estimated to be 3.7 days. The flowers showed diurnal rhythm in their opening, anther dehiscence, and pollination. Some flowers open in the morning between 9.30 h to 10.30 h while others open in the afternoon between 15 and 16 h. Anther dehiscence takes place between 10 and 11 h and then between

15.30 h and 16.30 h. The stigma is visually receptive at around 10 h in the morning and at 16 h in the afternoon when a clear drop of fluid is seen on it. The flowers remain receptive for 4–5 h. The flowers that are receptive in the morning were not so in the afternoon, and the latter were not so the next day. The flowers being cream in colour changes to light yellow on pollination. The colour of unpollinated flowers remains the same and gradually wilts away. The saddle-shaped ovary destined to develop into a pod emerges out of the corolla after 15–20 days of pollination. The ovary has 3 to 8 ovules with an average number per ovary of five. The pods emerge and become prominent in April and gain full size in July [11]. The green pods when matured turn brown by September end. The flat pod is 5–8 cm long, and contain 1–4 seeds. By the middle of December, the fully developed seeds are formed [9, 12], while there are also reports of maturation of pods and ripening of seeds occurring in December or in the initial weeks of January [13]. After maturation, pods remain on the trees for about 7–8 months [14].

2.1 Pollination Biology

The flowers are cream-colored and emit a pleasant fragrance due to the presence of nectar, attracting a variety of insects such as honeybees, butterflies, beetles, and thrips. These insects play crucial roles in the pollination process of the plant. Honeybees and butterflies begin their activity as soon as the flowers open, exhibiting a diurnal pattern, most active between 10:00 AM to 12:00 PM and 3:00 to 5:00 PM. Beetles are mostly active during the morning hours, while thrips remain inside the flowers throughout the day, occasionally emerging [15]. Honey bees are the most effective pollinators, feeding on both nectar and pollen. The flowering season aligns with the swarming season of honey bees, evidenced by the numerous beehives found on these trees. During their foraging, bees transfer pollen from one flower to another on the same or different trees, facilitating both self-pollination and cross-pollination. The structure of the flowers, with stamens and stigmas positioned at the same height, enhances the efficiency of this pollination process. Beetles also contribute to pollination, moving between flowers in search of nectar. They use their heads to access nectar, transferring pollen to the stigmas, thus causing both self- and cross-pollination. However, their activity is limited to the forenoon. Butterflies, though less aggressive in their foraging, use their long proboscis to extract nectar, collecting and transferring pollen between flowers, aiding in pollination. Thrips, being smaller in size, stay mostly at the base of the flower, and increase in number as the flowering season progresses. They move within flowers to promote self-pollination and occasionally fly to other flowers, facilitating cross-pollination as well [1, 2, 7, 12].

2.2 Pollination Pattern

For understanding self-pollination, artificial crossing, cross versus self-pollination, various studies were carried out by Bangarwa [7].

Self-Pollination: A single bud inflorescence was selected on each tree before it opened. One hundred buds were bagged to allow self-pollination, and data on pod setting was recorded. It was observed that 16.25% of the buds set pods under controlled self-pollination conditions. Additionally, some inflorescences were selected and tagged, with their buds counted to determine the natural pod setting percentage, which was found to be 35%. The lower pod setting rate under controlled self-pollination with bagging is attributed to the altered environment within the bags. Effective pollination was noted to occur while the buds were still in their closed condition.

Artificial Crossing: Young buds, in which the corolla had just become visible, were selected on marked trees. These buds were emasculated early in the morning before 8:00 AM to expose the gynoeciums. After emasculation, each bud was bagged to prevent contamination. Later, between 11:00 AM and 12:00 PM on the same day, the emasculated buds were pollinated with pollen from other trees and bagged again. Data on pod setting was recorded, revealing a pod setting rate of only 5–7%.

2.3 Cross-Pollination Versus Self-Pollination

It is evident from Fig. 1 that the 44% pod formation following cross-pollination and 6% due to self-pollination lead to infer it is a self-incompatible species. It is considered to be self-incompatible if its fruit formation due to cross-pollination is 5 times more than that due to self-pollination [16]. *D. sissoo* showed an Index of self-incompatibility of 0.14, confirming the presence of self-incompatibility. The low formation of only 15% pods (fruits) in nature is also indicative of the presence of self-incompatibility.

The observation of a wet stigma and the inhibition of self-pollen tube growth in the ovary of *D. sissoo* suggest the presence of a gametophytic self-incompatibility system [17, 18]. The wet-type stigmas in the Leguminosae family [19], and also the presence of small, papillate stigmas that produce stigmatic exudate indicate a gametophytic self-incompatibility mechanism [20].

As majority (56%) of experimentally cross-pollinated flowers fail to set fruit and the rejection of pollen tube sometimes occur in the ovary, post zygotic factors may be responsible for the abortion of fruits in *D. sissoo*. It also suggests that maternal resource constraints might contribute to the abortion of young fruits.

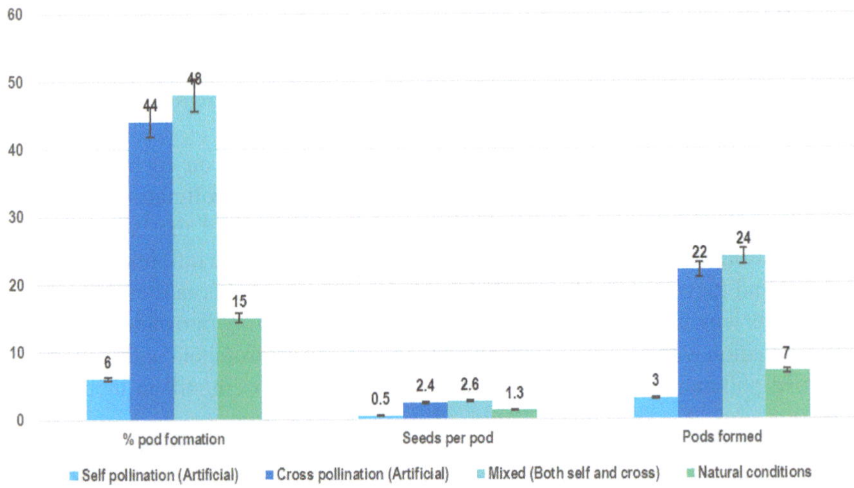

Fig. 1 Fruit and seed formation in a pollination experiment

3 Fruit Biology

Pods mature in November–December, developing a light brown colour and a flat shape. It has been observed that flowering and fruiting co-occur (Fig. 2), as young pods can be seen alongside flowers on the same inflorescences.

3.1 *Number of Pods per Inflorescence*

In naturally pollinated inflorescences, the number of pods per inflorescence ranged from 4 to 7.5, with an average of 6.01. In self-pollinated inflorescences, the range was 3.2 to 4.5, with an average of 4.0. Pods take approximately 6 to 7 months to fully

Fig. 2 Occurrence of flowers and pods in *Dalbergia sissoo* over a period of 90 days

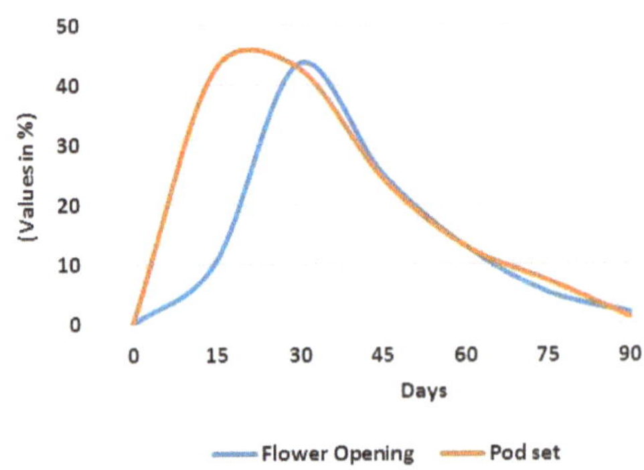

mature, and the difference in the average number of pods per inflorescence between naturally and self-pollinated plants was significant. It was observed that the fruit set of *D. sissoo* in Haryana averaged at 15%, with a range of 10% to 25.3% and most flowers and young pods were lost within 3 to 14 days after anthesis [21]. A study in Uttarakhand recorded fruit production between 11% and 17.2% [1] indicating that geography could play a role in fruit production.

3.2 *Pod and Seed Characteristics*

Light brown, indehiscent, and flat pods mature in November and December and can be seen alongside inflorescences. The number of pods per inflorescence differs significantly between natural and controlled self-pollination, with averages of 6.01 and 4.0, respectively. The average pod length was 5.13 cm and 4.54 cm in naturally pollinated inflorescences and in self-pollinated inflorescences, respectively and were found to be significantly different. The mean pod breadth in naturally and self -pollinated inflorescences, measured 7.42 mm and 7.40 mm, respectively, but they were not significantly different. However, there were significant differences within the natural and self-pollinated trees. Pods are indehiscent and remain on the tree for several months even after reaching maturity (Fig. 3). During pod development, seed development also occurs inside the pod. Depending on the number of fertilized ovules, the corresponding number of seeds (1–4) develop inside the pod. Seed development begins within pods, where they are attached to the pod wall through a funicle. Initially, seeds are green with endospermic cotyledons. Mature seeds are light brown, flat, thin, and kidney-shaped, with an average length of 0.83 cm and a breadth of 0.42 cm [12, 22].

The mean weight of 100 pods in naturally pollinated inflorescences ranged from 6.46 to 8.31 g, with an average of 7.05 g, while in self-pollinated pods, it ranged from 5.10 to 6.42 g, with an average of 5.65 g. These averages were found to be significantly different.

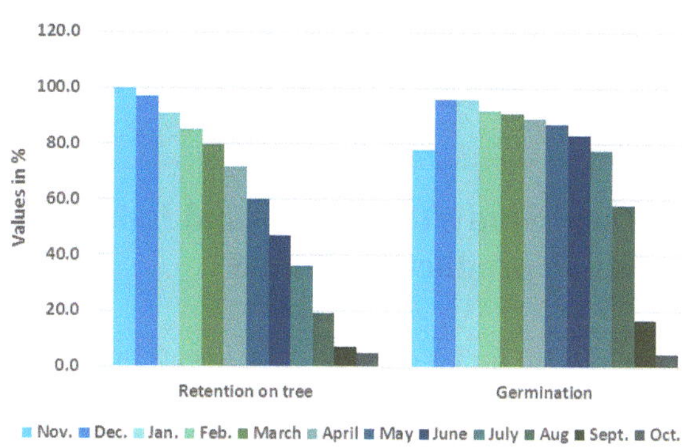

Fig. 3 Retention time of pods on trees in relation to viability of the seeds

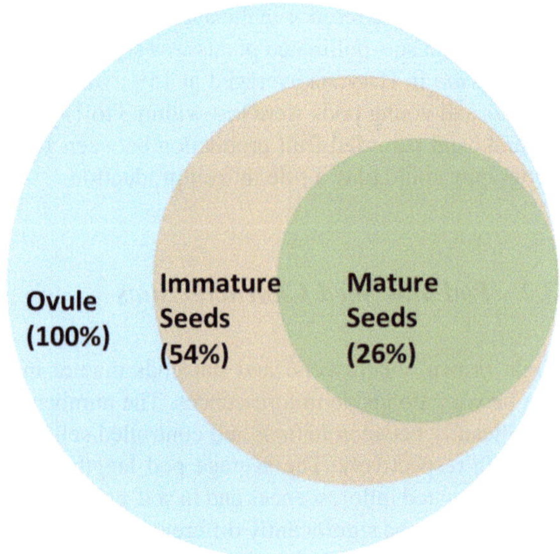

Fig. 4 Ovule to seed ratio in *Dalbergia sissoo*

Seed abortion in multiovulated ovaries, where only a small percentage of ovules mature into seeds, is common in flowering plants [23]. In *D. sissoo,* approximately 74% of ovules abort, leaving just 26% to develop into mature seeds (Fig. 4). Typically, a mature pod contains one to four seeds, averaging 1.30 seeds per pod from about five ovules per ovary. This low seed count is attributed to intra-fruit seed abortions [24, 25]. *D. sissoo*, which primarily outbreeds, exhibits a high rate of seed abortions, consistent with other tree species [26, 27]. The number of ovules and seeds produced is consistent within and between trees, suggesting strong canalizing selection for this trait, as seen in other species [26]. Low seed set is not due to pollen or resource limitations, nor the absence of fertilization [24, 28]. Species with low seed numbers in dispersal units, like pods, often have increased dispersal efficiency due to reduced wing loading [25, 29]. Hence, low seed production in *D. sissoo* enhances the efficient dispersal of pods by wind and water. This trait also benefits post-dispersal seedling establishment and survival. In single-seeded pods, competition is absent, while in multi-seeded pods, minimal competition occurs due to genetic relatedness [30].

4 Field Study Techniques and Laboratory Methods for Reproductive Studies

Bangarwa [9] for his crossing experiments in *Dalbergia sissoo* selected young buds on mother parent for emasculation just before the visibility of corolla during the early hours prior to 0800. Using a pointed forceps, the gynoecium was revealed towards the keel, removing the sepals and petals. The nine stamens containing anthers

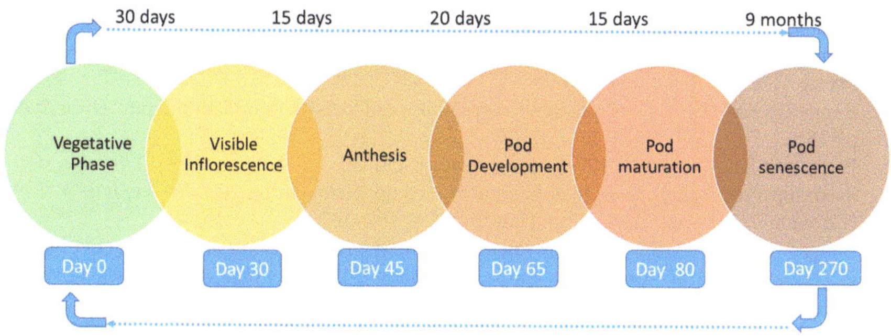

Fig. 5 Timelines of pod formation in *Dalbergia sissoo*

and surrounding the gynoecium were removed gently in a single operation, without causing injury to the stigma. The emasculated buds were pollinated with a fine brush between 1100 and 1200 h on the same day. The pod setting percent was reported to vary from 10 to 17 while under natural conditions, only 5–7% was observed [8, 9]. Adopting the same methodology, it was reported that the pod setting was 16 to 35 per cent [12]. The timelines for pod formation are depicted in Fig. 5.

5 Conclusion

The pollination mechanism of *Dalbergia sissoo* remains a topic of debate, as it can be either self-pollinated or outcrossing. The timing of pollination and fertilization is also uncertain. However, several factors suggest self-pollination: the stamens encase the ovary, anthers dehisce before anthesis, pollen has high fertility, and the stigma is receptive during the bud stage. On the other hand, cross-pollination cannot be dismissed, as the stigma remains receptive and pollen viable even after anthesis, the ovary emerges from the stamen whorls during the receptivity stage, and insect visits occur. Viable seeds can be collected almost throughout the year, enhancing opportunities for wide planting.

References

1. Tewari SK, Shubhanjana S, Pandey BS (2002) Reproductive behaviour of *Dalbergia sissoo* (Shisham). Indian Forester 128(3):336–340
2. Kumar A (2017) Know Your Trees—*Dalbergia sissoo* Roxb. Van Vigyan, ENVIS Newsletter 3(4):2–8
3. Tewari DN (1994) A monograph on *Dalbergia sissoo* Roxb. International Book distributors, Rajpur Road, Dehradun, India
4. Parker RN (1956) A forest flora for the Punjab with Hazara and Delhi. Government Printing Press, Lahore, Pakistan

5. Luna RK (1996) Plantation trees. International Book Distributors, Dehra Dun, India, pp 975
6. Bangarwa KS, Singh VP, Sangwan RS (1994) Breeding system in *Dalbergia sissoo*. Nitrogen Fixing Tree Res Rep 12:57–59
7. Bangarwa KS (1996) Sissoo breeding. Agriculture and forestry information centre, Hisar, India, pp 156
8. Bangarwa KS, Singh VP (1994) Floral biology and crossing techniques in *Dalbergia sissoo* Roxb. In: Proceedings of an international workshop. Hetauda, Nepal. May 31, June 4, 1993, Nitrogen Fixing Tree Association, Hawaii, USA, 162–167
9. Bangarwa KS (1993) Studies on provenance variation, progeny testing and reproductive biology in *Dalbergia sissoo* Roxb. Ph.D. thesis, Ch.Charan Singh Haryana Agricultural university, Hissar, India
10. Bakewell-Stone P (2022) *Dalbergia sissoo* (Shisham). CABI Digital Library, 17808. https://www.cabidigitallibrary.org/doi/https://doi.org/10.1079/cabicompendium.17808
11. Ijaz S, Haq IU (2021) *Dalbergia sissoo*: Biology, ecology and sustainable agroforestry. CRC Press, Boca Raton, USA
12. Sharma R, Chauhan SK, Khajuria HN (2009) Reproductive biology and variability studies in *Dalbergia sissoo* (Roxb). J Tree Sci 28(1&2):23–28
13. Haq R (1958) Monograph on *Dalbergia sissoo* Rox. (Shisham). M.Sc. Thesis, Forest Education Division, Pakistan Forest Institute Peshawar, Pakistan
14. Chauhan R, Chauhan S, Khajuria HN (2004) Reproductive biology and variability studies in *Dalbergia sissoo* (Roxb). Advances in Forestry Research in India 28, 24–37
15. Orwa C, Mutua A, Kindt R, Jamnadass R, Simons A (2009) Agroforestree database: a tree reference and selection guide. Version 4. Nairobi, Kenya: World Agroforestry Centre. http://www.worldagroforestry.org/sites/treedbs/treedatabases.asp
16. Bawa KS Breeding systems of tree species of a lowland tropical community. Evolution 28, 85–92
17. Richards AJ (1986) Plant breeding systems. George Allen and Unwin, London
18. Shivanna K, Owens S (1989) Pollen-pistil interactions (Papilionoideae). Advances in legume biology: Monograph of Systematic Botany Missouri, Missouri Botanical Garden 29:157–182
19. Heslop-Harrison J, Shivanna KR (1977) The receptive surface of angiosperm stigma. Ann Bot 41(6):1233–1258
20. Heslop-Harrison J, Heslop-Harrison Y, Barber J (1975) The stigma surface incompatibility responses. Proc. Royal Soc. London. B. 188:287–297
21. Vasudeva SP, Sareen TS (2011) Reproductive biology in *Dalbergia sissoo* Roxb. (Papilionodeae; Leguminosae). Indian Forester 137(4):411–425
22. Vakshasya RK, Rajora OP, Rawat MS (1992) Seed and seedling traits of *Dalbergia sissoo* Roxb: seed source variation studies among ten sources in India. For Ecol Manag 48(3–4):265–275 (1992)
23. Mohana GS, Uma Shaankar R, Ganeshaiah KN, Dayanandan S (2001) Genetic relatedness among developing seeds and intrafruit seed abortions in *Dalbergia sissoo* (Fabaceae). Amer. J. Bot. 88(7):1181–1188
24. Ganeshaiah KN, Shaanker RU (1988) Seed abortion in wind-dispersed pods of *Dalbergia sissoo*: maternal regulation or sibling rivalry? Oecologia 77(1):135–139. https://doi.org/10.1007/bf00380936
25. Uma Shaankar R, Ganeshaiah KN, Bawa KS (1988) Parent-offspring conflict, sibling rivalry and brood size patterns in plants. Ann Rev Eco Syst 19:177–205
26. Bawa KS, Webb CJ (1984) Flower, fruit and seed abortion in tropical forest trees: implications for the evolution of paternal and maternal reproductive patterns. Amer J Bot 71(5):736–751
27. Krishnamurty KS, Uma Shankar R, Ganeshaiah KN (1997) Seed abortion in an animal dispersed species. *Syzgium cumini* (L) Skeels (Myrtaceae): the Chemical basis. Current Sci 73(10):869–873
28. Mohan Raju B, Uma Shaanker R, Ganeshaiah KN (1996) Intra-fruit seed abortion in a wind dispersed tree, *Dalbergia sissoo* Roxb: Proximate mechanisms. Sexual Plant Reproduction 9(5):273–278

29. Ganeshaiah KN, Uma Shaanker R, Joshi NV (1991) Evolution of polyembryony: consequences to the fitness of mother and offspring. J Genet 70(2):103–127
30. Ganeshaiah KN, Uma Shankar R (1991) Seed size optimisation in a wind dispersed tree *Butea monosperma*: trade-off between seedling establishment and pod dispersal efficiency. Oikos 60:3–6

Chapter 23
Indicators of Seed Maturation and Deterioration in *Dalbergia sissoo*: Implications for Germination and Longevity

Geeta Joshi, S. S. Phartyal, J. S. Nayal, Rekha R. Warrier, A. N. Arunkumar, and R. C. Thapliyal

1 Introduction

In the current context of rapid deforestation and climate change, which are driving species loss, ex situ conservation through seed storage is an effective strategy for species preservation. Seed storage is the simplest and most cost-effective conservation method [1, 2], but understanding optimal storage conditions is essential for maintaining seed viability [3]. Orthodox seeds develop desiccation tolerance during the late maturation stage, allowing them to remain viable for extended periods in a dry state [4]. During maturation, seed development is marked by a series of physical, physiological, and biochemical transformations. The maturation process progresses

G. Joshi (✉)
Indian Council for Forestry Research and Education, Dehradun, Uttarakhand, India
e-mail: geeta@icfre.org

S. S. Phartyal
School of Earth Science and NRM, Mizoram University, Aizawal, Mizoram, India

J. S. Nayal
Uttarakhand State Seed and Organic Production Certification Agency, Dehradun, Uttarakhand, India

R. R. Warrier
ICFRE-Institute of Forest Genetics and Tree Breeding, Coimbatore, Tamil Nadu, India

A. N. Arunkumar
ICFRE-Institute of Wood Science and Technology, Bengaluru, Karnataka, India

R. C. Thapliyal
ICFRE-Forest Research Institute, Dehradun, Uttarakhand, India

R. R. Warrier et al. (eds.), *Handbook of Asian Rosewoods*, Materials Horizons: From Nature to Nanomaterials, https://doi.org/10.1007/978-981-96-9459-4_23

through three primary stages: cell division and histodifferentiation, cell expansion and reserve deposition, and maturation drying [5]. During the second phase, seeds undergo significant morphological and biochemical modifications, including increased dry weight and accumulating storage compounds such as starch, proteins, fats, and phytin in the endosperm or cotyledonary tissues. At maturity, orthodox seeds exhibit low moisture content and enter a metabolically quiescent state [6]. The final stage of development is marked by a decline in metabolic activity as moisture content decreases [7]. At this point, orthodox seeds acquire desiccation tolerance, making them viable for germination [8].

The longevity of seed, i.e. the duration from maturation to the loss of viability under storage conditions, varies across species and is influenced by a complex interplay of environmental and genetic factorsF [9]. In orthodox seeds, viability is better preserved under cool and dry conditions but declines with increased temperature and humidity [10]. Viability and vigour loss rate in orthodox seeds primarily depend on seed moisture content and storage temperature [11]. Additionally, the embryo cells of these seeds can endure extreme dehydration due to a specialised developmental programme [12].

During storage, seed viability declines due to the onset of seed ageing, which is both irreversible and inevitable [13]. The deterioration of metabolic activity leads to cellular damage, delayed seedling emergence, reduced vigour, and, ultimately, loss of viability [14]. In orthodox seeds, viability loss during ageing is associated with damage to macromolecules, including membrane phospholipids and proteins, as well as degradation of the nucleus and cytoplasmic organelles [13, 15–25]. Optimal storage conditions help slow down metabolic processes while preventing seed damage and extending longevity.

Effective seed storage is critical for any propagation activity or ex situ conservation programme as it restricts deterioration process which is crucial for maintaining seed viability. Improper seed storage leads to poor quality seeds, resulting in waste of natural and economic resources [26].

Dalbergia sissoo is an important leguminous tree species extensively used for afforestation purposes. Therefore, high-quality seeds are essential for the large-scale propagation of vigorous seedlings. However, research on the seed storage behaviour of *D. sissoo* remains limited [27–29]. This study discusses seed maturation and examines the impact of seed moisture content, various storage temperatures, and different containers on seed longevity, while documenting the changes occurring during storage in *D. sissoo* seeds.

2 Seed Maturity

Pods of *D. sissoo* were collected at 15-day intervals from the second week of April, starting at 135 days after anthesis (DAA) from ICFRE-FRI, Campus, Dehradun, Uttarakhand, India. The morphological parameters such as pod and seed length, width, fresh and dry weight, and moisture content were recorded. Biochemical changes in sugar, starch, phosphorus, magnesium, potassium, calcium, and nitrogen were assessed. Germination potential and seedling survival rates were evaluated under controlled conditions. The changes occurring during seed maturation are briefly discussed here, while the detailed methodology and results are published [30].

Pod length and width increased until 90 DAA, then declined steadily until 195 DAA. Seed length peaked at 210 DAA, while seed width stabilised at 195 DAA. Pod fresh weight increased until 60 DAA, followed by a decline until 180 DAA and a rapid drop until 225 DAA. Seed fresh weight peaked at 195 DAA and declined after 210 DAA. Pod dry weight stabilised at 180 DAA, while seed dry weight reached stability at 210 DAA. Moisture content declined steadily from 30 to 210 DAA, followed by a drop at 240 DAA. Soluble sugar content remained low throughout fruit development. Starch content declined, increased sharply, and then stabilised at 195 DAA. Phosphorus, magnesium, and nitrogen levels decreased with maturity, while potassium and calcium increased two- and three-fold, respectively. By 210 DAA, these nutrients had stabilised (Fig. 1). Pod colour changed from light green to yellow at 195 DAA, with seed colour transitioning at 180 DAA. By 225 DAA, both turned light brown. Chloroplasts, visible in immature seeds at 195 DAA, disappeared in mature seeds at 240 DAA. Seed germination began at 195 DAA, increasing with maturity, reaching maximum viability at 240 DAA. Desiccation further improved germination, reaching 92% in dried immature seeds.

In desiccation-tolerant seeds, moisture content decreases during the final stages of maturity, influencing dry weight accumulation [31, 32]. In *D. sissoo*, dry weight stabilises by 210 days after anthesis (DAA). Maximum dry weight coincides with

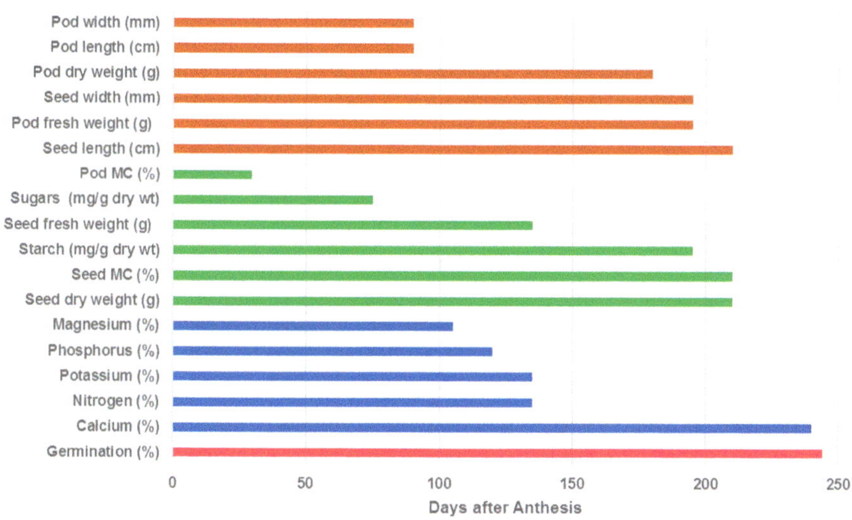

Fig. 1 Stabilising of different parameters (morphological—brown, physiological—green, and biochemical—blue) following different Days After Anthesis (DAA) in *Dalbergia sissoo*

a sharp drop in moisture at 210 DAA, but peak germination and lowest moisture content occur at 240 DAA. Similar patterns have been observed in *Atriplex cordobensis* [33], where viability is highest after mass maturity. This supports the model where mass maturity (maximum dry weight) precedes physiological maturity (maximum viability) [34, 35], contrasting with the idea that viability and dry weight peak simultaneously [2].

During seed ontogeny, soluble food reserves are translocated and stored as starch, proteins, and fats [36]. Organic and inorganic components fluctuated and stabilised at mass maturity. Starch content varied during growth, a trend also observed in *Eugenia jambolina* and *Avicennia marina* [37, 38]. While soluble sugars remain constant, starch serves as a temporary carbon reserve for seed development [39, 40]. Inorganic nutrients such as nitrogen, phosphorus, and magnesium declined, while calcium and potassium increased, as observed in other species like Quercus and Fraxinus [41, 42].

Physiologically, germinability was attained by 195 DAA. Immature seeds continued maturing post-harvest under dry conditions [43]. However, while germination rates improved with storage, seedlings from immature seeds (195 DAA) had poor survival, indicating incomplete metabolic development [44].

Maturation is accompanied by colour changes, with pods turning brown by 225 DAA. Degreening, the loss of chlorophyll, coincides with desiccation tolerance [45, 46]. Given these developmental milestones, 240 DAA is the optimal time for seed collection in *D. sissoo*.

3 Seed Longevity

Mature pods of *Dalbergia sissoo* were collected in December from Shakumbra in the Dehra Dun Forest Division, Uttarakhand, India. Each pod, containing one to four seeds, was stored intact. Before germination, a portion of the pod containing a single seed (hereafter referred to as "seed") was separated for use. The seeds were stored in different combinations of storage temperatures (40, 25, 15 + 5, –5 °C and room temperature), three seed moisture contents (12.5, 8.0, and 5.6%), and three storage containers (vacuum boxes, polythene bags –0.45 mm thick, and plastic boxes) for 665 days. The dried seeds initially had a moisture content of 12.5%, which was reduced through sun drying.

Moisture content was assessed following ISTA guidelines [47]. Prior to storage, seed viability was determined using a laboratory germination test. Germination studies were conducted under laboratory conditions at 30 °C, using four replications of 50 seeds each. A seed was considered germinated when the radicle reached approximately 1.0 cm in length. Initial germination was recorded before storage, and viability was monitored every two months during storage. The viability and vigour of seeds stored under different moisture content and temperature combinations were analysed for statistical significance using ANOVA after arcsine transformation of percentage values. Probit-transformed germination percentages were regressed against storage time to determine the time required for viability to decline to 50% of the original level (P50) [48]. The survival curve slope indicated the seed deterioration rate [49].

To assess the impact of storage temperature, container type, seed moisture content, storage duration, and their interactions on seed viability, germination data from seeds stored under various conditions were statistically analysed ($p = 0.05$). The results indicated that seed viability decreased with increasing temperature. The lowest mean germination (11.66%) was recorded at 40 °C, while the highest (84.09%) was observed at –5 °C (Table 1). Seeds with a moisture content of 12.50% exhibited a mean germination rate of 61.55%, whereas those with a moisture content of 5.60% had a higher mean germination rate of 71.85% (Table 1). Seeds with lower moisture content retained viability across all temperature conditions for longer periods.

The analysis of storage containers showed that seeds stored in plastic boxes had the highest mean germination rate (68.18%), making them the most effective for seed storage (Table 2). In contrast, the lowest mean germination rate (66.28%) was recorded in vacuum-sealed boxes, followed by 67.81% in polythene bags (Table 2). The combined effect of storage temperature and seed moisture content indicated that seeds with a moisture content of 5.60% stored at 5 °C or –5 °C were optimal for maintaining seed longevity.

The analysis of the interaction between storage temperature, container type, seed moisture content, and storage duration revealed a significant decline in germination percentage. At 40 °C, seeds with moisture contents of 12.50% and 8.00% lost viability completely by day 122 in all three storage containers. However, seeds with 5.60% moisture content at the same temperature (40 °C) retained viability until day 305 before experiencing complete viability loss. The lowest germination recorded was 4.00% on day 665 for seeds with 12.50% moisture content stored at ambient room temperature in a vacuum box. In contrast, the highest mean germination (86.00%) was observed in seeds with 5.60% moisture content stored in a vacuum box at –5 °C (Figs. 2, 3, and 4).

Table 1 Mean germination for interaction of storage temperature (40, 25, 15, 5, -5 °C, and RT), and moisture content (12.50, 8.00, and 5.60%)

Temp. (°C) / MC (%)	40 °C	RT °C	25 °C	15 °C	5 °C	–5 °C	Mean MC
12.50	07.50 (06.14)	56.83 (48.90)	65.47 (54.58)	73.71 (59.88)	82.60 (65.72)	83.19 (66.13)	61.55 (50.32)
08.00	08.28 (07.47)	73.98 (60 01)	79.46 (63.55)	80.76 (59.88)	83.10 (66.12)	83.93 (66 73)	68.25 (54.70)
05.60	19.21 (17.51)	77.06 (62.01)	82.39 (65.54)	83.14 (66.08)	84.17 (66.99)	85.15 (67.73)	71.85 (57.64)
Mean Temp	11.66 (10.57)	69.28 (56.97)	75.77 (61.22)	79.20 (63.44)	83.29 (66.28)	84.09 (66.86)	

* Values in parenthesis are arc sine transformed values

Table 2 Mean germination for interaction of storage temperature, container (vaccum box—V, polythene bag—P, and plastic box—B) and moisture content

Temp. (°C)	40 °C			RT			25 °C			15 °C			5 °C			–5 °C			Mean Cont
MC (%)	12.50	08.00	05.60	12.50	08.00	05.60	12.50	08.00	05.60	12.50	08.00	05.60	12.50	08.00	05.60	12.50	08.00	05.60	
Cont																			
V	07.54 (06.80)	08.42 (07.59)	18.29 (16.61)	54.92 (46.88)	71.50 (58.64)	69.79 (57.62)	64.58 (54.06)	80.04 (63.99)	84.13 (66.92)	71.92 (58.61)	80.21 (64.05)	80.83 (64.36)	83.04 (65.99)	81.96 (65.17)	85.08 (67.74)	81.13 (64.46)	83.38 (66.36)	86.33 (68.76)	66.28 (53.58)
	07.46 (06.67)	08.46 (07.61)	19.54 (17.80)	58.21 (50.17)	76.54 (61.65)	80.71 (64.34)	62.25 (52.54)	78.08 (62.71)	79.92 (63.76)	73.29 (59.70)	81.08 (64.42)	84.42 (67.06)	80.56 (64.19)	83.04 (66.16)	83.50 (66.55)	84.38 (67.04)	83.50 (66.28)	84.33 (66.93)	67.81 (54.20)
P	07.50 (06.74)	07.96 (07.21)	19.79 (18.14)	57.38 (49.64)	73.79 (59.75)	80.67 (64.16)	69.58 (57.15)	80.25 (63.84)	83.08 (65.85)	75.92 (61.32)	81.00 (64.58)	84.17 (66.83)	84.21 (66.97)	84.29 (67.04)	83.90 (66.69)	84.08 (66.89)	84.29 (67.53)	84.79 (67.49)	68.18 (54.89)
	07.54 (06.80)	08.42 (07.59)	18.29 (16.61)	54.92 (46.88)	71.50 (58.64)	69.79 (57.62)	64.58 (54.06)	80.04 (63.99)	84.13 (66.92)	71.92 (58.61)	80.21 (64.05)	80.83 (64.36)	83.04 (65.99)	81.96 (65.17)	85.08 (67.74)	81.13 (64.46)	83.38 (66.36)	86.33 (68.76)	66.28 (53.58)
B	07.46 (06.67)	08.46 (07.61)	19.54 (17.80)	58.21 (50.17)	76.54 (61.65)	80.71 (64.34)	62.25 (52.54)	78.08 (62.71)	79.92 (63.76)	73.29 (59.70)	81.08 (64.42)	84.42 (67.06)	80.56 (64.19)	83.04 (66.16)	83.50 (66.55)	84.38 (67.04)	83.50 (66.28)	84.33 (66.93)	67.81 (54.20)
	07.50 (06.74)	07.96 (07.21)	19.79 (18.14)	57.38 (49.64)	73.79 (59.75)	80.67 (64.16)	69.58 (57.15)	80.25 (63.84)	83.08 (65.85)	75.92 (61.32)	81.00 (64.58)	84.17 (66.83)	84.21 (66.97)	84.29 (67.04)	83.90 (66.69)	84.08 (66.89)	84.29 (67.53)	84.79 (67.49)	68.18 (54.89)

Variables	MC	Cont	Temp	MC*Temp	MC*Temp*Cont	MC*Temp*Cont*Days
S.Em ±	0.17	0.17	0.25	0.43	0.75	2.6
CD at 5%	0.34	0.34	0.49	0.84	1.47	5.09

* Values in parenthesis are arc sine transformed values

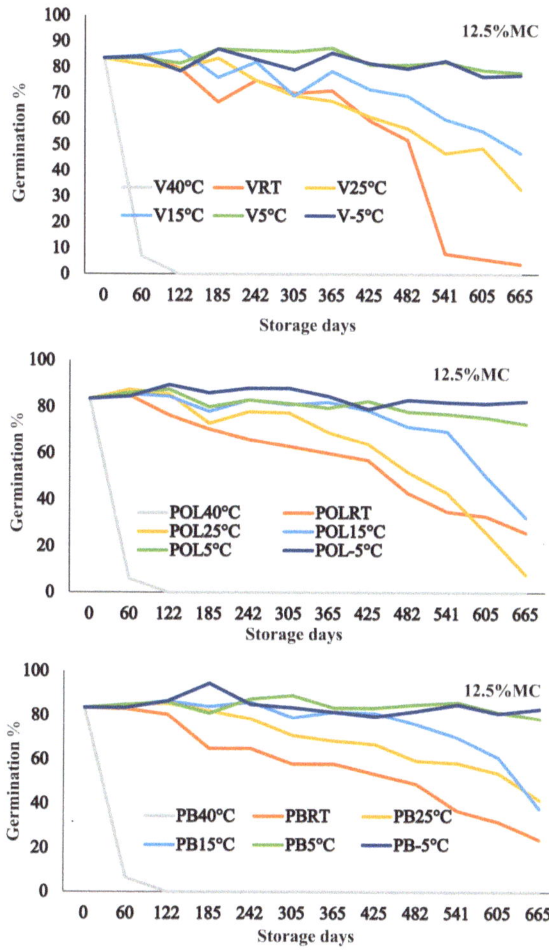

Fig. 2 Mean germination for the storage of seed stored at different temperatures (40, 25, 15, 5, –5 °C, and RT), containers (vacuum box (V), polythene bag (POL), and plastic box (PB)) at 12.50 percent moisture content for 665 days

Statistical analysis ($p = 0.05$) indicated that the mean germination of seeds stored under the following conditions was comparable to the maximum mean germination:

- Seeds with 5.60% moisture content stored in a vacuum box at 25 °C and 5 °C, in a polythene bag at ambient room temperature, 15 °C, and -5 °C, and in a plastic box at 25 °C, 15 °C, 5 °C, and –5 °C.
- Seeds with 8.00% moisture content stored in a vacuum box at –5 °C, in a polythene bag at –5 °C, and in a plastic box at 5 °C and –5 °C.
- Seeds with 12.50% moisture content stored in a polythene bag and plastic box at –5 °C.

After two years of storage, 70% germination was reported in *D. sissoo* seeds [27]. A study on the seed storability of twenty provenances of *D. sissoo* across northern India revealed that, in general, seeds from low- to medium-rainfall areas exhibited

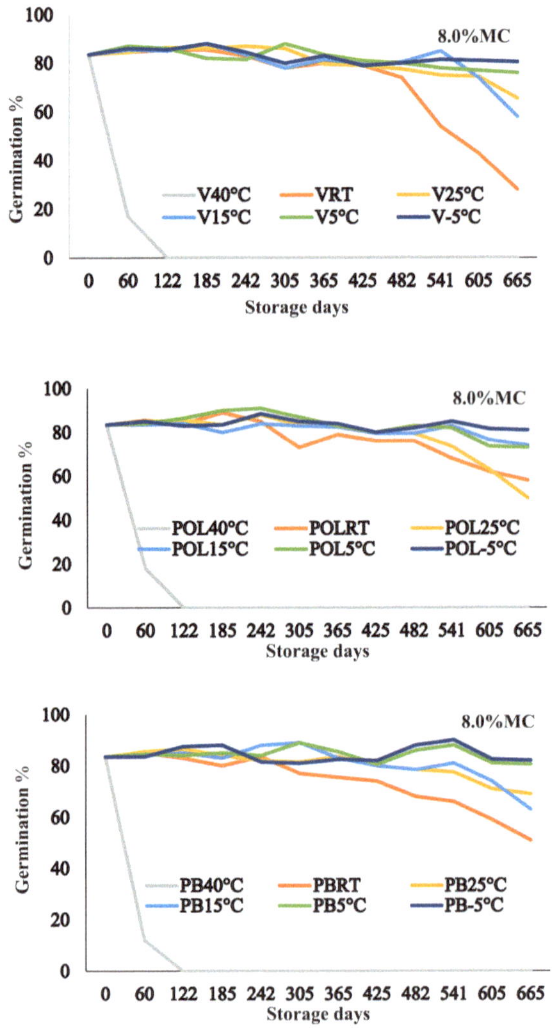

Fig. 3 Mean germination for the storage of seed stored at different temperatures (40, 25, 15, 5, –5 °C, and RT), containers (vacuum box (V), polythene bag (POL), and plastic box (PB)) at 8.0 percent moisture content for 665 days

relatively higher germination percentages [28]. After 18 months of storage, germination ranged from 75.3% in the Meerut provenance to 80.4% in the Hanumangarh provenance, with the latter showing the least reduction in germination percentage over this period [29]. For orthodox seeds, ultra-dried seeds can be stored effectively at room temperature under vacuum conditions in tightly sealed containers or bags impermeable to water vapour [11]. Storing seeds in airtight containers with a 3–7% moisture content is ideal for long-term preservation [50].

The survival curve indicated that seeds stored at 40 °C exhibited a steep slope, reflecting a rapid loss of viability across all seed moisture contents and storage containers. In general, seeds with a moisture content of 12.50% experienced the highest viability loss when stored at ambient room temperature, followed by those

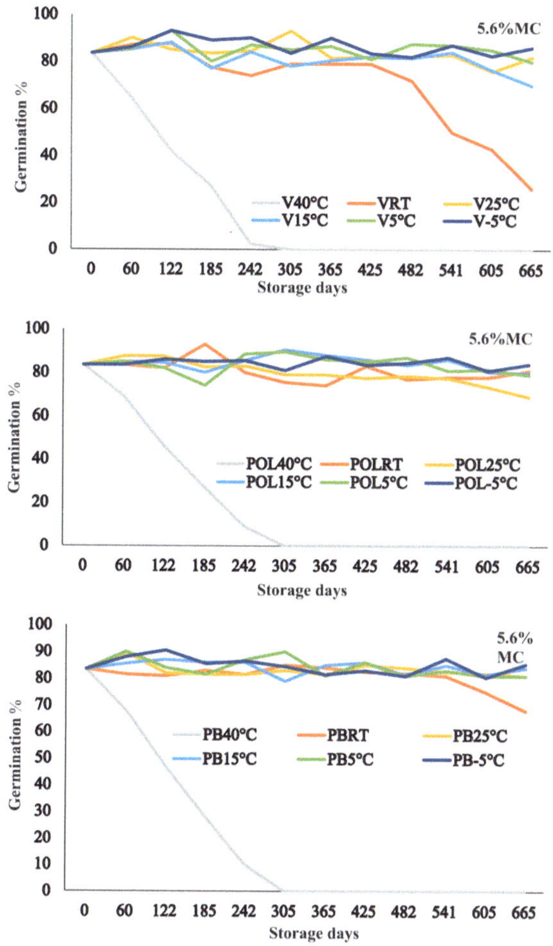

Fig. 4 Mean germination for the storage of seed stored at different temperatures (40, 25, 15, 5, –5 °C, and RT), containers (vacuum box (V), polythene bag (POL) and plastic box (PB)) at 5.6 percent moisture content for 665 days

stored at 25 °C and 15 °C. Conversely, seeds stored at 5 °C and –5 °C with a moisture content of 5.60% exhibited minimal viability loss, as reflected by an almost straight survival curve (Fig. 5). The rate of viability loss (-V/D) showed a clear trend: as storage temperature and seed moisture content decreased, the rate of viability loss also declined. The highest rate of viability loss was observed at 40 °C across all containers and moisture levels. For seeds with a 12.50% moisture content, the maximum viability loss rate (0.143) was recorded in vacuum-sealed storage at 40 °C, while the minimum (0.0004) was observed in seeds stored in polythene bags at –5 °C. Similarly, for seeds with a 5.60% moisture content, the maximum loss rate (0.135) occurred in vacuum-sealed storage at 40 °C, while the minimum (0.0002) was recorded for seeds stored in polythene bags at 5 °C. The overall trend confirmed that reducing storage temperature from 40 °C to –5 °C and seed moisture content from 12.50 to 5.60% significantly slowed viability loss (Table 3).

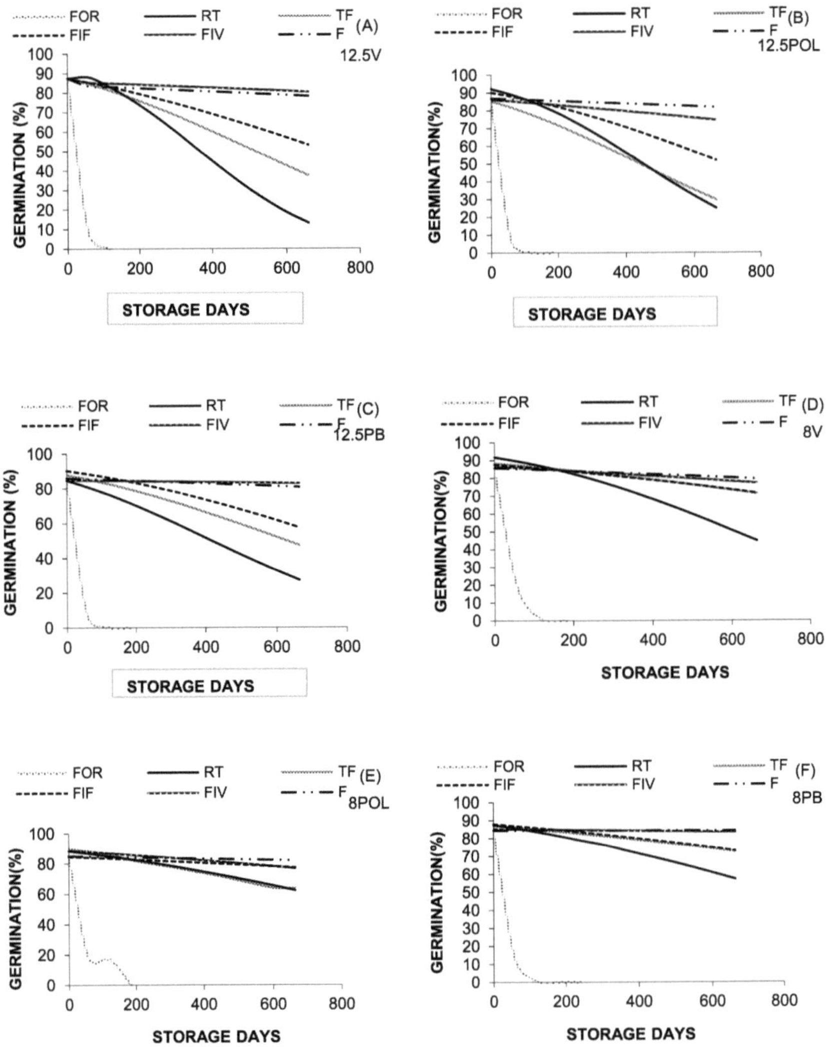

Fig. 5 **A–I** Survival curves of seed lot in vacuum box (V), polythene bag (POL), and plastic box (PB) at 12.50% MC (A, B, C), 8.00% MC (D, E, F), and 5.60% MC (G, H, I) at different temperatures. *Legend*: FOR = 40°C, RT = Room Temperature, FIF = 15°C, FIV = 5°C, and F = -5°C

The survival curve shape and the viability loss rate (-V/D) indicate that *D. sissoo* seeds exhibit increased longevity with decreasing seed moisture content and storage temperature. Based on these findings, it can be concluded that seed viability and vigour decline progressively with increasing storage duration, seed moisture content, and storage temperature. It is recommended that survival curve can be drawn portraying loss of germination from storage data under different storage conditions

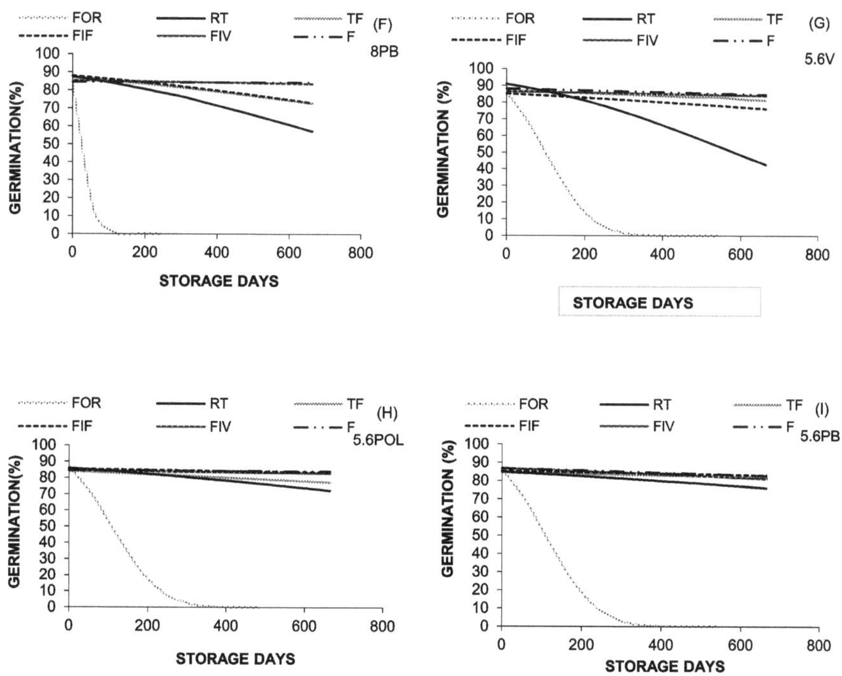

Fig. 5 (continued)

and storage period, it can further aid in extrapolating different parameters depicting longevity [51].

The storage behaviour of *D. sissoo* seeds aligns with the definition of "orthodox" seeds as proposed by [48], given their ability to maintain viability at low temperatures (up to –5 °C) and low moisture levels (as low as 5.6%). The ability to survive under these conditions underscores the importance of proper seed storage for the long-term conservation of valuable genetic resources in gene banks [52].

4 Seed Deterioration

Seed deterioration refers to an irreversible decline in seed quality once it has reached its maximum potential. A search for reliable deterioration indices revealed that, regardless of the underlying causes, certain quantifiable changes occur as seeds degrade. In *D. sissoo*, these changes were analysed to estimate the extent of deterioration.

Viability tests were conducted after 18 months of storage to assess the changes in stored *D. sissoo* seeds. Eight seed samples with varying viability percentages

Table 3 Rate of loss of viability (-V/D) of seed stored under different combinations of storage temperature, container, and seed moisture content

Temperature (°C)	Container	Moisture content (%)		
		12.5	8.0	5.6
40	Vacuum box	0.143	0.142	0.135
	Polythene bag	0.143	0.142	0.133
	Plastic box	0.143	0.143	0.133
RT	Vacuum box	0.066	0.030	0.033
	Polythene bag	0.051	0.018	0.006
	Plastic box	0.049	0.022	0.007
25	Vacuum box	0.040	0.011	0.001
	Polythene bag	0.049	0.016	0.010
	Plastic box	0.031	0.009	0.002
15	Vacuum box	0.026	0.010	0.007
	Polythene bag	0.025	0.006	0.0007
	Plastic box	0.020	0.008	0.001
+ 5	Vacuum box	0.002	0.006	0.002
	Polythene bag	0.008	0.004	0.0002
	Plastic box	0.0005	0.0008	0.001
-5	Vacuum box	0.006	0.001	0.003
	Polythene bag	0.0004	0.001	0.0009
	Plastic box	0.0007	0.005	0.0008

(96.0, 88.0, 76.0, 68.0, 54.0, 22.0, 16.0, and 0.0%) were selected for further analysis. The study documented changes in seed coat colour, seedling length (cm), the electrical conductivity of seed leachates (mhos g^{-1}) [53], sugar and starch content [54], protein content [55], and respiration rate (μmol mg^{-1} s^{-1}). The respiration rate was measured using a LI-Cor Model 6200n portable photosynthesis system, which included a Model 6250 infrared gas analyser and a 0.251-L chamber (LI-Cor). 100 seeds were carefully weighed and placed in the chamber for each sample. The respiration rate was recorded at a CO_2 concentration of approximately 360 ppm and 40% relative humidity (RH), calculated based on the rate of CO_2 increase within the chamber. Additionally, ultrastructural changes in seed cells at viability levels of 96.0%, 54.0%, and 0.0% were observed using Transmission Electron Microscopy (TEM) at the Central Drug Research Institute (CDRI), Lucknow, India.

In *D. sissoo*, physical seed deterioration can be assessed by visual examination of the seed coat colour—it transitions from light brown to dark brown as germination declines from 96.00% to 0.00% (Fig. 6). Previous studies have indicated that seed coat colour serves as an indicator of seed deterioration [56–59]. These colour changes are likely due to oxidative reactions in the seed coat, which are accelerated under high temperature and humidity conditions.

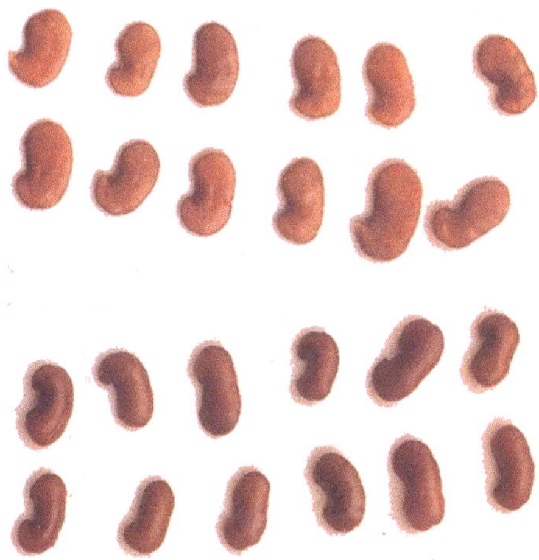

Fig. 6 Change in colour of seed from light brown to dark brown with loss in viability

A decline in seed viability was accompanied by a reduction in seedling length, decreasing from 9.07 cm at 96.00% germination to 2.45 cm at 16.00% germination (Fig. 7a). Reduced germination and delayed seedling emergence are among the earliest symptoms of deterioration followed by slower seedling growth and development [60]. As deterioration progresses, germination percentage decreases, weak seedlings are produced, and ultimately, seed viability is lost [61, 62].

Physiological assessment of viability loss was conducted through quantitative analysis of respiration and electrical conductivity. Similar methods has been used to evaluate seed quality loss during ageing [63, 64]. Respiration, measured as CO_2 release, increased from 0.049 μmol mg^{-1} s^{-1} in seeds with 96.00% viability to 0.268 μmol mg^{-1} s^{-1} in non-viable seeds (Fig. 7b). As seed deterioration advances, respiration gradually weakens, ultimately leading to complete germination failure [59].

Electrical conductivity is widely used as a vigour testing method and can also serve as a rapid viability assessment tool. In *D. sissoo*, conductivity increased from 0.10 to 0.58 mhos g^{-1} as seed viability declined from 96.00% to 0.00% (Fig. 7b). Previous studies have shown that deteriorated and damaged seeds leak more cellular material compared to vigorous and intact seeds [56, 65, 66]. A similar increase in conductivity with declining viability has been reported in *D. sissoo* [67].

Biochemical parameters serve as key indicators of seed viability and vigour. In *D. sissoo*, sugar, starch, and protein levels exhibited a declining trend with a loss of viability. Starch content decreased from 152.3 to 70.82 mg g^{-1} fresh weight, while sugar content declined from 256.2 to 108.2 mg g^{-1} fresh weight as viability dropped from 96.00% to 0.00% (Fig. 7c). Protein content also decreased, from 2.95 mg g^{-1} fresh weight in seeds with 96.00% viability to 2.34 mg g^{-1} fresh weight in non-viable seeds (Fig. 7c). Loss of viability is associated with protein degradation, a phenomenon previously documented in ageing seeds [11, 68, 69].

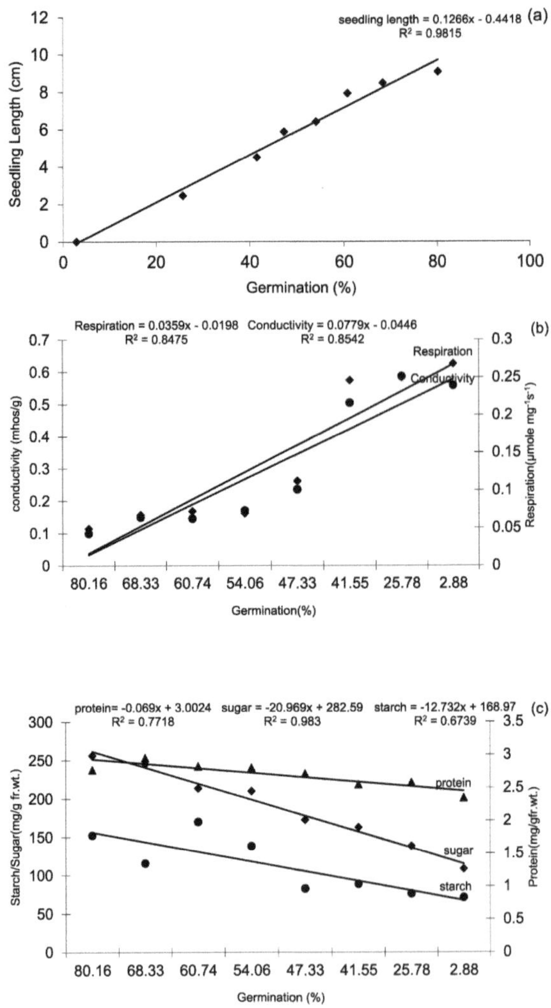

Fig. 7 a–c Regression equation showing relationships of germination with seedling length **a**, respiration and conductivity **b**, and protein, sugar, and starch **c** during deterioration

Physiological and biochemical deterioration in ageing seeds corresponds with ultrastructural changes at the cellular level (Figs. 8 and 9). As seeds aged, all organelles exhibited degradation. Cell walls appeared rigid and smooth in seeds with 96.00% viability, with a uniform deposition of granular material. With viability loss, cell walls became wavy, lost rigidity, and showed disrupted granular deposition. These structural changes contributed to the increased conductivity observed in aged seeds. The rise in conductivity is attributed to cell membrane disintegration, which is linked to a reduction in membrane phospholipids—a process commonly associated with seed deterioration, resulting from phospholipase enzyme activity or lipid peroxidation [70]. Membrane degradation during seed ageing has been documented across multiple species [59, 71–74].

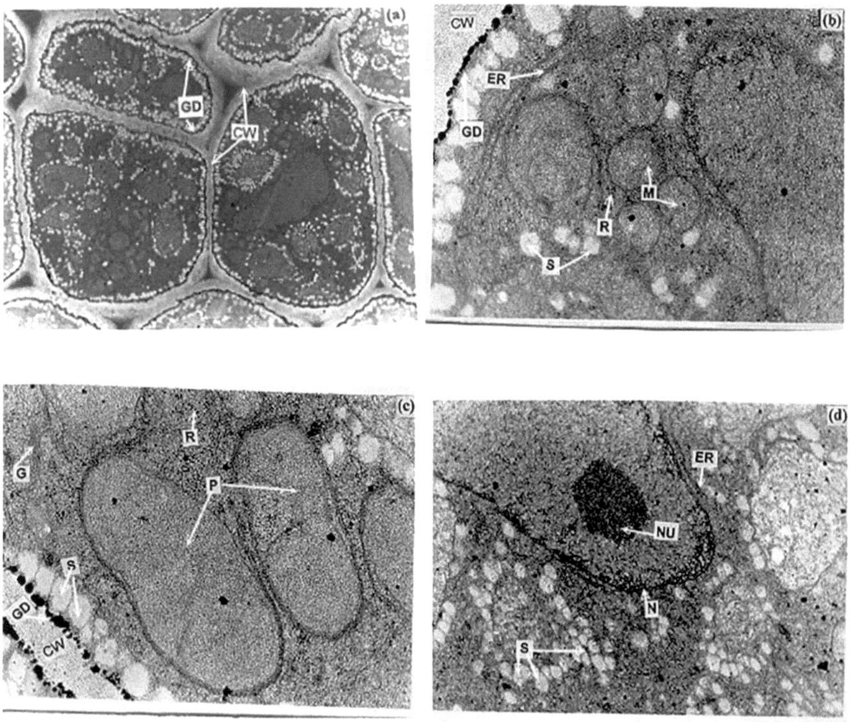

Fig. 8 **a–d** Radical tip of seed from seed lot with 96.00 per cent viability showing **a** general view of Cells (X 3000); **b** cell wall (CW) with granular deposition (GD), ribosomes (R), mitochondria (M), and starch (S) granules in clear matrix (X 21,000); **c** cell wall (CW) with granular deposition (GD), ribosomes (R), golgi bodies (G), Plastid (P), along with other cell organelles (X 21,000); **d** Nucleus (N), nucleolus (Nu) with clear nuclear matrix, mitochondria, plastids, endoplsmic retiulum, and starch granules (X 14,000)

Previous studies have primarily linked the increase in conductivity with viability loss to membrane disruption. However, our findings indicate that cell membrane and cell wall damage contribute to declining viability. Cytoplasmic metabolites leach into the intercellular space as the cell wall deteriorates alongside the membrane. Additionally, an increase in CO_2 release correlates with viability loss, reinforcing the close association between respiration rate and seed deterioration. This phenomenon is likely linked to the breakdown of membrane structures, particularly in mitochondrial cristae, which reduces overall respiration efficiency [74]. Ultrastructural analysis of aged seeds reveals swollen mitochondria and disintegration of cristae (Fig. 9b), consistent with the observation that mitochondria are among the first organelles to deteriorate during ageing [75]. The mitochondrial respiratory chain generates reactive oxygen species (ROS), leading to membrane damage, organelle structural deterioration, increased seed leachate loss, reduced respiratory rates and ATP production, and diminished enzymatic activity [76].

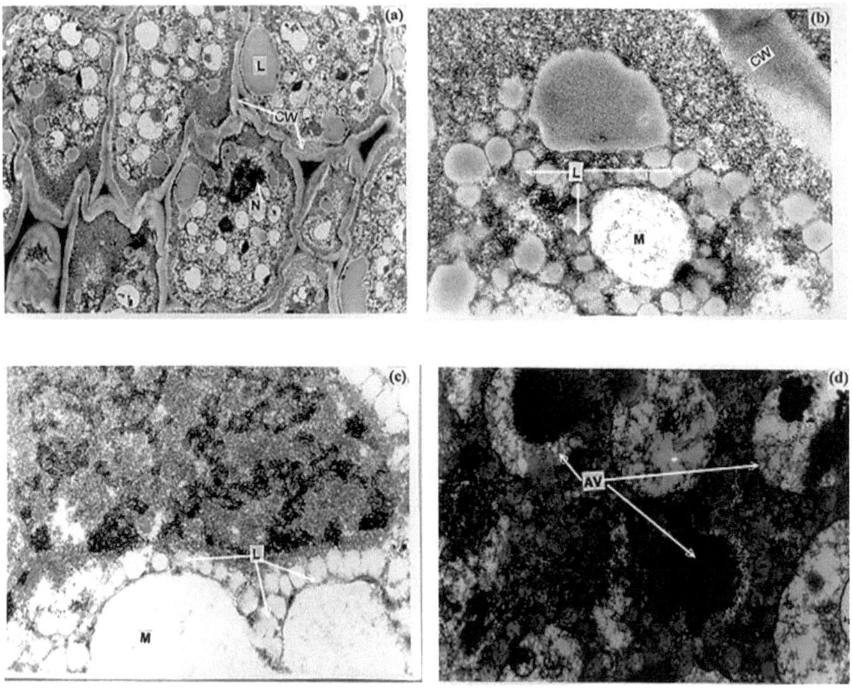

Fig. 9 **a–d** Radical tip of seed from non-viable seed lot showing **a** general view of tissue having disintegrated cell wall (X 3000); **b** cell wall without granular deposition, swollen mitochondria (M), dense lipid (L) bodies, and dense cytoplasmic matrix (X 21,000); **c** disintegrated nucleus with dense matrix, swollen mitochondria (M) surrounded with lipid bodies (X 21,000); **d** autophagic vacuole (AV), dense lipid bodies, and swollen mitochondria in dense cytoplasmic matrix (X 7100)

Along with mitochondria, plastids are among the earliest organelles exhibiting age-related changes. In seeds with 96.00% viability, plastids appear as large double-membraned structures with a clear matrix. However, in aged seeds, plastids gradually disappear. Other organelles, such as the endoplasmic reticulum, Golgi bodies, and ribosomes, were also absent in aged cells. The nucleus becomes irregular, exhibiting lobe formation in seeds with 54.00% viability, and eventually, nucleolus and nuclear membrane degeneration occur in non-viable seeds. Structural changes such as mitochondrial swelling, plastid disappearance, early disintegration of dictyosomes, distortion of the endoplasmic reticulum, and nuclear lobbing have been documented in ageing seeds [76, 14].

In aged cells, lipid bodies and autophagic vacuoles were dispersed throughout the cytoplasmic matrix (Fig. 9c and d). In seeds with 54.00% viability, lipid bodies appeared small, whereas in non-viable seeds, their coalescence was evident. Similar lipid body coalescence has been reported in pine [77]. Oleosins are the most common lipid protein that stabilises lipid bodies during the dehydration stage of seed maturation. Lack of these oleosins leads to a fatal coalescence of lipid bodies after germination [78]. Extensive vacuolation and vacuole fusion have also been observed in *Azadirachta indica* [76] and *Avicennia marina* [14]. This progressive membrane deterioration, beginning with mitochondrial and plastid damage and culminating in tonoplast dissolution, likely contributes to the decline in seed vigour and eventual loss of viability.

Acknowledgements The work was part of ICFRE-Forest Research and Extension Project (FREEP) on "Storage of Forest Tree Seed" funded by the World Bank, and the same is acknowledged. We thank CDRI Lucknow for support in performing TEM work.

References

1. Harrington JF (1970) Seed and pollen storage for conservation of plant gene resource. In: Genetic Resources in plants, their exploration and conservation, Handbook No. 11, International Biological Programme, London
2. Harrington JF (1972) Seed storage and longevity. In: Seed Biology, Kozlowski TT, ed, Academic Press, Chicago 145–245
3. FAO/IBPGR (1992) Report of the expert consultation on genebank standards. Rome: International Board for Plant Genetic Resources
4. Ranganathan U, Groot SPC (2023) Seed longevity and deterioration. In: Seed science and technology: biology, production, quality. Springer Nature, Singapore 91–108
5. Sripathy KV, Groot SPC (2023) Seed development and maturation. Seed science and technology: biology, production, quality. Springer Nature, Singapore 17–38
6. Leprince O, Pellizzaro A, Berriri S, Buitink J (2017) Late seed maturation: drying without dying. J Exp Bot 68(4):827–841
7. Shaheb MR, Islam MN, Nessa A, Hossain MA (2015) Effect of harvest times on the yield and seed quality of French bean. SAARC J Agriculture 13(1):1–13
8. Bewley JD, Bradford KJ, Hilhorst HWM, Nonogaki H (2013) Seeds: physiology of development, germination and dormancy, 3rd edition. Springer Science & Business Media, New York, USA 381
9. Zinsmeister J, Berriri S, Basso DP, Ly-Vu B, Dang T, Lalanne D, Da Silva EAA, Leprince O, Buitink J (2020) The Seed-specific heat shock factor A9 regulates the depth of dormancy in *Medicago truncatula* via ABA signalling. Plant, Cell Environ 43(10):2508–2522 https://doi.org/10.1111/pce.13853
10. Ellis RH, Hong TD (2006) Temperature sensitivity of the low-moisture-content limit to negative seed longevity–moisture content relationships in hermetic storage. Ann Bot 97(5):785–791
11. Corbineau F (2024) The effects of storage conditions on seed deterioration and ageing: How to improve seed longevity. Seeds 3(1):56–75
12. Waterworth W, Balobaid A, West C (2024) Seed longevity and genome damage. Biosci Reports 44(2)
13. McDonald MB (1999) Seed deterioration: physiology, repair and assessment. Seed Sci. Technol 27:177–237
14. Zhang K, Zhang Y, Sun J, Meng J, Tao J (2021) Deterioration of orthodox seeds during ageing: influencing factors, physiological alterations and the role of reactive oxygen species. Plant Physiol Biochem 158:475–485
15. Priestley DA (1986) Seed aging. In: Implications for seed storage and persistence in the soil. Cornell University Press- Ithaca, NY, USA 304

16. McDonald MB (2004) Orthodox seed deterioration and its repair. In: Handbook of Seed Physiology: Applications to Agriculture. The Haworth Reference Press, New York, USA 273–304
17. Nadarajan J, Walters C, Pritchard HW, Ballesteros D, Colville L (2023) Seed longevity—The evolution of knowledge and a conceptual framework. Plants 12(3):471
18. Fu YB, Ahmed Z, Diederichsen A (2015) Towards a better monitoring of seed ageing under ex situ seed conservation. Conserv Physiol 3(1): cov026
19. Walters C (1998) Understanding the mechanisms and kinetics of seed ageing. Seed Sci Res 8(2):223–244
20. Bailly C (2004) Active oxygen species and antioxidants in seed biology. Seed Sci Res 14:93–107
21. Come D, Corbineau F (2000) Storage of seeds. In: Improving postharvest technologies of fruits, vegetables and ornamentals. International Institute of Refrigeration, Paris, France 755–770
22. Roberts EH (1972) Storage environment and the control of viability. In Viability of Seeds. Chapman and Hall, London, UK, pp 14–58
23. Osborne DJ (1980) Senescence in seeds. In: Senescence in plants, CRC Press, Boca Raton, FL, USA, 13–37
24. Roberts EH (1988) Seed aging: The genome and its expression. In: Senescence and aging in plants. Academic Press, San Diego, CA, USA: 465–498
25. Li W, Niu Y, Zheng Y, Wang Z (2022) Advances in the understanding of reactive oxygen species-dependent regulation on seed dormancy, germination, and deterioration in crops. Front Plant Sci 13:826809
26. De Vitis M, Hay FR, Dickie JB, Trivedi C, Choi J, Fiegener R (2020) Seed storage: maintaining seed viability and vigor for restoration use. Restor Ecol 28:S249–S255
27. Robbins AMJ (1988) Storage of sissoo seeds. Bano Ki Janakari 2(1):57–59
28. Bangarwa KS (1993) Study on provenance variation, progeny testing and reproductive biology in *Dalbergia sissoo* Roxb. Ph.D. Thesis, CCS Haryana Agricultural University, Hissar, India
29. Khajuria S (2016) Studies on genotypic and phenotypic variations in *Dalbergia sissoo* Roxb. Ph.D. Thesis, CCS Haryana Agricultural University, Hissar, India
30. Joshi G, Thapliyal RC, Phartyal SS, Nayal JS (2008) Morphological and physiobiochemical changes in *Dalbergia sissoo* Roxb. Seed during its development. Indian Forester 134(4):563–573
31. Bhambri N, Malik CP (1982) Biochemical changes in developing seeds and pods of *Phaseolus vulgaris*. Plant Physiol 37:911
32. Bewley JD (2017) Seed maturation and germination. In: Reference module in life sciences. New York, USA: Springer: 1–9. https://doi.org/10.1016/B978-0-12-809633-8.05092-5
33. Aiazzi MT, Arguello JA, Rienzo JAD (1988) Physiological maturity in seeds of *Atriplex cordobensis* correlation with visual indicators. Seed Sci Tech 26(2):405–411
34. Ellis RH, Pieta-Filho C (1992) The development of seed quality in spring and winter cultivars of barley and wheat. Seed Sci Res 2:9–15
35. Finch-Savage WE, Bassel GW (2016) Seed vigour and crop establishment: extending performance beyond adaptation. J Exp Bot 67(3):567–591. https://doi.org/10.1093/jxb/erv490
36. Groot SPC (2022) Seed maturation and its practical implications. Seed Sci Techn 50(1):141–151
37. Rawat DCS (1998) Physiology and biochemical aspects of seed development in a recalcitrant tree seed, Doctoral Thesis H.N.B. Garhwal University, Srinagar, India
38. Farrant JM, Pammenter NW, Berjak P (1992) Development of recalcitrant (Homoiohydrous) seeds of *Avicennia marina*: anatomical ultra-structural and biochemical events associated with development from histodifferentiation to maturation. Ann Bot 70:75–86
39. Bewley JD, Black M (1985) Seeds: Physiology of development and germination. Plenum Press, New York
40. Oliveira GE, Pinho RG, Andrade T, Pinho EVR, Santos CD, Veiga AD (2013) Physiological quality and amylase enzyme expression in maise seeds. Ciência e Agrotecnologia 37:40–48
41. Bonner FT (1976) Maturation and collection of Yellow-Poplar seeds in the Midsouth. USDA Forest Service Research Paper SO-121

42. Bonner FT (1976) Maturation of shumard and white oak acorns. Forest Science 22:149–154
43. Adams CA, Fjerstad MC, Rinne RW (1983) Characteristics of soybean seed maturation: Necessity for slow dehydration. Crop Sci 23:265–267
44. Rasyad A, Van Sanford DA, Tekrony DM (1990) Changes in seed viability and vigour during wheat seed maturation. Seed Sci Tech 18:259–267
45. Long SR, Dale RMK, Sussex IM (1981) Maturation and germination of *Phaseolus vulgaris* embryonic axes in culture. Planta 153:405–415
46. Tanaka A, Ito H (2025) Chlorophyll degradation and its physiological function. Plant Cell Physiol 66(2):139–152. https://doi.org/10.1093/pcp/pcae093
47. ISTA (1993) International Society for Seed Testing Rules. Seed Sci and Techn 211–287
48. Roberts EH (1973) Predicting the storage life of seeds. Seed Sci Tech 1:499–514
49. Vertucci, C.W., Farrant, J.M.: Acquisition and loss of desiccation tolerance. Seed development and germination, 237–271 (2017)
50. FAO/IPGRI Genebank Standards (1994) Food and agriculture organization of the United Nations, Rome, International Plant Genetic Resources Institute, Rome (1994) ftp://ftp.fao.org/docrep/fao/meeting//aj680e.pdf
51. Buitink J, Leprince O (2024) A seed storage protocol to determine longevity. Methods Mol Biol 2830:63–69. https://doi.org/10.1007/978-1-0716-3965-8_6
52. Li DZ, Pritchard HW (2009) The science and economics of ex situ plant conservation. Trends Plant Sci 14:614–621
53. Bonner FT (1991) Estimating seed quality of southern pines by leachate conductivity. USDA Forest Service Research Paper SO-263. USDA Forest Service, Southern Forest Experiment Station, New Orleans
54. Dubois MK, Gills K, Hamiton JK, Rebers PA, Smith F (1951) A colorometric method for determination of sugars. Nature 168–167
55. Bradford MM (1976) A rapid and sensitive method for the estimation of microgram quantities of protein utilising the principle of protein-dye binding. Anal Biochem 72:248–254
56. Ghive DV, Barabde NP, Pote SR (2007) Seed viability and factors affecting seed storage. Asian J Bio Sci 2(2):202–208
57. Copeland LO, McDonald MB (2012) Principles of seed science and technology. Chapman and Hill, New York, USA
58. Sano N, Rajjou L, North HM, Debeaujon I, Marion-Poll A, Seo M (2016) Staying alive: molecular aspects of seed longevity. Plant Cell Physiol 57(4):660–674
59. Choudhury A, Bordolui SK (2023) Concept of seed deterioration: reason, factors, changes during deterioration and preventive measures to overcome seed degradation. Am Int J Agric Stud 7(1):41–56
60. Copeland LO, McDonald MB (1985) Seed longevity and deterioration. Principles of seed science and technology. Macmillan Publishing Company, New York
61. Maity RK, Limon SM, Wesche P, Beling E (2000) Responses of some crops to various abiotic stress factors and its physiological and biochemical basis of resistances: a review. Agric Rev 21(3):155–167
62. Tilebeni GH, Golpayegani A (2011) Effect of seed ageing on physiological and biochemical changes in rice seed (*Oryza sativa* L). Int J AgriScience 1(3):138–143
63. Egli DB, White GM, Tekrony DM (1978) Relationship between seed vigour and the storability of soyabean seeds. J Seed Techn 3(2):1–11
64. Marcos–Filho J, Carvalho RW, Cicero SM, Demetrio CGB (1988) Physiological quality and soyabean seed performance during storage and in the field. Anasi da E.S.A. 'Luiz de Querioz' 43:389–443
65. Mirdad Z, Powell AA, Matthews S (2006) Prediction of germination in artificially aged seeds of *Brassica* spp. using the bulk conductivity test. Seed Sci Technol 34(2):273–286
66. Powell AA (2022) Seed vigour in the 21st century. Seed Sci Technol 50(2):45–73
67. Thapliyal RC, Connor KF (1997) Effects of accelerated ageing on viability, leachate exudation, and fatty acid content of *Dalbergia sissoo* Roxb seeds. Seed Sci Technol 25(2):311–319

68. Nautiyal AR, Thapliyal AP (1984) Changes in the proteins and Isoenzymes during imbibition of viable and non-viable seeds of soyabean. Pl. Physical and Biochem. 11(1):20–25
69. Sugliani M, Rajjou L, Clerkx EJ, Koornneef M, Soppe WJ (2009) Natural modifiers of seed longevity in the *Arabidopsis* mutants abscisic acid insensitive3-5 (abi3-5) and leafy cotyledon1-3 (lec1-3). New Phytol 184(4):898–908
70. Vidhya CJ, Nethra N (2022) Seed deterioration: An overview. Seed Research 50(2):91–99
71. Jyoti J, Malik CP (2013) Seed deterioration: a review. Int J Life Sci Bot Pharm Res 2:374–385
72. Dawidowicz–Grzegorzewska A, Podstolski A (1992) Age–related changes in the ultrastructure and membrane properties of *Brassica rapus* L. seeds. Annals of Botany 69:39–46
73. Ebone LA, Caverzan A, Chavarria G (2019) Physiologic alterations in orthodox seeds due to deterioration processes. Plant Physiol Biochem 145:34–42
74. Wang Y, Li Y, Xue H, Pritchard HW, Wang X (2015) Reactive oxygen species-provoked mitochondria-dependent cell death during ageing of elm (*Ulmus pumila* L) seeds. Plant J 81(3):438–452
75. Berjak P, Campbell GK, Farrant JM, Omandi-Oloo W, Pammenter NW (1995) Responses of seeds of *Azadirachta indica* (neem) to short-term storage under ambient or chilled conditions. Seed Sci. and Technol. 23:779–792
76. Motete N, Pammenter NW, Bberjak P, Fredric JC (1997) Response of the recalcitrant seeds of *Avicennia marina* to hydrated storage: events occurring at the root primordial. Seed Sci Res 7:169–178
77. DeCastro MFG, Martinez-Hunduwilla CJ (1984) Ultrastructural changes in naturally aged in *Pinus pinea* seeds. Physiologia Plantarum 62:581–588
78. Murphy DJ, Hernández-Pinzón I, Patel K (2001) Role of lipid bodies and lipid-body proteins in seeds and other tissues. J Plant Physiol 158(4):471–478

Chapter 24
Mechanical Properties and Wood Drying Behaviour of *Dalbergia sissoo*

Shailendra Kumar, Shweta B. Kukreti, and Ismita Nautiyal

1 Introduction

Dalbergia sissoo, a multipurpose species known as shisham or sissoo, is native to Afghanistan, Bangladesh, Bhutan, India, Iran, Iraq, Myanmar, Nepal, Pakistan, the Philippines, and South Africa [1, 2]. The wood is hard, moderately heavy to very heavy, and having a specific gravity in the range of 0.63–0.83 (air dry). Its heartwood is golden brown to deep brown with darker streaks, distinct from the sapwood, which is yellowish-white or pale brownish. The grain is generally straight to shallowly interlocked, and the texture is medium coarse [3]. Its wood is strong and highly elastic, and some properties are comparable to teak, with some aspects even superior. The heartwood is exceptionally durable and requires no chemical treatment [4]. It exhibits excellent working characteristics and responds well to most machining operations, though sections with interlocked grain may tear out during surfacing. Despite this, Sissoo turns, glues, and finishes effectively. These exceptional properties make *D. sissoo* a first-class timber for cabinetry, furniture, panelling, and flooring, as well as for ornamental turnery, carving, and engraving, tool handles, and sporting goods [1, 4]. It is also used for marine and aircraft grade plywood, decorative plywood, and block boards. A material undergoes a significant transformation from its raw state to the final product state through various phases, including sawing, drying, treatment, woodworking, and finishing. It is essential to understand its behaviour throughout

S. Kumar
Forest Products Division, Forest Research Institute Dehradun, Dehradun, Uttarakand, India

S. B. Kukreti
College of Forestry, Veer Chandra Singh Garhwali Uttarakhand University of Horticulture and Forestry, Ranichauri, Uttarakhand, India

I. Nautiyal (✉)
Research and Planning Division, Indian Council of Forest Research & Education, Dehradun, Uttarakhand, India
e-mail: ismita06@gmail.com

R. R. Warrier et al. (eds.), *Handbook of Asian Rosewoods*, Materials Horizons: From Nature to Nanomaterials, https://doi.org/10.1007/978-981-96-9459-4_24

these phases, especially during drying and under different stresses, to make the best use of its exceptional properties. This chapter will discuss the mechanical properties of *D. sissoo* and its drying characteristics to comprehensively understand *D. sissoo* as a timber.

2 Mechanical Properties of Timber

Mechanical properties of wood define its behaviour under various loading conditions, which plays critical role in structural design, material selection, and product development. It is important to comprehend how timber reacts to external forces or different stress types since assessing stress amount and distribution within timber components is vital for evaluating its structural integrity, which further leads to the calculation of its suitability indices and safe working stress. These indices are the basis for further timber grading and classification. Stress is a distributed force (internal & external), basically, the force per unit area applied to timber through various forms like tensile stress (pulling), compressive stress (pushing), and shear stress (sliding). These three basic stresses can combine to create compound stresses in wood, as seen in flexure [5]. The mechanical properties of wood are well described in different editions of wood handbook [6–8]. Mechanical properties, typically expressed as strength properties, encompass various aspects such as modulus of rupture in bending, maximum stress in compression parallel to grain, compressive stress perpendicular to grain, and shear strength parallel to grain. Additional measurements may include maximum load in bending, impact bending strength, tensile strength perpendicular to the grain, and hardness [7, 8]. Being orthotropic in nature, wood exhibits distinct properties along three perpendicular axes: longitudinal, radial, and tangential [8]. This unique behaviour allows it for various end uses. These properties vary between species and within a species due to natural variability. Understanding these properties is crucial for applications where wood is being used for structural purposes.

2.1 *Methods for Mechanical Testing of* D. sissoo

The mechanical strength of sissoo (*D. sissoo*) has been extensively studied using various standardized testing methods used in various countries. ASTM D-143–94 was used to evaluate sissoo's mechanical properties, providing a comprehensive understanding of its strength characteristics [9]. The British Standard Institute BS-373, which focuses on the method of testing for small clear specimens of timber, was applied to provide further insights into the material properties [10]. In India, the testing follows the IS: 1708 [11], which covers a wide range of mechanical tests, ensuring a thorough assessment of sissoo's mechanical strength performed at small clear specimen free from any defect. Tariq et al. [12] followed ASTM D143-2014, and the Janka hardness test [13] was used to test the mechanical properties of Sissoo in Pakistan. It was mentioned that ASTM standard (1971) was used for timber testing in Bangladesh [14]. In Pakistan *D. sissoo* is taken as standard species for comparing other timbers [15], and strength properties of wood were determined using ASTM D-143–94 and ASTM D-2395–93 procedures [16]. In China, national standard GB1927-1943–91 (the testing method of physical and mechanical properties of test method) for timber testing was followed [17]. These diverse standards allow for a robust comparison and validation of Sissoo's mechanical properties across different regions and methodologies.

2.2 Mechanical Properties of D. sissoo

The mechanical properties of wood can vary significantly based on factors such as species, growth conditions, and geographic location. In wood, bending strength is typically represented by Modulus of Rupture (MOR), which indicates the maximum load-carrying capacity of a member in bending [7]. However, wood's combined strength and toughness under bending stresses can be measured as work to maximum load in bending. *D. sissoo* displayed differing stiffness levels across locations and moisture conditions (Table 1). Sample tested from India showed average MOR of 62,880–116,000 Kpa [18–20], whereas sissoo from Pakistan and Bangladesh exhibited average MOR of up to 138,953 kPa and 96,497 Kpa, respectively [10, 13] (Table 2). Several factors could contribute to these variations. The property of a material to revert to its original shape after the load's removal is termed as elasticity and expressed as the modulus of elasticity, or Young's modulus. The elastic properties (modulus of elasticity and Poisson's ratio) are crucial for predicting deflections, whereas strength properties (modulus of rupture and shear strength) ensure safe designs, and wood's Poisson's ratio indicates how it deforms laterally when compressed or stretched [7]. In sissoo, the average MOE was recorded in the range of 6549 to 11,790 MPa [10, 14, 18–20]. This property is crucial for applications where the wood will be subjected to bending forces, such as in construction and furniture making. Stiffness contrasts with elasticity, as wood maintains its natural size and shape, resisting deformation [5]. The wood that is easily bent is flexible and that difficult to bend is stiff.

Different stress types result in specific strains, such as elongation under tension or shortening under compression. Compression strength of wood is the ability to bear the compressive load, which can be measured parallel to grain or perpendicular to grain, crucial for use as columns, furniture, props, and chair logs. The studies on sissoo showed average compressive strength of 29,616–69,000 kPa when tested parallel to grain and 6668–14,824 kPa in perpendicular to grain (Table 1). Some compression tests at elastic limit and MOE have also been performed [18].

In various applications such as agricultural implements, lorry bodies, bearings, and packing, shear strength plays a crucial role, as these structures can experience high longitudinal shear stresses. It determines how well wood resists forces parallel to its grain. It is relevant for joints, fasteners, and connections. *D. sissoo* also reflects

Table 1 Mechanical properties of *Dalbergia sissoo* samples collected from different locations in India

Location of sample	Sample condition	Specific gravity	Static bending		Compressive strength (KPa)				Shear strength parallel to grain (kPa)	Tensile strength Perpendicular (kPa)	Hardness (N)			Reference
			MOR (kPa)	MOE (MPa)	Comp. stress at elastic limit	Max. Crushing stress or Compression strength parallel	Modulus of elasticity	Comp. perpendicular stress at elastic imit			Radial	Tangential	End	
Hosiyarpur Punjab,	Green	0.678	62,880	7,005	17,995	32,130	7722	7239	R-11238 T-11100	R-8515 T-7481	6116	6250	5872	[18]
	Air dry	0.713	90,804	9266	26,821	50,883	10,432	12,135	R-13824, T-12480	R-7722, T-6412	7962	8407	8496	
	Kiln dry	0.725	85,701	9425	27,786	52,193	10,087	14,824	R-11238, T-13858	R-6722, T-9135	8118	8874	8763	
Jalpaiguri, Bengal,	Green	0.643	75,980	9,404	22,132	38,197	10,204	10,204	R-9756 T-9894	R-7929 T-7205	6561	6272	6294	
	Air dry	0.668	95,492	11,472	24,752	52,469	12,204	10,342	R-11307, T-10273	R-6619, T-6033	5916	6050	5560	
	Kiln dry	0.693	108,523	11,790	31,164	55,089	12,900	9722	R-14272, T-11170	R-5723, T-6550	7050	7117	7762	
Uttar Pradesh,	Air Dry	0.695	103,559	10,618	31,302	60,191	12,390	15,582	R-14720, T-12204	R-7136, T-5861	8763	8585	8563	[19]
	Kiln dry	0.695	105,834	11,259	30,751	63,363	11,535	15,444	R-15099, T-13548	R-9032, T-4792	8296	8763	8140	
Andreta (Kangra), HP	Oven dry	0.56	89,000	8591		64,000		38,000		32,000				[20]

(continued)

Table 1 (continued)

Location of sample	Sample condition	Specific gravity	Static bending		Compressive strength (KPa)				Shear strength parallel to grain (kPa)	Tensile strength Perpendicular (kPa)	Hardness (N)			Reference
			MOR (kPa)	MOE (MPa)	Comp. stress at elastic limit	Max. Crushing stress or Compression strength parallel	Modulus of elasticity	Comp. perpendicular stress at elastic imit			Radial	Tangential	End	
Baroh (Kangra), HP		0.604	113,000	8921		69,000		31,000		94,000				
Galore (Hamirpur), HP		0.604	98,000	9283		62,000		33,000		78,000				
Kangu (Hamirpur), HP		0.561	104,000	10,369		62,000		29,000		66,000				
Nalagarh (Solan, HP)		0.517	97,000	6935		56,000		29,000		68,000				
Chowkiwala (Solan), HP		0.576	84,000	7188		61,000		23,000		47,000				
Dattowal (Solan), HP		0.644	99,000	8716		53,000		25,000		39,000				
Ghumarwin (Bilaspur, HP)		0.559	88,000	7891		63,000		23,000		30,000				

(continued)

Table 1 (continued)

Location of sample	Sample condition	Specific gravity	Static bending		Compressive strength (KPa)				Shear strength parallel to grain (kPa)	Tensile strength Perpendicular (kPa)	Hardness (N)			Reference
			MOR (kPa)	MOE (MPa)	Comp. stress at elastic limit	Max. Crushing stress or Compression strength parallel	Modulus of elasticity	Comp. perpendicular stress at elastic imit			Radial	Tangential	End	
Sundernagar (Mandi), HP		0.58	116,000	9199		65,000		22,000		55,000				
Sarahan (Sirmaur, HP)		0.546	81,000	9297		46,000		32,000		76,000				

Table 2 Mechanical properties of *Dalbergia sissoo* from other Asian countries

Sample details	Specific gravity	Static bending		Compression (kPa)		Tensile strength kPa	
		MOR (kPa)	MOE (MPa)	Parallel to grain	Perpendicular to grain	Parallel to grain	Perpendicular to grain
Bangladesh general Green	0.65	54,917	6.374	29,616	7845	78,453	R-5394 T-6080
Bangladesh general Oven dry	0.74	96,497	10.78	49,327	8630	11,780	R-6668 T-6472
China	0.746		7470	46,500			
Faisalabad	0.738	112,973		9905	6668	66,881	
Gujranwal	0.69			34,000		12,480	
Zafarwal	0.75			36,000		21,650	
Peshawar	0.679	101,450	6549	33,880	12,992		2790
Swabi	0.669	92,163	8109	38,420	9695		2486
Kohat	0.79	113,192	15,375	41,455	14,266		2167
Dera Ismail Khan	0.788	138,953	10,221	47,976	14,402		4488
Pakistan general sample (@ 12% moisture)		110,030	8413	54,917			3530

(continued)

Table 2 (continued)

Impact Bending kgm/cm^2	Shear strength parallel	Hardness N		Nail Withdrawal (kg)		Toughness cm-kg		Cleavage kg/cm		Flexural strength	Reference
		Side	End	Side	end	Radial	Tangential	Radial	Tangential		
	R-10782 T-13395	6462	5790	83	365	365	403	69	64		[14]
	R-10337 T-11572	6884	6894	119	360	360	393	95	79		
		7070	7880							87.89 Mpa	[17]
601 kg/cm					173 kg / cm						[9]
		7880	6980								[12]
		6280	6340								
		6622	6945								[10]
		6367	7377								
		7814	8231								
		10,634	10,673								
1.79		6377	7848					22			[15]

Table 3 Comparative suitability index of *D. sissoo* in terms of Teak as taken 100 [22]

Species	Strength as beam	Suitability as post	Shock resistance ability	Surface hardness	Refractoriness (Splitting coefficient)	Nail and screw holding property
Dalbergia sissoo	95	90	130	120	80	115

good shear strength in the range of 9756–15,099 kPa in radial direction and 9894–13,858 kPa in tangential direction (Table 1). Similarly, tensile strength parallel to grain indicates wood's maximum resistance to tensile stress in that direction. Sissoo showed average tensile strength of 5394–9032 kPa and 4792–9135 kPa in radial and tangential directions respectively perpendicular to grain (Table 1). The hardness of sissoo was also tested and showed average hardness of 5916–8118 kPa in radial, 6050–8874 kPa in tangential direction, 6462–10,634 kPa in side, and 5560–10,673 kPa in end direction [10, 12, 14, 15, 17–19]. In packing cases, construction, and joinery work, high nail and screw holding power is an essential requirement. Sissoo exhibited 83–119 kg average nail withdrawal capacity in side and 360–365 kg in end direction (Table 1). Apart from this, impact bending test evaluates wood's ability to absorb shocks beyond the proportional limit, representing its resilience to sudden loads. In sissoo, based on this test average impact bending of 601 kg/cm was recorded [9].

Sissoo is categorized under "Super Group Timbers" [21], which requires a strength coefficient of 75 or higher and has a weight ranging from 85 to 110, with teak being the reference at 100. The comparative suitability index of sissoo as compared to teak is given in Table 3. It shows that some properties like shock resistance ability, surface hardness, refractoriness, and nail/screw holding capacity of sissoo are more than that of teak.

3 Drying Behaviour

Wood's characteristics are significantly impacted by water. Depending on the air temperature, relative humidity, and moisture content (MC) of the wood, it constantly exchanges moisture with its environment. The wood contains three different types of water. The liquid form of free water can be found in the wood's lumens or cell cavities. The components of the cell wall include bound water, whereas the air in cell lumens may also contain water vapour. When wet wood dries, free water exits the lumens first before bound water does. Up to fibre saturation point (FSP), water can be extracted quite easily from wood. Wood doesn't shrink until it dries to a MC lower than its FSP. For the majority of wood species, the FSP ranges from 25 to 30% MC [23]. Wood experiences expanding and shrinking when its moisture content falls below FSP. Wood swells as the moisture content rises, and shrinks when it falls.

Not every wood direction experiences the same dimensional changes. The tangential direction has the largest dimensional change. While longitudinal shrinkage (along the grain) is so tiny that it is typically ignored (0.1% to 0.3%), radial shrinkage is significantly smaller than tangential shrinkage. In general, the wood shrinks by 10% in the tangential direction and 6% in the radial direction. Equilibrium moisture content (EMC) is the moisture content level at which the wood reaches environmental equilibrium. The wood will therefore attain dimensional stability at EMC. For the most part, the EMC falls between 8 and 12% on average worldwide [23]. As a result, below and over EMC, the wood is dimensionally unstable. Therefore, in order to ensure that there are as few dimensional changes as possible as a result of water absorption or desorption in wood, it is necessary to bring the MC percentage of the wood into the range of EMC's (8–12%) before producing any high-quality product. Wood drying, also known as seasoning, is the process of reducing the MC percentage of wood to EMC's or a desired level in controlled conditions.

3.1 Fibre Saturation Point, Equilibrium Moisture Content, and Hygroscopicity

The term "fibre saturation point," which is particularly used in the context of wood drying, refers to the stage of the drying process at which only water that is bonded to the cell walls is still present; all other water, referred to as "free water," has been forced out of the cell cavities. Additional drying of the wood strengthens its fibres and is accompanied by shrinkage.

Wood shrinks almost linearly up to 0% moisture content, beginning at the fibre saturation threshold, or approximately 30% moisture content. On the other hand, the wood's surface dries first during the typical drying process of timber. The surface shrinks as it dries below the fibre saturation point. However, the inside is still at higher MC and has not shrunk. As a result, timber may start to shrink before its overall average moisture content falls below the fibre saturation limit. This also implies that shrinkage will start early for the timbers having higher FSP as compared with those with lower FSP. Thus, lower FSP gives an indication of the stability of the timbers.

Fibre Saturation Point values of some common Indian timber species are depicted in Fig. 1 [23]. It may be seen that *Tectona grandis,* which is most sought-after timber has very low FSP (19%) indicating its stability. *Mangifera indica,* a common timber used in handicrafts, has FSP value of 22%. Other timbers like *Ailanthus excelsa, Toona ciliata,* etc. have higher FSP values. *D. sissoo*, compared with these timbers, has relatively lower FSP (20%) and is very close to that of teak. This indicates the stability of *D. sissoo*. Equilibrium moisture content of some common Indian woods along with *D. sissoo* at 35 °C and 96%, and 30% relative humilities is present in Fig. 2 [24]. Teak showed minimum EMC values at both the relative humidities, i.e. 16% and 4.3%. Lower EMC in a saturated air indicates that the wood will not shrink or

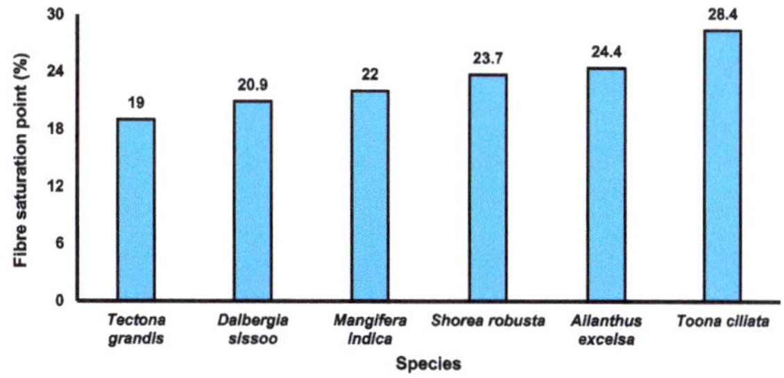

Fig. 1 Fibre saturation point of a few common Indian species

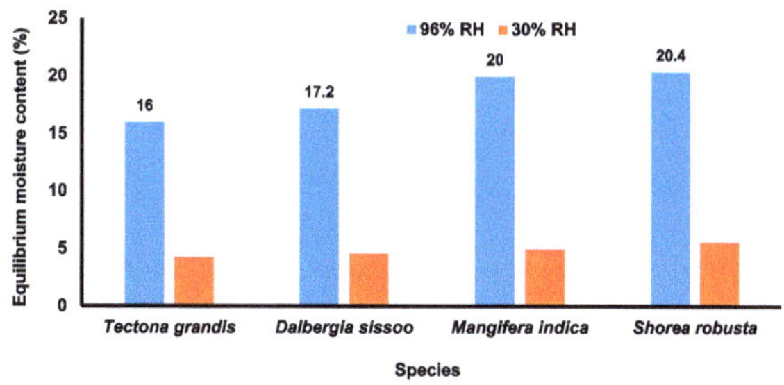

Fig. 2 EMC of some common Indian timbers at 35 ° C and 96% and 30% RH [24]

swell much when the temperature and humidity change. *D. sissoo* with EMCs 17.2% and 4.6% shows little higher EMC than teak but considerably lower than other two species (*M. indica* and S. *robusta*). Relatively lower EMC of *D. sissoo* indicates the species' suitability for producing high-quality products.

3.2 Anisotropy in Shrinkage and Swelling

Shrinkage behaviour of some common Indian woods is presented in Fig. 3 [23]. Out of these species, *T. grandis* wood shows least shrinkage in both the directions, i.e. radial (2.2%) and tangential (3.9%). The volumetric shrinkage of teak is 5.5%. Teak wood is considered to be most dimensionally stable as its total shrinkage is considerably lower than other species. As shown in Fig. 3, *D. sissoo* appears to be closest to the teak wood shrinkage characteristics. It has 2.8%, 4.9%, and 7.9% shrinkage in radial, tangential, and volumetric shrinkage, respectively. *Mangifera indica*, a common wood currently being used by Indian wood-based handicrafts industries, appears to be closest to *D. sissoo* shrinkage behaviour. Other wood species show relatively higher shrinkage properties. Thus, with the lower shrinkage properties, *D. sissoo* appears to be a stable wood.

The tangential–radial shrinkage ratio, or T/R ratio, has been used to characterize the dimensional stability of wood [25]. Many workers have noted that wood with a lower T/R ratio is more dimensionally stable.

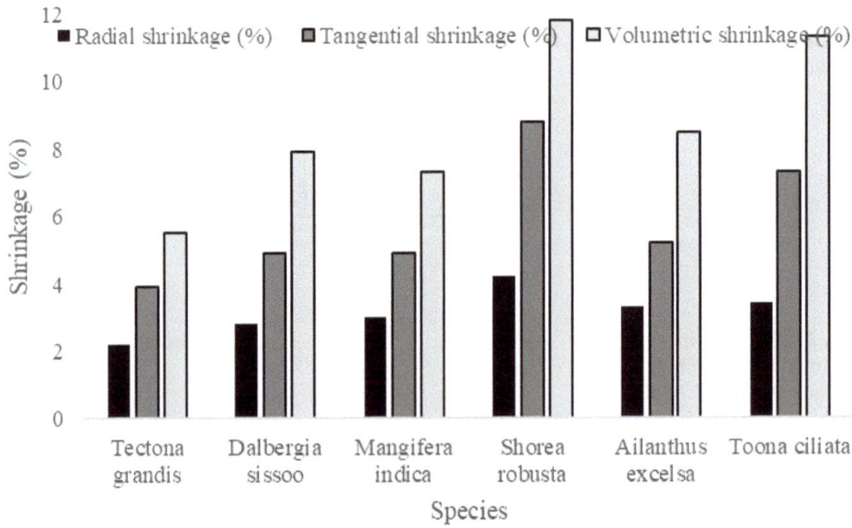

Fig. 3 Radial, tangential, and volumetric shrinkage of a few common Indian species

T/R ratio of *D. sissoo* along with *T. grandis, M. indica,* and *Ailanthus excelsa* is lower than 2 (Fig. 4), indicating the dimensional stability of the species. Anisotropic behaviour in dimensional changes of some Indian timber was evaluated by calculating anisotropic coefficient [26]. The Indian timbers were grouped into three major classes. The species which have anisotropic coefficient value below 60% have been considered to be dimensionally very stable (Class A), whereas above this value have been considered as less stable (Class B).

Figure 5 presents anisotropic coefficient of some common Indian wood species. As one can see, teak has the least anisotropic coefficient (58%) and next to it is *D. sissoo* (60%). The other two timbers show higher anisotropic coefficient. This fact explains why *D. sissoo* is traditionally preferred timber for high-quality products.

Fig. 4 Tangential–radial shrinkage ratio of a few common Indian species

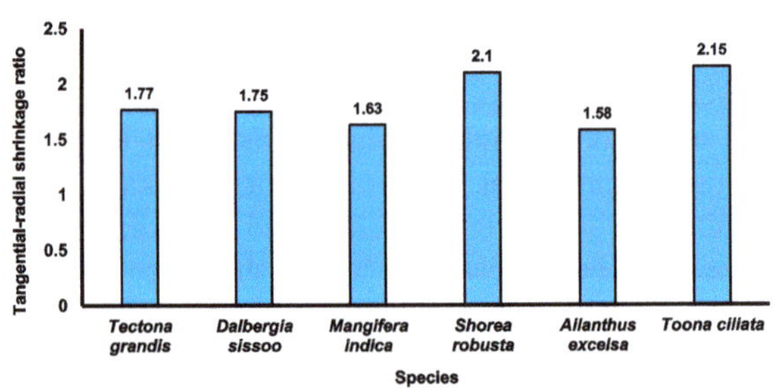

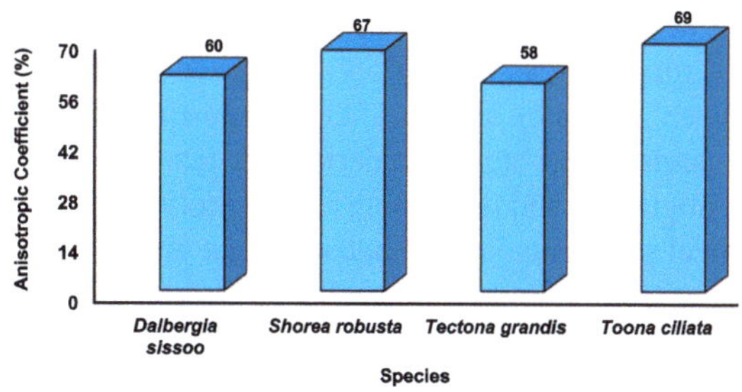

Fig. 5 Anisotropic coefficient of a few common Indian timbers [26]

3.3 Moisture Diffusion Coefficient of D. sissoo

The passage of water vapour through hardwoods occurs through a system of void spaces consisting of cell cavities, pit membrane pores, interstitial openings, and cell wall substance. Water vapour diffusion is a measure of the ease of conductance of moisture through wood. Physical and chemical properties and gross structure of wood play a major role in determining the magnitude of water vapour diffusion. Under steady-state conditions, one might anticipate that variations in the moisture content will also affect the diffusion coefficient in any given direction. Activated molecules diffuse along the cellulose molecules, passing through the molecular-dimensional interstices between them. The available space for diffusion decreases when the cellulose molecules are brought together in shrinkage during desorption, and the system's conductance decreases as a result. Using a moisture conductivity cell, the influence of diffusion coefficient on moisture content under steady-state conditions in *D. sissoo* wood was investigated [27]. Calculating the mathematical link between moisture content and water vapour diffusion was made possible. It was found that at lower moisture content (10.5%) the diffusion coefficient was 0.318×10^{-5} cm/s, and at 30.2% MC, it was 0.465×10^{-5} cm/s. At higher MC (52.2%), the diffusion coefficient was found to be 3.603×10^{-5} cm/s. This indicates the fact that the nature of the drying and water extraction at lower and higher MC are very different and complex.

3.4 Kiln Drying

Pre-steaming green wood at high temperatures has a significant impact on the behaviour of the wood drying process, in addition to the well-known benefits of steaming wood, like sterilization against mould, fungal stain, and decay, equalizing and conditioning, recovering from collapse (re-conditioning), toning the colour of some timbers, and occasionally reducing hygroscopicity. Pre-streaming has been found to increase the longitudinal permeability to gases in many woods. This is likely to directly impact the rates of "free" water movement during the drying process of green wood in the range above the fibre saturation point and potentially reduce drying degradation such as surface cracking and collapse. For various woods, average drying time savings of 20–25% have been documented. The permeability of wood can be enhanced by altering their composition or moving extractive location within the vessels or pit areas; rendering the insoluble extractives, which prevents them from migrating and depositing in and around the important pit areas; changing the chemical properties of the materials that make up the pit areas, which is said to be different from the rest of the cell wall; or actually rupturing the pit areas of the cell wall [28].

The effects of pre-steaming green timber under saturated conditions at 100 °C and 127 °C on the drying rates in subsequent drying for radial and tangential directions of moisture movement in two species of hardwoods, *Grewia tiliaefolia* and *D. sissoo*, were investigated after restricting the moisture movement in small laboratory-sized specimens to a single direction [29]. The findings show that pre-steaming considerably impacts drying rates, but only when dealing with more refractory wood. Pre-stressing at 127 °C has greatly increased refractory wood Grewia tiliaefolia drying rates in both moisture transport directions. However, the radial direction of moisture flow improves more than the tangential direction. Pre-steaming has only slightly increased drying rates in less refractory wood like *D. sissoo* in the tangential direction and not at all in the radial direction of moisture transport.

The permeability of the rays, their volume, and their relative contribution to radial moisture movement as compared to other tissues can likely be used to explain differences in the relative order of improvement in drying rates for the two directions of moisture movement in the two species. To overcome the problem of refractory wood experiencing severe first cracking during pre-steaming, pre-boiling in water at a temperature exceeding 100 °C (under pressure) has been proposed as an alternative to pre-steaming.

Kiln Drying Schedule for *D. sissoo*

A kiln schedule is a rough guide concerning the temperature and humidity conditions to be maintained in the kiln when the moisture content of the wettest samples on the entering air side has reached the moisture content stage specified in the schedule tables. The conditions in the kiln should be regulated as per the speed of drying, moisture distribution, case hardening, permissible degrade, quality, end products, thickness of the wood, etc. A detailed report on the drying behaviour of the Indian timbers was published [30]. The schedules are conservative, safe, and meant for 25 mm thick timbers. For higher thickness, relative humidity should be kept 5% higher at each stage during the drying. While starting the drying process, giving direct steaming treatment to the timber stack at 55 ºC temperature and 100% relative humidity is desirable. Seven major kiln drying schedules for the drying of Indian timbers were published [30]. Generally, light timber is easier and quicker to dry than heavy timber. *D. sissoo* is classified as Refractoriness Class B along with *T. grandis*, *Toona ciliata*, etc. Kiln drying schedule number IV is recommended for *D. sissoo*. Table 4 presents the kiln drying schedule for 25 mm thick *D. sissoo* [30]. As per the recommended schedule, dry bulb temperature varies from 42 ºC to 55 ºC and relative humidity progressively decreases from 80 to 39% towards end of the drying.

Table 4 Kiln drying schedule for *D. sissoo* [30]

Moisture content of wettest sample on the air inlet side	Temperature (°C)		Relative humidity (%)
	Dry bulb	Wet bulb	
Green	42	38.5	80
60%	42	38	76
40%	45	40	72
35%	47	40	64
30%	49	40	56
25%	52	40	47
20%	55	40	39

During the drying process, intermittent steaming (55 °C and 100%) for 2–4 h will help mitigate the drying stresses [30].

3.5 *Chemical Dehumidification and Solar Drying*

Timber drying is a very energy-intensive operation. Drying uses between 60 and 70 percent of the energy used to manufacture timber. A variety of traditional and unconventional techniques are popular for drying wood. The easiest drying technique is natural drying; however, it has drawbacks such as being unpredictable, requiring a lot of room and stock, and, most importantly, not being able to reach low moisture contents. Convective kilns have a low energy efficiency of approximately 45% and are energy-intensive [23], need trained operators to maintain and operate boilers, and handle pollution risks and high-temperature-related problems. In the past few decades, there has been a lot of interest in replacing traditional steam-heated kilns with refrigeration-style dehumidifiers, or heat pumps. This is mostly because, in comparison to steam-heated kilns, less energy is used. But there are two significant disadvantages. The current generation of refrigeration-style dehumidifiers is not designed to operate above 50–55 °C, resulting in a slow drying rate and high condensing pressure, frequently leading to unit breakdown. Second, in order to preserve the vapour pressure gradient, RH must be lower than 45%. The unit must run at a low suction temperature to maintain low RH, significantly increasing energy usage.

Drying of *D. sissoo* was investigated while testing an adsorption-based drier of 3.5 m^3 (125ft^3) [31]. Very few defects were detected in the dried materials, which were extremely high quality. The chemical dehumidification drier consumed 25–30% less energy than electrically heated kilns. Extensive studies on drying of shisham wood in the solar kiln at Dehradun have been carried out [32]. Shisham wood of cross section 75 mm × 100 mm requires approximately 18 days to dry in a steam-heated kiln which runs 24 × 7. In April, solar kiln drying took 25 days to bring MC of shisham from initial 55% to final 18%. The solar kiln-dried planks were found to develop only light to moderate drying stresses. No serious defects were observed during the drying.

4 Conclusions

Dalbergia sissoo is a valuable Indian timber species, classified under the Super Group Timber, with a strength coefficient of 75 compared to teak, which makes it highly suitable for various structural and decorative applications. Various standardized methods have been used to evaluate its mechanical properties, including MOE, MOR, compressive, shear, tensile strength, and hardness, consistently showing the highest strength in kiln-dried conditions compared to green and air-dried samples. However, its mechanical strength varies across locations due to moisture content, specific gravity, and testing methods. It is also known for its stability because it maintains equilibrium moisture content and avoids dimensional changes caused by water absorption or desorption. The tangential–radial shrinkage ratio (T/R ratio) of *D. sissoo* is lower than 2, indicating its dimensional stability. *D. sissoo* has a lower fibre saturation point (20%) and an EMC lower than teak, making it suitable for high-quality production. It is classified as refractoriness Class B, and kiln drying schedule number IV is recommended for its seasoning. *D. sissoo* also responds very well during dehumidification of kiln and solar kiln drying.

References

1. Anoop EV, Pasha MKS (2017) Timber identification manual: manual of timber used by wood-based handicrafts industry of Kerala, Uttar Pradesh, and Rajasthan. European Union /WWF India, New Delhi
2. Bhattacharjee A, Krishna G, Kumar A, Sengupta S, Chakraborty S, Dhanavate R, Sarkar S, Sahu RK, Soni AK, Chakraborty O, Mallick B, Prasad K (2018) Report on Non-Detriment Findings (NDFs) of Dalbergia latifolia Roxb and D sissoo DC. In India. Botanical Survey of India, Kolkata
3. ICFRE (2020) A handbook on field identification of selected timbers of India. Indian Council of Forestry Research and Education, Dehradun, p 102
4. Hegde M, Garg SL, Khatri R, Parihar PK, Gehlot N, Midde SN, Jayant BV, Tomar UK, Baloch MR (2024) Genetic Improvement of *Dalbergia sissoo* at ICFRE-AFRI Jodhpur and release of high yielding clones for planting in hot semi-arid and sub-humid tracts of Gujrat and Rajasthan. Wood is Good 4(4):128–133
5. Samuel JR (1914) The mechanical properties of wood. including a discussion of the factors affecting the mechanical properties and methods of timber testing. J Wiley and Sons, New York, p 141
6. Green DW, Winandy JE, Kretschmann DE (1999) Mechanical properties of wood. In: Wood handbook, FPL-GTR-113. U.S. Department of agriculture, forest service, forest products laboratory, Madison, pp 4.1–4.45

7. Kretschmann DE (2010) Mechanical properties of wood. In: Wood handbook. Wood as an engineering material, Forest Products Laboratories, Technical Report. Madison, WI, USA: USDA, 5, pp 1–44
8. Senalik CA, Farber B (2021) Commercial lumber, round timbers, and ties; General Technical Report FPL-GTR-282; Forest Service, U.S. Department of Agriculture 6:1−25
9. Awan AR, Chughtai MI, Ashraf MY, Mahmood K, Rizwan M, Akhtar M, Siddiqui MT, Khan RA (2012) Comparison for physico-mechanical properties of farm-grown *Eucalyptus camaldulensis* Dehn. with conventional timbers. Pakistan J Botany 44(6):2067–2070
10. Hussain M, Shah SMA, Ahmad S, Yousaf A, Shaukat S, Yousaf S, Ahmed SW, Rauf Z, Afzal QUA (2023) Physico-mechanical properties of shisham (*Dalbergia sissoo*) wood grown in Khyber Pakhtunkhwa, Pakistan. J Xi'an Shiyou University, Natural Sci Edition 66(3):38–62
11. IS:1708 (1986) Indian standard methods of testing of small clear specimens of timber. Indian Standards Institution, New Delhi, p 64
12. Tariq KA, Haroon M, Mughal AH, Ali, S (2022) Physical and mechanical characterization of structural wood used in Pakistan. Wood Research 67(1):134–146 doi.org/https://doi.org/10.37763/wr.1336-4561/67.1.134146
13. Janka G (1906) Die härt dês holzes. Centralblatt für das Gessamte Forstwessen. Wien 32(6):193–202
14. Sattar MA, Bhattacharjee DK, Kabir MF (1999) Physical and mechanical properties and uses of timbers of Bangladesh. Bangladesh Forest Research Institute, Chittagong, pp 61
15. Mahmood K, Awan AR, Chughtai MI (2016) Anatomical, physical and mechanical properties of salt tolerant tree species grown in Punjab. Pakistan. Pak. J. Bot. 48(5):1813–1818
16. Anonymous (1996) ASTM International Standard Methods (D143–94 and D-2395–93). American Society for Testing and Materials. Philadelphia
17. Shi L, Sun QF, Deng J (2011) Primary study on the anatomical construction and physical-mechanical-properties of introduced *Dalbergia sissoo*. Advanced Mat Res 179:1446–1452
18. Sekhar AC, Bhatia DN (1957) Physical and mechanical properties of woods tested at Forest Research Institute Report VI. Indian Forest Records (Timber Mechanics) 1(7):137–146
19. Sekhar AC, Bhatia DN (1960) Physical and mechanical properties of woods tested at the Forest Research Institute. Dehra Dun VIII. Indian Forest Records (Timber Mechanics) 1(9):155–157
20. Sunny PP, Dutt B, Sharma KR, Dhiman B, Sumthane YY (2019) Wood variation in physico-mechanical properties of *Dalbergia sissoo* Roxb. ex DC. from local markets of Himachal Pradesh. J Tropical Agricult 57(1):17–26
21. IS 13622 (1993) Indian Standard on Indian timbers for furnitures and cabinets – Classification. Bureau of Indian Standards, New Delhi
22. SP 33 (1986) Handbook on timber engineering. Bureau of Indian Standards, New Delhi
23. Pandey CN, Jain VK (1992) Wood seasoning technology. Indian council of Forestry Research and Education, Dehradun, p 200
24. Rehman MA (1953) The seasoning behaviour of Indian timbers, part II: Kiln drying schedules. Indian Forester 79(7):369–375
25. Basri E, Hadjib N, Saefudin S (2005) Basic properties in relation to drying properties of three wood species from Indonesia. Indonesian J Forest Research 2(1):49–56
26. Pandey CN, Dayal R, Rao V (1986) Evaluation of Anisotropic behaviour in dimensional changes of some Indian timbers. J Timber Develop Associat India 32(2):31–36
27. Pandey CN, Kambo AS (1990) Effect of moisture on water vapour diffusion in *Dalbergia sissoo*. J Timber Develop Associat India 36(4):26–31
28. Donaldson LA, Wong KKY, Mackie KL (1988) Ultrastructure of steam-exploded wood. Wood Sci Technol 22:103–114
29. Sharma SN, Bali BI (1969) Effect of pre-steaming on drying rates in subsequent seasoning of green and refractory hardwood. Indian Forest Bulletin No. 258, Forest Research Institute, Dehradun, pp 13
30. Rehman MA, Singh G (1969) Wood water relationship and equilibrium moisture content of Indian woods. J Timber Develop Associat India 15(1):30–34

31. Pandey CN, Gandhi BL, Singh D (1995) Technical evaluation and feasibility study of chemical dehumidification based wood dryer. J Timber Develop Associat India 41(3):5–13
32. Sharma SN, Nath P, Badoni SP (1981) Commercial trials on a 7.1 Cu.M. solar kiln. Indian Forest Bulletin No. 274, Forest Research Institute, Dehradun, pp 48

Chapter 25
Dalbergia tonkinensis

Phi Hong Hai, Nong Phuong Nhung, Nguyen Minh Chi, Pham Quang Thu, and Bernard Dell

1 Introduction

Dalbergia tonkinensis Prain, popularly referred to as yellow rosewood, is a medium-sized, slow-growing tree reaching a height of 20–25 m and a diameter of 70–90 cm at maturity [28, 30]. It is becoming increasingly rare in the wild, and the wood of mature trees has high commercial value in Vietnam and China [16, 30]. Despite a ban on harvesting [16, 46–48], wild populations in Vietnam are dwindling due to habitat destruction, timber extraction, and illegal trade. According to local Vietnamese statistics, trees with a diameter of over 40 cm exist only in protected areas such as office compounds, streetscapes, parks, historical sites, and pagoda gardens. There are only about 100 *D. tonkinensis* trees in Vietnam aged 60 years or older [16, 39], with relatively low genetic diversity [14].

Over the past two decades, *D. tonkinensis* has been grown in some smallholder plantations and home gardens in Vietnam, using seeds sourced mainly from mother trees in Vinh Phuc province [31, 39]. Mostly, these plantings have been a spontaneous activity of householders attracted by the very high value of *D. tonkinensis*. This enthusiasm has contributed to investment in small *D. tonkinensis* populations even though the quality of wood is unknown. The high demand for growing this tree has promoted activities to improve genetic sources [31]. Recently, some insect pests and diseases have also been recorded. Studies on propagation from seed and cuttings, conservation, and insect pests and diseases have been undertaken for over a decade in Vietnam to preserve and develop its genetic resources effectively. Though *D.*

P. H. Hai (✉) · N. P. Nhung
Vietnamese Academy of Forest Sciences, Hanoi, Vietnam
e-mail: phi.hong.hai@vafs.gov.vn; phihonghai2008@gmail.com

N. M. Chi · P. Q. Thu
Forest Protection Research Centre, Vietnamese Academy of Forest Sciences, Hanoi, Vietnam

B. Dell
Agriculture and Forest Sciences, Murdoch University, Murdoch, Australia

R. R. Warrier et al. (eds.), *Handbook of Asian Rosewoods*, Materials Horizons: From Nature to Nanomaterials, https://doi.org/10.1007/978-981-96-9459-4_25

tonkinensis is strictly protected by national law, illegal logging and trading of this species are still continuing with weak regulations.

2 Taxonomy

Kingdom: Plantae

Order: Fabales

Family: Fabaceae

Subfamily: Faboideae

Tribe: Dalbergieae

Genus: *Dalbergia*

Species: *Dalbergia tonkinensis* Prain

3 Distribution

Dalbergia tonkinensis is naturally distributed in Vietnam, from the Northern (Lang Son, Quang Ninh, Bac Ninh, Hoa Binh, Phu Tho, Hung Yen, Ha Noi, Ha Nam, and Ninh Binh provinces) to the Central (Nghe An, Quang Binh, Quang Tri, Quang Nam, Phu Yen, and Khanh Hoa provinces) regions, as well as the Central Highlands (Gia Lai, Dak Lak provinces) and the South (Dong Nai province). It is also reported from some locations in Southern China, concentrated on Hainan Island [27], where the closely related *D. odorifera* occurs. In nature, trees are scattered in evergreen, tropical forests, concentrated at altitudes below 500 m. The species prefers fertile, deep, and moist soil [27].

4 Morphology

Dalbergia tonkinensis is a seasonally deciduous tree. Tree trunks are axial and scattered. The bark is yellow–brown or pale-grey, and is longitudinally cracked. Young branches are green and have sparse fine hairs. Leaves are imparipinnately compound with 11–17 leaflets [16, 39]. The pinnules are alternate and elliptic, the apex is acute with a short point, the base is rounded, the underside of the blade is pale white, and there are 10 pairs of lateral veins (Fig. 1b). The inflorescence is an axillary panicle producing white, fragrant flowers, often appearing after the leaves have fully expanded. The legumes are flat, elliptic, 5–7 cm long, 1.6–2.5 cm wide, with 1–2

Fig. 1 Morphology of *Dalbergia tonkinensis*: **a** Seven- to eight-year-old trees planted in a garden; **b** leaves; **c** inflorescences; **d** recently harvested fruits; **e** seeds for commercial trade

seeds (Fig. 1d), and are indehiscent. The brown seeds are oblong, 9–13 mm long, 5–8 mm wide, and flattened (Fig. 1e).

Morphological variation among 12 provenances of *Dalbergia tonkinensis* in Vietnam (Lang Son, Bac Ninh, Hung Yen, Ha Noi, Vinh Phuc, Ninh Binh, Thanh Hoa, Nghe An, Quang Tri, Hue, Dak Lak, and Gia Lai) revealed notable differences in several traits. The number of pinnules per secondary rachis ranged from 11 to 17, with the lowest count observed in Lang Son and the highest in Hue. Significant variation was also found in tree height, stem cracking, bark colour, and stem form. Superior provenances featured tall trees (>17 m) with a long, straight bole and fissured bark, while inferior provenances were characterised by shorter, often multi-stemmed trees [39].

Distinct differences were also evident in legume size and hardness across provenances. For instance, trees from Ninh Binh produced long, hard legumes, whereas those from Thanh Hoa had shorter pods with soft walls. Seed traits are size, colour, and germination varied. Seeds from Vinh Phuc were typically yellowish-brown, oval, and somewhat flattened, while those from Ninh Binh were dark grey, elongated, and rounded.

Ma et al. [25] reported that seed size, seed weight, and germination potential differed ($P < 0.01$) among provenances and also between families within provenances. The Gia Lai provenance seeds had double the germination per cent from Hue, 72.6% and 30.0% [31]. Germination was positively correlated with 1,000-seed weight and seed width; altitude, and hours of sunshine [39], indicating the influence of the environment on the seed performance.

5 Phenology

Flowering occurs in February and March; the legumes mature from September to November of the same year [27]. It was observed that flowering was early and fruit maturation faster, with early leaf fall in Quang Tri, Hue, Dak Lak, Gia Lai, Ninh Binh, Thanh Hoa, and Nghe An, while the trend was different in Lang Son, Hung Yen, Ha Noi, and Vinh Phuc provenances. Fruit characteristics also show clear variation across provenances. Trees from Ninh Binh produce long and thick fruits, while those from Gia Lai yield very large round fruits. The fruit shells from Lang Son, Nghe An, and Gia Lai provenances are generally soft, with those from Thanh Hoa being hard. The seeds of *D. tonkinensis* are typically yellowish-brown, thin, and oval. However, seeds from the Hue and Ninh Binh provenances are distinct, being dark grey, thick, long, and tapering. The weight of 1,000 seeds varied significantly, ranging from 80 g in Hue to 137.3 g in Gia Lai. Additionally, the proportion of high-quality seeds differs across provenances. Lang Son, Ninh Binh, and Hue have relatively low percentages of quality seeds (33.2–51.1%), compared to higher rates in Hung Yen, Gia Lai, and Ha Noi , where over 81% of seeds are classified as high quality [16, 39].

6 Propagation

Generally, *D. tonkinensis* is propagated from seed in nurseries or home gardens, but it has challenges. The major difficulty is separating seeds from dry pericarps, a task currently performed manually, which is labour-intensive. Once the pericarp turns brown and dry, extracting seeds becomes even more challenging. To overcome this, seeds are removed when the pods are yellow or light brown, retaining some moisture. These seeds are then shade-dried in well-ventilated areas and stored in plastic bags in a refrigerator at 5–6 °C for 1–2 years [31]. However, some local collectors harvest unripe fruit while the pericarp is still green (Fig. 1d). This makes seed extraction easier but can significantly reduce seed quality.

Germination is done in moist cotton bags. Once the radicles appear, they are sown in trays with moist river sand as the substrate. They are transplanted into polybags containing soil when seedlings develop a pair of compound leaves. Seedlings are highly adaptive to various soil types, but propagation experiments have shown that using organic fertilisers supplemented with arbuscular mycorrhizal fungi promotes

Fig. 2 *Dalbergia tonkinensis* propagation. **a** Seedlings in plastic bag containers in nursery beds; **b** Rooted cuttings showing weak to vigorous root formation (left to right)

faster, more uniform growth and reduces disease incidence [7]. Improvements in commercial nursery practices could further enhance seedling health and reduce pest and disease issues. These practices include using sterile substrates, root trainers or trays, and raised containers. The seedlings are ready for outplanting, in six months, when they attain >50 cm height, >8 mm basal diameter, and are free from pests and diseases (Fig. 2a).

Clonal propagation using cuttings from superior *D. tonkinensis* trees is another viable approach, as cuttings root moderately well (Fig. 2b). Rooting success can be significantly enhanced using rooting hormones such as IBA (1.5 g/L), achieving a 92% rooting rate and producing an average of 12 adventitious roots per cutting [31]. Cuttings should ideally be taken in spring or autumn to maximise rooting success [41]. This increases the chance of obtaining field-ready plants twice a year. Young trees (less than five years old) are the best source of cuttings, as rooting ability declines with age, a phenomenon common in many tree species [2, 4, 23, 38, 44].

As the demand for planting *D. tonkinensis* is increasing in Vietnam, clonal propagation is expected to play a more prominent role in establishing commercial plantations. Cuttings from trees with superior traits—such as excellent form, early heartwood formation, and resistance to pests and diseases—are likely to replace seedlings for plantation establishment.

7 Current Practices

Dalbergia tonkinensis is becoming popular in home gardens in Vietnam using seeds from mother trees in Vinh Phuc province, supplemented with seed taken from scattered trees in other provinces. Seedlings produced in local nurseries reached about 50 cm in height and 5 mm in basal diameter at five months of age. Seedlings from selected seed sources had superior growth and less disease than other seed sources [16]. So far, most stands of *D. tonkinensis* have been planted in home gardens with 50–100 trees per garden (area of 500–2,000 m^2) or in plantations with density of 833–1,600 tree/ha. In plantations and home gardens, trees can attain 1.5 m/year in height and 1.6 cm/year in diameter up to 5 years of age [39]. Trees in Son La, Bac Ninh, and Ha Noi provinces had the best growth. Staking, pruning, and thinning have been widely applied to improve stem form and growth. Bamboo culms about 2–3 m long were used as stakes to fix the trees, one stake per tree, tied with polymer rope and replaced annually until 3 or 4 years after planting. Pruning was done 1–2 times per year, for 3 years after planting. Branches with a base diameter >1/3 of the trunk diameter at the branch location were removed. However, 38–58% of trees of Hoa Binh, Ninh Binh, and Nghe An provenances were multi-stemmed. Thinning was also done to ensure a remaining density of about 800–1,000 trees/ha. Poorly growing or trees with multiple stems were removed, while some others were transferred to a new location. Fertilisation was usually done only once when planting, and the trees were not watered after planting. Canker disease (see below) was present in all sites but the most serious impact was in Tuyen Quang and Phu Tho provenances [39]. Techniques to promote heartwood development are being used by cutting 100% of branches +50% of roots or cutting all branches followed by transplanting [30].

8 Pests and Diseases

Root rot caused by *Fusarium solani* and *Pythium vexans,* and *Colletotrichum* leaf spot disease have often been observed in *D. tonkinensis* seedlings [7]. Symptoms of these diseases become severe during the wet season (spring in northern Vietnam). The disease is more prevalent in nurseries with poor drainage or very high seedling density [32]. This highlights the need to improve nursery hygiene by adopting best practice protocols.

Fusarium decemcellulare and *F. lateritium* can cause canker disease in seedlings (Fig. 3a) and in 1–3-year-old trees (Fig. 3b and c). The morphological characteristics of these pathogens have been described by Nhung et al. [32] (Fig. 3d–k). These canker pathogens can damage seedlings grown in dark and humid conditions or areas heavily infested with weeds. Field observations indicate that these *Fusarium* species can kill 1–2-year-old trees (Fig. 3b) and cause dieback in older trees (Fig. 3c). In Vietnam, other *Fusarium* species cause severe damage in horticultural and field crops, such as *Fusarium oxysporum* and *F. solani* on *Piper nigrum* [42], and *F. graminearum*, *F. proliferatum*, and *F. verticillioides* on *Zea mays* [28].

Recently, *D. tonkinensis* trees have become prone to attack from wilt disease caused by *Ceratocystis manginecans*. The pathogen mainly infects trees through wounds on the trunk and lateral branches, causing trees to wilt and die. This occurs in all tree age cohorts (Fig. 4a). Morphological characterisation (Fig. 4d–g) and DNA analysis of this pathogen have been documented [6]. This pathogen, which has a wide host range, was first recorded in mango in Oman and Pakistan [45], later, on *D. sissoo* and *Prosopis cineraria* in Pakistan [1, 3], *Mimusops elengi* in Thailand [33], and *Acacia* spp. in Indonesia and Malaysia [15, 50]. In Vietnam, *C. manginecans* is a serious pathogen in commercial *Acacia* [39, 40], *Eucalyptus* [21, 43], *Chukrasia*

Fig. 3 Symptoms of canker disease in *Dalbergia tonkinensis* and appearance of the main fungal pathogens. **a** Seedlings with stem cankers in a nursery; **b** two-year-old tree in the field with stem canker on the trunk; **c** six-year-old tree with branch dieback; **d**, **e**, **h**, **i**. *Fusarium decemcellulare*; **f**, **g**, **j**, **k** *F. lateritium*; **d**, **e**, **f**, **g** Mycelium; **h**, **i**, **j**, **k** Conidia

Fig. 4 *Ceratocystis manginecans* wilt disease on *Dalbergia tonkinensis* trees, and spore forms of the pathogen: **a** diseased tree in a garden; **b** lesions on the trunk; **c** vascular discolouration of diseased tree; **d** hat-shaped ascospores; **e** chlamydospores; **f** barrel-shaped conidia; **g** cylindrical conidia

tabularis [8], and *Melia azedarach* [9] plantations. Operations such as pruning, root cutting, and stem drilling create entry points for the pathogen [7]. These practices should be refined to minimise the risk of infection.

Attack by ambrosia beetles in the genus *Euwallacea* can lead to branch dieback and tree death [5]. This was first observed in young stands of *D. tonkinensis* in 2018 [6]. The typical symptoms include small holes on the trunk (Fig. 5e) or branches with purple or brown sap flowing from those holes (Fig. 5). Ambrosia beetles are generally recognised as dangerous insect pests in Vietnam, especially *Euwallacea fornicatus*, *E. similis*, *Xyleborus perforans*, and *Xylosandrus crassiusculus* [20, 40]. They often attack trees that have been stressed, physically wounded (such as from branch/root pruning), or weakened by infection by plant pathogens. In the latter case, trees infected with canker (Fig. 3c) or wilt disease (Fig. 4c) are susceptible to ambrosia attack. Also, these beetles are carriers of spores of fungal pathogens such as *Fusarium* spp. [20, 22].

Fig. 5 Ambrosia beetle (*Xylosandrus crassiusculus*) damages *Dalbergia tonkinensis* trees in Vietnam: **a** adult; **b** egg; **c** larvae; **d** pupa; **e** trunk with exit holes. Gallery of split open, with pupae and black fungus

9 Uses

The wood of *D. tonkinensis* is heavy, hard, durable, resilient to weathering, beautifully veined, and has a unique aroma. It is highly valuable and popular for making high-quality furniture [19, 27, 30]. The heartwood is the main product and of core economic value, which sells for 13–152 USD/kg in local trade [30]. Large timber from old trees commands a higher price than small timber from young trees, making *D. tonkinensis* furniture expensive. A heartwood table, chair set, or bed can sell for thousands of USD. Small heartwood pieces are also valuable, fuelling a handcraft industry that produces bracelets (9–26 USD), statues (87–652 USD), pens (17–22 USD), and head supports (435–522 USD). In recent years, the sale of seeds and seedlings of *D. tonkinensis* has become a significant source of income for smallholders. The price of seeds can reach 87–130 USD/kg and seedlings are priced at 1–2 USD [30].

The heartwood of *D. tonkinensis* contains chemical compounds with anti-inflammatory properties [10, 39] and mono- and bifunctional vasodilators [12]. Some of these compounds are used in antidiabetic drug formulations [26, 35]. Additionally, the seeds show pharmaceutical potential due to their chemical composition. Studies by Cuong et al. [10, 11], Son et al. [35,

36, 37], and Nhan et al. [29] indicate the seed oil is lipid-rich, with a triacylglycerol content of 76%. The fatty oil content of *D. tonkinensis* seeds is approximately 13%, exceeding that of *D. sissoo* (4.8%) and *D. paniculata* (7.4%).

10 Present Threat Status

D. tonkinensis populations face threats from overexploitation, habitat loss, and illegal harvesting. Recognised as Vulnerable by the IUCN Red List and Vietnam Red Data Book, the species has been protected by Vietnamese government decrees (32/2006, 06/2019, 84/2021) and listed in CITES Appendix II since 2017 [26, 46–48]. Despite these measures, illegal trade persists, and new plantations will take decades to meet market demand. Smallholder plantings, initiated in northern Vietnam in the early 2000s, are being harvested, but their heartwood is of lower quality than mature trees. These plantings, typically 20–100 trees per garden [30], are expanding and the species is popular for amenity planting due to its canopy and fragrant flowers. However, cultivated populations exhibit low genetic diversity. Over the last decade, smallholder plantings of *D. tonkinensis* have increased to about 2000 ha [40], mainly using low-quality seedlings from parent trees [39]. Domestication has not followed any systematic or scientific approach. The mating system of the species has not been investigated.

11 Genetic Resources and Conservation

11.1 Genetic Resources and Diversity

As mentioned earlier (Sect. 4 of this chapter), there is considerable morphological variation in *D. tonkinensis* in Vietnam. Seeds from 109 healthy trees (DBH 41.4–84.6 cm, height 15.5–28.5 m, boles >50% tree height) (Fig. 6) from twelve provinces [13] were vacuum sealed in plastic bags and stored in a seed bank at 5–7 °C for both gene conservation and establishment of field gene banks. However, diversity assessment using inter-simple sequence repeat (ISSR) markers revealed low genetic diversity in *D. tonkinensis* [14]. The estimated values for the effective number of alleles, Shannon's information index, intralocus gene diversity and Nei's gene diversity were 1.198, 0.166, 0.526, and 0.123, respectively; and there was a correlation between genetic and geographic distance [14].

11.2 Gene Conservation in Field Gene Banks

There are no identified seed sources of *D. tonkinensis* in Vietnam [31, 39]. To address this, three field gene banks were established for conservation and future seed production: in Tan Son and Doan Hung (Phu Tho) in 2015 [13], and in Tan Bien (Binh Phuoc) in 2019 [16]. These banks will later serve as seed orchards designed as progeny tests, enabling family testing, selection, breeding, and seed production within a single stand [49]. Selective thinning ensures unrelated neighbours surround each seed tree,

Fig. 6 Examples of *Dalbergia tonkinensis* mother trees used for seed collection: **a** in the grounds of a Buddhist temple in Hanoi (Dbh = 70.2 cm, Ht = 25 m); **b** in a home garden in Vinh Phuc (Dbh = 55.3 cm, Ht = 19 m)

promoting outcrossing and minimising related cross-pollination. Further thinnings based on rigorous family and individual selection will drive genetic improvement. These banks also enable long-term monitoring of growth, disease resistance, and heartwood quality, with superior trees available for clonal propagation [31].

(a) Evaluation of Provenances in Field Gene Banks

After three years, a preliminary evaluation of the three stands revealed varying results. Bac Ninh and Hung Yen provenances performed well in terms of height (Fig. 7), whereas the Ha Noi and Gia Lai provenances had the maximum girth in Tan Son. In Doan Hung and Tan Son, the best provenance was Hanoi, with superior height, diameter growth, and survival (Fig. 8b), followed by the Vinh Phuc provenance (Fig. 8a).

(b) Evaluation of Families in Field Gene Banks

Mean values of growth traits, stem number, disease index, and survival for *D. tonkinensis* families are summarised in Table 1. Family significantly affected growth ($p < 0.001$) but not stem number or disease index. Growth varied widely across sites: in Tan Son, heights ranged from 1.43 to 4.64 m and diameters from 0.72 to 3.61 cm; in Doan Hung, heights were 2.36–5.24 m and diameters 2.37–4.56 cm. In Tan Bien, heights ranged from 4.88 to 6.69 m and diameters from 3.86 to 5.61 cm, though

Fig. 7 Three-year-old *D. tonkinensis* planted in Tan Son: **a** superior family from Hung Yen; **b** superior family from Bac Ninh

stem quality was lower than in northern Vietnam. Twelve families exceeded 3.5 m in height across all sites, 29–48% above the test average. Significant differences among provenances and families highlight the potential for substantial growth improvement through selection in progeny tests [13, 16].

At present, *D. tonkinensis* seeds are being collected mainly from trees grown in home gardens in Vinh Phuc [30, 31]. The Vinh Phuc provenance performed poorly in the Tan Son soil (Table 1) [16]. However, family DT66 from Vinh Phuc showed good early growth in both field sites and could be a potential candidate for breeding activities [13].

(c) Genetic Variation of the Field Gene Banks

High heritability and coefficient of additive variation of growth parameters in the Tan Son, Doan Hung [13], and Tan Bien sites suggest that these are useful traits to improve productivity in *D. tonkinensis* in northern Vietnam. The expected genetic gain (R_Y) of height and diameter at breast height were 25.7–40.0% and 36.5–71.2%, respectively, which shows the potential for selection of quality *D. tonkinensis* clones from these trials. In *D. sissoo*, 5 clones with good growth performance had heritability estimates of 87%, 83%, and 80% for height, crown width, and crown length, respectively [34]. In the field gene banks in Vietnam, the individual heritability ($\hat{h}^2$) of height and

Fig. 8 Three-year-old *D. tonkinensis* planted in Doan Hung: **a** superior family from Vinh Phuc; **b** superior family from Ha Noi

diameter at breast height of *D. tonkinensis* at age 3 years were 0.24–0.39 and 0.27–0.49, respectively [13, 16]. This value was higher than or similar to the individual heritability of fast-growing species such as *Acacia auriculiformis*, $\hat{h}^2 = 0.36–0.39$ [18], *A. mangium*, $\hat{h}^2 = 0.11–0.30$ [17], and *Eucalyptus urophylla*, $\hat{h}^2 = 0.10–0.31$ [24], suggesting promising results for improvement in *D. tonkinensis*. The interaction between the growth traits in the Tan Son and Doan Hung test was strongly correlated (0.88–0.90), indicating that families with good growth in one site will also perform well at the second site. Eleven families are reportedly growing well in both sites [13].

(d) Location of Mother Trees

Mother tree locations significantly affected the height and diameter of trees in the three field sites [13]. Families with superior height and diameter came from identified trees of natural forests. The poorest performing families were from mother trees in government office buildings' courtyards or city parks.

Table 1 Phenotypic variations of growth, number of stems, disease index, and survival between *D. tonkinensis* families at age 3 years in Tan Son, Doan Hung, and Tan Bien

Trait	Unit	Test mean	Variation	P value
Tan Son				
Diameter	cm	1.71	0.72–3.61	<0.001
Height	m	2.72	1.43–4.64	<0.001
Disease index	point	0.99	0.41–1.64	0.178
Number of stems	point	1.42	1.06–1.73	0.361
Survival	%	92.4	56.3–100	<0.001
Doan Hung				
Diameter	cm	3.01	2.37–4.56	<0.001
Height	m	3.55	2.36–5.24	<0.001
Disease index	point	0.91	0.33–1.40	0.190
Number of stems	point	1.43	1.06–1.84	0.260
Survival	%	95.16	78.1–100	<0.001
Tan Bien				
Diameter	cm	4.84	3.86–5.61	<0.001
Height	m	5.89	4.88–6.69	<0.001
Stem straightness	point	2.90	2.17–3.50	0.478
Survival	%	62.9	44.8–88.9	<0.001

12 Need for Conservation and Maintenance of Genetic Diversity

Illegal logging threatens the genetic diversity of *D. tonkinensis* in Vietnam, endangering both biodiversity and sustainability. Urgent population studies are needed to assess genetic diversity in the wild. Meanwhile, the approximately 100 mature mother trees in parks and pagodas and the three existing field gene banks [40] can support germplasm collection efforts. Expanding seed and field gene banks nationwide across diverse microclimates and soil types is crucial for conservation and sustainable development.

13 Conclusion

Dalbergia tonkinensis has been classified as Vulnerable under criteria A1cd in the IUCN Red List of Threatened Species and by the Vietnamese decree 84/2021/ND-CP, aiming to curb illegal harvesting. Identified seed sources exhibit low genetic diversity, and mature trees in natural populations are low. Field gene banks have been established to serve as seed orchards, support the development of clonal plantations, and act as genetic resources for a breeding program. Conservation and genetic diversity maintenance of the species are crucial for its long-term survival. While home-garden cultivation of *D. tonkinensis* is expanding across various provinces in Vietnam, no efforts have been made to provide growers with certified genetic resources. Until seeds from selected mother trees in commercial seed orchards or high-quality clones are available, growers should aim to source legal planting stock from plus trees to avoid adverse effects.

References

1. Al Adawi AO, Barnes I, Khan IA, Al Subhi AM, Al Jahwari AA, Deadman ML, Wingfield BD, Wingfield MJ (2013) *Ceratocystis manginecans* associated with a serious wilt disease of two native legume trees in Oman and Pakistan. Aust Plant Pathol 42(2):179–193. https://doi.org/10.1007/s13313-012-0196-5
2. Amri E, Lyaruu HVM, Nyomora AS, Kanyeka ZL (2010) Vegetative propagation of African blackwood (*Dalbergia melanoxylon* Guill. & Perr.): effects of age of donor plant, IBA treatment and cutting position on rooting ability of stem cuttings. New Forests 39(2):183–194. https://doi.org/10.1007/s11056-009-9163-6
3. Arif M, Zaidi NW, Haq QMR, Singh YP, Khan S, Singh US (2013) Molecular phylogeny and pathotyping of *Fusarium solani*: a causal agent of *Dalbergia sissoo* decline. Forest Pathol 43(6):478–487. https://doi.org/10.1111/efp.12060
4. Brix H (1974) Rooting of cuttings from mature douglas-fir. NZ J Forest Sci 4(2):133–139
5. Chi NM, Nhung NP, Hung TX, Quang DN, Hinh TX, Nam NV, Thuy PTT, Thu PQ (2019) First report of stem borer beetle associated with *Dalbergia tonkinensis* in Viet Nam. J Plant Prot 4:15–19
6. Chi NM, Nhung NP, Trang TT, Thu PQ, Hinh TX, Nam NV, Quang DN, Dell B (2019) First report of wilt disease in *Dalbergia tonkinensis* caused by *Ceratocystis manginecans*. Aust Plant Pathol 48(5):439–445. https://doi.org/10.1007/s13313-019-00643-1
7. Chi NM, Quynh DN, Thong NQ, Nam NV, Ngan DH, Hinh TX (2014) Study on effects of fertiliser to growth and disease of *Dalbergia tonkinensis* Prain in nursery. Sci Technol J Agric Rural Dev 23:137–142
8. Chi NM, Trang TT, Nhung NP, Quang DN, Son VM, Tuan TA, Mai LT, Hung TX, Nam NV, Thu PQ, Dell B (2021) *Ceratocystis* wilt in *Chukrasia tabularis* in Vietnam: identification, pathogenicity and host tolerance. Aust Plant Pathol 50(1):17–27. https://doi.org/10.1007/s13313-020-00754-0
9. Chi NM (2022) Wilt disease in *Melia azedarach* caused by *Ceratocystis manginecans*. Sci Technol J Agric Rural Dev 1+2:57–64
10. Cuong NM, Nhan NT, Son NT, Nghi DH, Cuong TD (2017) Daltonkins A and B, two new carboxyethylflavanones from the heartwood of *Dalbergia tonkinensis*. Bull Korean Chem Soc 38:1511–1514. https://doi.org/10.1002/bkcs.11313
11. Cuong NM, Son NT, Nhan NT, Khanh PN, Huong TT, Tram NTT, Sgaragli G, Ahmed A, Trezza A, Spiga O et al (2020) Vasorelaxing activity of R-(-)-3′-hydroxy-2,4,5-trimethoxydalbergiquinol from *Dalbergia tonkinensis*: involvement of smooth muscle $Ca_V1.2$ channels. Planta Med 86:284–293. https://doi.org/10.1055/a-1099-2929
12. Cuong NM, Son NT, Nhan NT, Fukuyama Y, Ahmed A, Saponara S, Trezza A, Gianibbi B, Vigni G, Spiga O, Fusi F (2022) Vietnamese *Dalbergia tonkinensis*: a promising source of mono-and bifunctional vasodilators. Molecules 27(14):4505. https://doi.org/10.3390/molecules27144505

13. Dell B, Thu PQ, Nghia NH, Hai PH, Nhung NP, Chi NM (2022) Early field performance and genetic variation of *Dalbergia tonkinensis*, a valuable rosewood in Vietnam. Forests 13:1977. https://doi.org/10.3390/f13121977
14. Dinh TP, Hien VTT, Thanh TTV, Van NT, Binh NQ (2011) Genetic diversity on the tropical rare wood species of Dalbergia in Vietnam revealed by inter-simple sequence repeat (ISSR) markers. Afr J Biotechnol 10(55):11397–11408
15. Fourie A, Wingfield MJ, Wingfield BD, Thu PQ, Barnes I (2016) A possible centre of diversity in South East Asia for the tree pathogen *Ceratocystis manginecans*. Infection Genetics and Evolution 41:73–83. https://doi.org/10.1016/j.meegid.2016.03.011
16. Hai PH (2021) Gene Conservation of endangered species in Vietnam. Project report in period of 2016-2020. Vietnamese Academy of Forest Sciences
17. Hai PH, Duong LA, Toan NQ, Ha TTT (2015) Genetic variation in growth, stem straightness, pilodyn and dynamic modulus of elasticity in second-generation progeny tests of *Acacia mangium* at three sites in Vietnam. New Forests 46: 577–591. https://doi.org/10.1007/s11056-015-9484-6
18. Hai PH, Jannson G, Harwood CE, Hannrup B, Thinh HH, Pinyopusarerk K (2008) Genetic variation in wood basic density and knot index and their relationship with growth traits for *Acacia auriculiformis* in northern Vietnam. New Zeal J For Sci 38:176–193
19. Hai TN (2006) Conservation and development of *Dalbergia tonkinensis* Prain and endangered species in Vietnam. Paper presented at the National Scientific and Biological Resource Conference, pp 34–36
20. Hung TX, Thu PQ, Chi NM, Binh LV, Dell B (2022) Impacts and trapping of ambrosia beetles *Euwallacea fornicatus* and *E. similis* in *Acacia* plantations in Vietnam. Southern Forests: a J Forest Sci 84(3):242–252. https://doi.org/10.2989/20702620.2022.2128931
21. Indrayadi H, Glen M, Halimah, Fahrizawati, Prihatini I, Beadle C, Tjahjono B, Mohammed C (2023) Recognising ceratocystis disease symptoms in a *Eucalyptus pellita* nursery. Aust Plant Pathol 52(6):625–636. https://doi.org/10.1007/s13313-023-00951-7
22. Kajii C, Morita T, Jikumaru S, Kajimura H, Yamaoka Y, Kuroda K (2013) Xylem dysfunction in *Ficus carica* infected with wilt fungus *Ceratocystis ficicola* and the role of the vector beetle *Euwallacea interjectus*. IAWA J 34(3):301–312. https://doi.org/10.1163/22941932-00000025
23. Khudhur SA, Omar TJ (2015) Effect of NAA and IAA on stem cuttings of *Dalbergia sissoo* (Roxb). J Biol Life Sci 6(2):208–220. https://doi.org/10.5296/jbls.v6i2.7445
24. Kien ND, Jansson G, Harwood C, Thinh HH (2009) Genetic control of growth and form in *Eucalyptus urophylla* in Northern Vietnam. J Trop For Sci 21:50–65
25. Ma YH, Yang ZJ, Xu DP, Liu XJ, Zhang NN (2016) Variations in morphology and sowing quality of *Dalbergia tonkinensis* seeds from different provenances in Vietnam. J South China Agric Univ 37(3):107–114. (Ref. 22)
26. Mulia R, Simelton E, Le TT, Pham TV, Do TH (2018) Native and endangered timber tree species in home gardens of Northeast and Northcentral Vietnam. Biodiversity Int J 2(1):40–43. https://doi.org/10.15406/bij.2018.02.00041
27. Nghia NH (2008) Atlas of Vietnam's forest tree species, vol 2. Cartographic Publishing House, Hanoi, Vietnam
28. Nguyen TTX, Dehne H-W, Steiner U (2016) Histopathological assessment of the infection of maise leaves by *Fusarium graminearum*, *F. proliferatum*, and *F. verticillioides*. Fungal Biol 120(9):1094–1104. https://doi.org/10.1016/j.funbio.2016.05.013
29. Nhan NT, Son NT, Cuong TD, Nguyen NPD, Khanh PN, Huong TT, Cuong NM (2018) Further study on chemical constituents from the heartwood of *Dalbergia tonkinensis* collected in Daklak province. Vietnam J Sci Technol 56:252–258. https://doi.org/10.15625/2525-2518/56/4A/12862
30. Nhung NP, Chi NM, Thu PQ, Thuong BH, Ban DV, Dell B (2020) Market and policy setting for the trade in *Dalbergia tonkinensis*, a rare and valuable rosewood, in Vietnam. Trees, Forests and People 1(1):100002. https://doi.org/10.1016/j.tfp.2020.100002
31. Nhung NP, Thu PQ, Chi NM, Dell B (2019) Vegetative propagation of *Dalbergia tonkinensis*, a threatened, high-value tree species in South-east Asia. Southern Forests: a J Forest Sci 81(3):195–200. https://doi.org/10.2989/20702620.2018.1542570

32. Nhung NP, Thu PQ, Dell B, Chi NM (2018) First report of canker disease in *Dalbergia tonkinensis* caused by *Fusarium lateritium* and *Fusarium decemcellulare.* Austr Plant Pathol 47(3):317–323. https://doi.org/10.1007/s13313-018-0562-z
33. Pornsuriya C, Sunpapao A (2015) A new sudden decline disease of bullet wood in Thailand is associated with *Ceratocystis manginecans.* Austr Plant Dis Notes 10(1):26–31. https://doi.org/10.1007/s13314-015-0176-z
34. Sharma A, Bakshi M (2011) Growth and heritability estimates among clones of *Dalbergia sissoo* Roxb. in a clonal seed orchard. For Stud China 13:211–217
35. Son NT, Fukuyama Y, Cuong NM (2019) Chemical constituents of the heartwood of *Dalbergia tonkinensis.* Chem Nat Compd 55:854–856. https://doi.org/10.1007/s10600-019-02831-3
36. Son NT, Kamiji M, Huong TT, Kubo M, Cuong NM, Fukuyama Y (2019) Chemical constituents of the Vietnamese plants *Dalbergia tonkinensis* Prain and *Cratoxylum formosum* (Jack) Dyer in Hook and their DPPH radical scavenging activities. Med Chem Res 28:1441–1447. https://doi.org/10.1007/s00044-019-02383-9
37. Son NT, Oda M, Hayashi N, Yamaguchi D, Kawagishi Y, Takahashi F, Harada K, Cuong NM, Fukuyama Y (2018) Antimicrobial activity of the constituents of *Dalbergia tonkinensis* and structural-bioactive highlights. Nat Prod Commun 13:157–161. https://doi.org/10.1177/1934578X1801300212
38. Swamy SL, Puri S, Singh AK (2002) Effect of auxins (IBA and NAA) and season on rooting of juvenile and mature hardwood cuttings of *Robinia pseudoacacia* and *Grewia optiva.* New Forest 23(2):143–157
39. Thu PQ, Chi NM, Quang DN, Dell B (2014) Study on phenology and morphological characteristics of some provenances of *Dalbergia tonkinensis* Prain in Vietnam. Sci Technol J Agric Rural Dev 1(1):247–253
40. Thu PQ, Hung TX, Dinh PX, Binh VD, Chi NM (2023) Insect pests and disease in *Acacia* plantations in North Central, Vietnam. Vietnam J For Sci Special issue, 132–140
41. Tran MD, Le TH (2012) Investigatory result of *Dalbergia tonkinensis* Prain and propagation status of this tree species in Thua Thien Hue province. Hue Univ J Sci 75A:19–28
42. Tran TTH, Nguyen HTT, Pham TTH, Nguyen TPA, Hoang THQ (2011) Morphological characteristics and pathogenicity of *Fusarium* caused yellowing disease of black pepper in Cam Lo, Quang Tri. J Plant Prot 6(11–16):2354–0710
43. Trang TT, Thu PQ, Khai PQ, Tuan TA, Hinh TX, Nam NV, Thuy PTT, Chi NM (2022) First report of canker and wilt disease in eucalypt caused by *Ceratocystis manginecans* in Vietnam. Indian Phytopathol 75(1):287–291. https://doi.org/10.1007/s42360-021-00435-y
44. Truong MH (2002) The influence of concentration of treating solution and treating time on rooting of *Dalbergia bariaensis* cuttings. Vietnam J Agric Rural Dev 5:446–447
45. Van Wyk M, Al Adawi AO, Khan IA, Deadman ML, Al Jahwari AA, Wingfield BD, Ploetz, R., Wingfield, M.J.: *Ceratocystis manginecans* sp. nov., causal agent of a destructive mango wilt disease in Oman and Pakistan. Fungal Divers 27:213–230
46. Vietnam Government: Decree No. 06/2019/NĐ-CP dated 10/3/2019 on management of endangered and rare forest plants and animals and implementation of the Convention on International Trade in Endangered Species of Wild Fauna and Flora (2019)
47. Vietnam Government: Decree No. 84/2021/NĐ-CP dated 22/9/2021 on Amending and supplementing a number of articles of Decree No. 06/2019/ND-CP on management of endangered and rare forest plants and animals and implementation of the Convention on International Trade in Endangered Species of Wild Fauna and Flora (2021)
48. Vietnam Government: Decree No. 32/2006/ND-CP dated 30/3/2006 on management of endangered and rare forest plants and animals and implementation of the Convention on International Trade in Endangered Species of Wild Fauna and Flora (2006)
49. White TL, Adams WT, Neale DB (2007) Forest Genetics. CABI International, Massachusetts

50. Wingfield MJ, Wingfield BD, Warburton P, Japarudin Y, Lapammu M, Rauf MRA, Boden D, Barnes I (2023) *Ceratocystis* wilt of *Acacia mangium* in Sabah. J Trop For Sci 35:51–66. https://doi.org/10.26525/jtfs2023.35S.SI.51

Part II
Advancements and Developments

Chapter 26
Chemometrics in the Analysis of *Dalbergia latifolia* Wood: Current Techniques, Applications, and Emerging Trends

Rakesh Kumar and Rohit Sharma

1 Introduction

1.1 Introduction to Dalbergia *Genus*

The genus *Dalbergia*, belonging to the *Fabaceae* family, consists of approximately 275 species [1, 2], with around 35 species recorded in India [3, 4]. *Dalbergia latifolia* Roxb., commonly known as Rosewood, predominantly occurs in southern India, whereas *Dalbergia sissoo* DC., or Shisham, is native to the northern and central regions [3]. Both species hold considerable economic importance: *D. latifolia* is prized for its use in high-end furniture and musical instruments [5, 6], while *D. sissoo* is favoured for its adaptability and widespread application in carpentry and furniture production [7, 8]. The increasing demand and over-harvesting, particularly of *D. latifolia*, have resulted in their inclusion in Appendix II of the CITES list, signifying that while it is not at immediate risk of extinction, its trade must be regulated to ensure its continued survival [9, 10]. Despite *D. sissoo* being abundant, its anatomical similarity to *D. latifolia* has led to its inclusion in Appendix II as well, complicating enforcement and identification [11, 12]. Consequently, trade restrictions have affected artisans and farmers who depend on wood of *D. sissoo*, fuelling calls for its delisting [13, 14].

R. Kumar (✉) · R. Sharma
Wood Properties and Processing Division, ICFRE-Institute of Wood Science and Technology, Bengaluru, Karnataka, India
e-mail: colourrakesh@gmail.com; rakesh@icfre.org

R. R. Warrier et al. (eds.), *Handbook of Asian Rosewoods*, Materials Horizons: From Nature to Nanomaterials, https://doi.org/10.1007/978-981-96-9459-4_26

1.2 Overview of Chemotaxonomy

Chemotaxonomy, also known as chemosystematics, is the practice of classifying and identifying organisms (particularly plants) based on detectable differences and similarities in their biochemical compositions [15, 16]. It is the study how chemical compounds or groups of chemicals produced by living organisms are distributed across different plant species, either related or potentially related [17, 18] primary and secondary metabolites (SMs), including fatty acids, essential oils, and volatile compounds such as terpenes (mono-, di-, and triterpenes), saponins, alkaloids, flavonoids, lignin, and phenolics [19]. Integrating multiple classification methods can enhance the robustness of the taxonomic system. A key emerging trend in chemotaxonomy is the development of metabolomics and chemoinformatics tools, along with the construction of relevant databases. Chemotaxonomy plays a crucial role in plant classification when traditional botanical methods face uncertainty. As an interdisciplinary field, it integrates morphological and molecular data to enhance plant systematics [15]. In recent decades, chemotaxonomy has become a vital tool in taxonomy, complementing traditional morphological classification and genetic analysis. By examining secondary metabolites unique to plant families, genera, or species, chemotaxonomy aids in clarifying phylogenetic relationships and distinguishing between closely related species [16, 20–27].

A notable study highlights that, since the European Deforestation Regulation (EUDR) 2023/1115 prohibits the trade of wood products from illegal logging within the EU, a study presented a chemotaxonomic method using thermal desorption-gas chromatography-mass spectrometry (TD-GC–MS) to identify tropical hardwood species in pulp and paper, complementing anatomical techniques and assisting in the detection of CITES-protected species [25]. However, chemotaxonomy also faces challenges due to genetic and environmental variations, which can lead to inaccurate conclusions. By applying a holistic, multivariate data analysis approach, such as 1H NMR-based metabolomics, has successfully identified chemotaxonomic biomarkers in Euphorbia species, optimizing plant metabolomics protocols for improved species differentiation [28]. Chemotaxonomy also plays a role in fields beyond taxonomy, such as ecology, pharmacognosy, and conservation biology, by helping to identify economically and medicinally valuable plants [29]. The chemical profiles of plants can reveal adaptations to environmental pressures, and provide a basis for understanding plant biodiversity at the biochemical level [30].

Traditional taxonomic methods that rely on morphological characteristics may not always be sufficient to differentiate closely related species within *Dalbergia* genus, such as *D. latifolia* and *D. sissoo*, which share many similar anatomical features. However, chemotaxonomic markers, such as flavonoids, quinones, and phenolic compounds, offer species-specific chemical profiles that improve the accuracy of classification. Chemotaxonomic analysis enables the differentiation of species by leveraging their unique chemical signatures, making it particularly valuable for identifying and distinguishing wood species such as *D. latifolia* and *D. sissoo*. This chapter explores the chemotaxonomy of *D. latifolia*, examining the biochemical markers that define its classification and exploring how chemotaxonomic tools help to resolve the taxonomic challenges within the *Dalbergia* genus, particularly with regard to *D. latifolia* and *D. sissoo*. By providing a detailed analysis of the chemometric techniques, this chapter enhances our understanding of how to distinguish *D. latifolia* from closely related species such as *D. sissoo* and others.

2 Phytochemical Profile of *D. latifolia*

Wood of *Dalbergia* species is an important species, and it has a wide variety of secondary metabolites. The flavonoid and neoflavanoid constituents of the *Pterocarpus* and *Dalbergia* species were reviewed together with related compounds [31]. The phytochemical profile of *D. latifolia* is rich and complex, encompassing various classes of secondary metabolites that are crucial for taxonomic differentiation, medicinal applications, and ecological adaptation. Chemotaxonomic analysis of *D. latifolia* primarily focuses on identifying these compounds, including flavonoids, benzofurans, quinones, phytosterol tannins, phenolic acids, lignans and terpenoids, which contribute to the unique biochemical identity of the species. Latinone, a substituted phenanthrene-1, 4-quinone from *D. latifolia* was isolated and characterized [32]. Latifolin, a natural flavonoid, was reported in *Dalbergia* species including *D. latifolia* [33].

Phytochemicals as Chemotaxonomic Markers in *D. latifolia*

The secondary metabolites of *D. latifolia* serve as effective chemotaxonomic markers. Some of the most prominent metabolites found in *D. latifolia* include the following:

- **Dalbergichromene and latifolin**: These quinones are unique to *D. latifolia* and are rarely found in other *Dalbergia* species. They are often used to differentiate *D. latifolia* from species like *D. sissoo*. *D. latifolia* wood also contains dalbergichromene, latinone, and dalcriodain [34]. Latifolin, O-methyl dalbergin, 3-hydroxy-2,4,5-trimethoxydalbergiquinol, 2,4,5-trimethoxydalbergiquinol, and dalbergin are the other major flavonoids found in *D. latifolia* wood [35, 36].
- **Isoliquiritigenin**: The flavonoids, found in high concentrations in *D. latifolia*, distinguish it from other Fabaceae members [37].
- **Flavonoids**: The primary flavonoids in *D. latifolia* wood include dalbergiphenol, 2,3-dimethoxy-2,4-dihydroxychalcone, an isomer of latifolin dimethyl ether, and latifolin dimethyl ether itself. Additionally, 5-O-methyl latifolin and an isomer of 2-O-methyl latifolin were also reported. Other significant flavonoids include latifolin, 2-O-methyl dalbergin, 3-hydroxy-2,4,5-trimethoxydalbergiquinol, 2,4,5-trimethoxydalbergiquinol, and dalbergin [36].

- **Benzofurans**: Three benzofurans from the heartwood of *D. latifolia* were obtained including a new benzofuran compound, named 2-[5-hydroxy-4-methoxy2-(3-phenyl-trans-allyloxy) benzyl]-5-hydroxy-6-methoxy-3-phenylbenzofuran, together with (+)-obtusafuran and isoparvifuran [38].
- **Tannins and phenolic acids**: Variations in the types and concentrations of tannins and phenolic acids are particularly valuable in distinguishing *D. latifolia* from related species, as these compounds are influenced by environmental conditions and genetic factors [35, 36, 39].
- **Phytosterols and triterpenes**: *D. latifolia* wood contains an important compound called lupeol. Additionally, other important compounds found in this species include β-sitosterol, β-eudesmol, β-sitostenone, and stigmasterol [39, 40]. The distinct chemical profile of *D. latifolia* may supports its classification within the *Dalbergia* genus, highlighting the chemotaxonomic role of secondary metabolites in refining species-level distinctions. The identification of these markers has implications for wood verification processes in trade, conservation efforts, and further taxonomic studies. The detailed phytochemical description has been discussed in the chapter titled "Phytochemistry of *Dalbergia latifolia*: Bioactive Compounds and Their Therapeutic Potential" Chap. 9 .

3 Challenges in Traditional Taxonomy of *D. latifolia*

Conventional methods for identification of wood include visual, physical, and anatomical characteristics of wood species, which takes more time as well as is labour-intensive. In recent years, advanced techniques in wood identification such as chemical isotopes and DNA markers have been developed [41, 42]. However, identifying timber from Dalbergia species based solely on wood anatomy remains challenging [43, 44], which hinders the enforcement of CITES regulations, as accurate timber identification is crucial for trade monitoring and control. Additionally, many Dalbergia species in the Indochina region are difficult to distinguish in the field, even when fruits or flowers are present [45]. These challenges are further compounded by the absence of a comprehensive and updated taxonomic revision. *D. latifolia* closely resembles other *Dalbergia* species, especially with *D. sissoo* both of which share similar geographic distributions and morphological characteristics. Due to these similarities, taxonomic misidentification is common, posing challenges for conservation, trade regulation, and species-specific research [43, 45]. Misidentification has also fuelled illegal logging and trade, particularly for high-value timber species like *D. latifolia.*

Traditional taxonomy is often insufficient in accurately identifying *Dalbergia* species due to the following:

- **Morphological convergence**: Many *Dalbergia* species exhibit convergent morphological traits, which are similar in appearance due to ecological adaptations.

- **Phenotypic variations**: Environmental factors can cause substantial phenotypic variation within a species, making field-based identification challenging.
- **Hybridization**: Natural hybridization among *Dalbergia* species contributes to genetic and morphological overlap, further complicating classification efforts.

4 Chemotaxonomy as a Tool in *Dalbergia* Classification

Chemotaxonomy offers an alternative approach to traditional methods by focusing on chemical compounds that serve as consistent and species-specific markers. Sharma et al. [46] explored the integration of ATR-FTIR spectroscopy and machine learning to enhance wood species identification for combating illegal logging and ensuring responsible sourcing. By analysing *Eucalyptus, Dalbergia*, and *Populus* species in India, it demonstrates effective species discrimination, supporting law enforcement and quality control in the timber industry. Shang, et al. [47] used gas chromatography with high-resolution mass spectrometry (GC/QToF) to identify *Dalbergia* species, with the objective of combating illegal timber trade. A Microsoft Excel® template and Kernel Discrimination Analysis (KDA) were developed for species identification. The approach ensures accurate identification, addresses mislabelling, and proposes creating an EI + spectral database for future reference. Liu et al., [48], developed a computer vision model, SE-ResNet, using a CITES-28 macroscopic image dataset for species identification. The model achieved 99.65% accuracy, with effective image pre-processing methods like cropping and rotation. Real-world testing on timber market specimens resulted in 82.3% accuracy, aiding illegal timber trade enforcement. Woods primarily consist of cellulose, hemicellulose, lignin, and extractives which contains secondary metabolites [49], with their composition varying based on factors like tree part, geographic origin, and environmental conditions [50]. Detailed tree identification is typically achieved through chemotaxonomy, focusing on minor constituent's confirmable differences and similarities in their biochemical compositions [15]. Since few chemicals in plants are specific to certain species or geographical regions, chemometrics can be used to differentiate them. By analysing chemical data through statistical methods, chemometrics helps to identify unique chemical signatures, enabling accurate identification of plant species and their origin. In a study, FT-IR and 2DCOS-IR spectroscopy were used to differentiate *D. cochinchinensis* from similar species. Specific peaks at 738 cm^{-1}, 755 cm^{-1}, and 701 cm^{-1} helped distinguish it from *D. retusa, D. bariensis*, and *D. oliveri*. Additionally, *D. cochinchinensis* exhibited more stable extractives, providing effective chemical markers for identification [51].

Key benefits of chemotaxonomy in the classification of *D. latifolia* include the following

1. **Accuracy in species differentiation**: Chemical markers allow for precise differentiation between *D. latifolia* and other *Dalbergia* species, especially those with similar morphology.

2. **Forensic applications**: Chemotaxonomic markers aid in the forensic identification of wood samples, supporting legal enforcement in the timber trade.
3. **Conservation efforts**: By accurately identifying species, chemotaxonomy helps conservationists prioritize protection efforts for vulnerable species like *D. latifolia.*

5 Comparative Chemotaxonomic Analysis Within the *Dalbergia* Genus

Chemotaxonomic comparisons between *D. latifolia* and other *Dalbergia* species reveal significant variations in secondary metabolite composition. Following are some examples:

- **Comparison with *D. sissoo*:** While both species share certain phenolic compounds, *D. latifolia* exhibits higher concentrations of quinones, particularly dalbergichromene and latifolin, making these key distinguishing markers [36]. Additionally, its flavonoid profile differs, with a higher presence of genistein compared to *D. sissoo* [52]. Dalbergichromene found in the stem bark and heartwood of *D. sissoo* [53] and a red crystalline solid compound obtained from the methanol extract of heartwood of *D. latifolia* known as latinone [32] may also help in identification. Isodalbergin present in the stem bark and heartwood of *D. sissoo* [53] may also be helpful for their differentiation.
- **Comparison with *D. melanoxylon***: *D. melanoxylon*, or African blackwood, contains distinct terpenoid profiles not found in *D. latifolia*, making terpenoids valuable markers for differentiating these species. Furthermore, quinones in *D. melanoxylon* differ significantly from those in *D. latifolia*, further aiding species-specific identification. Compounds namely *S*(+)-3′-hydroxy-4′,2,4,5-tetramethoxy dalbergiquinol (1), and a benzofuran, named (2*S*,3*S*)-5-hydroxy-6-methoxy-3-methyl-2-(4′-hydroxyphenyl)-2,3dihydrobenzofuran, was reported in *D. melanoxylon* [38].
- **Comparison with *D. nigra***: *D. nigra*, known for its highly prized Brazilian rosewood, is chemically distinguished from *D. latifolia* by its phenolic acid profile, which contains unique acids such as rosmarinic acid not present in *D. latifolia* wood. The flavonoid content in *D. nigra* also shows distinct patterns that separate it from other *Dalbergia* species including *D. latifolia* [54].

6 Analytical Methods in Wood Identification

Modern wood identification leverages several analytical techniques, each providing unique insights into the chemical and structural characteristics of wood. When combined with chemometrics, these methods can achieve high accuracy and efficiency (Fig. 1).

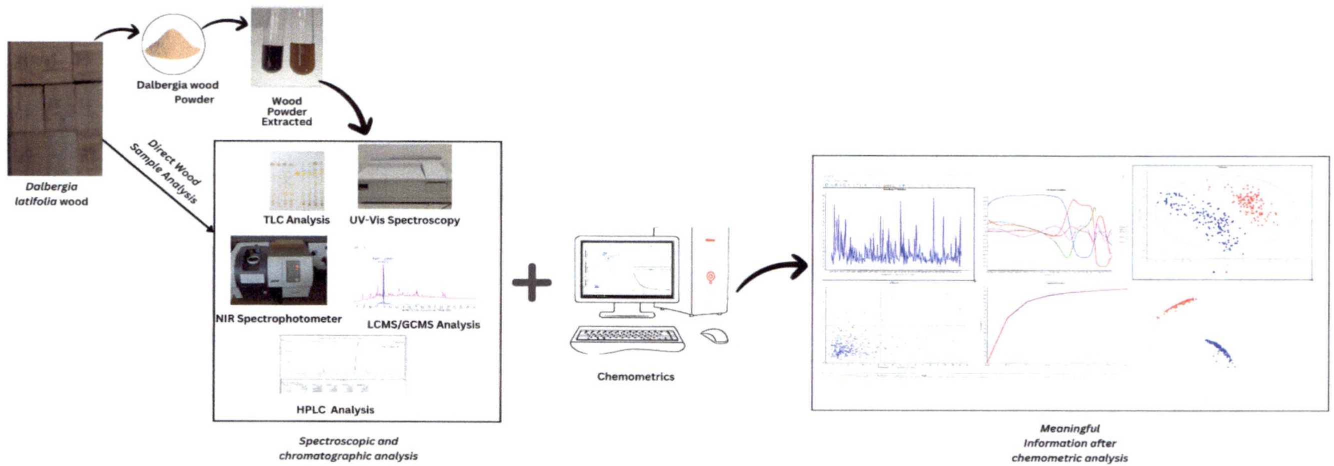

Fig. 1 Graphical representation of chemotaxonomic discrimination between *Dalbergia* species

6.1 Spectroscopic Methods

Many researchers around the globe have reported different approaches for the identification of *Dalbergia* species. For chemotaxonomic discrimination between these easily confused species, *D. cultrate, D. latifolia*, and *D. melanoxylon*, the total extractive content of the three wood species was determined [55], using four different organic solvents. They concluded that FTIR and GC–MS spectroscopy can be applied to identify and discriminate among *D. cultrate, D. latifolia*, and *D. melanoxylon.*

Further, fluorescence spectroscopy may be a valuable technique for the identification of timber [56]. The researchers conducted a study to measure fluorescence excitation and emission spectra of solid wood blocks, powders, and their methanol extracts from the heartwood of various tree species, including *Acacia nilotica* (Babul), *Albizia lebbek* (Kokko), *Pterocarpus marsupium* (Bijasal), *Toona ciliata* (Toon), *Tectona grandis* (Teak), and *Eucalyptus tereticornis* (Eucalypts). Fluorescence spectra obtained from solid dry wood were found to be independent of excitation wavelength, while those obtained from extracts exhibited excitation wavelength dependence, indicating the presence of multiple fluorescent chemicals, which would be beneficial for the identification of different wood species. FTIR and 2D correlated IR spectroscopy were used for the rapid identification of *Pterocarpus santalinus* and *Dalbergia louvelii* [51]. A tri-step identification method, consisting of conventional infrared spectroscopy (FT-IR), second derivative infrared spectroscopy (SD-IR), and two-dimensional correlation infrared spectroscopy (2DCOS-IR), was applied to investigate *P. santalinus* and *D. louvelii* wood in furniture. The aim was to develop a quick and efficient method for easily differentiating these commonly confused wood species in furniture. Further a study was developed using a three-step identification of infrared spectra for distinction of tree species closely related to *Pterocarpus santalinus* [57], thereby eliminating the difficulties faced to assess the wood species in the furniture market in resembling tree species such as *Dalbergia louvelii* and *Pterocarpus tinctorius.* It was concluded that tree species can be rapidly and accurately identified using the three-step process using infrared spectroscopy. In a study, multivariate analyses of UV–Visible absorption spectral data from cachaça wood extracts to classify varied aged Brazilian cachaças according to the wood species was used. The technique offered a low-cost substitute for techniques based on liquid chromatography and mass spectrometry to categorize cachaça aged in barrels made of various wood species. Six clusters of cachaça wood extracts (amburana, amendoim, balsamo, castanheira, jatoba, and oak) were identified by the use of Principal Components Analysis (PCA) and Hierarchical Cluster Analysis (HCA). Linear Discriminant Analysis (LDA) further classified 10 different wood species (amburana, amendoim, balsamo, cabreuva-parda, canela-sass afras, castanheira, jatoba, jequitiba-rosa, louro-canela, and oak) used in the cachaça extracts with an accuracy ranging from 80% (amendoim and castanheira) to 100% (balsamo and jequitiba-rosa) [58].

1. **Near-Infrared (NIR) Spectroscopy**: This technique is widely used due to its rapid and non-destructive nature. NIR spectroscopy analyses the overtones and combinations of molecular vibrations in wood components like cellulose, lignin, and hemicellulose. Studies demonstrate the effectiveness of NIR spectroscopy in determining wood species, moisture content, and chemical composition [59]. Different researchers have done considerable work to identify timber species by using various approaches including near-infrared spectroscopy [60–62] mass spectroscopy [1, 63, 64], radiocarbon dating, and stable isotope ratio analysis. Moreover, genetic marker-based approaches such as DNA profiling, DNA barcoding, metabolome profiling, and karyotyping have also been used for timber analysis [65–67].
2. **Hyperspectral Imaging (HSI)**: Hyperspectral imaging combines imaging and spectroscopy to capture spatial and spectral information. Near-Infrared HSI has been applied to identify wood species [62]. Many researchers have proposed innovative solutions for automated systems leveraging HSI technology for various aspects of wood analysis, including wood classification, species identification, fungal detection, and wood segmentation [68].
3. **Fourier Transform Infrared (FTIR) Spectroscopy**: FTIR provides detailed information on the functional groups in wood. FTIR spectroscopy combined with chemometrics tools like Partial Least Squares Regression (PLSR), has been used to identify wood origin and assess chemical modifications. This approach has applications in both the wood industry and forensic science [46]. FTIR spectroscopy was used to examine variations in functional groups within the extractive components of three species from *Dalbergia* genus (*Dalbergia cultrata, D. latifolia,* and *D. melanoxylon*), allowing for the deduction of the main chemical component types based on the positions, intensities, and shapes of characteristic peaks [55].
4. **UV–Visible Spectroscopy**: UV spectroscopy and chemometrics combined can be used for the identification of different species. Combination of the UV spectroscopy and chemometrics was used to identify six different commercial vegetable tannins [69]. The use of UV–visible spectroscopy for chemical fingerprinting is a valuable method for differentiating the different wood species based on the chemical components of wood samples [70, 71]. Similar looking *Pterocarpus* species was differentiated using UV–visible spectroscopy [70] in another study; the two look-alike species *Tectona grandis* and *Acacia auriculiformis* were distinguished using the UV–visible spectroscopy [71].
5. **Raman Spectroscopy**: This technique detects molecular vibrations to identify specific chemical bonds. Its non-invasive application, combined with multivariate analysis, allows for precise wood species discrimination. Research findings suggest that Raman spectroscopy holds great potential as a reliable method for identifying wood species [72]. Raman spectroscopy combined with pattern recognition techniques is utilized to develop a promising method for distinguishing and classifying wood based on its type [73]. Fourier-transform Raman spectroscopy can also be applied to analyse wood from various geographical

regions and time periods. One study examined specimens of *Caesalpinia echinata* Lam. (from Brazil, South America), *Caesalpinia sappan* L. (from the East Indies, Asia), *Haematoxylum brasiletto* Karsten (from Central America), and *Haematoxylum campechianum* L. (from North America) with a goal to identifying distinct Raman biomarker bands that could serve as the foundation for a reliable identification protocol [74].

6.2 Chromatographic Method

1. **High Performance Liquid Chromatography (HPLC) Method**: HPLC for identification and quantification of bioactive compounds has been a method of choice for researcher across the globe. Multilevel profiling and identification of *D. odorifera* and *D. stevensonii* was done using GC/MS [75]. A method was developed and validated [76] using reverse-phase HPLC to identify and quantify nine chemical markers in red propolis, including liquiritigenin, calycosin, isoliquiritigenin, formononetin, vestitol, neovestitol, medicarpin, biochanin A, and 7-O-methylvestitol. These markers are present in Brazilian red propolis extracts, as well as in *Dalbergia ecastaphyllum* and *Etlingera pavieana* [75, 76]. Similarly, [78] developed HPLC method to determine the presence of nine major flavonoids in *D. odorifera*, including 3'-hydroxymelanettin, melanettin, stevenin, butein, isoliquiritigenin, dalbergin, 2,4-dihydroxy-5-methoxybenzophenone, pinocembrin, and 4-methoxydalbergione, and it was successfully applied to analyse 32 commercial samples of *D. odorifera*. Development and validation of an analytical method for condensed tannin extracts, and phenol and its derivatives and catechin from different tree species using HPLC was reported [78–80]. Sterile and fungus-treated pine wood was analysed using HPLC with evaporative light scattering detection [80], and it was found that free fatty acids, sterols, and triglycerides are clearly distinguished. From *Osmanthus fragrans* root bioactive components were identified and quantified by HPLC–MS/MS [81]. Fingerprint evaluation using HPLC and HPTLC for leaf and stem wood extracts was effectively demonstrated for genus identification in *Artocarpus hirsutus* Lam, and it was reported that oxyresveratrol and artocarpin are the two important components to be present in the stem wood extract, but absent from the leaf extract [82]. The bioactivity of latifolin and its derivatives against termites and fungi was studied [83].
2. **Thin-Layer Chromatography (TLC)**: Researchers have been utilizing thin layer chromatography (TLC) for identification of wood and other plant parts. *Chamaecyparis obtusa* and *C. obtusa var. formosana* are significant species utilized in traditional Japanese architecture [84]. Chemometrics may also be used in fingerprinting by means of thin layer chromatography [85]. In order to develop a straightforward, inexpensive technique for identifying the colourants employed in the historic manufacturing of artefacts, TLC was the primary technology selected [86]. Cellulose thin-layer chromatography plate (Cell-TLC) was developed and its usefulness was showed in screening wood extracts that contained a

variety of chemicals [87]. Thirteen *Dalbergia* (rosewood) species were studied, and it was found that the species produces a purple quinonemethide pigment in the callus that appeared to be identical across the different species, which may serve as an excellent model system for investigating metabolic differentiation similar to that observed in heartwood [88]. Three different species of softwoods, namely Douglas fir, Radiata pine, and Sugi were classified with the help of thin-layer chromatography (TLC) combined with chemometric analysis thereby improving the reliability of near-infrared (NIR)-based classification via extractive content [89]. By applying chemometrics techniques, the TLC fingerprints were able to differentiate *Curcuma longa* samples from various origins [90]. The PCA score plot, based on the first two principal components (PCs), and the Cluster Analysis (CA) clearly identified two distinct clusters of samples. It was found that combination of TLC fingerprinting with PCA and Cluster Analysis (CA) proved effective in distinguishing *C. longa* rhizomes obtained from different geographical locations. In a study, TLC was used for distinguishing and developing fingerprints for different Thymus species [91]. Some recent developments in the field of chemotaxonomic techniques involve distinctive profiling of phytochemical markers. Direct electrospray ionization mass spectrometry (ESI–MS) analysis of a straightforward methanolic extract of a microscopic wood chip was used to complete this difficult task utilizing a straightforward, quick, and unambiguous technology. The distinctive polyphenols like catechin derivatives and cinchonain, as well as typical limonoids like khivorin, khayanolide A, and mexicanolide for African mahogany (*Khaya ivorensis*) and phragmalin-type limonoids for the native Brazilian mahogony (*Swietenia macrophylla*) species make up the distinguishing phytochemical marker pools for different species [22].

3. **DARTMS/QTOF, GC/QTOF, and LC/QTOF Method**: Keeping the view of the conventional methods of anatomical studies and DNA analysis have proven insufficient and/or not practical in identifying wood species to the necessary level for effective monitoring, and thus alternative approaches, such as chemotyping techniques, were necessary [92]. In their study, they extracted samples from endangered rosewood species (*Dalbergia* spp.) using a simple field method, which revealed distinct and recognizable hues suggestive of the specific wood species. To establish more conclusive and precise wood species identification methods, a thorough study was carried out utilizing high-resolution quadrupole time-of-flight (QTOF) mass spectrometers coupled with liquid chromatography (LC), gas chromatography (GC), and Direct Analysis in Real Time (DART) techniques [92]. Due to insufficient ability of anatomy, [23] developed direct analysis in real time connected to Fourier transform ion cyclotron resonance mass spectrometry (DART-FTICR-MS), and found distinct chemical signatures for the heartwood samples from the two species, *Pterocarpus santalinus* and *Pterocarpus tinctorius* which allowed for quick species-level identification when multivariate statistical analysis was used. Direct Analysis in Real Time (DART) Time-of-Flight Mass Spectrometry (TOFMS) was used for species identification of protected *Dalbergia* trees from Africa, Madagascar, and Asia [1]. They

concluded that *D. cochinchinensis* and *Dalbergia* spp. from Madagascar (both listed in CITES Appendix II) could be differentiated from each other, as well as from the non-protected *D. latifolia* and *D. melanoxylon*. In another study, DART-MS was found to be effective in assisting in the identification of various *Dalbergia* species [93]. Also [94, 95] reported DART-MS is suitable for identification of *Dalbergia* species. Mix-ups and identification issues have developed from the rising demand for *P. santalinus,* and woods with morphological similarities like *P. tinctorius* have been illegally traded in its name. Similar to *Dalbergia* wood, unlawful harvesting due to high value and commercial demand of *Pterocarpus santalinus* resulted in its inclusion in CITES Appendix II as an endangered timber species [96]. The researchers developed methods for precise differentiation between *P. santalinus* and *P. tinctorius* using direct analysis in real time and Fourier transform ion cyclotron resonance mass spectrometry (DART-FTICR-MS). The DART-FTICR-MS spectra revealed sizable variations with the most accurate prediction being made using orthogonal partial least square discriminant analysis (OPLS-DA). These results show that DART-FTICR-MS and multivariate statistical analysis can be used to authenticate wood types.

6.3 Chemometric Techniques In Wood Identification

Chemometric methods are essential for processing the complex datasets generated by analytical techniques. These methods include the following:

1. **Principal Component Analysis (PCA):** PCA reduces the dimensionality of large datasets while retaining the most significant variations. It is frequently used in exploratory data analysis to visualize differences among wood samples and identify species clusters. Near-infrared spectroscopy (NIRS) was utilized combined with PCA for identifying hardwood species in wood chip form, considering moisture as a key influencing factor [97]. Similarly, NIR spectroscopy was studied as a rapid, non-destructive method to distinguish *Abies balsamea* and *Abies lasiocarpa* from Pine and Spruce respectively by using multivariate-based PCA [98]. Wood samples collected from the market belonging to three wood genera, *Instia* spp., *Couratari* spp., and *Pometia* spp.were identified, and a model was developed using PCA and PLS-DA [99]. Their results demonstrated that NIR spectroscopy can accurately distinguish between different wood species. Similarly, NIRS was utilized to identify wood species from archaeologically and historically significant wooden artefacts in a non-invasive manner. Their findings indicated that an NIR-based PCA model is an effective tool for evaluating the historical or archaeological significance of these items [100]. When UV–Vis spectra combined with PCA and Discriminant Analysis (DA) for classification of species, the DA provided a more distinct classification, accurately grouping 95.5% of the samples through leave-one-out cross-validation [101].

2. **Partial Least Squares Regression (PLSR)**: PLSR correlates spectral data with reference measurements to predict properties such as moisture content, lignin concentration, and species classification. This method is particularly effective in NIR and FTIR spectroscopic analyses. [102] investigated and concluded the suitability of separation of wood with the NIR spectroscopy technique combined with PLS-R and orthogonal signal correction (OSC) filtering of spectra to differentiate heartwood and sapwood of *P. abies* (Norway spruce).
3. **Discriminant Analysis (DA)**: DA classifies wood samples into predefined groups (e.g., species or regions of origin) based on their chemical or physical attributes. Techniques like Linear Discriminant Analysis (LDA) and Quadratic Discriminant Analysis (QDA) have shown high accuracy in species identification. [103] used different models such as PLS-DA and artificial neural networks (ANN) to identify mixed population of *Tsuga heterophylla* and *Abies amabilis,* revealing that both modelling techniques can be effective tools for species differentiation. Multivariate analysis was used to investigate the ability of hand-held NIRS technology to distinguish the spectra of a few important *Dalbergia* species [104]. It was observed that PLS-DA approach proved to be better than soft independent modelling of class analogies (SIMCA) for generating a field-deployable NIRS technique. Themultivariate analysis such as PCA and LDA were used to classify wood species into distinct groups based on their chemical content. UV–Vis absorption spectra and chemometrics together were used for rapid detection of sappan wood raw material adulteration from three different dyes [105].
4. **Clustering and Machine Learning Algorithms**: Hierarchical clustering and algorithms like Support Vector Machines (SVM) and Random Forests are increasingly used for complex classifications. These approaches are particularly valuable when handling datasets with overlapping spectral features. The researchers analysed data using PCA and hierarchical cluster analysis (HCA) [77]. The analysis resulted in the formation of two distinct clusters, each containing a single species. The complete dataset was subsequently split into two subsets: a training set and a test set. Wood sample differentiation is achieved through multivariate statistical analysis, including Hierarchical Cluster Analysis (HCA) and Principal Component Analysis (PCA) [46]. PCA and HCA approaches were used to effectively classify the wood into softwoods and hardwoods [106].

7 Challenges and Future Directions in the Chemotaxonomy of *D. latifolia*

While chemotaxonomy has proven invaluable in studying and classifying *D. latifolia*, several challenges and future directions must be considered to enhance its applications further.

7.1 Challenges In Chemotaxonomy

1. **Limited availability of reference material**

   One of the primary challenges in chemotaxonomic research is the availability of authenticated reference materials. Establishing a reliable chemotaxonomic database for *Dalbergia* species requires access to well-preserved and genetically /botanically verified samples, which can be difficult to obtain for endangered species like *D. latifolia.*
2. **Environmental variability impact**

   While chemotaxonomy is generally more reliable than morphological methods, environmental factors like soil composition, climate, and altitude can still affect chemical profiles, leading to variations within a single species. To create accurate and representative chemical profiles, it is essential to sample extensively from multiple populations.
3. **Analytical limitations**

   Traditional methods like High-Performance Liquid Chromatography (HPLC) and Mass Spectrometry (MS) are highly effective but demand significant resources and expertise, which can limit their accessibility. Further, these techniques offer high accuracy, they may not identify all relevant compounds, highlighting the importance of multimodal analysis approaches.

7.2 Advances In Analytical Techniques

Technological advancements in analytical chemistry have opened new possibilities for chemotaxonomy in *D. latifolia.* Emerging methods include the following:

1. **Nuclear Magnetic Resonance (NMR) Spectroscopy**

   NMR offers a non-destructive analysis method that provides detailed information on the molecular structure of chemical compounds. Its use in *Dalbergia* research could enhance the accuracy of chemotaxonomic markers by identifying small chemical differences that distinguish closely related species.
2. **DNA-Metabolite Integration**

   Combining chemotaxonomic data with DNA analysis can improve the resolution of species classification and facilitate studies on the genetic basis of secondary metabolite production. This integrated approach could lead to a more robust taxonomic framework for *Dalbergia* species identification and improve species verification in forensic applications.
3. **Machine learning and data mining in chemotaxonomy**

   Machine learning algorithms can analyse complex chemotaxonomic datasets and identify patterns that might not be immediately apparent. Data mining could help classify *Dalbergia* species more accurately by correlating chemical markers with specific taxonomic traits, environmental conditions, and genetic factors.

7.3 Future Research Directions

1. **Establishment of a chemotaxonomic database for *Dalbergia* spp.**

   Developing an accessible, standardized chemotaxonomic database would aid in conservation efforts, trade regulation, and academic research. This database could gather information on chemical markers, geographic origin, and species status, facilitating more efficient species identification and conservation monitoring.
2. **Conservation-centric chemotaxonomy**

   Future research should place greater emphasis on applying chemotaxonomic methods within conservation biology. Studies could investigate how chemical markers fluctuate in response to environmental stress, providing valuable insights for conservationists in managing and restoring populations of *D. latifolia* and other threatened species of *Dalbergia* spp.
3. **Pharmacological exploration of *D. latifolia* compounds**

   Expanding research on the pharmacological properties of *D. latifolia* phytochemicals can help to develop sustainable applications in natural medicine. Studies could examine how chemotaxonomic markers correlate with bioactive properties, providing insights into potential health benefits and therapeutic uses.

8 Conclusion

The chemotaxonomy of *D. latifolia*, or Indian rosewood, plays a pivotal role in the taxonomy, conservation, and sustainable utilization of this valuable species. Traditional taxonomic approaches relying on morphological characteristics have proven insufficient to differentiate *Dalbergia* species variations. Chemotaxonomy, by analysing unique secondary metabolites like flavonoids, quinones, and polyphenols, may provide a more precise and reliable method for identifying *D. latifolia* and distinguishing it from other *Dalbergia* species. Chemotaxonomic markers not only aid in species verification but also serve crucial applications in conservation and legal frameworks. As *D. latifolia* faces pressures from illegal logging and habitat loss, chemotaxonomic identification has become crucial in regulating timber trade and enforcing protections, such as those under CITES. Forensic applications in wood science rely mainly on these chemical markers to authenticate species origin, aiding in the fight against illegal timber trade and supporting ethical sourcing in industries. Despite the advantages, challenges remain in chemotaxonomy due to environmental variability and analytical limitations. However, advancements in analytical techniques, such as NMR spectroscopy, DNA-metabolite integration, and machine learning, are opening new frontiers in chemotaxonomic research. The establishment of comprehensive chemotaxonomic databases and the integration of chemotaxonomy with molecular methods have potential to enhance accuracy in species identification and broaden applications in conservation and pharmacology. Hence, the chemotaxonomy of *D. latifolia* highlights the importance of chemical markers in modern taxonomy, supporting sustainable conservation and utilization of a valuable natural resource. Through continued research and technological advancements, chemotaxonomy will remain essential in the preservation, ethical management, and understanding of *D. latifolia* and the broader *Dalbergia* genus. The practical use of chemotaxonomy in *Dalbergia* identification, conservation, forensic wood science, and species-specific research is crucial for precise wood identification.

Acknowledgements The authors express their sincere gratitude to the Indian Council of Forestry Research and Education (ICFRE), Dehradun. The authors sincerely express their gratitude to the Director, ICFRE-Institute of Wood Science and Technology, Bengaluru for their continuous support and encouragement. We extend our appreciation to the Group Coordinator (Research), ICFRE-IWST, and the Head, Wood Properties and Processing, Division, for their valuable guidance and insights throughout this work.

References

1. McClure PJ, Chavarria GD, Espinoza E (2015) Metabolic chemotypes of CITES protected *Dalbergia* timbers from Africa, Madagascar, and Asia. Rapid Commun Mass Spectrom 29:783–788. https://doi.org/10.1002/rcm.7163
2. Cowell C, Williams E, Bullough LA, Grey J, Klitgaard B, Govaerts R, Andriambololonera S, Cervantes A, Crameri S, Lima HC, Lachenaud O, Li SJ, Linares JL, Phillipson P, Rakotonirina N, Wilding N, van der Burgt X, Vatanparast M, Barker A, Barstow M, Beentje H, Plummer J (2022) CITES *Dalbergia* Checklist: Commissioned by the CITES Secretariat
3. Arunkumar AN, Warrier RR, Kher MM, Teixeira da Silva JA (2022) Indian rosewood (*Dalbergia latifolia* Roxb.): biology, utilisation, and conservation practices. Trees 36(3):883–898. https://doi.org/10.1007/s00468-021-02243-3
4. Purkayastha SK (1997) A manual of Indian timbers. Sribhumi Publishing Company, Calcutta
5. Atikah T, Wardani W, Surya D (2021) Population, economic value and the trade of *Dalbergia latifolia* in Lombok, West Nusa Tenggara, Indonesia. In: Presented at the IOP conference series: earth and environmental science 762(1):012015. https://doi.org/10.1088/1755-1315/762/1/012015
6. Yulita KS, Atikah TD, Wardani W (2020) Unraveling genetic variations of *Dalbergia latifolia* (Fabaceae) from Yogyakarta and Lombok Island, Indonesia. Biodivers J Biol Div 21(2):833–841. https://doi.org/10.13057/biodiv/d210251
7. Bhattacharya M, Singh A, Ramrakhyani C (2014) *Dalbergia sissoo*-variability in morphology. J Med Plants Stud 2(3):8–13
8. Hossain S, Martin A (2013) Merchantable timber production in *Dalbergia sissoo* plantations across Bangladesh: regional patterns, management practices and edaphic factors. J Trop For Sci 25(3):299–309
9. IUCN, TRAFFIC: IUCN/TRAFFIC analyses of the proposals to amend the CITES, appendices. Prepared by IUCN Global Species Programme and TRAFFIC for the 18th Meeting of the, Conference of the Parties to CITES, Gland, Switzerland (2019)

10. Dormontt EE, Boner M, Braun B, Breulmann G, Degen B, Espinoza E, Gardner S, Guillery P, Hermanson JC, Koch G, Lee SL (2015) Forensic timber identification: it's time to integrate disciplines to combat illegal logging. Biol Cons 191:790–798. https://doi.org/10.1016/j.biocon.2015.06.038
11. Espinoza EO, Wiemann MC, Barajas-Morales J, Chavarria GD, McClure PJ (2015) Forensic analysis of CITES-protected *Dalbergia* timber from the Americas. IAWA J 36(3):311–325. https://doi.org/10.1163/22941932-20150102
12. Mahatara D, Acharya AK, Dhakal BP, Sharma DK, Ulak S, Paudel P (2021) Maxent modelling for habitat suitability of vulnerable tree *Dalbergia latifolia* in Nepal. Silva Fennica 55(4). https://doi.org/10.14214/sf.10441
13. Nautiyal S, Pal M, Sagta H, Rawat R, Negi D (2003) Clonal variation in growth, flowering and seed production of *Dalbergia sissoo* in a clonal seed orchard. J Trop For Sci 15(1):97–108
14. Ashutosh S, Lakhchaura P, Pandey K, Ghosh S, Tripathi S, Tripathi HK (2020) Trees outside forest resources in India. In: FSI technical information series. Forest Survey of India, Ministry of Environment, Forest & Climate Change, Government of India, pp. 1–30
15. Hao D, Gu X, Xiao P (2015) Chemotaxonomy: a phylogeny-based approach. Medicinal plants. Woodhead Publishing, Cambridge, UK, pp 1–48
16. Flaig ML, Berger J, Helmling S, Olbrich A, Schaffrath HJ, Zahn D, Saake B (2024) Chemotaxonomic and anatomic wood species identification in bleached pulp: blind test and method validation. Holzforschung 78(9):487–502. https://doi.org/10.1515/hf-2024-0025
17. Elbalola AA, Abbas ZK (2024) Chemotaxonomy, antibacterial and antioxidant activities of selected aromatic plants from Tabuk region-KSA. Heliyon 10(1):e23641. https://doi.org/10.1016/j.heliyon.2023.e23641
18. Bhargava VV, Patel SC, Desai KS (2013) Importance of terpenoids and essential oils in chemotaxonomic approach. Int J Herbal Med 1(2):14–21
19. Hao D, Gu X, Xiao P (2015) Taxus medicinal resources: a comprehensive study. Med Plants 1:97–136
20. Seikel MK, Hall SS, Feldman LC, Koeppen RC (1965) Chemotaxonomy as an aid in differentiating wood of eastern and western white pine. Am J Bot 52(10):1046–1049. https://doi.org/10.1002/j.1537-2197.1965.tb07283.x
21. Herout V (1971) Chemotaxonomy of the family Compositae (Asteraceae). Presented at the Pharmacognosy and Phytochemistry: 1st International Congress Munich, pp 93–110. https://doi.org/10.1007/978-3-642-65136-6_5
22. Fasciotti M, Alberici RM, Cabral EC, Cunha VS, Silva PR, Daroda RJ, Eberlin MN (2015) Wood chemotaxonomy via ESI-MS profiles of phytochemical markers: the challenging case of African versus Brazilian mahogany woods. Anal Methods 7(20):8576–8583. https://doi.org/10.1039/C5AY01725D
23. Zhang M, Zhao G, Guo J, Wiedenhoeft AC, Liu CC, Yin Y (2019) Timber species identification from chemical fingerprints using direct analysis in real time (DART) coupled to Fourier transform ion cyclotron resonance mass spectrometry (FTICR-MS): comparison of wood samples subjected to different treatments. Holzforschung 73(11):975–985. https://doi.org/10.1515/hf-2018-0304
24. Tuncer FD (2023) Comparing modelling performance of chemometric methods for wood discrimination by near infrared spectroscopy. Wood Mat Sci Eng 18(2):422–433. https://doi.org/10.1080/17480272.2022.2039960
25. Flaig ML, Berger J, Wenig P, Olbrich A, Saake B (2023) Identification of tropical wood species in paper: a new chemotaxonomic method based on extractives. Holzforschung 77(11):860–878. https://doi.org/10.1515/hf-2023-0048
26. Abdelfattah HM, Hussein HA, Teleb SS, El-Demerdash MM, George NM (2024) Chemotaxonomy compared to morphological and anatomical taxonomy of five *Hibiscus* species. J Plant Res 137(6):967–984. https://doi.org/10.1007/s10265-024-01566-9
27. Zhiponova M, Yordanova Z, Ivanov G, Gašić U, Mišić D, Alipieva K, Mantovska D, Cholakova G, Kapogianni A, Tsacheva I (2025) Chemotaxonomy and bioactive potential of high-mountain *Plantago atrata*. Chem Biodivers e202402433. https://doi.org/10.1002/cbdv.202402433

28. Sofrenić I, Anđelković B, Gođevac D, Ivanović S, Simić K, Ljujić J, Tešević V, Milosavljević S (2023) Metabolomics as a potential chemotaxonomical tool: application on the selected *Euphorbia* species growing wild in Serbia. Plants 12:262. https://doi.org/10.3390/plants12020262
29. Davis CC, Choisy P (2024) Medicinal plants meet modern biodiversity science. Curr Biol 34:158–173. https://doi.org/10.1016/j.cub.2023.12.038
30. Sharma MK (2024) Exploring the biochemical profiles of medicinal plants cultivated under stressful environmental conditions. Current Agric Res J 12(1):81–103. https://doi.org/10.12944/CARJ.12.1.07
31. Seshadri TR (1972) Polyphenols of *Pterocarpus* and *Dalbergia* woods. Phytochemistry 11(3):881–898. https://doi.org/10.1016/S0031-9422(00)88430-7
32. Criodain TO, O'Sullivan M, Meegan MJ, Donnelly DM (1981) Latinone, a phenanthrene-1, 4-quinone from *Dalbergia latifolia*. Phytochemistry 20(5):1089–1092. https://doi.org/10.1016/0031-9422(81)83033-6
33. Yun H-M, Park JE, Lee JY, Park K-R (2022) Latifolin, a natural flavonoid, isolated from the heartwood of *Dalbergia odorifera* induces bioactivities through apoptosis, autophagy, and necroptosis in human oral squamous cell carcinoma. Int J Mol Sci 23(21):13629. https://doi.org/10.3390/ijms232113629
34. Yadav SK, Nagarathna P, Yadav CK (2015) Research article of evaluation of immunomodulatory activity of *Dalbergia latifolia* on Swiss albino mice. J Pharm Biol Sci 10:58–64. https://doi.org/10.9790/3008-10135864
35. Khalid M, Siddiqui H, Freed S (2011) In-vitro assessment of antioxidant activity of *Dalbergia latifolia* barks extract against free radicals. Am Eurasian J Sci Res 6(3):172–177
36. Masendra M, Purba BAV, Irawati D, Lukmandaru G (2021) Investigation of flavonoid extractives and their contribution to color of *Dalbergia latifolia* Roxb. Wood. Wood Res J 12(2):53–60. https://doi.org/10.51850/wrj.2021.12.2.53-60
37. Sutrapu S, Pal RS, Khurana N, Vancha H, Mohd S, Chinnala KM, Kumar B, Pilli G (2024) Diabetes warriors from heart wood: unveiling dalbergin and isoliquiritigenin from *Dalbergia latifolia* as potential antidiabetic agents in-vitro and in-vivo. Cell Biochem Biophys 82:1–16. https://doi.org/10.1007/s12013-024-01285-x
38. Liu RH, Mei DY, Lin S, Wang DQ, Shao F, Chen LY, Guo SL (2019) A new benzofuran from the heartwood of *Dalbergia latifolia*. Nat Prod Res 33(1):47–52. https://doi.org/10.1080/14786419.2018.1431639
39. Deshmukh VP, Lunge MS, Rajurkar AV, Dharkar NS, Raut SR, Dhoran VS (2021) Chemical characterization and therapeutics of *Dalbergia latifolia* Roxb.: a review. J Pharmacogn Phytochem 10(4):340–345. https://doi.org/10.22271/phyto.2021.v10.i4d.14174
40. Khalid M, Shoaib A, Akhtar J, Alqarni MH, Badruddeen, Siddiqui HH, Foudah AI (2020) Anti-obesity prospective of *Dalbergia latifolia* (Roxb.) hydroalcoholic bark extract in high fat diet induced obese rats. 3 Biotech 10:1–12. https://doi.org/10.1007/s13205-020-02491-z
41. Keppler F, Harper DB, Kalin RM, Meier-Augenstein W, Farmer N, Davis S, Schmidt H, Brown DM, Hamilton JT (2007) Stable hydrogen isotope ratios of lignin methoxyl groups as a paleo-climate proxy and constraint of the geographical origin of wood. New Phytol 176(3):600–609. https://doi.org/10.1111/j.1469-8137.2007.02213.x
42. Degen B, Fladung M (2008) Use of DNA-markers for tracing illegal logging. In: Proceedings of the international workshop–Fingerprinting methods for the identification of timber origins. Bonn, Germany, pp 6–14
43. Gasson P (2011) How precise can wood identification be? Wood anatomy's role in support of the legal timber trade, especially CITES. IAWA J 32(2):137–154
44. Yu M, Liu K, Zhou L, Zhao L, Liu S (2016) Testing three proposed DNA barcodes for the wood identification of *Dalbergia odorifera* T. Chen and *Dalbergia tonkinensis* Prain. Holzforschung 70(2):127–136. https://doi.org/10.1515/hf-2014-0234
45. Hartvig I, Czako M, Kjær ED, Nielsen LR, Theilade I (2015) The use of DNA barcoding in identification and conservation of rosewood (*Dalbergia* spp.). PLoS One 10(9):e0138231. https://doi.org/10.1371/journal.pone.0138231

46. Sharma V, Yadav J, Kumar R, Tesarova D, Ekielski A, Mishra PK (2020) On the rapid and non-destructive approach for wood identification using ATR-FTIR spectroscopy and chemometric methods. Vib Spectrosc 110:103097. https://doi.org/10.1016/j.vibspec.2020.103097
47. Shang D, Brunswick P, Yan J, Bruno J, Duchesne I, Isabel N, VanAggelen G, Kim M, Evans PD (2020) Chemotyping and identification of protected *Dalbergia* timber using gas chromatography quadrupole time of flight mass spectrometry. J Chromatogr A 1615:460775. https://doi.org/10.1016/j.chroma.2019.460775
48. Liu S, Zheng C, Wang J, Lu Y, Yao J, Zou Z, Yin Y, He T (2024) How to discriminate wood of CITES-listed tree species from their look-alikes: using an attention mechanism with the ResNet model on an enhanced macroscopic image dataset. Front Plant Sci 15:1368885. https://doi.org/10.3389/fpls.2024.1368885
49. Ramage MH, Burridge H, Busse-Wicher M, Fereday G, Reynolds T, Shah DU, Wu G, Yu L, Fleming P, Densley-Tingley D (2017) The wood from the trees: the use of timber in construction. Renew Sustain Energy Rev 68(1):333–359. https://doi.org/10.1016/j.rser.2016.09.107
50. Pettersen RC (1984) The chemical composition of wood. In: The chemistry of solid wood. American Chemical Society, Washington DC, pp 57–126
51. Zhang F-D, Xu C-H, Li M-Y, Chen X-D, Zhou Q, Huang A-M (2016) Identification of *Dalbergia cochinchinensis* (CITES Appendix II) from other three *Dalbergia* species using FT-IR and 2D correlation IR spectroscopy. Wood Sci Technol 50:693–704. https://doi.org/10.1007/s00226-016-0815-3
52. Sen A, Kaur A, Johar V, Singh V (2023) Indian rosewood: a versatile tree for economic prosperity. Int J Phytol Res 3(4):16–21
53. Dhingra V, Mukerjee S (1971) Chemical examination of the bark and sapwood of *Dalbergia latifolia*. Phytochemistry 10(10):2551. https://doi.org/10.1016/S0031-9422(00)89912-4
54. Kite GC, Green PW, Veitch NC, Groves MC, Gasson PE, Simmonds MS (2010) Dalnigrin, a neoflavonoid marker for the identification of Brazilian rosewood (*Dalbergia nigra)* in CITES enforcement. Phytochemistry 71(10):1122–1131. https://doi.org/10.1016/j.phytochem.2010.04.011
55. Yin X, Huang A, Zhang S, Liu R, Ma F (2018) Identification of three *Dalbergia* species based on differences in extractive components. Molecules 23(9):2163. https://doi.org/10.3390/molecules23092163
56. Pandey K, Upreti N, Srinivasan V (1998) A fluorescence spectroscopic study on wood. Wood Sci Technol 32:309–315. https://doi.org/10.1007/BF00702898
57. Liu M, Peng L, Lyu S, Lyu J (2020) Properties of common tropical hardwoods for fretboard of string instruments. J Wood Sci 66:1–11. https://doi.org/10.1186/s10086-020-01862-7
58. da Silva AA, De Keukeleire D, Cardoso DR, Franco DW (2012) Multivariate analyses of UV-Vis absorption spectral data from cachaça wood extracts: a model to classify aged *Brazilian cachaças* according to the wood species used. Anal Methods 4(3):642–646. https://doi.org/10.1039/C2AY05670D
59. Yadav A, Sharma S, Singh V, Kapoor M, Singh R (2024) Forensic identification and differentiation of some protected timber species using ATR-FTIR spectroscopy and chemometrics. Front Anal Sci 4:1508509. https://doi.org/10.3389/frans.2024.1508509
60. Adedipe OE, Dawson-Andoh B, Slahor J, Osborn L (2008) Classification of red oak (*Quercus rubra*) and white oak (*Quercus alba*) wood using a near infrared spectrometer and soft independent modelling of class analogies. J Near Infrared Spectrosc 16(1):49–57. https://opg.optica.org/jnirs/abstract.cfm?URI=jnirs-16-1-49
61. Braga JWB, Pastore TCM, Coradin VTR, Camargos JAA, da Silva AR (2011) The use of near infrared spectroscopy to identify solid wood specimens of *Swietenia macrophylla* (Cites Appendix II). IAWA J 32(2):285–296
62. Ma T, Inagaki T, Ban M, Tsuchikawa S (2019) Rapid identification of wood species by near-infrared spatially resolved spectroscopy (NIR-SRS) based on hyperspectral imaging (HSI). Holzforschung 73(4):323–330. https://doi.org/10.1515/hf-2018-0128

63. Musah RA, Espinoza EO, Cody RB, Lesiak AD, Christensen ED, Moore HE, Maleknia S, Drijfhout FP (2015) A high throughput ambient mass spectrometric approach to species identification and classification from chemical fingerprint signatures. Sci Rep 5:11520. https://doi.org/10.1038/srep11520
64. Kite GC, Green PW, Veitch NC, Groves MC, Gasson PE, Simmonds MS (2010) Dalnigrin, a neoflavonoid marker for the identification of Brazilian rosewood (*Dalbergia nigra*) in CITES enforcement. Phytochemistry 71:1122–1131. https://doi.org/10.1016/j.phytochem.2010.04.011
65. Tsuchikawa S, Siesler H (2003) Near-infrared spectroscopic monitoring of the diffusion process of deuterium-labeled molecules in wood. Part I: softwood. Appl Spectrosc 57(6):667–674. https://doi.org/10.1366/000370203322005364
66. Pastore TCM, Braga JWB, Coradin VTR, Magalhães WLE, Okino EYA, Camargos JAA, de Muñiz GIB, Bressan OA, Davrieux F (2011) Near infrared spectroscopy (NIRS) as a potential tool for monitoring trade of similar woods: Discrimination of true mahogany, cedar, andiroba, and curupixá. Holzforschung 65(1):73–80. https://doi.org/10.1515/hf.2011.010
67. Ramalho FM, Andrade JM, Hein PR (2018) Rapid discrimination of wood species from native forest and plantations using near infrared spectroscopy. For Syst 27(2):e008. https://doi.org/10.5424/fs/2018272-12075
68. Htun PP, Htwe NH, Kyi NNN, Phyu PE (2024) A review on hyperspectral imaging and its applications in wood domain. Indian J Comput Sci Eng 15(4):433–439. https://doi.org/10.21817/indjcse/2024/v15i4/241504027
69. dos Santos Grasel F, Ferrão MF, Wolf CR (2016) Ultraviolet spectroscopy and chemometrics for the identification of vegetable tannins. Ind Crops Prod 91:279–285. https://doi.org/10.1016/j.indcrop.2016.07.022
70. Bisht S, Mishra K, Pandey KK, Chandra G, Nayak MM, Shashikala S (2022) Chromatographic and spectrophotometric analysis of heartwood extracts of four *Pterocarpus* species. Wood Sci Technol 56(2):459–476. https://doi.org/10.1007/s00226-022-01368-z
71. Sharma R, Kumar R (2024) Distinguishing between *Tectona grandis* and *Acacia auriculiformis* woods. Microchem J 202:110788. https://doi.org/10.1016/j.microc.2024.110788
72. Gerasimov V, Gurovich A, Kostrin D, Selivanov L, Simon V, Stuchenkov A, Paltcev A, Uhov A (2016) Raman spectroscopy for identification of wood species. J Phys Conf Ser 741:012131. https://doi.org/10.1088/1742-6596/741/1/012131
73. Lavine BK, Davidson C, Moores AJ, Griffiths P (2001) Raman spectroscopy and genetic algorithms for the classification of wood types. Appl Spectrosc 55(8):960–966. https://doi.org/10.1366/0003702011953108
74. Edwards HG, de Oliveira LF, Nesbitt M (2003) Fourier-transform Raman characterization of brazilwood trees and substitutes. Analyst 128(1):82–87. https://doi.org/10.1039/B209052J
75. Wang F, Huang A, Yin X, Wang W, Chen J (2018) Multilevel profiling and identification of *Dalbergia odorifera* and *Dalbergia stevensonii* by FTIR, NMR and GC/MS. Chin Chem Lett 29(9):1395–1398. https://doi.org/10.1016/j.cclet.2017.11.007
76. Aldana-Mejia JA, Ccana-Ccapatinta GV, Ribeiro VP, Arruda C, Veneziani RC, Ambrósio SR, Bastos JK (2021) A validated HPLC-UV method for the analysis of phenolic compounds in Brazilian red propolis and *Dalbergia ecastaphyllum*. J Pharm Biomed Anal 198:114029. https://doi.org/10.1016/j.jpba.2021.114029
77. Wang L, Wang X, Kong L (2012) Automatic authentication and distinction of *Epimedium koreanum* and *Epimedium wushanense* with HPLC fingerprint analysis assisted by pattern recognition techniques. Biochem Syst Ecol 40:138–145. https://doi.org/10.1016/j.bse.2011.10.014
78. Liu R-X, Li L, Wang Q, Wang W, Bi K-S, Guo D-A (2005) Simultaneous determination of nine flavonoids in *Dalbergia odorifera* by LC. Chromatographia 61:409–413. https://doi.org/10.1365/s10337-005-0520-0
79. Ruiz-Aquino F, Santiago-García W, Suárez-Mota M, Esquivel-Reyes H, Feria-Reyes R, Rutiaga-Quiñones J (2021) Development and validation of an analytical method for condensed tannin extracts obtained from the bark of four tree species using HPLC. Wood Res 66(2):171–182. https://doi.org/10.37763/wr.1336-4561/66.2.171182

80. Claassen FW, van de Haar C, van Beek TA, Martínez-Iñigo MJ, Dorado J, Sierra-Alvarez R (1998) HPLC analysis of low molecular weight constituents in wood. In: Proceedings 7th International conference on biotechnology in the pulp and paper industry. Canadian Pulp and Paper Association, Quebec, Canada, June (1998)
81. Liao X, Hong Y, Chen Z (2021) Identification and quantification of the bioactive components in *Osmanthus fragrans* roots by HPLC-MS/MS. J Pharm Anal 11(3):299–307. https://doi.org/10.1016/j.jpha.2020.06.010
82. Nayak M, Nagarajan A, Majeed M (2017) Pharmacognostic evaluation of leaf and stem wood extracts of *Artocarpus hirsutus* Lam. Pharmacogn J 9(6):887–894. https://doi.org/10.5530/pj.2017.6.139
83. Sekine N, Ashitani T, Murayama T, Shibutani S, Hattori S, Takahashi K (2009) Bioactivity of latifolin and its derivatives against termites and fungi. J Agric Food Chem 57(13):5707–5712. https://doi.org/10.1021/jf900719p
84. Narita H, Ando N, Kuga S (2013) Simple identification of weathered wood of *Chamaecyparis obtusa* var. formosana. Mokuzai Gakkaishi 59(4):203–210
85. Komsta Ł (2012) Chemometrics in fingerprinting by means of thin layer chromatography. Chromatogr Res Int 2012(1):893246. https://doi.org/10.1155/2012/893246
86. Bischoff JJ, Nolley SW, Thornton J, Street N (1996) The use of thin layer chromatography in the identification of historic artificially colored varnishes and glazes. In: Proceedings of the objects specialty group session, Norfolk, Virginia, pp 14–22
87. Yusiasih R, Yoshimura T, Umezawa T, Imamura Y (2003) Screening method for wood extractives: direct cellulose thin-layer chromatography plate. J Wood Sci 49:377–380. https://doi.org/10.1007/s10086-002-0482-x
88. Czakó M, Márton L (2001) A heartwood pigment in Dalbergia cell cultures. Phytochemistry 57(6):1013–1022. https://doi.org/10.1016/S0031-9422(01)00111-X
89. Park SY, Kim JH, Kim JC, Yang SY, Yeo H, Choi IG (2021) Classification of softwoods using wood extract information and near infrared spectroscopy. BioResources 16(3):5301
90. Kartini K, Jayani NIE, Hadiyat MA (2021) Validating and developing TLC-based fingerprinting for *Curcuma longa* L. J Pharm Pharmacogn Res 9(5):704–715. https://doi.org/10.56499/jppres21.1062_9.5.704
91. Hajimehdipoor H, Khanavi M, Zahedi H, Abedi Z, Kalantari Khandani N, Adib N, Pirali Hamedani M (2009) Fingerprint study of *Thymus* spp. by TLC. J Med Plants 8(31):24–19. https://dor.isc.ac/dor/20.1001.1.2717204.2009.8.31.3.8
92. Brunswick P, Cuthbertson D, Yan J, Chua CC, Duchesne I, Isabel N, Evans PD, Gasson P, Kite G, Bruno J (2021) A practical study of CITES wood species identification by untargeted DART/QTOF, GC/QTOF and LC/QTOF together with machine learning processes and statistical analysis. Environ Adv 5:100089. https://doi.org/10.1016/j.envadv.2021.100089
93. Lancaster C, Espinoza E (2012) Analysis of select *Dalbergia* and trade timber using direct analysis in real time and time-of-flight mass spectrometry for CITES enforcement. Rapid Commun Mass Spectrom 26(9):1147–1156. https://doi.org/10.1002/rcm.6215
94. Wiemann MC, Espinoza EO (2017) Species verification of *Dalbergia nigra* and *Dalbergia spruceana* samples in the wood collection at the forest products laboratory. Research Paper FPL–RP–690. US Department of Agriculture, Forest Service, Forest Products Laboratory, Madison, WI, vol 690, pp 1–10
95. Evans PD, Mundo IA, Wiemann MC, Chavarria GD, McClure PJ, Voin D, Espinoza EO (2017) Identification of selected CITES-protected Araucariaceae using DART TOFMS. IAWA J 38(2):266-S3
96. Zhang M, Zhao GJ, Liu B, He T, Guo J, Jiang X, Yin Y (2019) Wood discrimination analyses of *Pterocarpus tinctorius* and endangered *Pterocarpus santalinus* using DART-FTICR-MS coupled with multivariate statistics. IAWA J 40(1):58–74
97. Russ A, Fišerová M, Gigac J (2009) Preliminary study of wood species identification by NIR spectroscopy. Wood Res 54(4):23–32
98. Cooper P, Jeremic D, Radivojevic S, Ung Y, Leblon B (2011) Potential of near-infrared spectroscopy to characterize wood products. Can J For Res 41(11):2150–2157. https://doi.org/10.1139/x11-088

99. Yang Z, Liu Y, Pang X, Li K (2015) Preliminary investigation into the identification of wood species from different locations by near infrared spectroscopy. BioResources 10(4):8505–8517
100. Abe H, Kurata Y, Watanabe K, Ishikawa A, Noshiro S, Fujii T, Iwasa M, Kaneko H, Wada H (2020) The separation of softwood and hardwood in historical wooden statues of the Nazenji-temple in Japan using NIR spectroscopy. IAWA J 41(4):740–750
101. Jannah R, Rafi M, Heryanto R, Kautsar A, Septaningsih DA (2018) UV-Vis spectroscopy and chemometrics as a tool for identification and discrimination of four *Curcuma* species. Int Food Res J 25(2):p643
102. Sandberg K, Sterley M (2009) Separating Norway spruce heartwood and sapwood in dried condition with near-infrared spectroscopy and multivariate data analysis. Eur J For Res 128:475–481. https://doi.org/10.1007/s10342-009-0296-0
103. Lazarescu C, Hart F, Pirouz Z, Panagiotidis K, Mansfield S, Barrett J, Avramidis S (2017) Wood species identification by near-infrared spectroscopy. Int Wood Prod J 8(1):32–35. https://doi.org/10.1080/20426445.2016.1242270
104. Snel FA, Braga JW, da Silva D, Wiedenhoeft AC, Costa A, Soares R, Coradin VT, Pastore TC (2018) Potential field-deployable NIRS identification of seven *Dalbergia* species listed by CITES. Wood Sci Technol 52:1411–1427. https://doi.org/10.1007/s00226-018-1027-9
105. Batubara I, Husna S, Rafi M, Sumaryada T, Uchiyama S, Juliandi B, Putri SP, Fukusaki E (2022) A combination of UV-Vis spectroscopy and chemometrics for detection of Sappan-wood (*Caesalpinia sappan*) adulteration from three dyes. Sains Malaysiana 51(3):775–781. https://doi.org/10.17576/jsm-2022-5103-11
106. Chen H, Ferrari C, Angiuli M, Yao J, Raspi C, Bramanti E (2010) Qualitative and quantitative analysis of wood samples by Fourier transform infrared spectroscopy and multivariate analysis. Carbohyd Polym 82(3):772–778. https://doi.org/10.1016/j.carbpol.2010.05.052

Chapter 27
Rosewood in Guitars: Balancing Acoustic Excellence and Ecological Survival

A. N. Arunkumar

1 Introduction

Wood, nature's endowment to humanity, has been an intrinsic material of various musical instruments throughout the world since time immemorial [14]. As an essential component of the diversified instruments, wood has been continuously evolving and helping in the unique designs of musical instruments. The distinct performance of the wood instruments has been one of the strong factors in defining the nature, heritage, and cultural identity of places and people across the globe. Therefore, wood and musical instruments remain inseparable.

As a by product of nature, wood as a material is enigmatic and complicated, resembling the differences between people. The inherent variability within and between tree species can have a direct bearing on the performance of any musical instrument.

Some of the traits making a tree species applicable for use in musical instruments must have some prerequisite traits. The trees must be abundantly available, facilitating sustained harvest. A typical tone wood species should have a larger girth, with a clean, straight, clear bole height and sufficient heartwood. The quality attribute of the wood is complex but plays a crucial role. Quantitative traits are strength and mechanical properties, while qualitative are colour, grain, porosity, and figure. The wood has to be durable, workable, bendable, and glueable-friendly. The chemical constituents in terms of secondary metabolites can also impact the sound quality as well as be friendly towards the human skin. Other factors include the sawing, drying, and storage of the wood [2].

A. N. Arunkumar (✉)
ICFRE-Institute of Wood Science and Technology, Malleswaram, Bengaluru, India
e-mail: arun@icfre.org

R. R. Warrier et al. (eds.), *Handbook of Asian Rosewoods*, Materials Horizons: From Nature to Nanomaterials, https://doi.org/10.1007/978-981-96-9459-4_27

2 Acoustics and Wood

During pre-industrial times, wood was the only available material and therefore became indispensable for string instruments. However, the natural uniqueness and versatility of wood have resulted in it being unable to be replaced by the industrial materials such as metals and plastics even after a century. A good example to prove this is considering 'weight' as a trait. A spruce violin top weighing 70 g, if replaced, requires a 440 g acrylic top or 400 g of aluminium, which makes the instrument heavy. Considering resonance as the quality factor, compared to wood, the resonance of acrylic is much less, while that of aluminium is much higher. Therefore, in the case of string instruments, wood as nature's time-tested material cannot be easily replaced [27].

Factors contributing to the uniqueness of wood and its acoustic properties are the tree's age, genetic makeup, growing conditions, growth rate and pattern, and orientation of the grain, colour, and mechanical and strength properties. The orthotropic nature of the wood materials in terms of unique mechanical properties in the three perpendicular axes, namely longitudinal (along the tree length), radial (perpendicular to the tree), and tangential (perpendicular to the tree but tangential to the fibre or growth rings), sets it apart from all the other synthetic or human-made materials.

It is generally agreed that a musical instrument made from the same maker can respond differently because of all the innate variable traits of the wood. Some of the other wood properties having a direct bearing on the acoustic response are the combination of mechanical strength, tactile properties, and aesthetic appeal, all influencing the overall experience of the music. A distinct trait of wood is that when exposed to different environments, it responds and aligns accordingly, which every musical instrument needs [8, 26, 28].

The material properties that define the acoustical quality and the instruments' performance are density, Young's modulus, damping, and shrinkage, which are dependent on the moisture content of wood. These three properties determine the speed, frequency, and intensity of the sound from the musical instrument [3, 15, 16, 25, 32].

3 Association of Guitar and Rosewood

Among many musical instruments in the world, the guitar is considered one of the most popular and extensively used instruments worldwide [13]. The etymology of the word'guitar' is derived from the Persian language 'sihtar' through the Greek derivative 'kithara' or through the Spanish name 'guittara' [3, 12]. Guitar has a historical relevance of more than five centuries. Over time, several modifications in design, string numbers, size, and use of material occurred, and the present-day guitar has evolved. Guitar can be related to some of the ancestral string instruments, such as the oud, lute, Renaissance guitars, and Chitarra battente [7]. The artisans or craftsmen who make stringed instruments in general are known as luthiers, and the artist who performs using a guitar is known as a guitarist. The sole purpose of a guitarist is to generate sound from the acoustic instruments at a specific frequency that is pleasant to the ears of the listener.

Guitar as an instrument has an illustrative history in terms of trends in materials used and has been categorized into five eras, which have been commonly accepted periods of music history. They are the Renaissance era (1400–1600), the Baroque era (1600–1750), the Classical era (1750–1820), the Romantic era (1820–1900), and the Twentieth Century (1900–2000) [20]. The oldest guitar, which had six strings, was built in 1779 by Italian luthier Gaetano Vinaccia [19]. In the mid-1850s, Spanish luthier Antonio Torres Jurado built a revolutionary modified guitar that was bigger in size and proportions, and fan bracing was optimized [17]. This modified classical guitar enabled the modern dimensions, enhancing the tone, volume, range, and projection [9, 22].

Every component of the guitar has an impact on the tone of the instrument. However, soundboards and the backs and sides have a greater impact. The rigidity of the neck assists in the vibrations being transferred through the bridge to the soundboard. Fretboards can indirectly impact the sound quality, and heavier headstocks provide longer sustainment at the cost of treble and volume [9]. The guitar's bridge looks simpler but plays an important role as it anchors the string and is critical for the instrument's performance. As a gateway for string energy, it manoeuvres the string vibration to the soundboard of the guitar.

Prominent tree species used for guitars were as follows: for back and sides, it was maple, walnut, cherry, mahogany, American black walnut, and Fraxinus, and for soundboards, it was European and Adirondack spruce. Spruce was the species extensively used for the main soundboard as it showed high efficiency in transferring string vibration to acoustic sound [3, 5, 11] along with greater preference culturally and musically [4]. Gradually, by the end of the nineteenth century, except for spruce soundboards, other components of guitars that used local species were replaced by tropical hardwoods [1].

Some of the traditional and long-lasting woods used in various parts of guitars are back and sides (rosewood, mahogany, maple, and koa); soundboard (spruce, cedar, mahogany, koa); neck (mahogany, maple, rosewood, Spanish cedar); fret (rosewood, ebony); bridge (rosewood, ebony); the headstock (rosewood, mahogany, and maple) and brace are made up of the same wood as that of the soundboard [3] (Fig. 1).

Rosewood became an essential component of guitars. One of the important rosewood species, *Dalbergia nigra,* also known as Brazilian rosewood, was the most desired wood for making guitars till the 1960s and was considered the 'Holy Grail' of tonewood [10]. However, due to overutilization, the population of *D. nigra* started severely dwindling in its natural habitat, and post-1969 onwards, *D. latifolia* gradually started being used, which had nearly 90% of the same attributes (Table 1). Though the purists still prefer or opine that Indian rosewood cannot match *D. nigra,* slowly the acceptance level reached higher once some of the globally reputed guitar manufacturers started producing the premium guitars of *D. latifolia.*

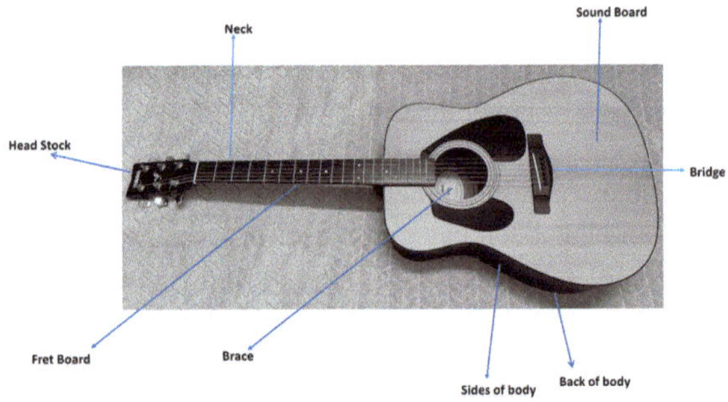

Fig. 1 Different parts of guitar in which rosewood is used

Table 1 Wood properties of *Dalbergia nigra* (Brazilian rosewood) and *Dalbergia latifolia* (Indian rosewood)

Sl.no	Wood traits	Values	
		Brazilian rosewood (*Dalbergia nigra*)	East Indian rosewood (*Dalbergia latifolia*)
1	Average dried weight	835 kg/m^3	855 kg/m^3
2	Specific gravity (Basic)	0.68	0.73
3	Specific gravity (12% moisture content)	0.84	0.86
4	Janka hardness	12,410 N	10,440 N
5	Modulus of rupture	135 MPa	115 MPa
6	Elastic modulus	13.93 GPa	12.30 GPa
7	Crushing strength	67.2 MPa	61.3 MPa
8	Radial shrinkage	2.9%	2.6%
9	Tangential shrinkage	4.6%	5.9%
10	Volumetric shrinkage	8.5%	8.6%
11	Tangential to Radial shrinkage ratio	1.6	2.3

(*Source* [18])

The key developments in the rise of rosewood (*Dalbergia* spp.) were enhanced improvement in woodworking tools due to the industrial revolution. The increase in global trade due to Spanish and Portuguese colonies across various parts of the world resulted in the availability of exotic woods. Mahogany, which was still used for backs and sides, was replaced by rosewood. Thus, rosewood became the premium choice in the top-end guitars.

In 1916, *D. nigra* was used in the making of one of the famous iconic guitars Martin D-28 from Martin and Company, USA. Steadily from 1930 onwards, dreadnought guitars also became popular. It is a typical guitar used by singer and songwriter that has a large, broad body with an iconic shape used both for strumming and fingerstyle playing. Thus began the golden era of rosewood between 1920 and 1960, resulting in *D. nigra* becoming the standard for high-end/premium guitars. Even now these guitars are considered prized possessions by the guitar collectors. With a reduction in the availability of Brazilian rosewood along with the premium cost for the wood, in 1969, Martin and Company (a 190-year-old guitar manufacturing company in the USA) replaced *D. nigra* (Brazilian rosewood) with *D. latifolia* (Indian rosewood) for the first time in one of its premium guitars, the D-28 [10]. This paved the popular entry of *D. latifolia* wood in the guitar industry, considering its availability, sustainability, and the tonal quality, which was similar but not identical to the Brazilian rosewood. *D. nigra* was added to the CITES Appendix I list in 1992, which resulted in stringent measures for trade.

Acoustic properties of Brazilian rosewood (*Dalbergia nigra*) and Indian rosewood (*Dalbergia latifolia*) when used for the top and back plates of guitars are compared in Fig. 2. The average specific gravity was 0.864 and 0.775 for *D. nigra* and *D. latifolia*, respectively. It was 1.1 times more for *D. nigra* compared to *D. latifolia.* Young's modulus (E), or elastic modulus (measured in Giga Pascal—GPa), in other words, is the ratio of stress and strain. It increases with specific gravity in wood and provides an idea about the stiffness of wood.

The average value for E along the longitudinal and cross-longitudinal direction was 1.81 and 2.32 GPa for *D. nigra,* and it was 17.4 and 3.00 GPa for *D. latifolia.* Differences were also observed for sound damping (tan δ) across longitudinal (L) and cross-longitudinal directions (CL). For *D. nigra,* the L and CL were 5.83 and 1.47, while they were 7.37 and 1.46, respectively, for *D. latifolia.* The data provided in the table and the figure clearly indicate that *D. latifolia* has nearly 90% of the traits of *D. nigra* and can be the ideal replacement of wood for making high-quality guitars. Considering the wood grains or the arrangement of fibres, *D. nigra* is compact with interlocked grains, which enhances stiffness and reflection of sound waves; *D. latifolia* has straight grains and less interlocked, which leads to quicker response but less overtone complexity. Wood of *D. nigra* being dense is susceptible to cracks,

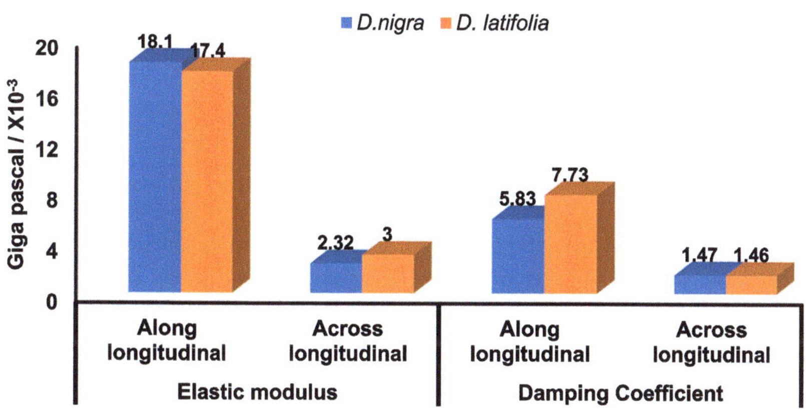

Fig. 2 Elastic modulus and damping coefficient values of *D. nigra* and *D. latifolia* (Adopted and modified from [30, 31])

Table 2 Qualitative assessment for tonal characteristics in *D. nigra* and *D. latifolia*

Tonal characteristic	*Dalbergia nigra*	*Dalbergia latifolia*
Bass response	Deep, powerful, and resonant	Strong, but slightly less boomy
Midrange	Warm and complex	Balanced and slightly drier
High frequencies	Smooth and rounded treble	Clear and articulate
Harmonic overtones	Exceptionally rich	Rich but not as layered
Projection	Extra strong and sustains longer	Strong, but sustains less longer

while *D. latifolia* wood is comparatively stable under different humid conditions. The elastic modulus and damping coefficient across and along longitudinal grains are not significantly different between the *D. nigra* and *D. latifolia* (Fig 2). When considering the qualitative assessment for tonal characteristics (Table 2), the bass response is deep, and the harmonic overtones are exceptionally rich in *D. nigra.* Interestingly, looking into the discussion by the guitar connoisseurs, many terms used are highly qualitative and equally inadequate as judging sound can be very personal. Some of those terms used are balance, boomy, bright, complex, crispy, dark, dry, rich, round, smooth, sparkling, warm, wide, woody, and woolly.

D. latifolia clearly articulates high frequencies. Considering the mechanical, acoustic, and tonal properties and the cheaper wood cost, *D. latifolia* is an ideal replacement for *D. nigra.* The demand for the guitar made out of *D. latifolia* remains unabated and has been accepted to a large extent by luthiers and guitarists. The guitar has become a globally mobile instrument, influencing various musical genres and impacting form, tonal textures, and playing techniques. It suits both a wide variety of musical contexts that seamlessly merge with the local specific and wide international music. The ample proof for this is the demand and requirement of guitar. Guitar sales were $8 billion in 2019 and increased by 15% to $9.2 billion in 2020. It further increased to $10.07 billion by 2022 and is projected to be ~$19.9 billion by the end of 2025 [21]. Globally, the relevance of guitar as instrument and music culture is in the increasing trend, and the intricate association of rosewood and guitar continues on a strong note.

4 Future Perspective

In guitar making, apart from the *Dalbergia* species, it is mostly other tropical woods that are extensively used. From a long-term perspective, it may not be a viable option as many of these species are either in the Red List of the International Union for Conservation of Nature or in the process of being listed soon. Research on tree improvement, conservation, sustainable harvest, marketing, and utilization of renowned species used for guitar has to become a research priority. Though efforts are already initiated, options of using lesser-known and new species, encouraging usage of engineered woods as suitable replacements for traditional hardwood, need a rigorous study along with foolproof research [23, 24, 29]. The stubbornness and unknown fear by the luthiers that the quality of other alternative woods may not match with the familiar tropical hardwood species will always be a hidden threat. They need to be educated [6] and become a bit lenient in accepting the new genre of tree species, as each wood has a unique tonal quality.

5 Conclusion

It is necessary to have a holistic approach of converging species and music culture. To achieve this, bringing together the luthiers, guitarists, audience, customers, traders, conservationists, and policymakers is the crucial step. Developing strategies to make them understand, accept, and commit to a multilevel conservation and sustainability of species and music ecosystems.

References

1. Applegate B (2021) The rise and fall of iconic guitar tonewoods and evaluation of alternate species. Doctoral Thesis, University of Edinburgh, UK, 2021
2. Bennett BC (2002) Forest products and traditional peoples: economic, biological, and cultural considerations. Nat Res Forum 26:293–301
3. Bennett BC (2016) The sound of trees: wood selection in guitars and other chordophones. Econ Bot 70:49–63. https://doi.org/10.1007/s12231-016-9336-0
4. Brémaud I (2012) Acoustic properties of wood in string instruments soundboards and tuned idiophones: biological and cultural diversity. J Acoust Soc Am 131(1):807–818
5. Buksnowitz C, Teischinger A, Müller U, Pahler A, Evans R (2007) Resonance wood [*Picea abies* (L.) Karst.]–evaluation and prediction of violin makers' quality-grading. J Acoust Soc Am 121:2384–2395
6. Calvano S, Negro F, Ruffinatto F, Zanuttini-Frank D, Zanuttini R (2023) Use and sustainability of wood in acoustic guitars: an overview based on the global market. Heliyon 9(4):1–15
7. Chapman R (2000) Guitar: music, history, players. Dorling Kindersley, New York
8. Fletcher NH, Rossing TD (1998) The physics of musical instruments. Springer, New York, p 756
9. Gerken T, Johnston R, Ford F, Simmons M (2003) Acoustic Guitars: The composition, construction, and evolution of one of the world's most beloved instruments. Hal Leonard Corporation, Milwaukee, USA
10. Gibson G, Warren A (2021) The guitar: tracing the grain back to the tree. University of Chicago Press, USA
11. Gore T (2011) Wood for guitars. J Acoust Soc Am 129(4):2519
12. Harper D (2015) Online etymology dictionary. http://www.etymonline.com
13. Jahnel F (2000) Manual of guitar technology. The Bold Summer Ltd., Westport, UK
14. Morley I (2013) The prehistory of music. Human evolution, archaeology, and the true origins of musicality. Oxford University Press New York
15. Negro F, Cremonini C, Fringuellino M, Meloni F, Prina S, Zanuttini R (2021) Sound absorbing systems made of wood cross sections paired with void cavities-a first investigation. Wood Fiber Sci 53(1):69–76. https://doi.org/10.22382/wfs-2021-07

16. Quintavalla M, Gabrielli F, Canevari C (2021) Grading materials for stringed instruments soundboard: an approach considering the orthotropic elastic and damping properties. Appl Acoust 187:108521. https://doi.org/10.1016/j.apacoust.2021.108521
17. Romanillo JL (2006) Antonio De Torres: Guitar maker—his life and work. The Bold Strummer, Westport, CT, USA
18. Ross RJ (2021) Wood handbook—wood as an engineering material. General technical report FPL-GTR-282. U.S. Department of Agriculture, Forest Service, Forest Products Laboratory Madison, WI, p 543
19. Rossing TD, Caldersmith G (2010) Guitars and lutes. In: The science of string instruments. Springer, New York, p 19–46
20. Sachs C (2006) The history of musical instruments. Dover Publications Inc., Mineola, USA
21. Sam P (2025) Guitar sales statistics–most recent guitar data. https://musicstrive.blog/guitar-sales-statistics/. Accessed 20 Mar 2025
22. Sandberg L (2000) The acoustic guitar guide: everything you need to know to buy and maintain a new or used guitar. A Capella Books, Chicago, USA
23. Shirmohammadi M, Faircloth A, Redman A (2021) Assessment of sound quality: Australian native hardwood species for guitar fretboard production. Eur J Wood Wood Product 79:487–497
24. Sproßmann R, Zauer M, Wagenführ A (2017) Characterization of acoustic and mechanical properties of common tropical woods used in classical guitars. Results Phys 7:1737–1742
25. Viala R, Pérez MA, Placet V, Manjon A, Foltête E, Cogan S (2020) Towards model-based approaches for musical instruments making: validation of the model of a Spanish guitar soundboard and characterization features proposal. Appl Acoust 172:107591. https://doi.org/10.1016/j.apacoust.2020.107591
26. Von Hornbostel EM, Sachs C (1914) Systematik der musikinstrumente. Ein versuch. Zeitschrift für ethnologie 46(4/5):553–590
27. Waltham C, Yoshikawa S (2018) Construction of wooden musical instruments. In: Springer handbook of systematic musicology. Springer, Germany pp 63–79
28. Wegst UGK (2006) Wood for sound. Am J Bot 93:1439–1448
29. Yano H, Furuta Y, Nakagawa H (1997) Materials for guitar back plates made from sustainable forest resources. J Acoust Soc Am 101(2):1112–1119
30. Yano H, Kyou K, Furuta Y, Kajita H (1995) Acoustic properties of Brazilian rosewood used for guitar back plates. Mokuzai Gakkaishi 41:17–24
31. Yano H, Oonishi K, Mukudai J (1990) Acoustic properties of wood for the top plate of guitar. J Soc Mater Sci Jpn 39:1207–1212
32. Yoshikawa S (2007) Acoustical classification of woods for string instruments. J Acoust Soc Am 122(1):568–573. https://doi.org/10.1121/1.2743162

Chapter 28
Artistry in Rosewood: The Craft of Inlay, Handicrafts, and Furniture

A. N. Arunkumar and Leonard J. Mills

1 Wood Inlay

Mysore Rosewood Inlay is a traditional craft from Mysore, India, known for its intricate decorative techniques. Artisans embed contrasting materials such as ivory (historically), mother-of-pearl, horn, shells, sandalwood, acrylic, brass, resin, or colored wood into carved depressions in rosewood surfaces. These inlays create elaborate patterns or pictorial compositions that seamlessly blend with the wood [1]. This centuries-old craft has flourished in Karnataka, particularly in Mysore, and remains deeply rooted in the region's history. Renowned for its artistry, Mysore Wood Inlay continues to be a significant part of India's cultural heritage, reflecting the exceptional skill of its craftsmen [2].

2 History and Origin

Mysore Rosewood Inlay is a centuries-old craft that has evolved into one of India's most celebrated artistic traditions. The use of rosewood, primarily sourced from the forests surrounding Mysore, dates back to the late eighteenth century during the rule of Tipu Sultan. However, the craft is believed to have originated much earlier during the Vijayanagara Empire and was later refined under the patronage of the Wadiyar (Wodeyar) king. His highness Nalwadi Krishna Raja Wadiyar, whom Mahatma Gandhi called Rajarishi or Saintly King, established the Chamarajendra Technical Institute in 1892 and included wood carving as a subject of study in

A. N. Arunkumar (✉)
ICFRE-Institute of Wood Science and Technology, Malleswaram, Bengaluru, India
e-mail: arun@icfre.org

L. J. Mills
Coorg Rosewood Timbers, Kattugollahalli, Bengaluru, Karnataka, India

R. R. Warrier et al. (eds.), *Handbook of Asian Rosewoods*, Materials Horizons: From Nature to Nanomaterials, https://doi.org/10.1007/978-981-96-9459-4_28

curriculum which was a major step toward the revival of the craft. In 1914, the Maharaja of Mysore showcased this intricate art on a global platform by sending a rosewood casket with ivory inlay and a photo frame to the British Empire Exhibition, where it won a gold medal. Based on this, the Elephant Cigarette Company placed major orders, increasing the economic position of the wood carvers. These pieces attracted the foreign tourists who placed huge orders in about 1930 [3]. These recognitions helped Mysore Rosewood Inlay gain international fame, establishing it as a significant cultural symbol and a major tourist attraction. The sales of inlay articles rose significantly after 1960 and now there are about 3,000 workers attending to its background cutting and finishing work. The bulk of these craftsmen are Muslims and Vishwakarmas.

3 Manufacturing Process

Creating a Mysore Wood Inlay masterpiece is an intricate and time-consuming process that demands exceptional precision, patience, and craftsmanship. This centuries-old art form primarily utilizes rosewood, yellow wood, and ebony, with designs often inspired by nature, Hindu mythology, and traditional motifs. Historically, ivory was the preferred inlay material, but modern artisans now use sustainable alternatives such as gold, silver, acrylic, resin, sandalwood, and brass [1].

The inlay process is highly refined, with up to individual pieces of wood and metal meticulously arranged within a single cubic centimeter. The tradition of inlay work traces its roots back to the Safavid era (one of the significant ruling dynasty in Iran), where it was extensively used to embellish furniture, doors, windows, mirror frames, Quran boxes, and ornate shrines. Today, Mysore artisans follow a detailed, multi-step process to craft exquisite inlay pieces:

1. **Selecting the Base and Design**
    - The artisan chooses a high-quality wooden base, typically rosewood, teak, or sheesham, which is shaped into various forms such as squares, rectangles, circles, and ovals.
    - The surface is smoothed and buffed to prepare it for the design.
2. **Transferring the Design**
    - The artist either sketches the design freehand or transfers it from paper using carbon paper or stencils.
3. **Carving the Recesses**
    - Using a marking knife, the design is outlined on the wood.
    - Chisels are used to carve grooves 2–3 mm deep, creating recesses for the inlay materials.
4. **Inlaying the Design**

- Contrasting materials such as brass, resin, acrylic, and colored wood are carefully inserted into the carved grooves.
- Glue is applied to hold the pieces in place, and a hammer is used to secure them firmly.

5. **Finishing Touches**

- Any gaps are filled with a mixture of glue, rosewood dust, and sawdust.
- Sandpaper smooths the surface, removing excess inlay.
- The final piece is polished with beeswax and lacquered for durability and shine.

The inlay artisans use both traditional and modern tools for precision. Essential tools include hand drills for holes, *tagad* chisels for detailing, *hathodi* hammers for securing inlays, *ambor* pliers and *bowor* hex saws for cutting, and flint papers of various grades for polishing and finishing [4].

4 Geographical Indication (GI) Tag

To preserve the authenticity and exclusivity of Mysore Rosewood Inlay, the Karnataka State Handicrafts Development Corporation Ltd. applied for its registration under the Geographical Indications of Goods Act, 1999. The application was submitted to the Office of the Controller-General of Patents, Designs, and Trademarks in Chennai, ensuring that only artisans from the Mysore region could use the name "Mysore Rosewood Inlay." The GI status was officially granted in 2005 [5], further cementing its legacy as a significant cultural and artistic tradition [4].

5 The Artisan Cluster and Government Support

Unlike many other crafts that have faded over time, Mysore's wood inlay tradition remains vibrant. The Mysore Fine Arts College offers specialized courses in wood inlay, attracting fine arts students to this discipline. Numerous workshops in Mysore's old quarters bustle with activity, providing livelihoods to over 5,000 artisans. The Regional Design and Marketing Center, a government organization, implements various schemes to support these craftsmen.

Recognizing the significance of this art form, the Karnataka government established a Multicraft Complex, providing artisans with housing and workspace. The Cauvery Karnataka State Arts & Crafts Emporium, a government procurement center, supplies artisans with raw materials such as sandalwood and markets their finished products. Additionally, training programs sponsored by the Development Commissioner of Handicrafts help artisans refine their skills. The government also provides health care and insurance to registered artisans, ensuring their well-being and livelihood.

Mysore Wood Inlay is not just a decorative art—it is a reflection of the region's rich cultural heritage. The artworks often depict scenes from Hindu mythology, nature, and rural life, each piece serving as a testament to the skill and dedication of Mysore's artisans. Today, Mysore Wood Inlay products continue to captivate art lovers and collectors worldwide. Whether adorning furniture, doors, or standalone art pieces, these inlays preserve a 400-year-old tradition that remains one of India's most cherished handicrafts.

6 Tarkashi of Rajasthan

Wood Inlay, known as *Tarkashi* in Rajasthan, is a traditional craft that involves embedding fine metal—brass, copper, silver, or gold wires—into wood to create intricate decorative patterns. This centuries-old art form flourished under royal patronage and continues to be practiced by skilled artisans, primarily in Udaipur, Jaipur, and Jodhpur. A highly skilled art form that is more than 500 years old, it is also practiced along the east coast in Odisha.

Tarkashi Inlay originated in Mainpuri, Uttar Pradesh. It was patronized by the Jaipur royals, who invited master craftsmen to Jaipur, to train the locals in the craft. Today, artisans from the Jangid community continue this rich tradition [6]. The craft begins with selecting a high-quality wooden base, often made from sheesham (*Dalbergia sissoo*). Artisans carefully design intricate patterns inspired by nature, Mughal motifs, and geometric shapes. These designs are then transferred onto the wood, either freehand or using stencils. Thin strips of brass, copper, or silver are then meticulously inlaid into these grooves, creating stunning visual contrasts against the wood. The metal is secured using adhesive, and excess material is carefully filed down to ensure a seamless finish. Once the inlay work is completed, the surface is polished with flint paper and waxed to enhance the shine and durability of the piece. Traditional *Tarkashi* designs often adorn furniture, decorative boxes, mirror frames, and jewelry, reflecting Rajasthan's rich artistic heritage [7].

7 Wood Craft Industry in Saharanpur

Renowned as the "Sheesham Wood Village," Saharanpur is home to some of India's most skilled wood carvers. The city has gained international acclaim for its exquisite craftsmanship, with artisans who have been transforming wood into works of art for generations [8].

Saharanpur's renowned woodwork industry has long been celebrated for its craftsmanship. Traditionally, artisans carved intricate door designs using sheesham wood [9] available in the surrounding Shivalik hills, but modern demand has shifted toward household furniture made from sheesham, tun, and *dudhi* wood. Fret-saw techniques have largely replaced traditional carving, leading to mass production and exports, though at the cost of craftsmanship.

The Saharanpur wooden handicraft industry dates back to 1856, founded by Atta Hussain, who moved from Multan and started carving wooden objects in 1857 with just five rupees [10]. Known for setting up the first organized wood carving unit in Saharanpur, the industry generates around ₹400 crore annually and supports approximately 150,000 artisans [11].

Inlay work was introduced by European residents, incorporating floral designs in brass, copper, and German silver on sheesham wood. Brass is widely used for inlaying floral, geometric, and typographic patterns into wood [10]. This process involves cutting metal strips, shaping them using die-pressing techniques, and then selling them by weight to wood carvers. The artisans carefully carve grooves into the wood, embed the brass pieces by hammering them in place, secure them with wood adhesive, and finally nail them for added durability. Any leftover metal strips are recycled and repurposed. In addition to brass, materials like bone and plastic are sometimes incorporated into inlay work. Beyond brass inlays, Saharanpur's artisans have begun blending wrought iron with wood. This serves both functional and aesthetic purposes—adding structural strength to furniture like chairs, tables, and beds, or enhancing decorative items such as vases, lanterns, and candle holders [12].

Regardless of the application, these artisans earn well-deserved recognition for their innovative designs. Even today, woodcraft designs of Saharanpur are inspired by naturalistic motifs such as trees, flowers, creepers, and leaves, creating a unique identity in the prevailing market [6]. While skilled artisans still create custom pieces, their numbers are dwindling, and the industry faces challenges in preserving its artistic heritage.

The artistry of Saharanpur's wood carving enhances spaces while embodying resilience, adaptation, and the enduring beauty of handmade craftsmanship. Each intricate detail reflects a deep connection to heritage and a continuous tradition of craftsmanship. These masterpieces stand as testaments to a legacy that shapes the artistic landscape of Saharanpur and extends far beyond (Figs. 1, 2, 3, 4, 5, 6, 7, 8, 9 and 10).

Fig. 1 Rosewood inlay works. The top, front, and back views of a table with inlay

The top, front and back views of a table with inlay
courtesy: Gurudev Arts, Malleswaram

Fig. 1 (continued)

Fig. 2 Types of cots made of Rosewood

Fig. 3 Types of sofas made of Rosewood

Fig. 4 Rosewood dining tables

Fig. 5 Carved Rosewood chairs

Picture Credit: Gurudev Arts, Malleswaram, Bengaluru

Fig. 6 Wooden swings made of Rosewood

Courtesy: Gurudev Arts, Bengaluru

Fig. 7 Rosewood artefacts in home Prayer Room (Pooja Room)

Fig. 8 Religious idols made of Rosewood

Courtesy: Gurudev Arts, Bengaluru

Fig. 9 Elephants carved out of Rosewood

Courtesy: Gurudev Arts,
Malleswaram, Bengaluru

Fig. 10 Miscellaneous Handicrafts made from Rosewood

References

1. Roy S, Bairagi N (2021) A study on craft practice and creativity among Mysore rosewood inlay artisans of India. In: With design: reinventing design modes. Springer Nature, Singapore (2021) https://doi.org/10.1007/978-981-19-4472-7_16
2. Dagrains.com (2024). https://dagrains.com/latest-blog/exploring-the-beauty-of-mysore-rosewood-inlay/1010
3. Kamath SU (1988) The Karnataka gazetteer: Mysore District. Government Press, Bangalore, pp 399–400
4. Roy S, Bairagi N (2022) Rethinking resilience: COVID-19 and creativity among Mysore rosewood inlay artisans. NIFT J Fashion 149. https://www.nift.ac.in/sites/default/files/2022-12/NJF%20Volume%201%20FINAL.pdf#page=160
5. Manjunatha NK (2016) Status of geographical indications in India, especially Karnataka. Third Concept 30:32–39
6. Ansari MA, Jahan B (2017) Role of Saharanpur in wooden craft tradition in contemporary Indian art. In: Proceedings of international conference on role of arts, culture, humanities, religion, education, ethics, philosophy, spirituality and science for holistic societal development. Krishi Sanskriti Publications, New Delhi, India, pp 93–96
7. Singh H (2022) Handicraft community and current practices for Rajasthan. Neuroquantology 20(12):744
8. Verma A, Gupta I (2019) Saharanpur's heritage of woodcraft: prototyping a future. J Herit Manag 4(1):36–72
9. Nevil HR (1909) The Saharanpur district Gazeetteer. Superintendent, Government Press, United Provinces, Allahabad
10. Hāṇḍā O, Jain M (2000) Wood handicraft: a study of its origin and development in Saharanpur. Indus Publishing, New Delhi, India, p 96
11. Yadav R, Mahara T (2018) An exploratory study to investigate value chain of Saharanpur wooden carving handicraft cluster. Int J Syst Assur Eng Manag 9:147–154. https://doi.org/10.1007/s13198-016-0492-5
12. Ansari MA (2021) Role of wood in contemporary Indian art with special reference to Nagina and Saharanpur. Int J Literacy Educ 1(2):79–82

Chapter 29
Rosewoods in Agroforestry: A Sustainable Approach to Conservation and Economic Development

Rekha R. Warrier, Geeta Joshi, and A. N. Arunkumar

1 Introduction

Rosewood, a group of tropical hardwood trees under the genus *Dalbergia*, is highly prized for its dense, richly hued wood, used in fine furniture, musical instruments, and decorative items. The global demand for rosewood, however, has led to extensive deforestation and illegal trade, pushing many species, such as *Dalbergia nigra* and *Dalbergia latifolia*, to the brink of extinction [1]. Agroforestry, the practice of integrating trees with agricultural crops and livestock, presents a viable strategy for conserving rosewood species and increasing global demand for sustainable and high-quality timber products, positioning rosewood as an attractive option in agroforestry systems. By fostering rosewood within diversified farming systems, agroforestry can support local economic growth, too [2]. Integrating rosewood into agroforestry practices may support economic growth and promote environmental sustainability, especially in tropical and subtropical regions.

The *Dalbergia* genus, part of the Fabaceae family, comprises over 100 species distributed in tropical and subtropical regions, particularly in Asia, Africa, and Latin America. Notable species, including *Dalbergia sissoo* (Indian rosewood), *D. latifolia* (East Indian rosewood), and *D. melanoxylon* (African blackwood), are renowned for their hardwood, which is highly valued in timber and furniture markets [3]. These species provide high-quality wood and serve essential roles in soil nitrogen fixation, significantly improving soil fertility in agroforestry systems [3].

R. R. Warrier (✉)
ICFRE-Institute of Forest Genetics and Tree Breeding, Coimbatore, India
e-mail: rekha@icfre.org

G. Joshi
Indian Council of Forestry Research and Education (ICFRE), Dehradun, India

A. N. Arunkumar
ICFRE-Institute of Wood Science and Technology, Bengaluru, India

R. R. Warrier et al. (eds.), *Handbook of Asian Rosewoods*, Materials Horizons: From Nature to Nanomaterials, https://doi.org/10.1007/978-981-96-9459-4_29

2 Agroforestry and Rosewood: A Symbiotic Relationship

Agroforestry, defined as the integration of trees and shrubs into agricultural landscapes, aims to increase biodiversity, improve soil fertility, and enhance water retention. Rosewood's adaptability to various environmental conditions, coupled with its nitrogen-fixing capability, makes it a suitable candidate for agroforestry systems. By cultivating rosewood alongside crops, farmers can potentially increase their income through diversified production while also contributing to ecological sustainability.

3 Ecological Benefits of Rosewood in Agroforestry

Rosewood trees contribute numerous ecological benefits to agroforestry systems. They serve as nitrogen-fixing plants, enriching the soil and enhancing fertility for surrounding crops [4]. This function is particularly advantageous in regions with poor soil quality, where rosewood's nitrogen fixation can enhance productivity across diverse plantings. In terms of biodiversity, rosewood provides habitat for various flora and fauna, promoting a more complex ecosystem structure. Rosewood plantations have been shown to support avian diversity, with species using these trees for shelter, nesting, and food sources [5]. Additionally, the dense canopy of mature rosewood trees helps mitigate soil erosion, further protecting nearby vegetation and crops in mixed-species agroforestry setups.

4 Environmental Benefits of Rosewood Cultivation in Agroforestry

4.1 Soil Fertility, Erosion Control, and Nutrient Cycling

Dalbergia species are known for their ability to fix atmospheric nitrogen through symbiotic relationships with Rhizobium bacteria in root nodules. This nitrogen fixation improves soil nutrient availability, benefiting both the trees and associated crops [6]. Research shows increased soil nitrogen levels in fields where trees intercropped with crops like wheat, maize, and millet lead to higher crop yields and reduced need for synthetic fertilizers [7]. Furthermore, *Dalbergia* trees contribute organic matter to the soil through leaf litter decomposition, enhancing soil structure, moisture retention, and overall productivity.

Dalbergia species, like other leguminous plants, form symbiotic relationships with Rhizobium bacteria, allowing them to fix atmospheric nitrogen in the soil. This process improves the nitrogen content of the soil, benefiting crops interplanted with *Dalbergia* trees. Studies on *Dalbergia sissoo* reveal that nitrogen levels and crop yields of wheat, maize, and millet significantly increase when these crops are grown in close proximity to *D. sissoo* trees [8]. This natural fertilization reduces the dependence on synthetic fertilizers, promoting sustainable farming practices.

4.2 Biodiversity Conservation

Agroforestry systems that incorporate *Dalbergia* species provide habitats and food sources for diverse organisms, promoting biodiversity within the system. These trees support various insects, birds, and mammals, which in turn enhance ecosystem functions like pollination and pest control [8]. In regions where *Dalbergia* species are prevalent, these trees contribute to creating biodiverse landscapes that support wildlife and help conserve genetic resources. Agroforestry systems that incorporate *Dalbergia* species provide habitats and food sources for diverse wildlife, including birds, pollinators, and soil organisms. These trees contribute to landscape heterogeneity, which fosters a variety of species and promotes ecological resilience [8]. This increased biodiversity supports ecosystem functions like pest control and pollination, which are essential for crop productivity.

4.3 Carbon Sequestration and Climate Resilience

Dalbergia trees in agroforestry systems enhance resilience to climate change through their role in carbon sequestration and soil conservation. They sequester significant amounts of carbon due to their large biomass, helping mitigate greenhouse gas emissions and contributing to climate change mitigation efforts [9]. Additionally, *Dalbergia* species help prevent soil erosion and retain water in the soil, which is critical in regions experiencing water scarcity and variable climate conditions.

5 Economic Scope of Rosewood in Agroforestry

5.1 Market Demand and Timber Value

Rosewood is a valuable hardwood sought for furniture, veneer, and musical instruments. Overexploitation has led to CITES restrictions, boosting demand for sustainably cultivated wood. Agroforestry systems integrating rosewood provide economic opportunities for farmers, ensuring legal and environmentally friendly trade. However, its slow growth demands patience, with harvestable maturity taking decades. Farmers may require incentives and financial support during early phases to sustain plantations, offering a steady supply and reducing illegal logging pressures on wild populations [10].

5.2 *Income Diversification and Stability*

Intercropping rosewood with crops like maize, beans, and medicinal plants offers short-term income while rosewood grows into a valuable long-term asset. *Dalbergia* timber, prized for its strength and fine grain, is ideal for furniture, instruments, and crafts, supporting rural livelihoods. In India, *Dalbergia sissoo* contributes significantly to local and global markets, enhancing rural economies [11]. These trees also provide fuelwood, fodder, and medicinal products, diversifying income and boosting resilience for smallholder farmers through agroforestry systems.

6 Social and Community Impacts

Rosewood agroforestry offers several social benefits, particularly in rural and impoverished areas where communities depend on agriculture for their livelihood. When rosewood is grown as part of a diversified farming system, it can act as a long-term economic security measure, potentially providing income from rosewood sales for future generations [12]. Moreover, agroforestry practices encourage traditional knowledge systems and the participation of local communities in forest management, fostering a sense of ownership and stewardship over natural resources.

Education and capacity building are crucial for communities involved in rosewood agroforestry, as they need training in sustainable harvesting practices, legal compliance, and agroforestry management. International and local NGOs play an essential role in providing this training, assisting communities in developing sustainable practices that can protect rosewood populations and offer economic returns [13].

7 Agroforestry with Rosewood Species

Involving local farmers in rosewood cultivation can facilitate, governments and organizations to promote sustainable livelihoods and strengthen community resilience. Training farmers in sustainable rosewood management practices and agroforestry techniques may further enhance the socioeconomic benefits associated with its cultivation. As mentioned earlier, different species have been tested in agroforestry. Two main species in the Asian region have gained popularity. The details are presented below.

7.1 *Latifolia Agroforestry*

7.1.1 India

D. latifolia is mostly found in non-native/artificial forest stands, mixed plantations, and agroforestry land. This species is planted with other tree species and as understory plants In India, in agroforestry systems, in the earlier stages of growth, the tree is intercropped with rice, maize, beans, or cassava and later, when the canopies begin to close, shade-tolerant crops like coffee, turmeric, and ginger are practiced. It is also grown with fruit trees like mango, annona, jackfruits, and guava [14]. It is also planted as a shade tree in coffee plantations and on roadsides [15]. Rosewood is suitable for agri-silvi, silvi-horti, and silvi-horti-agri systems. The horticulture species planted along with Rosewood includes mango, annona (custard apple), jackfruit, aonla (gooseberry), and guava. Rosewood is also used as a permanent host tree in Sandalwood plantations and as a shade tree in coffee plantations of Karnataka, Tamil Nadu, and Kerala [16]. During the initial three years millets like ragi are suitable. Afterwards when the canopies begin to close, shade-tolerant crops like ginger and turmeric are planted. A spacing of 3 × 2 m or 6 × 2 m is suitable for growing intercrops. In pure plantations, a spacing of 2.5 × 1.2 m or 1.8 × 1.0 m is adopted to prevent forking and induce clear bole formation. The tree is usually harvested at a rotation age of 30–40 years [17] (Fig. 1).

Rosewood is suited for boundary and block plantation. However, it is not preferred by tree growers due to stringent felling and transit regulations, hence it has not gained popularity as a farmland tree [18].

Fig. 1 **a** Rosewood-based Silvi-Horti-Agri system. **b** Irrigated Rosewood-based agroforestry system

7.1.2 Indonesia

In the planting systems of *D. latifolia* in Java and West Nusa Tenggara in Indonesia, the species is planted as mixed plantation and agroforestry, planted as hedges or mixed with annual crops. It is commonly planted on the edges of the gardens as boundary trees, or in the pathway or rice field bunds. Furthermore, it is also planted along the sides of residential or village roads as a reinforcing plant to prevent soil erosion/landslide. This is because its roots are strong in retaining the soil structure, especially on the steep contours or the side ravine of the road [19, 20].

The Government introduced a taungya system, *tumpangsari*, on government land as early as 1873. In this system, farmers grow food crops with timber trees for two to three years and then move to new sites when the tree canopy closes. Over time, the system has been improved to provide more benefits. Farmers plant *D. latifolia* at a spacing of 3 m between rows and 1–2 m between trees within rows. They grow annual food crops between the rows of trees, such as upland rice, maize, beans (peanut, soybean), and cassava. The system comprises three layers of vegetation—the ground layer with crops such as rice and beans, the second layer with crops such as maize and cassava, and the top layer of trees (Fig. 2).

In an effort to increase farm income and thus make agroforestry more attractive, the state forest corporation introduced another system in 1986—the whole-rotation agroforestry (Fig. 3). Under this system, farmers grow food crops and fruit trees throughout the entire rotation of timber trees. The farmers are allowed to keep all the food crops plus fruit and prunings from the fruit trees. The government keeps all the wood from thinning and harvesting operations, including wood from the fruit trees.

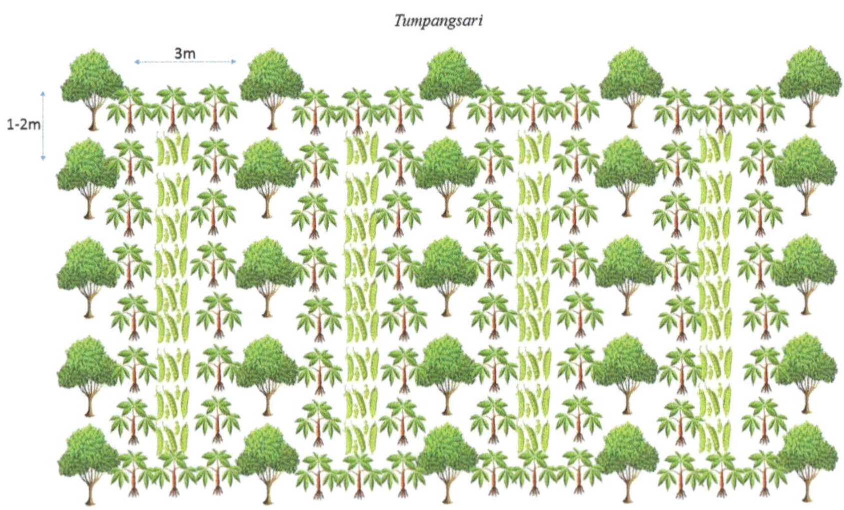

Fig. 2 A pictorial representation of *tumpangsari*

Fig. 3 A pictorial representation of a model of whole-rotation system agroforestry

Under this system, trees are spaced at 4, 5, or 6 m between rows and 2 m between trees within rows. In addition to annual food crops, farmers plant fruit trees such as mango, jackfruit, annona, and guava. In some places, they plant tobacco instead of food crops. As the canopies of timber and fruit trees increase, farmers replace food crops with shade-tolerant species such as ginger and turmeric. This system comprises five vegetation layers: a ground layer of rice and beans, a second layer of maize and cassava, a third layer of annona and guava, a fourth layer of jackfruit and mango, and a fifth layer of *D. latifolia.*

7.2 Shisham Agroforestry

7.2.1 India

Farmers have been planting *D. sissoo* for at least 100 years, primarily from seeds. Propagation and tree establishment techniques are well documented: they reflect diverse environmental conditions and management objectives. Planting strategies to improve productivity are increasingly based on vegetative reproduction and the use of clonal material. The species survives well in porous soil with adequate moisture, mostly along rivers and streams in the Himalayan tract. In its native countries of India, Nepal, Bangladesh, and Pakistan, *D. sissoo* has been widely planted in areas outside of its natural range. It is used as a windbreak in tea, coffee, and mango plantations, and these shade-loving crops also benefit from improved soil fertility under *D. sissoo* [21]. It has been grown under both temporary/taungya and permanent cultivation. Every year before the onset of or during winter, farmers carry out pollarding, lopping, individual branch pruning, trimming, or cutting of trees to promote coppicing as well as for fuelwood and fodder [22].

Many crops, including wheat, sugarcane, maize, and cotton, can be grown with *Dalbergia sissoo* [23], though there are reports on affected yields. Three rows of 25-year-old trees reduced wheat yield up to 5 m from the shelterbelt but increased yield beyond this distance [24, 25]. Reduced yields in wheat and paddy were observed near tree lines [26]. Cutting lateral roots of *D. sissoo* mitigated shading effects, boosting cotton and tobacco yields by 21% and 68% [23]. A Sissoo-rice agroforestry system evaluated at Jabalpur recommended a 50 percent high net monetary returns [27], while in a wheat-sissoo-based system, 25% pruning was reported to be the most profitable [28]. It was observed that closer spacing of sissoo reduced agri crop yield recommending a spacing of at least 10 × 10 m [29]. Similarly, significant yield reductions in rice were recorded when intercropped with closely spaced sissoo trees in Bangladesh, attributing it to tree canopy effects [30]. Pruning in agroforestry effectively manages competition for light and nutrients between trees and crops, enhancing system productivity [31]. The use of *D. sissoo* as an agroforestry component has been shown to be financially superior to monocropping. Agricultural crop and trees combined generate a higher net present value than other alternatives over a period of 20 years [32]. Assessment of compatibility of intercrops indicates that the horti-silvicultural system along with *D. sissoo* performed well [33].

Under agroforestry it can be grown without irrigation. However, irrigation is required during the first summer season to ensure better survival and growth. Crops like wheat, maize, mustard, millet, sesame, soybean, etc. can be grown during the initial stages [34]. After 2 or 3 years, shade-bearing crops like mung, ginger, turmeric, etc. can be grown in block plantations. The rotation of the natural crop is about 60 years, though under irrigated conditions it is about 25 years. For a 25-year rotation period of these clones, the total wood yield is expected to be 284 cum/ha (cubic metres per hectare) and a merchantable log yield of 84 cum/ha. A total income of Rs. 29.40 lakhs is estimated from the merchantable logs. An additional income from minor timber and fuelwood is estimated at Rs. 7.0 lakhs (at Rs. 5/Kg for 140 tons/ha fuelwood). The mean annual increment (MAI) for a 20-year-old, non-irrigated plantation in the eastern plain regions of India was 11.2 m^3/ha on good sites, 8.4 m^3/ha on fair sites, and 5.9 m^3/ ha on poor sites. At the end of 20 years, the system yielded a B:C ratio of 2.35.

The gain from planting these improved clones developed by ICFRE-AFRI is estimated at Rs. 12.52 lakhs/ha with a mean annual benefit of Rs. 50,000/ha. ICFRE-AFRI, Jodhpur, has recently released three high yielding clones of Shisham—AFRI-DS1, AFRI-DS2, and AFRI-DS4. These clones are suitable for plantation in hot semi-arid and sub-humid tracts of Gujarat and uplands of Rajasthan. ICFRE-FRI released a variety viz. FRI-DS-14 resistant to dieback disease [17] (Fig. 4).

Fig. 4 *D. sissoo* with wheat and mustard in Punjab

7.2.2 Bangladesh

Sissoo is a valuable multipurpose tree species in Bangladesh [35, 36]. It has been widely cultivated in community forests, private forests, woodlot plantations, and homesteads across various districts. This species has significantly contributed to the socioeconomic development of the region [36]. Estimates indicate that *D. sissoo* constitutes approximately 60% of plantations in the northern and southern regions of Bangladesh and 20% in the central and eastern regions [37] with combination of different crops (e.g., paddy, wheat, pineapple, and vegetables) [38]. It is predominantly cultivated in regions along the riverbanks of the Padma, Jamuna, Meghna, and Brahmaputra. The northern districts, such as Rajshahi and Rangpur, have embraced sissoo plantation, given their semi-arid characteristics and increasing demand for agroforestry models. *D.* sissoo accounts for a major share of woodlot trees in most of these areas due to its value as a high-value timber species. However, the preference has reduced to issues such as dieback resulting in mortality [39]. Considering the economic criteria for Bangladesh, 12 or 13 years for timber and 8 years for fuel wood are suggested as rotation ages in agroforestry [40].

8 Policy Recommendations and Conservation Strategies

Despite their ecological and economic benefits, many *Dalbergia* species are under threat due to overexploitation, habitat loss, and illegal logging. High demand for *Dalbergia* timber has led to severe population declines, prompting several species to be listed under Convention on International Trade in Endangered Species of Wild Fauna and Flora (CITES) Appendix II to control their trade [41]. Sustainable agroforestry practices, including regulated harvesting, replanting, and community-based conservation efforts, are essential to ensuring the continued availability of *Dalbergia* species for agroforestry. Agroforestry models offer a way to produce rosewood legally and sustainably, aligning with international conservation efforts. Farmers who adopt these models could benefit from policy incentives, such as subsidies or tax relief for sustainable practices, further encouraging rosewood's cultivation. Integrating rosewood into agroforestry systems requires supportive policies and legal frameworks to ensure both conservation and economic viability. Governments in rosewood-producing regions should offer incentives, such as tax breaks or grants, for farmers engaging in rosewood agroforestry. Certification schemes, like the Forest Stewardship Council (FSC) and Sustainable Forest Management & Forest Certification in India, could help establish market value for sustainably produced rosewood, promoting responsible management of rosewood agroforestry plantations [42].

Another important aspect is the establishment of traceability and legality in the rosewood trade. Given the challenges of illegal logging, it is essential to implementing transparent certification systems that can verify the legal origin of rosewood from agroforestry systems, reducing illegal harvesting from natural forests. Partnerships among local governments, international conservation organizations, and local communities could enable better enforcement of existing regulations and the development of sustainable, community-based rosewood agroforestry systems.

9 Challenges

Despite the promising scope, challenges remain, including the slow growth rate of rosewood, initial costs, and potential competition for light and resources between rosewood and intercropped species. Furthermore, illegal logging and poaching of rosewood continue to threaten its cultivation. Overexploitation and habitat loss from agricultural expansion and urban development also threatens *Dalbergia* species, which require specific conditions for optimal growth. *Dalbergia* species are susceptible to various pests and diseases, such as leaf-eating insects, fungal infections such as dieback disease, and stem borers, which can damage trees and impacts tree health and productivity. Implementing integrated pest management (IPM) strategies, such as biological control and resistant varieties, is critical for successfully establishing *Dalbergia* in agroforestry systems. Smallholder farmers may lack the knowledge and resources to effectively cultivate *Dalbergia* trees. Establishing sustainable agroforestry systems requires training on tree management, selective harvesting, and soil maintenance practices. Without adequate technical support, the benefits of incorporating *Dalbergia* species may not be fully realized. Other major challenges include Allelopathy: The release of certain biochemicals can inhibit the germination and growth of specific crops. Water Requirements: Although drought-tolerant, optimal growth demands consistent water availability, posing challenges in arid regions.

10 Sustainable Management Strategies

10.1 Community Involvement and Awareness Programs

Engaging local communities in *Dalbergia* species management is crucial for conservation. Awareness programs on sustainable harvesting and ecological benefits, along with community involvement in replanting efforts, can enhance stewardship. In India, community forestry initiatives have successfully supported *Dalbergia sissoo* replanting and protection.

10.2 Agroforestry Policy and Institutional Support

National and regional policies supporting agroforestry, including incentives for planting and conserving *Dalbergia* species, can promote sustainability. Financial support for smallholder farmers and policies ensuring tree tenure rights, financial aid, and market access for sustainably harvested timber can help reduce illegal logging and encourage sustainable *Dalbergia* management.

10.3 Research and Development on Dalbergia *Varieties*

Continued research on *Dalbergia* varieties with improved growth rates and disease resistance can enhance agroforestry. Collaboration between research institutions and farmers is crucial for testing and tailoring these varieties to local needs [43]. Future research should focus on optimizing intercropping, breeding for better growth rates, and developing policies to support sustainable rosewood agroforestry. Breeding pest-resistant varieties will help farmers maintain productive, long-term systems.

11 Conclusion

Integrating rosewood into agroforestry systems offers a promising solution to balance economic needs and ecological conservation. It provides a sustainable timber source, improves soil health, supports biodiversity, and mitigates climate change. Agroforestry promotes reforestation, biodiversity, and rural development by cultivating rosewood alongside crops or livestock. Success depends on supportive policies, financial incentives, and community involvement. Additionally, incorporating *Dalbergia* species in agroforestry enhances soil fertility, biodiversity, and carbon sequestration while providing valuable timber. However, balancing demand with conservation requires coordinated efforts from local communities, policymakers, and researchers. Sustainable agroforestry practices that protect *Dalbergia* species can support environmental and economic stability, benefiting both ecosystems and rural livelihoods.

References

1. Ghazoul J, Burivalova Z, Garcia-Ulloa J, King LA (2015) Conceptualizing forest degradation. Trends Ecol Evol 30(10):622–632
2. Nair PKR (1989) Agroforestry systems in the tropics. Kluwer Academic Publishers, Dordrecht, The Netherlands
3. Lemmens RHMJ, Soerianegara I, Wong WC (1995) Timber trees: minor commercial timbers; Plant Resources of South-East Asia Backhuys Publishers, Leiden, Netherlands, No. 5(2)
4. Méndez VE, Lok R, Somarriba E (2013) Interactions between carbon sequestration and shade coffee certification programs: a case study from a Nicaraguan cooperative. Ecol Econ 89:35–43
5. Porter-Bolland L, Ellis EA, Guariguata MR, Ruiz-Mallén I, Negrete-Yankelevich S, Reyes García V (2012) Community managed forests and forest protected areas: an assessment of their conservation effectiveness across the tropics. Forest Ecol Manag 268(Suppl C):6–17
6. Rao MR, Nair PKR, Ong CK (1997) Biophysical interactions in tropical agroforestry systems. Agrofor Syst 38:3–50
7. Tewari VP, Arya R, Prasad R (2021) Some insights into adaptive silviculture and forest management under climate change uncertainties. Indian J Agrofor 23(1):1–4
8. Harvey CA, Villanueva C, Villacís J, Chacón M, Muñoz D, López M, Ibrahim M, Gómez R, Taylor R, Martinez J, Navas A (2005) Contribution of live fences to the ecological integrity of agricultural landscapes. Agr Ecosyst Environ 111(1–4):200–230
9. Nair PKR, Mohan Kumar B, Nair VD (2009) Agroforestry as a strategy for carbon sequestration. J Plant Nutr Soil Sci 172(1):10–23
10. Gullison R, Rice R, Blundell A (2000) Marketing species conservation. Nature 404:923–924
11. Naqvi SAH, Mushtaq S, Malik MT, Umar UUD, Fareed S, Zulfiqar MA (2019) Factors leading towards *Dalbergia sissoo* decline (syndrome) in Indian sub-continent: a critical review and future research agenda. Pak J Agric Res 32(2):302–316
12. Mukul SA, Herbohn J, Firn J (2016) Co-benefits of biodiversity and carbon sequestration from regenerating secondary forests in the Philippine uplands: implications for forest landscape restoration. Biotropica 48(6):882–889
13. Swamy L, Drazen E, Johnson WR, Bukoski JJ (2018) The future of tropical forests under the united nations sustainable development goals. J Sustain For 37(2):221–256
14. Sukhadiya M, Dholariya CA, Behera LK, Nayak D, Patel SM, Mehta AA (2020) *Dalbergia latifolia* Roxb: biography of an indigenous multipurpose tree species of India. MFP News 10(1):4–7

15. Nath CD, Pélissier R, Ramesh BR, Garcia C (2011) Promoting native trees in shade coffee plantations of southern India: Comparison of growth rates with the exotic *Grevillea robusta*. Agrofor Syst 83:107–119
16. Arunkumar AN, Joshi G, Warrier RR (2022) Growth assessment of young plantation of *Santalum album* with *Dalbergia latifolia* as host. Indian For 148(8):785–790
17. Kumar S, Joshi G, Warrier RR, Arunkumar AN (2023) FAQs on Agroforestry. ICFRE, Dehradun, India, p 164
18. Sarath S, Sandra PR, Nikhil PV (2023) Farmland trees in India: perceptions, potentials and problems. *Authorea Preprints* (2023). https://authorea.com/users/667597/articles/667766-farmland-trees-in-india-perceptions-potentials-and-problems
19. Sunardi S (2013) The technique of agroforestry at communal forest areas in Pejarakan Village, District of Gerokgak, Regency of Buleleng, Province of Bali. J Trop Life Sci 3(1):34–42
20. Yulita KS, Wardani W, Atikah TD, Pratama BA, Subiakto A, Nurjanah A, Nopiansyah F, Nugroho A, Kamal I, Arrofaha A (2022) The non-detriment findings (NDF) report for *Dalbergia latifolia* in Java and West Nusa Tenggara, Indonesia
21. Orwa C, Mutua A, Kindt R, Jamnadass R, Anthony S (2009) Agroforestry database: a tree reference and selection guide, version 4.0. World Agroforestry Centre, Nairobi, Kenya. http://www.worldagroforestry.org/sites/treedbs/treedatabases.asp
22. Von Carlowitz PG (1991) Multipurpose trees and shrubs: sources of seeds and inoculants. In: International council for research in agroforestry (ICRAF), Nairobi, Kenya pp 328
23. Luna RK (1996) Plantation trees. International Book Distributors, Dehra Dun, India pp 975
24. Shah H, Bakhsh MI, Amjad M (1991) Tree growth on farmlands of NWFP. Pak J For 41(2):74–81
25. Wahid R, Hafeez M (1991) Effect of *Dalbergia sissoo* trees on the yield of wheat: crop 1. Pak J For 41(2):82–88
26. Dhillon MS, Singh S, Atwal AS, Dhillon GS (1984) Developing agri-silvicultural practices: effects of *Dalbergia sissoo* and *Acacia nilotica* on the yield of adjoining crops. Indian J. Ecol. 11(2):249–253
27. Nayak H, Koshta LD, Bhol N (2014) Economics of *Dalbergia sissoo*-rice based agroforestry system under different levels of pruning. Indian For 140(6):598–603
28. Meena MB (2008) Effect of different pruning intensities and doses of fertilizers on the growth and productivity of wheat under Agri-silviculture (Shisham+Wheat) practice. M.Sc. Thesis, College of Agriculture, JNKVV, Jabalpur
29. Nandal DP, Rana P, Kumar A (1999) Growth and yield of wheat (*Triticum aestivum*) under different tree spacings of *Dalbergia sissoo*-based agrisilviculture. Indian J Agron 44(2):256–260
30. Islam KK, Hoque ATMR, Mamun MF (2006) Effect of level of pruning on the performance of rice-sissoo based agroforestry system. Am J Plant Physiol 1(1):13–20
31. Frank E, Eduardo S (2003) Biomass dynamics of *Erythrina lanceolata* as influenced by shoot pruning intensity in Costa Rica. Agrofor Syst 57:19–28
32. Subhan F (1990) Financial analysis of selected shelterbelt systems in Pakistan. Pak J For 40(3):247–252
33. Sivakumar K, Ramah K, Surendran U, Kaleeswari RK, Jayakumar M (2022) Assessment of compatibility of intercrops in *Dalbergia sissoo*-based Hortisilvicultural system in Mondipatty, Manapparai block of Trichy district, Tamil Nadu. J Appl Nat Sci 14(1):94–101
34. Sharma KK, Khanna P, Gulati A (1996) The growth and yield of wheat and paddy as influenced by *Dalbergia sissoo* Roxb. boundary plantation. Indian For 122(12):114–126
35. Zabala NQ (1991) Plantation silviculture. Development of professional education in the forestry sector, Bangladesh. UNDP/FAO/BGD/85/011, Field Document No. 19. IFCU, FAO, UN, pp 234
36. Nawshadul H (1994) Production, management and genetic improvement of *Dalbergia sissoo* Roxb. Institute of Forestry, Chittagong University
37. Baksha MW, Basak AC (2003) Mortality of sissoo (*Dalbergia sissoo* Roxb.) and its management in Bangladesh. In: Mortality of Sissoo (*Dalbergia sissoo*) and top dying of Sundri (*Heritiera fomes*) in Bangladesh. Bangladesh Forest Research Institute, Chittagong, pp 1–6

38. Lodhiyal N, Lodhiyal LS (2003) Biomass and net primary productivity of Bhabar Shisham forests in Central Himalaya, India. For. Ecol. Manag. 176:217–235
39. Hocking D, Hocking A, Islam K (1996) Trees on farms in Bangladesh: 3. Farmers' species preferences for homestead trees, survival of new tree planting, and main causes of tree death. Agrofor Syst 33:231–247
40. Yunus M, Hossain MAT, Yasmin F, Milon M (2014) Economic rotation of Sissoo tree plantations (strip) in Bangladesh. Asian J Agric Rural Dev 4(9):449–457
41. CITES (2013) Appendices I, II and III. www.CITES.org
42. Bray DB, Ellis EA, Armijo-Canto N, Beck CT (2004) The institutional drivers of sustainable landscapes: a case study of the 'Mayan Zone'in Quintana Roo, Mexico. Land Use Policy 21(4):333–346
43. Dobhal S, Kumar A, Ravichandran S (2016) Evaluation of clonal divergence in *Dalbergia* sissoo Roxb. for developing production populations. Indian For 142(5):451–458

Chapter 30
Mapping the Vulnerability of Asian Rosewood Species to Multiple Threats: A Guide for Prioritizing Conservation and Restoration Areas Considering Ecoregional Diversity

Hannes Gaisberger, Tobias Fremout, Chris Kettle, Rekha R. Warrier, Thea So, Sineath Sreng, Chaloun Bounithiphonh, Huang Ping, Zheng Yongqi, Vivi Yuskianti, and Riina Jalonen

1 Introduction

Rosewoods (*Dalbergia* spp.) have long been revered for the beauty and exceptional properties of their wood, capturing the imagination of artisans, musicians, and collectors alike. With their rich hues, intricate grain patterns, and remarkable durability, rosewoods have found myriad applications in the crafting of fine furniture, musical instruments, and decorative objects [1]. However, the appeal of these hardwoods has also made them prime targets for exploitation, leading to a precipitous decline in their populations across tropical and subtropical regions worldwide [2], a situation that is further exacerbated by climate change [3, 4].

H. Gaisberger (✉) · C. Kettle
Bioversity International, Rome, Italy
e-mail: h.gaisberger@cgiar.org

C. Kettle
e-mail: c.kettle@cgiar.org

T. Fremout
Bioversity International, Lima, Peru
e-mail: t.fremout@cgiar.org

R. R. Warrier
ICFRE-Institute of Forest Genetics and Tree Breeding, Coimbatore, Tamil Nadu, India
e-mail: rekha@icfre.org

T. So
Institute of Forest and Wildlife Research and Development, Phnom Penh, Cambodia

R. R. Warrier et al. (eds.), *Handbook of Asian Rosewoods*, Materials Horizons: From Nature to Nanomaterials, https://doi.org/10.1007/978-981-96-9459-4_30

The genus *Dalbergia* includes a variety of species commonly referred to as rosewoods, many of which are native to Asia. Among these, five species—*Dalbergia cochinchinensis, D. cultrata, D. latifolia, D. oliveri*, and *D. sissoo*—hold particular significance due to their widespread distribution and economic importance. Siamese Rosewood (*Dalbergia cochinchinensis*), Burma Blackwood (*Dalbergia cultrata*), and Burmese Rosewood (*Dalbergia oliveri*) are found in various countries of the Greater Mekong Subregion, while Indian Rosewood (*Dalbergia latifolia*) and North Indian Rosewood (*Dalbergia sissoo*) are native to the Indian subcontinent, particularly in the Western and Eastern Ghats and in the Indo-Gangetic Plain, respectively. *D. latifolia* is also found in Indonesia, but its origin in the country is debated. Despite their cultural and ecological value, these species face existential threats ranging from habitat loss, illegal logging, and overharvesting [2]. On the IUCN Red List of Threatened Species [5], *Dalbergia* species were listed as threatened as early as 1998, but since then, their exploitation has only increased [2]. The entire genus *Dalbergia* was placed on Appendix II of the Convention on International Trade in Endangered Species of Wild Fauna and Flora (CITES) in 2017, placing restrictions on international trade to lessen pressure on the remaining natural populations. Population genetic studies have shown high levels of intraspecific diversity between populations [6–8], which may indicate limited gene flow and a high degree of local adaptation. Therefore, it is likely that the local extinction of populations of these species due to the threats will result in a permanent loss of genetic diversity and adaptive capacity.

The interest in planting *Dalbergia* spp. in Asia is growing [9–11]. Planted tree populations can potentially both relieve pressure on natural populations and help maintain their genetic diversity [12]. However, their productivity and conservation value depend on the genetic diversity of the planting material and its suitability to the current and future environmental conditions [13, 14]. Unfortunately, current restoration of forest and tree populations often relies on planting material with low genetic diversity collected from few parental trees in degraded and fragmented natural populations [15, 16], leading to reduced survival and productivity [17–19]. There is a clear need to integrate conservation and restoration efforts more closely to promote resilient restoration and improved biodiversity outcomes.

S. Sreng
Forest Development and Botanical Research Centre, IRD, Phnom Penh, Cambodia

C. Bounithiphonh
Forest Research Center, National Agriculture and Forestry Research Institute, Xaythany District, Lao PDR

H. Ping · Z. Yongqi
Research Institute of Forestry, Chinese Academy of Forestry, Beijing, China
e-mail: pippin09@163.com

V. Yuskianti
Research Center for Biota Systems, National Research and Innovation Agency, Bogor, Indonesia

R. Jalonen
Bioversity International, Penang, Malaysia
e-mail: r.jalonen@cgiar.org

The objectives of this chapter are to (i) assess the vulnerability of five economically important Asian rosewood species to key threats, (ii) identify species-specific and ecoregional conservation and restoration priorities to maintain adaptive variation within species, and (iii) explore opportunities for cooperation and synergies between countries in their conservation efforts across the species' ranges. We examined the vulnerability of *Dalbergia cochinchinensis, D. cultrata, D. latifolia, D. oliveri,* and *D. sissoo* to overexploitation, fire, overgrazing, habitat conversion, and climate change, across their modeled ranges in twelve countries (Bangladesh, Bhutan, Cambodia, China, India, Indonesia, Laos, Myanmar, Nepal, Pakistan, Thailand, and Vietnam). The results are based on a spatially explicit framework that has been applied to a total of 63 socioeconomically important tree species in South and Southeast Asia [20] and that has also been successfully implemented in other regions and biomes [21, 22]. The methodology includes habitat suitability modeling, threat exposure layers, and a trait-based sensitivity scoring approach [21] to create species—and threat-specific vulnerability maps. Based on the vulnerability levels, priority areas for conservation and restoration for each species are derived. We refined the results by partitioning the priority areas by ecoregions [23], with ecoregions serving as a proxy for genetic diversity, capturing the existing adaptive variation within species.

2 Consolidation of Occurrence Data

Dalbergia occurrence records were compiled from the literature, freely accessible databases, and through the network of the Asia Pacific Forest Genetic Resources Programme (APFORGEN, www.apforgen.org). The compiled occurrence dataset underwent a thorough cleaning process to ensure quality control and a spatial and environmental filtering process to reduce the negative impact of uneven sampling before being used for spatial analysis. More details on these steps can be found in Gaisberger et al. [20]. The final dataset consisted of 411 occurrence points for the five *Dalbergia* species cross nine countries, as shown in Table 1.

3 Habitat Suitability Modeling

We selected a set of 15 bioclimatic [24], edaphic [25], and topographic [26] variables with variance inflation factors (VIFS) <10 at a spatial resolution of 2.5 arc minutes (approximately 4.5 km at the equator) and modeled the habitat suitability of the five *Dalbergia* species using the maximum entropy algorithm (MaxEnt) [27]. As default settings can lead to overfitting [28], we executed MaxEnt with tuned parameter settings, using the ENMeval package [29], and evaluated the models with spatially independent partitions, using the blockCV package [30]. Background points were randomly selected from a geographic extent like the one of the occurrence points,

Table 1 Number of occurrence records of the five Dalbergia species used in habitat suitability modeling by country

Country	*Dalbergia cochinchinensis*	*Dalbergia cultrata*	*Dalbergia latifolia*	*Dalbergia oliveri*	*Dalbergia sissoo*
Bhutan					1
Cambodia	23			20	
China		22		1	
India			47		49
Laos	32	7		23	
Myanmar		5		1	
Nepal			2		3
Thailand	58	41		46	
Vietnam	8	2		20	
Total	**121**	**77**	**49**	**111**	**53**

which is aimed at improving the discriminatory power of models in the core distribution area [31], as well as the transferability in place and time [32]. The modeling extent was restricted to the convex hull around the occurrence points, extended with a buffer corresponding to 20% of the longest axis between them. We converted the habitat suitability maps into presence maps using the 10th percentile training presence omission threshold and validated the resulting maps through an online consultation with species experts and a review of the literature and existing databases. As part of this process, modeled ranges of the species in states or provinces and in ecoregions that were not confirmed as part of their natural range (e.g., due to barriers to dispersal such as mountain ranges) were identified and excluded. Further details can be found in Gaisberger et al. [20].

4 Threat Exposure Estimation

For each of the five threats (overexploitation, fire, overgrazing, habitat conversion, and climate change), threat exposure estimates were constructed using freely accessible spatial datasets, according to the methodology described in Gaisberger et al. [20]. The exposure estimates were created as raster layers with values between 0 (no threat) and 1 (maximum threat) and reflect the current exposure, except climate change, which reflects the predicted future exposure in 2050 (Table 2). To assess the exposure to climate change, we projected the species distribution models (SDMs) of the five *Dalbergia* species to predicted climate conditions for 2050 (2041–2060 period), using the following downscaled General Circulation Model (GCM) predictions from the sixth Coupled Model Intercomparison Project (CMIP6) [33], obtained from the WorldClim website (https://worldclim.org/): MIROC6, BCC-CSM2-MR,

IPSL-CM6A-LR, CNRM-ESM2-1, and MRI-ESM2-0. The climate change exposure maps were created by calculating the number of different GCMs predicting presence of suitable habitat across the validated native distribution ranges. All raster layers used to construct the exposure maps had a spatial resolution of 2.5 arc minutes (approximately 4.5 km at the equator). To assess the robustness of the threat mapping methodology, we carried out a sensitivity analysis by creating reference-, best-, and worst-case scenario exposure maps for each of the considered threats.

5 Sensitivity and Vulnerability Assessment

To estimate the sensitivity of the five *Dalbergia* species to the five key threats, we applied a sensitivity scoring approach [21] in which explicit relations between tree functional traits and resistance against threats were established based on the literature and expert judgement. We focused on a subset of 10 key traits with "medium" to "very high" importance, which we compiled through a separate literature study. Details of the data used to estimate species sensitivities and the final trait-based sensitivity values can be found in Gaisberger et al. [20].

Species-specific threat maps were created cell by cell, as the product of the species-specific threat sensitivity value (0–1) and the threat exposure value (0–1), restricted to the native species range corrected by experts. The grid cells of the threat maps were divided into five classes (no threat, low, medium, high, and very high) using the threshold values 0.01, 0.25, 0.50, and 0.75. The combined vulnerability of an area, corresponding to an individual pixel, was set to be equal to the highest vulnerability among the individual threat layers. To identify the overall vulnerability of a species, we calculated the proportion of distribution area under "medium" to "very high" vulnerability to current and predicted climate change threats individually and in combination.

6 Identifying Priority Areas for Conservation and Restoration

Based on the vulnerability to current threats (overexploitation, fire, overgrazing, and habitat conversion) and to climate change, we present the following conservation and restoration planning strategy [21]. We recommend (i) in situ conservation of populations in areas where both current and climate change threats are low, (ii) restoration activities such as active planting or assisted regeneration of populations in areas where current threats are high, but climate change threats are low, and (iii) ex situ conservation of populations in areas where climate change threats are high, in order to safeguard the genetic resources that might disappear due to climate change by 2050. The remaining combinations of current and climate change threats

Table 2 Summary information on the five key threats, the expected impact on tree populations, and the spatial layers used to estimate the exposure to these threats

Key threat	Impact on tree populations	Indicators	Spatial layers
Overexploitation	Reduces tree density and genetic diversity, fragments populations	Human population density	Human population density [34]
		Travel time to cities	Travel time to cities [35]
		Presence/ absence of protected areas	Designated protected areas [36], updated with protected areas from Bangladesh, Cambodia, and India by contributing species experts
Fire	Causes mortality, reduces tree density, inhibits natural regeneration	Fire frequency	NASA Fire Information for Resource Management System (FIRMS). MODIS Active Fire Detections from 2013 to 2017 [37]
Overgrazing	Inhibits natural regeneration	Tropical livestock unit (TLU) density (cattle, goats, and sheep)	Gridded Livestock of the World v2.0 [38]
Habitat conversion	Eliminates and fragments natural populations; promotes tolerated, cultivated species	Percentage of cropland coverage	Cropland cover data [39]
		Percentage of planted forests and tree crops	Tree Plantations [40] for Cambodia and Indonesia
Climate change	Has a negative or positive effect on the survival and reproduction of individuals	Number of General Circulation Models (GCMs) predicting suitable habitat	SDMs projected to future climate conditions (2041–2060 period). We used five high-performing and independent GCMs [41] for SSP126, SSP245, and SSP585, downloaded from the WorldClim website (https://worldclim.org/)

were classified as (iv) "no single priority activity" areas and may benefit from a combination of conservation and restoration approaches (e.g., passive restoration) depending on the site context.

We restricted the prioritized area for in situ conservation to areas where the modeled distribution showed highly suitable habitat (values ≥ 0.7) as predicted by the Maxent models, with the aim to identify populations that are likely to have maximal fitness and adaptive capacity [42]. In addition, the priority conservation maps were overlaid with a global protected areas map [36], updated with protected areas from Bangladesh, Cambodia, and India by contributing species experts [20], to assess the proportion of species' ranges that are currently under some type of protection. In addition, we added a land cover map [39] to identify areas that are "natural" or have been converted to agricultural land use. We defined "natural" as all areas that were not classified as "cropland" or "artificial surfaces." Distinguishing between converted and unconverted areas is important because these land use types may require different conservation and restoration strategies. Large-scale biodiversity restoration projects are more likely to be implemented in non-converted areas whereas agroforestry may be a more appropriate restoration strategy in areas that have already been converted to agricultural use [21].

7 Ecoregional Diversity Assessment

We applied the method described in Gaisberger et al. [43] and subdivided the modeled ranges of the five *Dalbergia* species by ecoregion [23], as surrogate for adaptive genetic variation [44], and combined them with the priority conservation and restoration maps to obtain distribution areas likely to reflect patterns of intraspecific diversity. We considered the modeled ranges that were not prioritized for recovery or ex situ conservation as "stable" and classified ecoregions as a priority for improving conservation of adaptive variation if <10% of these stable areas were located within protected areas.

Three possible follow-up activities were identified for these priority ecoregions: (i) designating additional protected areas or conservation units in areas prioritized for in situ conservation that remain outside the existing protected area network, (ii) restoring the species' populations in areas prioritized for restoration (degraded or converted lands but remaining suitable for the species under climate change), and (iii) conserving the species ex situ through seed collection and field gene banks, especially when conservation targets could not be met through the other means (i.e., >90% of the potential distribution area becoming unsuitable for the species under climate change). A combination of these activities could be applied in each ecoregion to meet the conservation threshold. All area calculations were performed after transformation into the cylindrical equal area projection.

8 Impacts of Current Threats and Projected Climate Change

The habitat suitability maps we created for the five *Dalbergia* species, which were validated by experts in their native ranges in South and Southeast Asia, covered a total area of 3.6 million km^2. The proportions of species ranges varied greatly between different biome types [23], as indicated in Table 3. On average, over half (54%) of the species ranges occurred in tropical and subtropical moist broadleaf forests, about 30% in tropical and subtropical dry broadleaf forests, and about 14% in xeric shrublands.

The spatially explicit threat assessment revealed that all five *Dalbergia* species exhibit medium to very high vulnerability to at least one of the five threats in more than 85% of their potential ranges (Fig. 1 and Table 4). Overexploitation emerged as the single most important threat, affecting 65–72% of the range of the three species (*D. cochinchinensis, D. cultrata,* and D. *oliveri*) in the Greater Mekong Subregion and 98–99% of the range of the two species (*D. latifolia* and *D. sissoo)* in the Indian subcontinent, and in Java Island in Indonesia for *D. latifolia.* The vulnerability to the other four threats varies even more between the five species. Habitat conversion poses an important risk to *D. sissoo* (76%), *D. latifolia* (62%), *D. cochinchinensis* (45%), and *D. oliveri* (41%) but is less of a risk for *D. cultrata* (18%). The species in the Indian subcontinent are more vulnerable to land degradation by overgrazing (71–78%), whereas the Greater Mekong species are much less affected (12–15%). In return, the *Dalbergia* species occurring in the Greater Mekong Subregion are more threatened by fire (32–39%), which is predicted to have less effect on the *Dalbergia* species in the Indian subcontinent (5–6%). More than 20% of the distribution ranges of *D. cultrata* and *D. oliveri* are projected to be under medium to very high threat from

Table 3 Percentage of modeled distribution area for each of the five *Dalbergia* species across South and Southeast Asia by biome type as defined by Dinerstein et al. [23]

Biome name	*Dalbergia cochinchinensis* (%)	*Dalbergia cultrate* (%)	*Dalbergia latifolia* (%)	*Dalbergia oliveri* (%)	*Dalbergia sissoo* (%)
Tropical and Subtropical Moist Broadleaf Forests	26	79	42	48	47
Tropical and Subtropical Dry Broadleaf Forests	74	21	33	52	28
Deserts and Xeric Shrublands			25		22

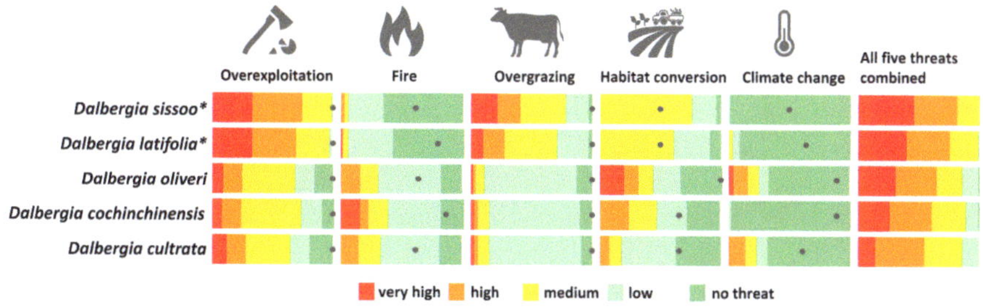

Fig. 1 Threat sensitivity and vulnerability estimates for the five *Dalbergia* species relative to five threats and the five threats combined (gray dots indicate sensitivity values; bars indicate relative share of suitable habitat for each species by level of threat [very high, high, medium, low, and no threat]; *, widely cultivated species). The species are in descending order of share of distribution range under high or very high vulnerability to combined threats

Table 4 Potential distribution area with medium to very high threat vulnerability to five threats and the five threats combined. The range from best-case to worst-case scenario is indicated in brackets. *, widely cultivated species

Species	Overexploitation (%)	Fire (%)	Over grazing (%)	Habitat conversion (%)	Climate change (%)	Combined threats (%)
D. cochinchinensis	72 (42–90)	39 (24–50)	12 (6–26)	45 (25–67)	<1 (<1–<1)	88 (69–95)
D. cultrata	65 (35–82)	33 (16–46)	15 (8–26)	18 (9–44)	23 (20–29)	86 (61–95)
*D. latifolia**	98 (90–100)	5 (2–9)	71 (43–89)	62 (34–88)	3 (3–4)	99 (93–100)
D. oliveri	67 (38–85)	32 (18–44)	9 (4–21)	41 (21–61)	24 (16–28)	85 (65–93)
*D. sissoo**	99 (94–100)	6 (3–9)	78 (54–91)	76 (52–93)	<1 (<1–<1)	100 (96–100)
Average	**80 (60–91)**	**23 (13–32)**	**37 (23–51)**	**48 (28–71)**	**10 (8–12)**	**91 (77–97)**

climate change, 3% for *D. latifolia,* and <1% for *D. cochinchinensis* and *D. sissoo.* Vulnerability maps of each species to the combined threats of overexploitation, fire, overgrazing, habitat conversion, and climate change are shown in Fig. 2.

9 Priority Areas for Conservation and Restoration

Based on the vulnerability levels to current threats and climate change, we mapped priority areas for conservation and restoration activities for the five *Dalbergia* species (Fig. 3). On average 6% (range from best-case to worst-case scenario 1–10%) of the potential native range was prioritized for in situ conservation and 56% (range

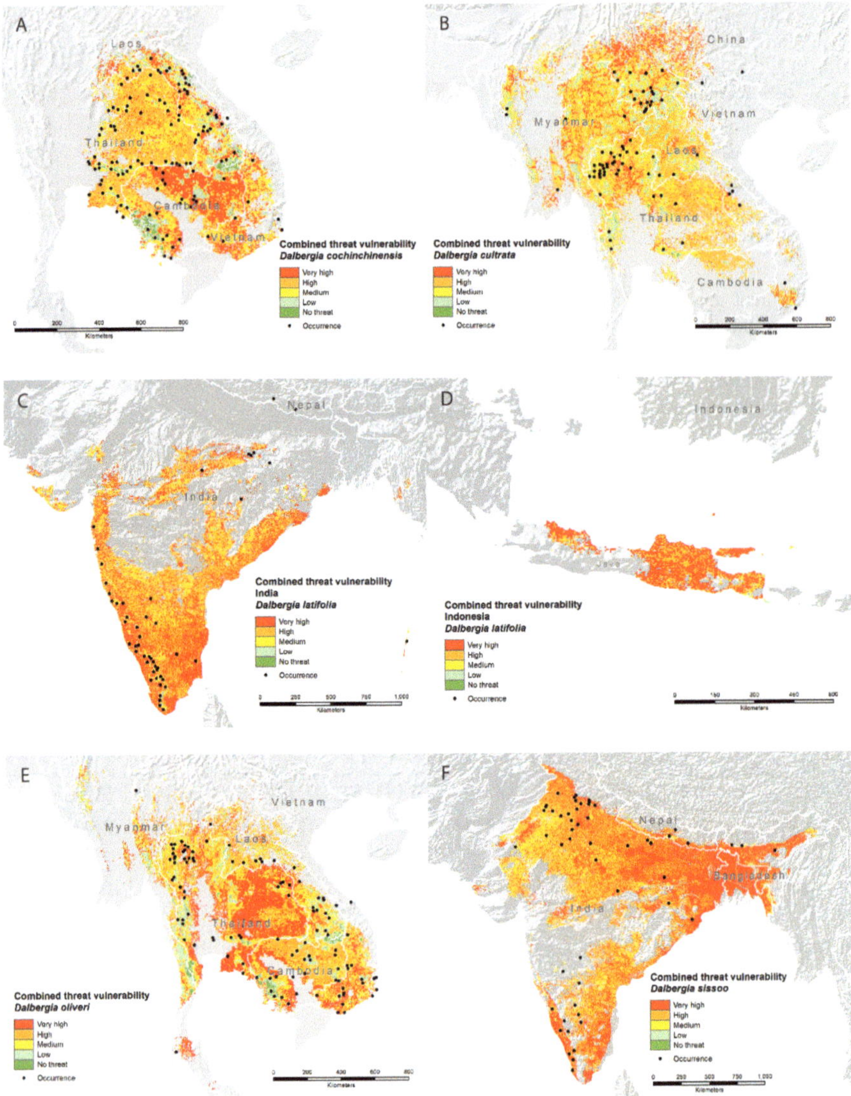

Fig. 2 Vulnerability to the combined threats of overexploitation, fire, overgrazing, habitat conversion, and climate change of **A** *D. cochinchinensis*, **b** *D. cultrata*, and **c** *D. latifolia* in India, **d** *D. latifolia* in Java, Indonesia, **e** *D. oliveri*, and **f** *D. sissoo* across their modeled native ranges

35–79%) for restoration activities, whereas on average, 6% (range <1–15%) of the distribution area was predicted to become unsuitable by 2050 as a result of climate change. The sensitivity analysis showed that, compared to the reference maps, an average of 34% (±5) of the grid cells of the priority area maps changed from one category to another for the best-case priority maps and 27% (±6) for the worst-case priority maps.

The 12–21% of the predicted ranges of *D. cochinchinensis*, *D. cultrata,* and *D. oliveri* fall within currently protected areas, whereas for *D. latifolia* and *D. sissoo* this proportion is 5 and 3%, respectively (Table 5). Cambodia has the largest proportion of the species' potential areas under protection, ranging from 37 to 38%. The

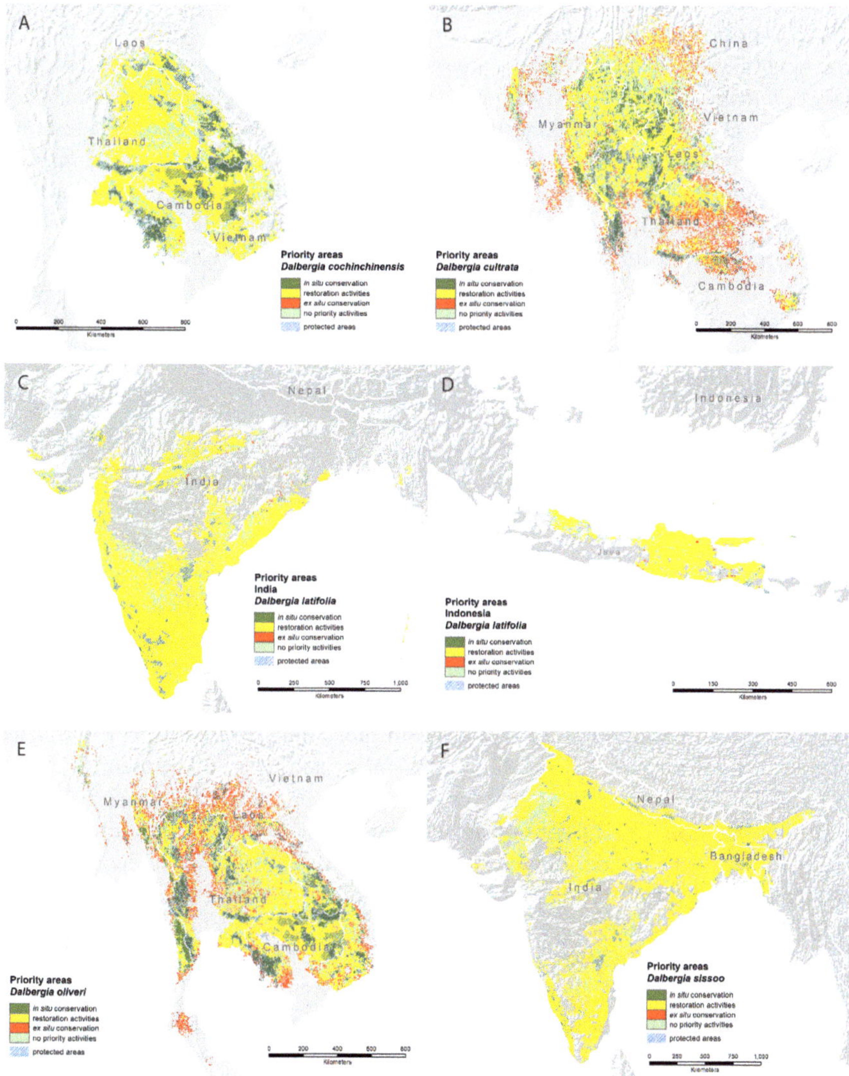

Fig. 3 Priority areas for conservation and restoration of **a** *D. cochinchinensis*, **b** *D. cultrata,* and **c** *D. latifolia* in India, **d** *D. latifolia* in Java, Indonesia, **e** *D. oliveri*, and **f** *D. sissoo* across their modeled native ranges

countries with the lowest proportion of protected species ranges are China (1%), Bangladesh (1–2%), Myanmar (0–4%), and India (2–5%). However, some species are also severely threatened within the protected area network: for example, 47% of the current distribution of *D. cochinchinensis* within protected areas in Cambodia is prioritized for restoration due to high vulnerability to current threats. Similarly, 33% of *D. cultrata*' s distribution range in protected areas in Cambodia is predicted to become unsuitable for the species due to climate change and therefore prioritized for ex situ conservation.

Areas prioritized for in situ conservation account for 1–10% of the total distribution ranges. More than 95% of these areas occur outside of the current protected areas for *D. cultrata* in China and Myanmar, for *D. latifolia* in Bangladesh, Myanmar, and Nepal, for *D. oliveri* in Myanmar and China, and for *D. sissoo* in Bangladesh and Myanmar. India is the country with the predominant distribution ranges of *D. latifolia* and *D. sissoo*, but due to severe current threats, only a maximum of 1% is prioritized for in situ conservation. At country level, more than 60% of the species' potential range is prioritized for restoration for *D. cochinchinensis* in Cambodia (62%) and Thailand (63%); for *D. latifolia* in Bangladesh (61%), India (71%), and Indonesia (81%); and for *D. sissoo* in India (78%), Nepal (81%), and Bangladesh (88%). The largest portion of the species' range prioritized for ex situ conservation is located in Cambodia (39%) for *D. cultrata* (Table 5).

10 Ecoregional Diversity

The conservation status of the five *Dalbergia* species varies depending on the ecoregion. For *D. cochinchinensis*, 6 of the 11 ecoregions (55%) have <10% of the predicted native range in stable protected areas. For *D. cultrata*, the proportion is 11 of 20 ecoregions (55%), for *D. latifolia* 18 of 22 ecoregions (82%), for *D. oliveri* 8 of 18 ecoregions (44%), and for *D. sissoo* 22 of 24 ecoregions (92%) (Table 6, Figs. 4 and 5). Seven large ecoregions with more than 100,000 km^2 of modeled species distribution area have <2% of stable protected areas, all of them occurring in India: Deccan thorn scrub forests for *D. latifolia* (Fig. 5c) and *D. sissoo*, Aravalli west thorn scrub forests, Chhota-Nagpur dry deciduous forests, Lower and Upper Gangetic Plains moist deciduous forests, Narmada Valley dry deciduous forests, and Thar desert for *D. sissoo* (Table 5). In 9 other ecoregions, the stable protected areas for individual species are very small (<10 km^2) or completely absent: Northern Indochina subtropical forests and Northern Thailand-Laos moist deciduous forests for *D. cochinchinensis;* Irrawaddy dry forests in Myanmar for *D. cultrata* and *D. oliveri;* Northern Triangle subtropical forests in Myanmar for *D. cultrata;* Western Java rain forests for *D. latifolia*; Mizoram-Manipur-Kachin rain forests in Myanmar for *D. oliveri*; Western Himalayan broadleaf forests in India and Nepal for *D. sissoo*.

For one of the total 65 species-ecoregion combinations prioritized for additional conservation measures, expanding the current protected area network to areas prioritized for in situ conservation is sufficient to meet the 10% threshold for stable

Table 5 Priority areas for in situ conservation, restoration activities, and ex situ conservation for the five *Dalbergia* species by country

Species	Country	Modeled distribution			Priority for in situ		Priority for restoration		Priority for ex situ
		km^2	% protected	% protected and stable	% of total area	% in protected areas	% of total area	% in natural habitats	% of total area
D. cochinchinensis	**All**	**460,076**	**21**	**14**	**8**	**69**	**58**	**46**	**< 1**
	Cambodia	140,210	37	20	12	84	62	75	< 1
	Laos	89,435	22	18	15	46	40	85	< 1
	Thailand	171,944	10	9	3	91	63	7	–
	Vietnam	58,487	10	8	3	46	57	52	–
D. cultrata	**All**	**856,402**	**12**	**9**	**9**	**34**	**35**	**76**	**14**
	Cambodia	25,828	38	14	1	47	28	90	39
	China	152,658	1	<1	8	<1	37	92	10
	Laos	134,591	12	8	10	27	36	94	13
	Myanmar	254,760	2	1	9	3	36	84	11
	Thailand	266,354	26	21	9	82	33	46	17
	Vietnam	22,211	12	9	<1	75	29	83	25
*D. latifolia**	**All**	**967,266**	**5**	**4**	**1**	**30**	**72**	**17**	**<1**
	Bangladesh	1,598	2	2	–	–	61	96	1
	India	899,594	5	4	1	31	71	16	<1
	Indonesia	66,074	2	2	< 1	13	81	35	1
D. oliveri	**All**	**794,463**	**21**	**15**	**10**	**60**	**39**	**51**	**15**
	Cambodia	142,147	37	21	15	86	56	73	7
	China	538	1	1	–	–	6	100	38

(continued)

Table 5 (continued)

Species	Country	Modeled distribution			Priority for in situ		Priority for restoration		Priority for ex situ
		km^2	% protected	% protected and stable	% of total area	% in protected areas	% of total area	% in natural habitats	% of total area
	Laos	131,702	18	13	12	37	23	82	25
	Myanmar	81,157	3	2	18	1	19	96	29
	Thailand	356,008	22	18	8	86	42	27	13
	Vietnam	82,911	10	8	6	48	46	58	13
D. sissoo	**All**	**1,899,013**	**3**	**2**	**1**	**9**	**79**	**14**	**–**
	Bangladesh	113,492	1	1	4	<1	88	16	–
	Bhutan	2,771	34	29	1	100	44	81	–
	India	1,740,772	2	2	<1	11	78	13	–
	Nepal	41,978	14	8	1	88	81	43	–

Table 6 Priority ecoregions with least stable protected areas (≤2%) and more than 10,000 km^2 of modeled *Dalbergia* distribution area in the Greater Mekong or more than 100,000 km^2 of modeled *Dalbergia* distribution area in India and Indonesia

Ecoregion	Country	Species	Modeled distribution			Priority for in situ		Priority for restoration		Priority for ex situ
			km^2	% protected	% protected and stable	% of total area	% in protected areas	% of total area	% in natural habitats	% of total area
Aravalli west thorn scrub forests	India	*D. sissoo*	162,899	1	<1	1	66	75	5	–
Chhota-Nagpur dry deciduous forests	India	*D. sissoo*	111,160	1	<1	1	–	88	10	–
Deccan thorn scrub forests	India	*D. latifolia*	229,580	<1	<1	<1	–	71	4	–
		D. sissoo	149,729	<1	<1	<1	–	75	3	–
Lower Gangetic Plains moist deciduous forests	Bangladesh, India	*D. sissoo*	244,017	1	<1	2	9	92	9	–
Narmada Valley dry deciduous forests	India	*D. sissoo*	120,867	2	1	<1	–	76	9	–
Northern Khorat Plateau moist deciduous forests	Laos, Thailand	*D. cochinchinensis*	16,362	2	1	2	23	69	29	–
		D. cultrata	12,609	1	1	1	22	48	22	20
		D. oliveri	11,960	2	1	1	36	51	22	16

(continued)

Table 6 (continued)

Ecoregion	Country	Species	Modeled distribution			Priority for in situ		Priority for restoration		Priority for ex situ
			km^2	% protected	% protected and stable	% of total area	% in protected areas	% of total area	% in natural habitats	% of total area
Thar desert	India	*D. sissoo*	106,308	<1	<1	<1	–	47	50	–
Upper Gangetic Plains moist deciduous forests	India	*D. sissoo*	261,994	2	1	1	4	89	2	–
Yunnan Plateau subtropical evergreen forests	China	*D. cultrata*	44,856	2	1	–	–	42	93	27

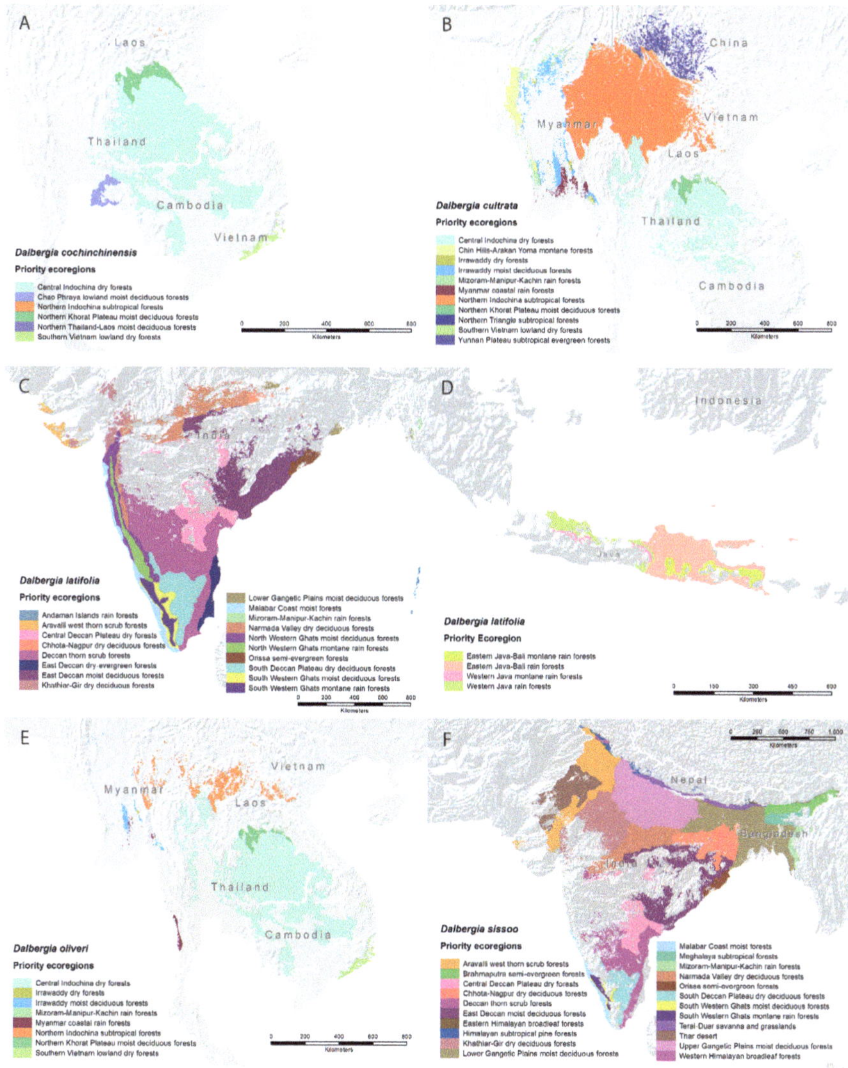

Fig. 4 Prioritized ecoregions with <10% of the predicted species range in stable protected areas for **a** *D. cochinchinensis*, **b** *D. cultrata* and **c** *D. latifolia* in India, **d** *D. latifolia* in Java, Indonesia, **e** *D. oliveri*, and **f** *D. sissoo* as part of their modeled native ranges

protected areas: *D. cultrata* in the Northern Indochina subtropical forests (Fig. 5b). The remaining 64 ecoregions (or species-ecoregion combinations) would require restoration outside of the current protected area network to meet the conservation target, including the Northern Khorat Plateau moist deciduous forests between Thailand and Laos for *D. cochinchinensis* (Fig. 5a), the Central Indochina dry forests for *D. oliveri* (Fig. 5c), the Eastern Java-Bali rain forests for *D. latifolia* (Fig. 5d), and

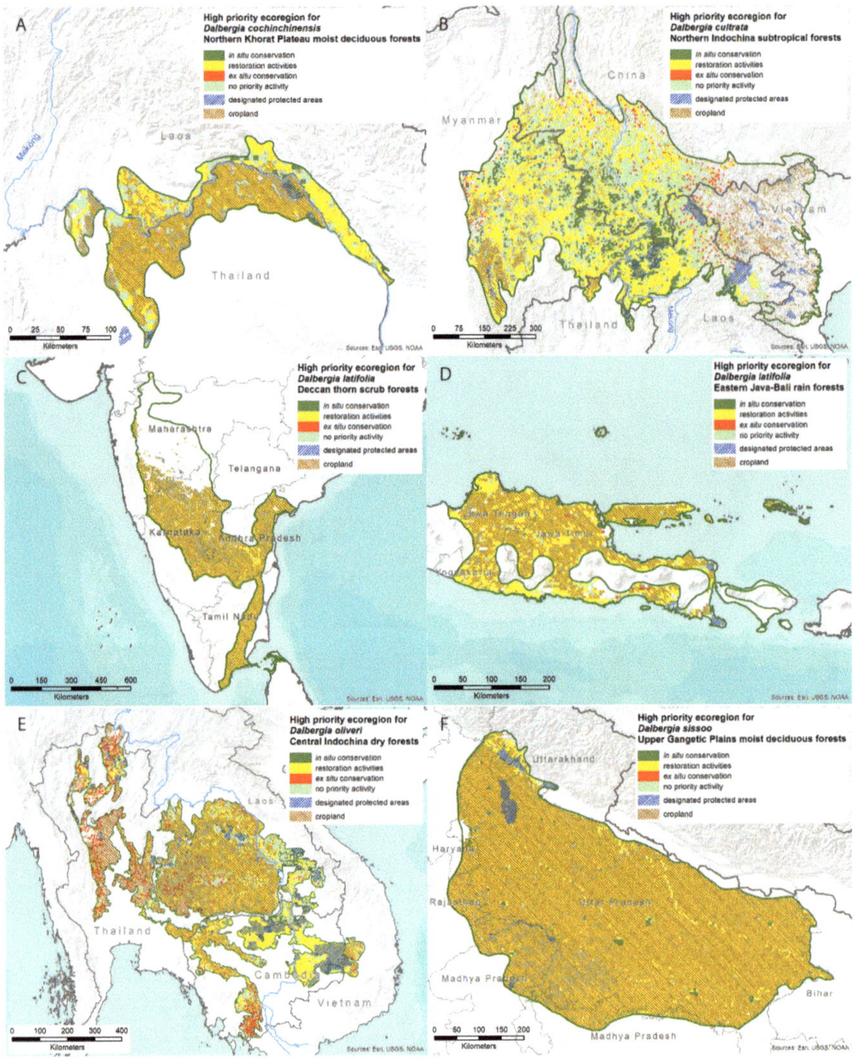

Fig. 5 Examples of priority ecoregion with <10% of the predicted species range in stable protected areas and prioritized conservation and restoration activities for **a** *D. cochinchinensis*, **b** *D. cultrata* and **c** *D. latifolia* in India, **d** *D. latifolia* in Java, Indonesia, **e** *D. oliveri*, and **f** *D. sissoo* as part of their modeled native ranges

the Upper Gangetic Plains moist deciduous forests for *D. sissoo* (Fig. 5e). This corresponds to an estimated restoration need of about 290,000 km^2 across the species' combined ranges.

Most of the priority areas for restoration in the Greater Mekong Subregion remain natural habitats and offer potential to restore the biodiversity of the three present *Dalbergia* species—except for three ecoregions that have experienced high land conversion: the Chao Phraya lowland moist deciduous forests, the Central Indochina dry forests, the Northern Khorat Plateau moist deciduous forests, and the Southern Vietnam lowland dry forests. In contrast, in most ecoregions in India and Indonesia (Java), much of the species' former natural habitat has already been converted to agricultural land, which is why planting *D. latifolia* or *D. sissoo* as part of agroforestry restoration projects has the greatest potential. Figure 3 shows maps of other areas within the landscapes where the species would benefit from restoration activities, independently of the conservation status within the ecoregion or country.

None of the ecoregions is a priority for ex situ conservation with the established (high) threshold of >90% of the range becoming unsuitable by 2050. The ecoregions most threatened by climate change are the Irrawaddy moist deciduous forests and the Mizoram-Manipur-Kachin rain forests (60–68% becoming unsuitable for *D. oliveri*), both located in Myanmar. Three other ecoregions are predicted to lose over 50% of the species' current distribution area: Northern Annamites rain forests in Laos for *D. cultrata*, and Irrawaddy dry forests in Myanmar and Northern Indochina subtropical forests in Myanmar, Thailand, Laos, and Vietnam for *D. oliveri*. Ex situ conservation may also be necessary in areas with little protected areas and few natural habitats remaining, independently of potential threat from climate change, such as the Chao Phraya lowland and the Northern Khorat Plateau moist deciduous forests in Thailand and Laos, the Yunnan Plateau subtropical evergreen forests in China, the Aravalli west thorn scrub forests, the Lower and Upper Gangetic Plains moist deciduous forests, and the Deccan dry deciduous and thorn scrub forests in India (Table 6).

11 Overall Assessment of the Conservation Status of Dalbergias Through Threat Mapping

In this chapter, we applied a comprehensive spatially explicit framework for quantitatively assessing the conservation status of tree species and their intraspecific diversity across their natural distribution area, considering multiple threats. Our method uses freely available spatial datasets, species distribution and climate change models, and occurrence and trait data combined with expert knowledge, making it well suited for tropical countries with high tree diversity and limited published information.

The study reveals that all five *Dalbergia* species (*D. cochinchinensis, D. cultrata, D. latifolia, D. oliveri*, and *D. sissoo*) are at risk across large parts of their native ranges in the Greater Mekong Subregion and in the Indian subcontinent and Java. Overexploitation is the most important threat, followed by habitat conversion and overgrazing. Part of the results presented here were used to update the IUCN Red List assessments for *D. cochinchinensis* and *D. oliveri* in collaboration with the Global Tree Assessment project [45, 46] and classify them as "critically endangered." The fact that the other three *Dalbergia* species are classified as less threatened is probably due either to a lack of population decline figures (*D. cultrata* is classified as "near threatened") or to widespread cultivation (*D. latifolia* is classified as "vulnerable" *and D. sissoo* as "least concern"). Nevertheless, the high level of vulnerability across their native ranges is of concern from a genetic conservation perspective. The widely cultivated species *D. latifolia and D. sissoo* often exhibit low levels of genetic diversity and disease resistance [47–49], whereas remaining natural and successfully introduced tree populations may still contain valuable genetic variation for tree improvement [49, 50], which is particularly important as climate change imposes new demands on species adaptation [51]. Therefore, we support the recommendations of a recent study that the IUCN extinction risk assessment should also include the impacts of climate change and additional aspects influencing species vulnerability such as functional traits, regional habitat threats, and genetic diversity [52].

Protected areas cover between 3 and 21% of the predicted range of the five species, but these areas may still be under pressure from both current threats and future climate change. Moreover, the existing protected areas are highly concentrated, leaving many parts of the species' ranges without protection. *Dalbergia* populations outside the existing protected area network, such as those found in agricultural lands and production forests in Laos [53], are declining rapidly, and in many of the studied ecoregions, areas identified as priority for in situ conservation are insufficient to meet the conservation goal of 10% of stable protected areas. Both the conservation and restoration are urgent and require cooperation between countries and stakeholders. To fill the gaps in the existing network of protected areas, the priority areas for in situ conservation presented here should be surveyed to confirm the presence of viable populations of the species and to delineate genetic conservation units, starting with the least protected ecoregions. High modeled habitat suitability values have been associated with high genetic diversity [42], so populations that grow in the most suitable areas can be particularly valuable for in situ conservation. However, and especially in the case of high genetic differentiation between populations [6], conservation measures should also be expanded to underrepresented areas to avoid loss of unique adaptative variation. Importantly, *Dalbergia* populations at the species' range margins may have unique, important traits, e.g., for climate change adaptation [54–56] and should be further investigated.

As low levels of overexploitation are required for grid cells to be classified as priority in situ conservation area, these are generally relatively remote areas. This results in lower opportunity costs for conservation compared to areas with higher accessibility and more land use pressure [57]. However, the remoteness of these areas can also lead to higher monitoring costs for tree populations and increased risks of illegal logging of valuable species [58]. Protected areas that are both remote and lack resources are at a higher risk of encroachment, as they may limit the informal access rights of local land users who depend on the forest [59]. Areas selected for in situ conservation may be well-suited for community forestry, as they face fewer threats to target species and can still allow for sustainable use while conserving genetic diversity. Employment opportunities and sources of income may arise from seed collection and seedling production, especially with the growing interest in planting

Dalbergia spp. in the region. Ensuring communities have access to seed markets is crucial for maintaining the benefits and commitment to protecting seed sources [60].

Restoration of *Dalbergia* spp. populations outside of the current protected area network is necessary for recovering and maintaining adaptive variation within species across predominant parts of their ranges. Mainstreaming these and other threatened, socioeconomically important tree species in broader land restoration programs would be important to align resources, help meet conservation targets, and improve biodiversity benefits of restoration initiatives [14, 61]. Restoration priority maps can be used to identify areas where species distribution overlaps with ongoing or planned restoration and tree planting programs. Capacities of farmers, other land managers, and restoration practitioners need to be strengthened to ensure that planted populations are established using genetically diverse and adapted seed to contribute to genetic conservation and result in resilient populations that can subsequently serve as sources of quality seed [16]. In contrast, intensive vegetative reproduction can contribute to low genetic diversity of populations, as studies of *D. latifolia* in Java have demonstrated [49]. The results of our study suggest that most priority ecoregions have few or no seed source populations. Remaining viable seed sources, including conservation stands in natural forests and ex situ collections, are therefore very important and need to be mapped and conserved across the species range to enable successful restoration. Seed orchards with material from multiple natural populations can contribute to the conservation of genetic diversity and provide locally adapted seeds for restoration projects [62], which underlines the importance of considering and planning conservation and restoration together.

Our results indicate that ex situ conservation is important for the long-term conservation of species' adaptive variation in ecoregions that are predicted to be severely affected by climate change. Responses to heat and drought stress vary even between closely related species [63], illustrating the need for species-specific analyses before synergies for conservation action can be identified. *D. cultrata* and *D. oliveri* in particular are at risk at the margins of their current distribution because of the combined effects of climate change and current threats. These areas should be selected for germplasm collection to avoid potential loss of adaptive variation. Our maps can be used to focus collection efforts on areas where climate threats are high but current threats are low, making it more likely that viable remnant populations can still be found. Ex situ conservation of tree genetic resources does not preclude sustainable use, and, apart from seed banks, conservation units can include plantations, provenance trials, seed orchards, agroforestry, and trees on farms. Common garden studies and genomic studies can help to understand species' capacity to adapt to climate change and identify opportunities for selection and breeding to support adaptation. Climate change should be considered in management plans for protected areas, which would benefit both the species under study and others.

24 of the 51 ecoregions (47%) in the study area extend across national borders, with protection status sometimes differing drastically across borders. Within individual ecoregions in individual countries, none of the species are sufficiently protected. Synergies between countries can therefore improve conservation status and help share resources [64]. Each ecoregion should have several protected populations to

avoid the risk of genetic erosion [65]. For highly threatened species such as *Dalbergia* spp., the designation of such reserves in cooperation between countries within an ecoregion is particularly important, as the remaining populations are scarce. Regional collaboration would also support efforts to address illegal cross-border trade, the main cause of loss of *Dalbergia* spp. in the region [66]. To improve the conservation status of *Dalbergia* ssp. in existing protected areas, countries must tackle ongoing issues in the planning [67] and management [68, 69] of these areas. This includes the enforcement of laws [53] and combating corruption [70]. Our results help to better target conservation efforts within and between protected areas, for example by highlighting protected areas where *Dalbergia* populations need to be better monitored and restored.

Transboundary collaboration for conservation actions is constrained by a lack of methods for identifying national roles, responsibilities, and complementarities [71]. Species distribution modeling contributes to such analyses, but the lack of distribution data in a standardized form limits its potential and is a major bottleneck for many native Asian tree species [20, 72]. We found that collaborative projects on species of common interest facilitate the compilation and standardization of previously unpublished data to enable range-wide assessments. Collaboration among national experts is also crucial for validating the resulting distribution maps and for identifying opportunities and recommendations for concrete collaborative actions. The results of the spatial analysis help focus field investigations on areas with high predicted conservation potential and the most synergies between species or countries, which is crucial in the race against local extinction of *Dalbergia* spp. and other endangered species.

12 Conclusion

The high vulnerability levels of *Dalbergia* species across Asia underscore the urgent need for concerted conservation and restoration efforts to protect these species and their habitats. By prioritizing their conservation and restoration, we can not only preserve biodiversity but also safeguard the cultural heritage and economic livelihoods associated with these iconic hardwoods. Moving forward, it is imperative to strengthen further research on ecology, population dynamics, and conservation strategies for *Dalbergia* spp. In addition, sustainable forestry and land use practices, international cooperation, promotion of capacity building, community involvement, and market-oriented approaches are crucial to ensuring the long-term sustainability of *Dalbergia* ecosystems. By fostering collaboration between governments, conservation organizations, local communities, and other stakeholders involved, we can work toward a future in which *Dalbergia* spp. thrive in their natural habitats, enriching ecosystems and enhancing human well-being for generations to come.

References

1. Ono T (2016) Musical instruments: materials. In: Book: reference module in materials science and materials engineering. https://doi.org/10.1016/B978-0-12-803581-8.02233-5
2. Winfield, K., Scott, M. and Grayson, C., 2016. Global status of Dalbergia and Pterocarpus rosewood producing species in trade. 17 Conference of the Parties, CITES. 24 Sept.–5 Oct. 2016.
3. Rakotonirina N, Nowak MM, Lowry PP, Rakouth HN, Rakouth BR (2024) Will Dalbergia species survive climate change? Predicting the potential future distribution of threatened species in Madagascar. Glob Ecol Conserv 52:e02936
4. Javed Muhammad A, Ul Haq I, Ghafoor N, Ur Rehman F. Climate Change Exacerbated *Dalbergia sissoo* Dieback Under Water Stress and Ceratocystis fimbriata Infection. Agri Res Tech Open Access J. 2023;27.
5. IUCN (2023) The IUCN red list of threatened species. Version 2023-1. [Internet]. 2023 [cited 2024 May 22]. https://www.iucnredlist.org/
6. Hartvig I, So T, Changtragoon S, Tran HT, Bouamanivong S, Theilade I et al (2018) Population genetic structure of the endemic rosewoods *Dalbergia cochinchinensis* and *D. oliveri* at a regional scale reflects the Indochinese landscape and life-history traits. Ecol Evol 8:530–545
7. Vasudevan R, Thangamani D, Lalitha S, Sasidharan KR (2023) Population structure and genetic diversity analysis of *Dalbergia latifolia* and *Dalbergia sissoides* using RAPD markers, [cited 2024 Apr 2]. http://www.jabonline.in
8. Dobhal S, Sharma S, Ahmed N, Kumar A (2019) Genetic polymorphism in Dalbergia sissoo Roxb. using RAPD markers. Indian J Biotechnol 18:164–173
9. CITES (2022) Viet Nam considers conservation plan for two of its most valuable tree species [Internet], [cited 2024 Apr 2]. https://cites.org/eng/news/viet-nam-considers-conservation-plan-two-valuable-tree-species
10. Asian Forest Cooperation Organization (AFoCO) (2020) Supporting the conservation of major timber species in Cambodia [Internet], [cited 2024 Apr 2]. https://afocosec.org/newsroom/news/project-highlights/feature-supporting-the-conservation-of-major-timber-species-in-cambodia/
11. Tewari DN (1994) A monograph on *Dalbergia sissoo* Roxb. Dehra Dun
12. Carnus J-M, Parrotta J, Brockerhoff E, Arbez M, Jactel H, Kremer A et al (2006) Planted forests and biodiversity. J For [Internet], [cited 2024 Apr 2] 104:65–77. https://academic.oup.com/jof/article/104/2/65/4599206
13. Thomas E, Jalonen R, Loo J, Boshier D, Gallo L, Cavers S et al (2014) Genetic considerations in ecosystem restoration using native tree species. For Ecol Manage [Internet] 333:66–75. https://doi.org/10.1016/j.foreco.2014.07.015
14. Gaisberger H, Jalonen R, Vinceti B, Elias M, Kettle Bioversity International CJ, Evert Thomas Bioversity International I et al (2023) Delivering tree genetic resources in forest and landscape restoration. A guide to ensuring local and global impact [Internet]. Food and Agriculture Organization of the United Nations, Rome. https://hdl.handle.net/10568/132089
15. Bosshard, E., Jalonen, R., Kanchanarak, T., Yuskianti, V. and Tolentino, E., 2022. Are tree seed systems for forest landscape restoration fit for purpose? An analysis of four Asian countries. TIDEE: TERI Information Digest on Energy and Environment, 21(1), pp.102-103
16. Jalonen R, Valette M, Boshier D, Duminil J, Thomas E (2018) Forest and landscape restoration severely constrained by a lack of attention to the quantity and quality of tree seed: insights from a global survey. Conserv Lett 11:e12424. https://doi.org/10.1111/conl.12424
17. Lengkeek AG, Jaenicke H, Dawson IK (2005) Genetic bottlenecks in agroforestry systems: results of tree nursery surveys in East Africa. Agrofor Syst 632 [Internet], [cited 2022 Sep 27] 63:149–155. https://link.springer.com/article/10.1007/s10457-004-9155-7
18. Banin LF, Raine EH, Rowland LM, Chazdon RL, Smith SW, Rahman NEB et al (2023) The road to recovery: a synthesis of outcomes from ecosystem restoration in tropical and sub-tropical Asian forests. Philos Trans R Soc B Biol Sci 378

19. Tito de Morais C, Kettle CJ, Philipson CD, Maycock CR, Burslem DFRP, Khoo E et al (2020) Exploring the role of genetic diversity and relatedness in tree seedling growth and mortality: a multispecies study in a bornean rainforest. J Ecol [Internet], [cited 2024 Apr 10] 108:1174–1185. https://onlinelibrary.wiley.com/doi/full/10.1111/1365-2745.13331
20. Gaisberger H, Fremout T, Kettle CJ, Vinceti B, Kemalasari D, Kanchanarak T et al (2021) Tropical and subtropical Asia's valued tree species under threat. Conserv Biol 36:e13873. https://doi.org/10.1111/cobi.13873
21. Fremout T, Thomas E, Gaisberger H, Van Meerbeek K, Muenchow J, Briers S et al (2020) Mapping tree species vulnerability to multiple threats as a guide to restoration and conservation of tropical dry forests. Glob Chang Biol 26:3552–3568
22. Ceccarelli V, Ekué M, Fremout T, Gaisberger H, Kettle C, Taedoumg H, et al (2022) Vulnerability mapping of 100 priority tree species in Central Africa to guide conservation and restoration efforts. Biol Conserv [Internet] 270:109554. https://www.sciencedirect.com/science/article/pii/S0006320722001070
23. Dinerstein E, Olson D, Joshi A, Vynne C, Burgess ND, Wikramanayake E et al (2017) An ecoregion-based approach to protecting half the terrestrial realm. Bioscience 67:534–545
24. Fick SE, Hijmans RJ (2017) WorldClim 2: new 1-km spatial resolution climate surfaces for global land areas. Int J Climatol 37:4302–4315
25. Hengl T, de Jesus JM, Heuvelink GBM, Gonzalez MR, Kilibarda M, Blagotić A et al (2017) SoilGrids250m: Global gridded soil information based on machine learning. PLoS One 12:e0169748
26. Danielson J, Gesch D (2010) Global multi-resolution terrain elevation data 2010 (GMTED2010). US Geol Surv Open-File Rep 2011–1073(2011):26
27. Phillips SJ, Dudik M, Schapire RE (2018) Maxent software for modeling species niches and distributions (Version 3.4.1) [Internet], [cited 2020 Sep 18]. https://biodiversityinformatics.amnh.org/open_source/maxent/
28. Radosavljevic A, Anderson RP (2014) Making better MAXENT models of species distributions: complexity, overfitting and evaluation. J Biogeogr 41:629–643
29. Muscarella R, Galante PJ, Soley-Guardia M, Boria RA, Kass JM, Uriarte M et al (2014) ENMeval: an R package for conducting spatially independent evaluations and estimating optimal model complexity for MaxEnt ecological niche models. Methods Ecol Evol 5:1198–1205
30. Valavi R, Elith J, Lahoz-Monfort JJ, Guillera-Arroita G (2019) BlockCV: an r package for generating spatially or environmentally separated folds for k-fold cross-validation of species distribution models. Methods Ecol Evol 10:225–232
31. Acevedo P, Ferreres J, Escudero MA, Jimenez J, Boadella M, Marco J (2017) Population dynamics affect the capacity of species distribution models to predict species abundance on a local scale. Divers Distrib [Internet], [cited 2021 Oct 5] 23:1008–1017. https://onlinelibrary.wiley.com/doi/full/10.1111/ddi.12589
32. Phillips SJ (2008) Transferability, sample selection bias and background data in presence-only modelling: a response to Peterson et al. (2007). Ecography (Cop) 0:080227084236895–0
33. Eyring V, Bony S, Meehl GA, Senior CA, Stevens B, Stouffer RJ et al (2016) Overview of the coupled model intercomparison project phase 6 (CMIP6) experimental design and organization. Geosci Model Dev 9:1937–1958
34. Lloyd CT, Sorichetta A, Tatem AJ (2017) High resolution global gridded data for use in population studies. Sci Data 41 [Internet], [cited 2024 May 31] 4:1–17. https://www.worldpop.org/geodata/summary?id=24777
35. Weiss DJ, Nelson A, Gibson HS, Temperley W, Peedell S, Lieber A et al (2018) A global map of travel time to cities to assess inequalities in accessibility in 2015. Nature [Internet] 533:333–336. http://www.nature.com/finder/10.1038/nature25181
36. UNEP-WCMC, IUCN (2018) World database on protected areas, WDPA, Version October 2018.www.protectedplanet.net. www.protectedplanet.net
37. NASA EOSDIS (2024) Fire information for resource management system (FIRMS). Archive download [Internet], [cited 2024 Feb 20]. https://firms.modaps.eosdis.nasa.gov/download/

38. Robinson TP, Wint GRW, Conchedda G, Van Boeckel TP, Ercoli V, Palamara E et al (2014) Mapping the global distribution of livestock. PLoS One [Internet], [cited 2014 Jul 9] 9:e96084. http://www.pubmedcentral.nih.gov/articlerender.fcgi?artid=4038494&tool=pmcentrez&rendertype=abstract
39. Latham J, Cumani R, Rosati I, Bloise M (2014) Global land cover SHARE (GLC-SHARE) database Beta-Release Version 1.0. Food Agric Organ United Nations (FAO), Rome, Italy [Internet], 1–39. http://www.fao.org/uploads/media/glc-share-doc.pdf
40. Harris N, Goldman ED, Gibbes S (2019) Spatial database of planted trees (SDPT Version 1.0) [Internet], [cited 2021 Aug 25]. https://www.wri.org/research/spatial-database-planted-trees-sdpt-version-10
41. Brunner L, Pendergrass A, Lehner F, Merrifield A, Lorenz R, Knutti R (2020) Reduced global warming from CMIP6 projections when weighting models by performance and independence. Earth Syst Dyn Discuss, 1–23
42. Nagaraju SK, Gudasalamani R, Barve N, Ghazoul J, Narayanagowda GK, Ramanan US (2013) Do ecological niche model predictions reflect the adaptive landscape of species?: a test using *Myristica malabarica* Lam., an endemic tree in the Western Ghats, India. PLoS One 8:1–13
43. Gaisberger H, Fremout T, So T, Thammavong B, Bounithiphonh C, Hoa TT et al (2022) Range-wide priority setting for the conservation and restoration of Asian rosewood species accounting for multiple threats and ecogeographic diversity. Biol Conserv [Internet], [cited 2022 Apr 28] 270:109560. https://doi.org/10.1016/j.biocon.2022.109560
44. Miller SA, Bartow A, Gisler M, Ward K, Young AS, Kaye TN (2011) Can an ecoregion serve as a seed transfer zone? Evidence from a common garden study with five native species. Restor Ecol 19:268–276
45. Barstow M, Boshier D, Bountithiponh C, Changtragoon S, Gaisberger H, Hartvig I et al (2022) *Dalbergia cochinchinensis*. The IUCN Red List of Threatened Species 2022 [Internet]. https://www.iucnredlist.org/species/215342548/2822125
46. Barstow M, Boshier D, Bountithiponh C, Changtragoon S, Gaisberger H, Hartvig I, Oliveri D et al (2022) The IUCN red list of threatened species 2022 [Internet]. https://www.iucnredlist.org/species/215341339/2813403
47. Atif S, Naqvi H, Mushtaq S, Malik MT, Umar U, Fareed S et al (2019) Factors leading towards *Dalbergia sissoo* Decline (Syndrome) in Indian Sub-continent: a critical review and future research Agenda. Pak J Agric Res 32(2):302–316. https://doi.org/10.17582/journal.pjar/2019/32.2.302.316
48. Shah KK, Tiwari I, Modi B, Pandey HP, Subedi S, Shrestha J (2021) Shisham (*Dalbergia sissoo*) decline by dieback disease, root pathogens and their management: a review. J Agric Nat Resour 4:255–272
49. Yulita K, Susila, Rachmat HH, Dwiyanti FG, Atikah TD, Subiakto A et al (2022) Population genetic of the Indonesian rosewood (*Dalbergia latifolia*) from Java and West Nusa Tenggara revealed using sequence-related amplified polymorphism. Forest Sci Technol [Internet] 18:172–181. https://doi.org/10.1080/21580103.2022.2123051
50. Lohr VI, Kendal D, Dobbs C (2016) Urban trees worldwide have low species and genetic diversity, posing high risks of tree loss as stresses from climate change increase. Acta Hortic 1108:263–270
51. Keenan RJ (2015) Climate change impacts and adaptation in forest management: a review. Ann For Sci 2015 722 [Internet], [cited 2021 Sep 18] 72:145–167. https://link.springer.com/article/10.1007/s13595-014-0446-5
52. Palacio RD, Abarca M, Armenteras D, Balza U, Dollar LJ, Froese GZL et al (2023) The global influence of the IUCN Red List can hinder species conservation efforts. TechRxiv [Internet], [cited 2024 May 27]. https://www.authorea.com/doi/full/10.22541/au.169945445.50394320?commit=8d4672045a5a15ac1eac181ca29dabc3ca9e046f
53. National Agriculture and Forestry Research Institute (NAFRI) (2023) Draft-Long-term rosewood conservation for natural forest. Developing CITES Non-Detriment Findings for Rosewoods in Lao PDR

54. Angert AL, Bontrager MG, Aringgren J (2020) What do we really know about adaptation at range edges? Ann Rev Ecol Evol Syst 51:341–361. https://doi.org/10.1146/annurev-ecolsys-012120-091002
55. Hartvig I, So T, Changtragoon S, Thi H, Somsanith T (2020) Conservation genetics of the critically endangered Siamese rosewood (*Dalbergia cochinchinensis*): recommendations for management and sustainable use. Conserv Genet [Internet] 21:677–692. https://doi.org/10.1007/s10592-020-01279-1
56. Hung TH, So T, Thammavong B, Chamchumroon V, Theilade I, Phourin C et al (2023) Range-wide differential adaptation and genomic offset in critically endangered Asian rosewoods. Proc Natl Acad Sci USA [Internet], [cited 2024 Apr 29] 120:e2301603120. https://www.pnas.org/doi/abs/10.1073/pnas.2301603120
57. Carrasco, L.R., Webb, E.L., Symes, W.S., Koh, L.P. and Sodhi, N.S., 2017. Global economic trade-offs between wild nature and tropical agriculture. PLoS Biology, 15(7), p.e2001657
58. Xue X, Nan C, Deng C, Liu X, Wang Z, Dao B (2019) Analysis on the characteristics of illegal cutting and deforestation of *Pinus kesiya* in Lancang County of Yunnan Province. In: E3S web conference
59. Geldmann J, Manica A, Burgess ND, Coad L, Balmford A (2019) A global-level assessment of the effectiveness of protected areas at resisting anthropogenic pressures. Proc Natl Acad Sci USA 116
60. Valette M, Vinceti B, Gregorio N, Bailey A, Thomas E, Jalonen R (2020) Beyond fixes that fail: Identifying sustainable improvements to tree seed supply and farmer participation in forest and landscape restoration. Ecol Soc 25
61. Strassburg BBN, Iribarrem A, Beyer HL, Cordeiro CL, Crouzeilles R, Jakovac CC et al (2020) Global priority areas for ecosystem restoration. Nature [Internet] 586:724–729. https://doi.org/10.1038/s41586-020-2784-9
62. Dudley N, Jones T, Gerber K, Ross-Davis AL, Sniezko RA, Cannon P et al (2020) Establishment of a genetically diverse, disease-resistant *Acacia koa* a. Gray seed orchard in Kokee, Kauai: Early growth, form, and survival. Forests 11:1–13
63. Hang Hung T, Rosemary H, Gabriele G, Thea R, Thammavong B, Thi H et al (2020) Physiological responses of rosewoods *Dalbergia cochinchinensis* and *D. oliveri* under drought and heat stresses. Ecology and Evolution10(19):10872–10885
64. Wang L, Yang B, Bai Y, Lu X, Corlett RT, Tan Y et al (2021) Conservation planning on China's borders with Myanmar, Laos, and Vietnam. Conserv Biol 1–12
65. Hoban S, Bruford M, D'Urban Jackson J, Lopes-Fernandes M, Heuertz M, Hohenlohe PA et al (2020) Genetic diversity targets and indicators in the CBD post-2020 Global Biodiversity Framework must be improved. Biol Conserv [Internet] 248:108654. https://doi.org/10.1016/j.biocon.2020.108654
66. Environmental Investigation Agency (EIA) (2016) CITES CoP17—Analysis of the demand-driven trade in Hongmu timber species: impacts of unsustainability and illegality in source region
67. Kukkonen MO, Tammi I (2019) Systematic reassessment of Laos' protected area network. Biol Conserv 229
68. Cuong C Van, Dart P, Dudley N, Hockings M (2017) Factors influencing successful implementation of Biosphere Reserves in Vietnam: challenges, opportunities and lessons learnt. Environ Sci Policy 67
69. Graham V, Geldmann J, Adams VM, Grech A, Deinet S, Chang HC (2021) Management resourcing and government transparency are key drivers of biodiversity outcomes in Southeast Asian protected areas. Biol Conserv 253
70. Milne S (2015) Cambodia's unofficial regime of extraction: illicit logging in the shadow of transnational governance and investment. Crit Asian Stud [Internet] 47:200–228. https://doi.org/10.1080/14672715.2015.1041275

71. Schmeller DS, Gruber B, Bauch B, Lanno K, Budrys E, Babij V et al (2008) Determination of national conservation responsibilities for species conservation in regions with multiple political jurisdictions. Biodivers Conserv 17
72. Serra-Diaz JM, Enquist BJ, Maitner B, Merow C, Svenning JC (2017) Big data of tree species distributions: how big and how good? For Ecosyst 4

Part III
Perspectives

Chapter 31
Sustainable Utilisation of Rosewood: An Industry Perspective

Leonard J. Mills

1 Introduction

The journey into rosewood furniture began unexpectedly in 2014, sparked by a newspaper article announcing that individuals could now purchase rosewood. This news led to a visit to Kushalnagar in Coorg, a decision that would eventually lay the foundation for a full-fledged manufacturing unit. Initially, there was no intention of starting a business—the first batch of rosewood was acquired purely for personal use, crafted into furniture and carvings for the joy of creation. However, the admiration of friends and family transformed this passion into something more significant. What began as a personal endeavour soon evolved into a pursuit of crafting masterpieces for others.

Driven by a deep appreciation for artistry, the business grew organically, with each piece reflecting dedication and skill. Customer satisfaction became the cornerstone of its success, ensuring that every creation was more than just furniture—it was a testament to craftsmanship.

In 2014, a small workshop was set up in Bengaluru, unaware of the complex legal challenges ahead. The forest department imposed stringent regulations, including the Property Mark Registration (PMR) requirement, which necessitated extensive documentation and frequent visits to the Range Forest Office for approvals at every stage. Additionally, timber sales were restricted without a permit, with each application taking over a week to process. These bureaucratic hurdles made sustaining a rosewood business impractical. As a result, operations were relocated to Kushalnagar, Coorg, in 2015, where the business flourished and became a recognised name.

Rooted in the local community, the company remains committed to ethical craftsmanship and employee well-being, fostering a positive and supportive work environment. Every piece is custom-made with meticulous attention to detail, ensuring

L. J. Mills (✉)
Coorg Rosewood Timbers, Kattugollahalli, Bengaluru, Karnataka, India
e-mail: crwtblr@gmail.com

R. R. Warrier et al. (eds.), *Handbook of Asian Rosewoods*, Materials Horizons: From Nature to Nanomaterials, https://doi.org/10.1007/978-981-96-9459-4_31

seamless integration into any space. Witnessing a rough log transform into a finely crafted creation is more than just furniture-making—preserving a dying art. With skilled artisans carrying forward a rich legacy of wood carving, each handcrafted piece is a tribute to tradition, artistry, and timeless craftsmanship.

2 Craftsmanship and Quality

Rosewood furniture is a timeless and valuable addition to any home, offering an ideal combination of durability, elegance, and ease of maintenance. As one of the most resilient hardwoods, rosewood is highly sought after by homeowners and furniture makers due to its strength and resistance to wear and tear, ensuring exceptional longevity. Its distinct grainy texture and rich red and brown hues exudes sophistication and charm, making it a striking choice for various interior settings. Beyond aesthetics, rosewood possesses practical advantages, such as natural termite resistance due to its interlocked grain pattern, making it an excellent choice for Indian households where termite infestations are a prevalent concern.

The maintenance of rosewood furniture is effortless, requiring only an initial varnish or polish to enhance its resistance to dust and pests. A periodic wipe-down is sufficient to retain its pristine appearance, making it a hassle-free option for busy homeowners. Additionally, rosewood emits a pleasant, soothing fragrance from natural essential oils, further enriching the ambience of any space. Its adaptability allows for intricate carvings, making it suitable for both traditional and modern furniture designs. Whether in elaborate, handcrafted pieces or sleek, contemporary furnishings, rosewood seamlessly blends with diverse decor styles, adding a touch of luxury.

Beyond its functional and aesthetic appeal, rosewood furniture is a valuable investment. Many pieces are cherished as family heirlooms, passed down through generations, carrying historical and cultural significance. Due to the meticulous craftsmanship involved in creating these exquisite furniture pieces, they hold high resale value and remain prized possessions. Combining strength, beauty, and long-term worth, rosewood furniture is an excellent choice for those seeking elegance, reliability, and investment value in their home decor.

However, working with rosewood presents unique challenges. Its extreme hardness requires patience and specialised tools to prevent cracking or damage during shaping. The wood's natural variations, such as unexpected colour shifts, often demand on-the-spot design adjustments. Due to its high density and interlocked grain pattern, cutting, shaping, and carving rosewood is labour-intensive, requiring expert handling to avoid tear-out and ensure a precise finish. Additionally, natural oils in rosewood can interfere with glue adhesion and finishing processes, necessitating careful preparation. Furthermore, prolonged exposure to rosewood dust can trigger allergic reactions or respiratory issues in some workers, highlighting the need for safety precautions in the workshop.

Maintaining craftsmanship quality while scaling up production poses another challenge—balancing artisanal techniques with increased efficiency. Heavy-duty machines are used for initial processing, while skilled hands ensure the final touches. Eco-friendly preservatives are applied, avoiding harmful shortcuts. Over time, design trends have evolved; while British-era styles remain an inspiration, they are now blended with modern customisations. Technological advancements like CAD assist in the design process, but craftsmanship remains central to production. While mass production has enhanced accessibility, preserving artisanal quality remains a priority. The company has upheld its legacy of superior craftsmanship by balancing tradition with innovation. The future lies in smart, multifunctional designs, and sustainable practices, merging heritage craftsmanship with cutting-edge technology to cater to evolving consumer needs.

3 Business and Market Insights

The growing demand for rosewood has significantly affected its availability, making sustainable harvesting essential for its long-term viability. Responsible sourcing practices help maintain ecological balance while preserving this valuable resource. Rosewood used in production is directly procured from the Karnataka Forest Department under strict legal regulations, ensuring compliance with sustainability standards and eliminating reliance on third-party suppliers.

Sustainability is central to operations. Every piece of rosewood is legally sourced with valid 2024 permits, ensuring no forests are destroyed, and no wildlife is harmed. Clients range from families seeking custom furniture to businesses furnishing luxury spaces, with over 700 metric tons shipped to destinations like China and Turkey. While figures may vary, the focus remains on bespoke and high-end markets, catering to those prioritising quality and ethical sourcing.

Handcrafted furniture stands superior to factory-made options due to its health benefits, durability, and long-term value. Unlike mass-produced furniture, which often contains MDF with formaldehyde—a known carcinogen—handcrafted pieces are made from solid wood, reducing health risks associated with toxic emissions. Handmade pieces retain authenticity, can be passed down as heirlooms, and are valuable long-term investments. Additionally, handmade furniture is more environmentally friendly, involving no harmful emissions and using biodegradable materials. The combination of superior craftsmanship, sustainability, and longevity makes handcrafted furniture a worthwhile choice over factory-produced options. The process also supports artisans who preserve India's centuries-old woodworking traditions.

The case study of Coorg Rosewood Timbers, which has earned the India 500 Startup Awards 2024, is a working model of how a timber company can venture into entrepreneurship through responsible sourcing and processing of rosewood timber, recognising its commitment to sustainability and high-quality craftsmanship, moving beyond mass production. This sets the standards for the timber trade, where sourcing could be from legal sources such as the forest departments, which auction wood regularly, ensure transparency and compliance with regulations, and promote sustainable forestry practices while supporting the livelihoods of local communities and artisans.

The major handcrafted works for which the company is unique include the following.

Rosewood Elephants: Elephants symbolise India's rich heritage and cultural elegance, making them a timeless motif in art and decor. Throughout Indian scriptures, the elephant represents strength and positive energy, enhancing its appeal. Rosewood elephants are crafted with precision and are highly sought after for their aesthetic and symbolic significance.

Rosewood Pooja Mantap: A Pooja Mantap holds deep spiritual significance, serving as the sacred centrepiece of a home. Thoughtful designs incorporate ample storage for idols, sculptures, photos, and essential accessories. Crafted from the finest rosewood and shaped by skilled artisans, these mantaps are available in various sizes and styles, blending functionality with artistic excellence.

Custom Rosewood Furniture: Handcrafted to meet individual preferences, custom rosewood furniture is tailored to unique designs and specifications. Depending on complexity and size, crafting each piece takes 4–8 weeks, ensuring high-quality craftsmanship and durability.

Inlay articles: Rooted in craftsmanship over generations, each creation follows a precise process, starting with selecting high-quality rosewood known for its rich colour and unique grain. Patterns are carefully designed, featuring floral, animal, or geometric motifs that reflect cultural traditions. The wood is then shaped, and shallow recesses are carved to accommodate inlaid materials like brass or bone. These elements are meticulously fitted to form seamless, intricate designs. Finally, the surface is sanded and polished to achieve a smooth, lustrous finish, enhancing the wood's natural beauty and ensuring a refined, timeless piece (Fig. 1).

4 Sustainability and Legal Aspects

With rosewood listed under CITES, exporting it—especially to markets like China, where it is valued as Hongmu—requires strict compliance with regulations. Adhering to Karnataka's legal framework and maintaining airtight permits ensures transparency and builds trust with buyers worldwide. While stringent regulations help protect the species, they can also create challenges for legitimate businesses, while illegal traders often find ways to bypass them. Improved policies, such as replanting subsidies or a global timber-tracking system, could enhance sustainability efforts, ensuring responsible trade which benefits both the environment and ethical enterprises.

Sustainable harvesting methods, such as selective cutting, allow only mature trees that have completed their natural life cycle to be harvested. This approach prevents

Fig. 1 Handcrafted rosewood items. Clockwise: 1. Elephant; 2. Mantap; 3. Dining set with inlay; 4. Chair

overexploitation and promotes forest regeneration. Additionally, managed plantations provide a controlled environment for rosewood cultivation, using responsible forestry practices and regular monitoring to prevent overharvesting. By following these sustainable methods, rosewood remains an eco-friendly and renewable material (Figs. 2, 3, 4, 5 and 6). Ethical sourcing and conservation-focused forestry practices ensure its quality and availability are maintained without compromising the environment.

Fig. 2 Selection of woodlots and procurement

Fig. 3 Processing of the procured woodlots

Fig. 4 Varied hues of rosewood

Fig. 5 Fine-grained rosewood logs ready for processing

Fig. 6 Processing of sawn logs of rosewood

5 Future Vision

In India, the goal is to bring rosewood craftsmanship into more homes beyond just the elite. Globally, the demand for sustainable luxury is rising, with markets like Europe as the next focus. Advancements in AI could play a role, potentially optimising timber supply and minimising waste, though the heart of this craft will always remain human. Emerging trends like 3D printing for intricate details bring excitement, but the tradition of hand-carving will never be replaced. Success in this field comes from mastering the wood, starting small, and maintaining integrity. The future looks promising when rosewood's heritage is honoured while innovation drives progress.

Chapter 32
Effective Conservation of Asian Rosewoods (*Dalbergias*) and Their Genetic Resources Through Regional Collaboration

Riina Jalonen, Ida Theilade, Ida Hartvig, Tin Hang Hung, John MacKay, Vivi Yuskianti, Chaloun Bounithiphonh, Hannes Gaisberger, Huang Ping, Syneath Sreng, Zheng Yongqi, and Rekha R. Warrier

1 Introduction

The genus *Dalbergia* in the Asia–Pacific region comprises approximately 80 species. Among these, 33 are native to India, 44 to Indochina, and eight to the New Guinea floristic region [1]. The list includes several valuable timber species, also known as the Asian Rosewood species, mostly endemics and species with extensive distributions across multiple countries (Table 1). They are highly valued for their beautiful and dense wood, often used in furniture, musical instruments, and decorative items.

S. Sreng
Institute of Forest and Wildlife Research and Development, Phnom Penh, Cambodia

R. R. Warrier (✉)
ICFRE-Institute of Forest Genetics and Tree Breeding, Coimbatore, India
e-mail: rekha@icfre.org

I. Hartvig
Center for Evolutionary Hologenomics, Globe Institute, University of Copenhagen, København K, Denmark

H. Ping · Z. Yongqi
Institute of Forestry, Chinese Academy of Forestry, Beijing, China

R. Jalonen
Bioversity International, c/o WorldFish Headquarters, Jalan Batu Maung, 11960 Bayan Lepas, Penang, Malaysia

I. Theilade
Department of Geosciences and Natural Resource Management, University of Copenhagen, Copenhagen, Denmark

R. R. Warrier et al. (eds.), *Handbook of Asian Rosewoods*, Materials Horizons: From Nature to Nanomaterials, https://doi.org/10.1007/978-981-96-9459-4_32

Table 1 Distribution of selected *Dalbergia* (timber) species across the Asia–Pacific

Scientific name	Distribution	Conservation status (as of 2024)
Dalbergia assamica	Vietnam, China, Lao PDR, Cambodia, Thailand, Myanmar, Bhutan, Bangladesh, and India	Least concern (2012)
Dalbergia cochinchinensis	Cambodia, Lao PDR, Thailand, and Vietnam	Critically Endangered (2020)
Dalbergia cultrata	Myanmar, Cambodia, China, Indonesia, Thailand, Lao PDR, Vietnam, and India	Near Threatened (2010)
Dalbergia latifolia	India, Indonesia, Nepal, Myanmar, Sri Lanka, and Vietnam	Vulnerable (2020)
Dalbergia oliveri	Cambodia, Lao People's Democratic Republic, Myanmar, Thailand, and Vietnam	Critically Endangered (2020)
Dalbergia odorifera	China	Vulnerable (1998)
Dalbergia sissoo	Afghanistan, Bangladesh, Bhutan, North India, Iran, Myanmar, Nepal, and Pakistan	Least Concern (2019)
Dalbergia tonkinensis	Vietnam and China	Vulnerable (1998)

Source [5, 6]

Many Asian Rosewood species are under threat due to overexploitation and illegal logging, driven by high demand for their valuable timber. A study to assess the extent of the illegal trade of Dalbergia species in Indonesia showed an annual increasing trend, with almost a fifth of rosewood being extracted from conservation areas rather than plantations, causing more significant damage than individual tree felling [2]. The high demand for rosewood timber, coupled with slow growth and a decline in natural habitat, has led to dwindling populations throughout the species' ranges

I. Hartvig
Department of Geosciences and Natural Resource Management, University of Copenhagen, Copenhagen, Denmark

Smithsonian Environmental Research Center, Smithsonian Institute, Washington, USA

T. H. Hung · J. MacKay
Department of Biology, University of Oxford, Oxford OX1 3EL, UK

V. Yuskianti
Research Center for Biota Systems, National Research and Innovation Agency, Bogor, West Java, Indonesia

C. Bounithiphonh
Forest Research Center, National Agriculture and Forestry Research Institute, Vientiane, Laos
H. Gaisberger
Bioversity International, Rome, Italy

[3]. Although there are scattered reports on the commercial cultivation of these species, slow growth rates and long rotation periods constrain opportunities to meet the demand for rosewood timber from plantations in the near future.

Previously, each country had different regulations for Dalbergia. For example, in Indonesia, *Dalbergia latifolia* and *D. sissoo,* which were successfully introduced and well-adapted in various geographical areas, have long been used in trade and did not evoke the Indonesian laws concerning protected plant and animal species. Since 2017, however, the entire genus except *D. sissoo* has been listed in the Convention on International Trade in Endangered Species of Wild Fauna and Flora (CITES) Appendices, which regulate and restrict species' international trade to prevent overexploitation and ensure sustainable use. The genus *Dalbergia* is included in CITES Appendix II, meaning that it requires export permits, reflecting the need for conservation measures across the genus [4]. While not all species of *Dalbergia* are threatened, the difficulty of distinguishing closely related species means that threatened *Dalbergia* species could be traded as non-threatened species, justifying a blanket decision to protect the genus. As an exception, specific regulations are in place for *D. sissoo* to facilitate trade of the species, which is abundantly found and cultivated in India and is important for small industries producing handicrafts.

Some of the commercially most valuable—and most threatened—species such as *Dalbergia cochinchinensis*, *D. oliveri,* and *D. latifolia* (Table 1) are distributed across several countries in Indochina, and so the effective conservation of the species and their genetic resources require collaboration between countries, not only in trade regulation and enforcement but also in research and conservation action. This chapter discusses the benefits of regional collaboration in conservation and research, delves into challenges in conserving species with cross-border distributions, and explores potential solutions to enhance regional cooperation for the protection and sustainable use of rosewoods.

2 Benefits of Regional Collaboration

Genetic diversity is the foundation for population viability, productivity, and adaptation to a changing climate. Developing strong research abilities in forest genetic resources (FGR) is crucial for protecting biodiversity, improving forest productivity and opportunities for plantation forestry, and ensuring ecosystem resilience. This requires a multidisciplinary approach, substantial investment, and strong collaboration among various stakeholders. Many Asian nations have developed significant capacities in FGR research, leveraging their rich forest biodiversity and advanced scientific infrastructure. The FGR research landscape in Asia is diverse and dynamic, characterised by strong national initiatives, international collaborations, and a growing emphasis on integrating modern genetic techniques with traditional forest management practices. Regional collaboration also creates capacity for research translation by providing guidance in support of conservation programmes.

Yet, research capacities across the region are unevenly distributed, especially considering the extremely high species diversity and the lower middle-income status of many countries within the distribution ranges of *Dalbergia* species. Further, many national initiatives are project-driven with a short time frame.

Through regional collaboration, it is possible to achieve synergies between national efforts, some of which are described below.

2.1 Developing Novel Research Methods and Approaches

Using 95 samples of 31 Southeast Asian *Dalbergia* species spanning 14 countries, DNA barcoding to support species conservation was successfully applied [7]. The study allowed identifying *D. cochinchinensis*, the first CITES-listed *Dalbergia* species from Asia, supporting efforts to curb the illegal trading of the species. Building on the same network of national partners a few years later, a regional research initiative led by the University of Oxford developed reference transcriptomes for six *Dalbergia* species, using samples from multiple Southeast Asian countries across the species distributions [8]. The study demonstrated more diverged disease resistance proteins in widely distributed species, implying higher genetic diversity and higher adaptive potential in species with wider ecological niches. The reference transcriptomes are expected to foster further studies on population genomics and gene-environment associations in Dalbergia. Later, a comparative greenhouse experiment characterised the different strategies of *D. cochinchinensis* and *D. oliveri* towards drought [9], where *D. cochinchinensis* tends to maximise photosynthetic assimilation by maintaining stomatal conductance during drought, but at the risk of hydraulic failure. *D. oliveri*, on the other hand, is more conservative in terms of water use. Their strategies are well-related to their ecological niche, as *D. cochinchinensis* is a pioneering species that prioritises growth. In a follow-up, a range-wide genomic study of both these species showed that the two species are adapted to different temperature and rainfall patterns and detected particularly strong local adaptation in the coastal ranges of the species, indicating that populations in those areas are of high genetic conservation priority [10]. Plant species, especially threatened species, are currently underrepresented in genomic databases, highlighting the importance of such studies.

2.2 Developing Range-Wide Conservation Assessments

Using species occurrence data contributed by 60 national research institutions and experts across Southeast Asian countries, species distribution models and vulnerability assessments for *D. cochinchinensis*, *D. cultrata*, and *D. oliveri* for current and future climate conditions and other threats such as overexploitation and habitat loss were developed [3]. The results were validated by the experts and helped to update the conservation status of *D. cochinchinensis* and *D. oliveri* on the IUCN Red List to Critically Endangered [10, 11], and are expected to help mobilise more resources for species conservation. Out of the 166,000 species on the IUCN Red List, only a handful of others are evaluated using quantitative conservation assessment methods.

2.3 Designing Effective Genetic Conservation and Restoration Strategies

Different conservation approaches are required based on the conservation status of and threats to each species. Knowledge of genetic differentiation between *D. cochinchinensis* populations across the species range was used to identify genetic management units to help ensure that local adaptations are effectively conserved through in situ and ex situ conservation units [12]. Building on range-wide, expert-validated distribution and threat maps, spatially explicit conservation priorities for *D. cochinchinensis, D. cultrata,* and *D. oliveri* by ecoregion, as a proxy for genetic differentiation, were identified [3]. Some ecoregions where the species are predicted to be the most threatened are transboundary, highlighting the need for practical collaboration between countries to conserve the species' unique genetic resources. In some cases, the species populations were predicted to have practically disappeared in one country while ample suitable habitats still remained on the other side of the border. In such cases, germplasm exchange can importantly support the restoration of viable, genetically diverse, and resilient populations [3]. Yet currently, regulations commonly prevent such exchange. In Indonesia, a genetic analysis of *D. latifolia* using 95 samples from eight populations across the country showed relatively low to medium levels of genetic diversity and complex/mixed origins [13]. The results imply the need to clarify the origin of the species, which would require analysing samples from countries of known natural distribution of the species, such as India.

Countries can also collaborate in exchanging experiences of working models of conservation and management. In Indonesia, the Indonesian State Forestry Company (Perum Perhutani) has since long implemented a management plan for *D. latifolia*, with planned rotation cycles and replanting to ensure the species' persistence. Since then, the practice has also been informally adopted on community-owned lands [14]. To our knowledge, similar detailed species management plans are not widely applied in other Asian countries within the *Dalbergia* spp. distributions, yet there is growing interest for sustainable plantation management in some countries. In Cambodia, *D. cochinchinensis* is the second most planted tree species nationwide (Institute of Forest and Wildlife Research and Development, Cambodia, unpublished report), and forestry department collaborates with smallholder farmers to supply seeds and seedlings from farmland seed sources for planting needs, an excellent model of decentralised tree seed supply chain.

2.4 Mobilising Funding

In recognition of the importance of regional collaboration for conserving threatened Dalbergias, regional research and development initiatives have received support from countries both within and beyond South and Southeast Asia. Notable donors include UK Darwin Initiative, DANIDA, and the National Science Foundation of China. Through the National Institute of Forest Sciences of the Republic of Korea, Korea Forest Service has supported regional exchanges and workshops to help refine regional research and conservation priorities and strategies. With its vast and diverse forest ecosystems, India has supported regional collaborations through Institutions such as the Indian Council of Forestry Research and Education (ICFRE) and various state forest research institutes. Research related to conservation, sustainability uses, and provision of superior seeds through tree breeding programmes has been carried out by research institutions and universities in Indonesia with funding from within the country and abroad, such as the CITES Tree Species Programme (CTSP). Many of the research initiatives and exchanges have been facilitated by the Asia Pacific Forest Genetic Resources Programme (APFORGEN), a regional network of 15 countries (Box 1).

Asia Pacific Forest Genetic Resources Programme (APFORGEN) APFORGEN works to enhance the conservation and sustainable use of forest genetic resources in the Asia-Pacific region. It aims to support and link national forest programmes, research institutions, non-government organizations, and individuals interested in forest genetic resources in the region [15]. The network was established in 2003 by the FAO, the Alliance of Bioversity International and the International Center for Tropical Agriculture (CIAT) (then International Plant Genetic Resources Institute, IPGRI), and the Asia -Pacific Association of Forestry Research Institutions (APAFRI). Network members are from 15 countries in the Asia-Pacific region: Bangladesh, Cambodia, China, India, Indonesia, Lao PDR, Malaysia, Myanmar, Nepal, Pakistan, the Philippines, Republic of Korea, Sri Lanka, Thailand, and Vietnam.

Since 2014, APFORGEN has had three regional strategy development and review workshops (2014, 2017, and 2023) where the National Coordinators of member countries identified and prioritized regional objectives and targets for joint research and capacity development [15]. These were developed into regional strategy documents (2014, 2017, and 2023), where the objectives and targets are explicitly linked to global conservation goals of the Global Plan of Action on Forest Genetic Resources, and the Kunming-Montreal Global Biodiversity Framework, to demonstrate how the network activities contribute to achieving these in the Asia-Pacific Region.

The Asia-Pacific Forestry Commission, FAO's highest forest decision- making body in the region, endorsed the APFORGEN's regional research and capacity development strategies in 2017 [3, 16–18]. Since then, Chairpersons of APFORGEN, elected from among the National Coordinators, attend the biennial Commission meetings to provide updates about the network's activities and achievements to the Commission members [19].

The recently completed Strategy 2018–2022: Implementing the Global Plan of Action on Forest Genetic Resources in Asia and the Pacific [3] helped the network attract funding for implementing four strategic objectives and genetic conservation assessments for over 60 species, restoration plans in 3 countries, and training workshops with over 150 participants [10, 11, 16]. It has significantly bolstered collaborative efforts between countries in forest genetic resource management in Asia and the Pacific region.

Challenges in Collaboration in Asia for Rosewoods

Collaboration in the conservation and sustainable use of *Dalbergia* spp. across Asia presents numerous challenges. Rosewoods face significant pressure from both legal and illegal logging. The exchange of germplasm, managing high species diversity, and dealing with the trafficking and regulatory landscape are critical factors that impact collaborative efforts.

Varying distributions of different species across multiple countries pose unique challenges for collaboration. Each species of rosewood may have distinct ecological requirements and threats, necessitating tailored conservation strategies. Coordinating these efforts across borders can be complex, as conservation measures effective in one country may not be transferable to other contexts. Despite recent research efforts, comprehensive and collaborative data collection efforts are still urgently needed on the distribution, population status, and ecological needs of many rosewood species. This lack of information hampers the development of effective regional conservation strategies and complicates collaborative efforts. Rosewoods are often found in fragmented habitats due to deforestation and land-use changes. Cross-border collaboration is essential to managing these fragmented landscapes effectively, but it requires coordination and resource-sharing between countries [20].

Illegal logging and trafficking of rosewoods pose severe threats to their conservation and sustainable use. High market demand drives these activities, resulting in significant ecological and economic losses. Effective monitoring and enforcement of environmental laws and regulations require substantial resources and coordination between countries. Weak enforcement mechanisms and corruption can exacerbate the problem, allowing illegal activities to flourish. Illegal logging and trafficking are often facilitated by well-organised transnational criminal networks in collusion with government authorities [21]. These networks exploit regulatory loopholes and differences in enforcement capabilities between countries, making it difficult to curb illegal activities.

The significant economic value of rosewood fuels powerful incentives for illegal logging. Tackling organised crime, corruption, and the underlying economic drivers is essential for successfully conserving *Dalbergia* species. Despite numerous communities reporting illegal logging activities to authorities, their concerns often go unaddressed. Those who push for enforcement frequently face threats, false accusations, or even fatal repercussions. A recent case in Cambodia highlights this alarming trend, where prominent environmental defenders documenting illegal logging were unjustly arrested, underscoring the dangers faced by those advocating for forest protection [22].

The regulatory environment surrounding rosewood conservation and trade is complex and varies significantly between countries. This complexity poses several challenges for collaboration. Differing national regulations on the harvesting, trade, and transport of rosewoods can create barriers to effective collaboration. Efforts to harmonise these regulations are essential but require significant political will and coordination. The Convention on International Trade in Endangered Species of Wild Fauna and Flora (CITES) regulates the trade of rosewoods to ensure their survival. However, compliance with CITES regulations can be challenging, especially for countries with weak governance and a lack of impartial courts. Ensuring consistent implementation of CITES provisions across countries is critical for effective collaboration. In response to overexploitation, some countries have implemented trade bans on rosewoods. While these bans are intended to protect the species, they can have unintended consequences, such as driving the trade underground and increasing illegal activities. Finding a balance between regulation and sustainable use is essential.

Germplasm exchange, the process of sharing genetic resources between different regions, is essential for the conservation and breeding of rosewoods. However, this process is fraught with challenges in Asia. Different countries have varying regulations regarding the exchange of genetic material. These laws are designed to protect national biodiversity but can inadvertently hinder collaborative conservation efforts. For instance, the Nagoya Protocol on Access and Benefit-sharing requires stringent compliance, which can complicate the process of germplasm exchange between countries. The movement of germplasm across borders carries the risk of spreading pests and diseases. Each country has its own phytosanitary regulations, which can create additional bureaucratic hurdles. Ensuring that exchanged germplasm is free from pests and diseases requires rigorous screening and certification processes. The ownership and use rights of genetic resources can be contentious. Countries and local communities want to ensure that they are fairly compensated for the use of their genetic resources, which can lead to protracted negotiations and delays in germplasm exchange agreements.

3 Way Forward

Addressing the challenges of collaboration in rosewood conservation requires a multi-faceted approach. Establishing regional platforms for dialogue and cooperation can help harmonise regulations, share best practices, and coordinate conservation efforts including environmental compliance assurance. Organisations such as the Association of Southeast Asian Nations (ASEAN) can play a pivotal role in facilitating such cooperation.

Enhancing the capacity of national and local authorities to enforce regulations, monitor rosewood populations, and manage conservation programmes is crucial.

This can be achieved through training, resource allocation, and technical assistance from international organisations. In addition, because not all countries have regulations governing the protection of Dalbergia in their country, reviewing the status of Dalbergia in CITES based on the utilisation, conservation status, benefits, and priorities of each country may facilitate the management of Dalbergia at the regional level.

Engaging local communities in conservation efforts is essential, especially where most Dalbergia wood is obtained from smallholder or community-owned land, such as in Indonesia where an estimated 80% of rosewood is supplied by local communities [14]. Efforts inspired by Indonesia's practices in sustainably managing Dalbergias on community lands focus on planned harvesting to maximise utilisation without harming the environment. These initiatives prioritise maintaining seed availability through stand management designated as seed sources, enhancing propagation, and providing seeds for various purposes, including conservation and reforestation programmes. Community monitoring also plays a pivotal role in the conservation of *Dalbergias*. With their deep ecological knowledge and direct connection to their surrounding environment, local communities become vital stakeholders in conservation efforts. Community-led actions strengthen the monitoring and protection of *Dalbergias* and empower communities, fostering a sense of stewardship and ensuring more sustainable and culturally attuned conservation outcomes. Such approaches ensure the sustainable management of *Dalbergias* along with benefits for local communities who contribute to their conservation. Providing economic incentives for sustainable use and involving communities in decision-making processes can help reduce illegal activities and ensure the success of conservation initiatives.

Investing in research to fill data gaps on rosewood species and promoting the sharing of scientific information between countries can support evidence-based conservation strategies. Collaborative research projects and databases can facilitate this process. Efforts are also directed towards improving tree species for timber, non-timber forest products, and agroforestry systems in India. Developing and promoting certification schemes for sustainably sourced rosewood can help balance conservation goals with economic interests. Encouraging consumers to choose certified products can reduce the demand for illegally sourced wood.

Collaboration in the conservation and sustainable use of rosewoods in Asia is a complex and challenging endeavour. By addressing the barriers to germplasm exchange, managing the high species diversity, combating illegal logging and trafficking, and navigating the regulatory landscape, stakeholders can enhance regional cooperation. Through strengthened regional cooperation, capacity building, community engagement, research, and the promotion of sustainable trade, it is possible to develop effective strategies for the conservation and sustainable use of rosewoods. Such efforts are crucial not only for the survival of these valuable species but also for the preservation of the ecological and economic benefits they provide to the region.

References

1. Lewis GP, Schrire B, Mackinder B, Lock M (2005) Legumes of the World. Royal Botanic Gardens, Kew, Surrey, 577
2. Nijman V (2024) The illegal trade in rosewood in Indonesia. Eur J Forest Res 143(3):1047–1055
3. Gaisberger H, Fremout T, Kettle CJ, Vinceti B, Kemalasari D, Kanchanarak T, Thomas E, Serra-Diaz JM, Svenning JC, Slik F, Eiadthong W (2022) Tropical and subtropical Asia's valued tree species under threat. Conservation Biology 36(3): e13873. https://conbio.onlinelibrary.wiley.com/doi/full/https://doi.org/10.1111/cobi.13873
4. Lakhey P, Pathak J, Adhikari B (2020) Dalbergia latifolia. The IUCN Red List of Threatened Species, e. T32098A67777757. https://doi.org/10.2305/IUCN.UK.2020-3.R.
5. Cowell C, Williams E, Bullough LA, Grey J, Klitgaard B, Govaerts R, Andriambololonera S, Cervantes A, Crameri S, Lima HC, Lachenaud O, Li SJ, Linares JL, Phillipson P, Rakotonirina N, Wilding N, van der Burgt X, Vatanparast M, Barker A, Barstow M, Beentje H, Plummer J (2022) CITES Dalbergia Checklist. Commissioned by the CITES Secretariat. Royal Botanic Gardens, Kew, Surrey
6. CITES Secretariat (2024). https://cites.org/sites/default/files/documents/E-PC27-27.pdf
7. Hartvig I, Czako M, Kjær ED, Nielsen LR, Theilade I (2015) The use of DNA barcoding in identification and conservation of rosewood (*Dalbergia* spp). PLoS ONE 10(9):e0138231. https://doi.org/10.1371/journal.pone.0138231
8. Hung TH, So T, Sreng S et al (2020) Reference transcriptomes and comparative analyses of six species in the threatened rosewood genus *Dalbergia.* Sci Rep 10:17749. https://doi.org/10.1038/s41598-020-74814-2
9. Hung TH, So T, Thammavong B, Chamchumroon V, Theilade I, Phourin C, Bouamanivong S, Hartvig I, Gaisberger H, Jalonen R, Boshier DH (2023) Range-wide differential adaptation and genomic offset in critically endangered Asian rosewoods. Proc Natl Acad Sci 120(33):e2301603120
10. Barstow M, Boshier D, Bountithiponh C, Changtragoon S, Gaisberger H, Hartvig I, Hung H, Jalonen R, Kanchanarak T, Mackay J, Ping H, Thammavong B, Theilade I, Tran T, Win P, Zheng Y (2022aa) *Dalbergia cochinchinensis.* The IUCN Red List of Threatened Species 2022: eT215342548A2822125. https://doi.org/10.2305/IUCN.UK.2022-1.RLTS.T215342548A2822125.en
11. Barstow M, Boshier D, Bountithiponh C, Changtragoon S, Gaisberger H, Hartvig I, Hung H, Jalonen R, Kanchanarak T, Mackay J, Ping H, Thammavong B, Theilade I Tran T, Win P, Zheng Y (2022b) *Dalbergia oliveri.* The IUCN Red List of Threatened Species 2022: eT215341339A2813403 . https://doi.org/10.2305/IUCN.UK.2022-1.RLTS.T215341339A2813403.en
12. Hartvig I, So T, Changtragoon S, Tran HT, Bouamanivong S, Ogden R, Senn H, Vieira FG, Turner F, Talbot R, Theilade I (2020) Conservation genetics of the critically endangered Siamese rosewood (*Dalbergia cochinchinensis*): Recommendations for management and sustainable use. Conserv Genet 21(4):677–692. https://doi.org/10.1007/s10592-020-01279-1
13. Yulita K, Susila Rachmat HH, Dwiyanti FG, Atikah, TD, Subiakto A, Pratama B, Setyawati T, Wardani W, Fambayun RA, Arrofaha N (2022a) Population genetic of the Indonesian rosewood (*Dalbergia latifolia*) from Java and West Nusa Tenggara revealed using sequence-related amplified polymorphism. For Sc Technol 18(4):172–181
14. Yulita KS, Wardani W, Atikah TD, Pratama BA, Subiakto A, Nurjanah S, Nopiansyah F, Nugroho A, Kamal I, Arrofaha N (2022b) The non-detriment findings (NDF) report for *Dalbergia latifolia* in Java and West Nusa Tenggara, Indonesia. Directorate of Biodiversity Conservation of Species and Genetic, the Ministry of Environment and Forestry, and the National Research and Innovation Agency, Jakarta, Indonesia
15. APFORGEN (2023) Strategy 2023–2030: conserving, restoring, and sustainably managing forest and tree genetic resources in Asia and the Pacific. Asia-Pacific Forest Genetic Resources Programme. Serdang: Bioversity International. https://www.apforgen.org/about/strategy-2023-2030

16. Gaisberger H, Fremout T, So T, Thammavong B, Bounithiphonh C, Hoa TT, Yongqi Z, Kanchanarak T, Changtragoon S, Sreng S, Ping H et al (2022) Range-wide priority setting for the conservation and restoration of Asian rosewood species accounting for multiple threats and ecogeographic diversity. Biological Conservation 270:109560. https://www.sciencedirect.com/science/article/pii/S0006320722001136?via%3Dihub6
17. Strengthening Seed Supply. Project website. Asia-Pacific Forest Genetic Resources Programme (APFORGEN). https://www.apforgen.org/initiatives/strengthening-seed-supply
18. Jalonen R, Yongqi Z, Warrier RR (2024) APFORGEN secures $1.48M to protect vital Asian tree species. Rome (Italy): Bioversity International; Cali (Colombia): CIAT 4p. https://alliancebioversityciat.org/publications-data/apforgen-secures-148m-protect-vital-asian-tree-species
19. FAO (2018) Report of the twenty seventh session of the Asia-pacific forestry commission–Colombo, Sri Lanka 24–27 October 2017. Regional Office for Asia and the Pacific, Food and Agriculture Organization of the United Nations. Bangkok: FAO. 6p. https://openknowledge.fao.org/items/c0d626b0-9fcd-42d0-b999-ac2ba7779fb5
20. Jalonen R, Gaisberger H, Fremout T, Bounithiphonh C, Kettle C (2024) Asian rosewoods now critically endangered on IUCN Red List. Rome (Italy): Bioversity International; Cali (Colombia): CIAT 4p. https://alliancebioversityciat.org/publications-data/asian-rosewoods-now-critically-endangered-iucn-red-list
21. GI-TOC (2022). https://globalinitiative.net/wp-content/uploads/2022/09/Cambodia-Logging-Report-web-v2.pdf
22. Mongabay (2024). https://news.mongabay.com/2024/11/six-activists-arrested-in-cambodia-while-investigating-illegal-logging/

Chapter 33
Epilogue: Charting the Future for *Dalbergia* Conservation and Utilization

Rekha R. Warrier, Geeta Joshi, and A. N. Arunkumar

The conservation and management of forest genetic resources (FGRs) of Asian rosewoods stand at a critical crossroads. As discussed throughout this handbook, the *Dalbergia* genus includes several species of immense cultural, ecological, and economic importance. However, these species are increasingly threatened by habitat destruction, overexploitation, and climate change. This volume provides a comprehensive overview of their distribution, genetic diversity, propagation techniques, pest and disease management, wood properties, and potential for sustainable utilization. As we conclude this exploration of Asian rosewoods, ensuring their survival requires a holistic approach that integrates scientific research, policy interventions, and community participation. This epilogue synthesizes key insights from the handbook and outlines a roadmap for future conservation and sustainable use. The challenges can be broadly categorized into scientific, social, and economic aspects. Addressing these issues through targeted strategies will support both the conservation and sustainable utilization of *Dalbergia* species.

Understanding the genetic resources of *Dalbergia* species and studying their distribution, habitat suitability, and timber properties highlight the need for targeted conservation and sustainable forestry practices. The presence of *Dalbergia* species across Asia (Warrier et al., Chap. 1) demonstrates their ecological importance and commercial value. Their varied habitats call for region-specific management strategies to mitigate threats and enhance sustainable utilization.

R. R. Warrier (✉)
ICFRE-Institute of Forest Genetics and Tree Breeding, Coimbatore, India
e-mail: rekha@icfre.org

G. Joshi
Indian Council of Forestry Research and Education (ICFRE), Dehradun, India

A. N. Arunkumar
ICFRE-Institute of Wood Science and Technology, Bengaluru, India

R. R. Warrier et al. (eds.), *Handbook of Asian Rosewoods*, Materials Horizons: From Nature to Nanomaterials, https://doi.org/10.1007/978-981-96-9459-4_33

The first section of this volume provided an in-depth understanding of various *Dalbergia* species, particularly those found across Asia. Rosewoods, represented by species such as *Dalbergia cochinchinensis*, *Dalbergia latifolia*, *Dalbergia oliveri*, *Dalbergia sissoo*, and *Dalbergia tonkinensis*, are highly valued for their premium-quality timber, medicinal properties, and ecological importance. Their distribution spans across Asia, with notable populations in India, China, Indonesia, Myanmar, Sri Lanka, Laos, Cambodia, Thailand, and Vietnam. However, these species face multiple threats, including illegal logging, land-use change, and unsustainable harvesting practices. The studies in this book emphasize the urgent need for a coordinated and science-driven approach to safeguard these genetic resources.

With its distribution in Laos, Thailand, Cambodia, and Vietnam (Chap. 2) and cultivation prospects in China (Yongqi and Yongxiang, Chap. 3), *Dalbergia cochinchinensis* exemplifies both the opportunities and challenges of commercial forestry. Research on its habitat and growth potential informs conservation priorities and future plantation strategies. Studies on distribution of *D. latifolia* in India (Chap. 4, Arunkumar et al.) and its naturalization in Indonesia (Yulita et al., Chap. 5), along with habitat modeling (Lalitha et al., Chap. 6), provide insights into its ecological resilience. Sri Lanka's *Dalbergia* species (Subasinghe et al., Chap. 13) are valued for their timber quality, but conservation challenges persist. Understanding their distribution and economic potential supports informed policy decisions and trade regulations. *Dalbergia oliveri* (Chaps. 14, 15 and 16, Chaloun et al., Yongqi and Yongxiang, Aye and Aung) requires region-specific management approaches. Its conservation must consider habitat protection and the prevention of illegal trade. Research on *Dalbergia tonkinensis* in Vietnam (Phi et al., Chap. 25) sheds light on its ecological role and economic value. Conservation measures tailored to its habitat and trade regulations will be essential for the survival of these species. Research into their genetic diversity, reproductive biology, and tissue culture has paved the way for improved conservation and cultivation strategies.

Investigations into wood properties (Sinha et al., Chap. 7; Sethy et al., Chap. 8), phytochemistry (Sharma and Kumar, Chap. 9), and disease resistance (Muthukumar et al., Chap. 10) of *D. latifolia* help optimize its commercial use while ensuring sustainability. The chapters on wood structure, mechanical properties, and working qualities provide a scientific basis for optimizing the use of rosewood in various industries.

Dalbergia sissoo, widely cultivated in India for its high-quality timber, and exempted from the CITES list, plays a crucial role in agroforestry systems due to its nitrogen-fixing properties and resilience in varied climatic conditions. However, challenges such as disease susceptibility, genetic variability, seed deterioration, and wood quality issues necessitate a multifaceted approach to its conservation and sustainable utilization. This species has been widely studied due to its multiple utilities. In Chap. 17, Sharma et al. provide a foundational overview of *D. sissoo*, discussing its botanical characteristics, ecological significance, distribution, and economic importance. It outlines the key threats facing the species, including overexploitation, habitat loss, and vulnerability to pests and diseases, while emphasizing the need for scientific interventions in genetic improvement, breeding, and disease management. Genetic variability is a cornerstone of sustainable forestry and breeding programs. Chapter 18 (Geeta et al.) explores the natural genetic diversity within *D. sissoo* populations and its implications for tree improvement. The authors highlight breeding strategies aimed at enhancing growth performance, disease resistance, and wood quality. Findings suggest that selective breeding and clonal propagation can significantly improve the species' adaptability and yield. *D. sissoo* adapts to diverse environmental conditions. Chapter 20 highlights integrated disease management strategies such as biological control, soil health improvement, and resistant genotype selection to reduce losses and sustain productivity and also takes the readers through a systematic review of the causes and management of these major threats, including pathogens, soil conditions, and climate factors. Flowering patterns, pollination mechanisms, and seed viability to enhance breeding programs and genetic quality, the role of insects, wind, and self-pollination in reproductive success have been detailed in Chaps. 21 and 22. These chapters also explore hybridization potential for developing superior cultivars. They also emphasize the need for increasing pollinator diversity to improve seed production and genetic diversity, ensuring healthy, natural, and plantation populations.

Factors affecting seed maturity and longevity, including moisture, storage conditions, and fungal infections (Chap. 23, Geeta et al.) along with best practices for preserving seed viability, essential for long-term conservation efforts are described.

Chapter 24 explores the mechanical properties, workability, and drying characteristics of *D. sissoo* wood. It also provides insights into the wood strength, workability, and resistance to deformation, which are key factors for applications in furniture making, construction, and musical instruments. Understanding optimal drying techniques is critical for preventing cracks, shrinkage, and warping during processing. Understanding optimal drying techniques helps prevent cracking, shrinkage, and warping, ensuring better use in furniture, construction, and musical instruments as the commercial value of *D. sissoo* largely depends on its mechanical properties and drying characteristics.

One of the key contributions of this handbook is the extensive information on the ecological aspects of *Dalbergia* species. Studies utilizing predictive habitat modeling (e.g., MaxEnt modeling for *D. latifolia* in India; Chap. 6) offer valuable insights into current and future habitat suitability. Such methodologies pave the way for proactive conservation planning. Additionally, advancements in micropropagation and tissue culture techniques have shown promise in addressing the challenge of propagating difficult-to-root species. The comprehensive review of micropropagation challenges in *D. latifolia* and *D. sissoo* (Chaps. 12 and 19) highlights the need for refining propagation protocols (Tewari, Chap. 11) to enhance genetic diversity and adaptability.

Integrating *Dalbergias* into agroforestry systems (Chap. 29, Warrier et al.) offers both economic and environmental benefits. Their cultivation on private and community lands can reduce pressure on wild populations, ensuring a sustainable timber supply. These trees enhance soil health, sequester carbon, and support biodiversity, making agroforestry a viable conservation strategy. By providing economic incentives, this approach encourages farmers to adopt sustainable land use while protecting natural forests.

The role of certification and traceability in the timber trade, as discussed in the industry-oriented chapter (Chap. 31, Mills), is crucial in balancing economic interests with conservation goals. The need for the timber industry to embrace sustainable practices to ensure long-term profitability, certification systems, legal frameworks, and responsible sourcing are key to the sustainable future of Rosewoods.

Advancements in technology, such as chemotyping for species differentiation (Chap. 26, Kumar and Sharma) have emerged as powerful tools for sustainable forest management. Chemotyping provides a scientific basis for species identification, helping combat illegal logging and ensuring the legal trade of rosewood products. This advancement is crucial for conservation enforcement. The ability to identify species through biochemical markers enhances traceability in trade, preventing illegal logging and promoting ethical sourcing.

Beyond their biological significance, rosewoods have been deeply embedded in human culture for centuries. Their use in musical instruments, furniture, and handicrafts highlights the intersection of tradition and commerce. The chapters dedicated to rosewood's role in the guitar industry (Arunkumar, Chap. 27) and artisanal crafts (Chap. 28, Arunkumar and Mills) emphasize the importance of sustainable sourcing to preserve both the trees and the industries they support. Developing responsible trade practices and promoting alternative materials can mitigate pressure on natural populations.

Illegal trade habitat degradation, and climate change remain major obstacles to conservation efforts. The mapping of threat status and vulnerability of different Dalbergia species (Gaisberger et al., Chap. 30) has highlighted the urgent need for targeted conservation efforts and prioritizing conservation areas. Regional collaboration, as emphasized in this book (Jalonen et al., Chap. 32), is essential for addressing these challenges effectively. Cross-border cooperation in genetic resource management, law enforcement, and policy harmonization will be crucial in safeguarding these species for future generations. Strengthening the enforcement of international regulations, such as CITES, alongside community-driven conservation initiatives, is essential to curbing illegal exploitation. The chapter on effective conservation emphasizes the necessity of transboundary cooperation among range countries. Harmonizing policies on trade, conservation, and research will enhance the impact of individual national efforts.

The way forward lies in a holistic approach that combines scientific research, sustainable industry practices, community involvement, and policy support. Governments, researchers, industries, and local communities must work together to balance conservation with utilization. Investing in restoration projects, strengthening legal frameworks, and promoting sustainable forestry initiatives will be key steps in this journey.

Capacity building and stakeholder engagement are critical components of sustainable management. Training programs for foresters, researchers, and policymakers can facilitate the implementation of best practices. Additionally, community involvement in conservation initiatives, including seed banking and habitat restoration, can ensure long-term success. This handbook serves as both a testament to the richness of *Dalbergia* species and a call to action for their preservation. The responsibility to protect these invaluable trees rests with all of us—by fostering responsible trade, advancing scientific research, and championing conservation efforts, we can ensure that rosewoods continue to thrive for generations to come.

Moving forward, several key actions are recommended:

1. **Enhanced Research and Innovation**: Continued investment in biotechnological approaches, such as genetic markers for species differentiation and climate-resilient breeding programs.
2. **Strengthening Conservation Networks**: Establishing a global *Dalbergia* conservation network to facilitate knowledge exchange and joint conservation efforts.
3. **Sustainable Trade and Certification**: Encouraging responsible sourcing and consumer awareness to reduce demand for illegally harvested rosewood.
4. **Community-Centered Conservation**: Empowering local communities through sustainable livelihoods linked to rosewood conservation.
5. **Policy Integration and Implementation**: Strengthening legal frameworks and international agreements to ensure the protection of Dalbergia species.

This book serves as a vital resource for researchers, policymakers, conservationists, and industry stakeholders committed to the preservation of Dalbergia genetic resources. The collective insights presented herein underscore the need for an interdisciplinary and collaborative approach. By bridging scientific research with practical conservation measures, we can secure the future of rosewoods for generations to come. The responsibility lies with all stakeholders—scientists, governments, industries, and local communities—to translate these findings into meaningful action.

The conservation of *Dalbergia* species is not merely a scientific endeavor but a moral, ethical, and ecological imperative. As we advance into an era of unprecedented climate change, the choices we make today will determine the fate of these precious species and the ecosystems they support. Let this work serve as a catalyst for sustained efforts toward the conservation and sustainable use of rosewood genetic resources worldwide.

MIX
Papier aus verantwortungsvollen Quellen
Paper from responsible sources
FSC® C105338

If you have any concerns about our products,
you can contact us on
ProductSafety@springernature.com

In case Publisher is established outside the EU,
the EU authorized representative is:
Springer Nature Customer Service Center GmbH
Europaplatz 3, 69115 Heidelberg, Germany

Printed by Libri Plureos GmbH
in Hamburg, Germany